Introduction to Dynamics and Control of Flexible Structures

John L. Junkins
Youdan Kim

AIAA

EDUCATION SERIES
J. S. Przemieniecki
Series Editor-in-Chief
Air Force Institute of Technology
Wright-Patterson Air Force Base, Ohio

Published by
American Institute of Aeronautics and Astronautics, Inc.
370 L'Enfant Promenade, SW, Washington, DC 20024-2518

American Institute of Aeronautics and Astronautics, Inc., Washington, DC

Library of Congress Cataloging-in-Publication Data

Junkins, John L.
Dynamics and control of flexible structures / John L. Junkins and Youdan Kim.
p. cm.—(AIAA education series)
Includes bibliographical references and index.
1. Large space structures (Astronautics) 2. Structural dynamics.
3. Structural control (Engineering) I. Kim, Youdan. II. Title. III. Series.
TL910.J86 1993 629.47′1—dc20 92-46384
ISBN 1-56347-054-3

Third Printing.

Texts Published in the AIAA Education Series

Re-Entry Vehicle Dynamics
Frank J. Regan, 1984
Aerothermodynamics of Gas Turbine and Rocket Propulsion
Gordon C. Oates, 1984
Aerothermodynamics of Aircraft Engine Components
Gordon C. Oates, Editor, 1985
Fundamentals of Aircraft Combat Survivability Analysis and Design
Robert E. Ball, 1985
Intake Aerodynamics
J. Seddon and E. L. Goldsmith, 1985
Composite Materials for Aircraft Structures
Brian C. Hoskins and Alan A. Baker, Editors, 1986
Gasdynamics: Theory and Applications
George Emanuel, 1986
Aircraft Engine Design
Jack D. Mattingly, William Heiser, and Daniel H. Daley, 1987
An Introduction to the Mathematics and Methods of Astrodynamics
Richard H. Battin, 1987
Radar Electronic Warfare
August Golden Jr., 1988
Advanced Classical Thermodynamics
George Emanuel, 1988
Aerothermodynamics of Gas Turbine and Rocket Propulsion, Revised and Enlarged
Gordon C. Oates, 1988
Re-Entry Aerodynamics
Wilbur L. Hankey, 1988
Mechanical Reliability: Theory, Models and Applications
B. S. Dhillon, 1988
Aircraft Landing Gear Design: Principles and Practices
Norman S. Currey, 1988
Gust Loads on Aircraft: Concepts and Applications
Frederic M. Hoblit, 1988
Aircraft Design: A Conceptual Approach
Daniel P. Raymer, 1989
Boundary Layers
A. D. Young, 1989
Aircraft Propulsion Systems Technology and Design
Gordon C. Oates, Editor, 1989
Basic Helicopter Aerodynamics
J. Seddon, 1990
Introduction to Mathematical Methods in Defense Analyses
J. S. Przemieniecki, 1990

Space Vehicle Design
Michael D. Griffin and James R. French, 1991

Inlets for Supersonic Missiles
John J. Mahoney, 1991

Defense Analyses Software
J. S. Przemieniecki, 1991

Critical Technologies for National Defense
Air Force Institute of Technology, 1991

Orbital Mechanics
Vladimir A. Chobotov, 1991

Nonlinear Analysis of Shell Structures
Anthony N. Palazotto and Scott T. Dennis, 1992

Optimization of Observation and Control Processes
Veniamin V. Malyshev, Mihkail N. Krasilshikov, and Valeri I. Karlov, 1992

Aircraft Design: A Conceptual Approach
Second Edition
Daniel P. Raymer, 1992

Rotary Wing Structural Dynamics and Aeroelasticity
Richard L. Bielawa, 1992

Spacecraft Mission Design
Charles D. Brown, 1992

Introduction to Dynamics and Control of Flexible Structures
John L. Junkins and Youdan Kim, 1993

Dynamics of Atmospheric Re-Entry
Frank J. Regan and Satya M. Anandakrishnan, 1993

Practical Intake Aerodynamic Design
E. L. Goldsmith and J. Seddon, Editors, 1993

Published by
American Institute of Aeronautics and Astronautics, Inc., Washington, DC

FOREWORD

An Introduction to Dynamics and Control of Flexible Structures by John L. Junkins and Youdan Kim provides a comprehensive introduction to two important disciplines indispensable in the design of spacecraft and aeronautical structures: *Engineering Mechanics* and *Control Engineering*. The rationale behind their novel unified approach in teaching dynamical modeling, analysis, and control design for flexible structures, according to the authors, is that "effective control laws for mechanical systems are best designed by one who understands both the basic mechanics of the system under consideration and the control methodology being used to design the control law." A better understanding of the interface between these two disciplines is indeed the crucial development needed to better designs for actively controlled aerospace structures. The authors have clearly provided an outstanding text to strengthen this important interface for practicing engineers and students of aeronautical/astronautical engineering.

The text starts with a review of essential mathematical preliminaries which consists of an efficient review of computational linear algebra. Also included in this discussion and throughout the text, are numerous applications of the authors' MATLAB operators. The MATLAB operators, used to do the computational examples in the text, provide an excellent supplement of the text, but their detailed discussion is relegated to an appendix and the documentation on the diskette itself. The mathematical preliminaries are followed by chapters on Lyapunov stability theory and application, and mathematical models for the dynamics of structural systems, represented both by systems of partial differential equations and systems of ordinary differential equations obtained by using the finite element method. The final three chapters deal with analysis and design of linear feedback controllers for structural systems. Each of these chapters provide excellent introductions to basic concepts for the design of control laws for flexible structures subjected to dynamic loading. While these chapters are tutorial, significant new material from recent research has been integrated throughout.

The AIAA Education Series embraces a broad spectrum of theory and application of different disciplines in aerospace, including aerospace design practice. The Series has now been expanded to include defense science, engineering, and technology. The basic philosophy for the Series is to develop both teaching texts for students and reference materials for practicing engineers and scientists.

J. S. PRZEMIENIECKI
Editor-in-Chief
AIAA Education Series

PREFACE

This text is the outgrowth of our work over the past decade on the interface of the two traditional disciplines: *Engineering Mechanics and Control Engineering.* Our work has been an interesting mixture of theoretical, computational, laboratory experimental, and applications to aerospace vehicle dynamics and control. From this range of basic and applied research, along with our teaching activities, a perspective has evolved which is reflected throughout this text.

The kernel of our viewpoint is that *effective control laws for mechanical systems are best designed by one who understands both the basic mechanics of the system under consideration and the control methodology being used to design the control law.* This idea is so obvious that it seems redundant to state it, however, we state it nonetheless in the hope that we will win as many converts as possible to what we feel is a right-thinking view of the universe. It is a trivial observation that there are numerous, highly productive specialists presently in the field who lack this perspective.

Certainly narrowly focused specialists can and do make enormous and widely useful contributions and we recognize that specialization has its merits. However, we believe our perspective is the correct one to foster among the next generation of structural dynamics and control engineers. We believe that a very significant and increasing fraction of the most important research contributions of the past few years have been made precisely because the researcher had gained broad competence in mechanics and control. Clearly there are a sufficient number of individuals who serve as counter-examples to conclude that technical depth and breadth are not mutually exclusive! Since most of the crucial research problems lie on the interface of the traditional disciplines of mechanics and control, it is extremely important that we develop a strong school of engineers who have both the requisite depth and breadth. In chapter one, we overview the history of these historically disjoint but recently merging disciplines. Our fondest hope is that this text will contribute to accelerating this highly desirable merger, especially amongst the new generation of engineers just entering this field.

In undertaking a book-writing project of this scope, one must decide a priori that it is worth it in some sense. The need for this book was not a serious issue in our thinking, because we knew that there exist very few books written from this general perspective and none with the blend of theory, computation, and experiment which we conceived as the format for this text. It was easy to undertake this writing project as a labor of love; our collaboration on the book was made easier by the fact that we have historically established a close technical impedance match while the second author studied for his Ph. D. under the direction of the first. While writing this book has been a labor of love, it was still a lot of work!

The approach we have taken is to concentrate on a central set of ideas in dynamics and control and the most important methods which flow from them. We have attempted to gain a higher degree of closure between analysis, computation, and experimental results than has heretofore been achieved in textbook form. Since a comprehensive text with this degree of closure would necessarily run several more years of effort and thousands of pages, it is apparent that we have exercised considerable selectivity to arrive at the set of topics considered. While this is an introductory text, we have introduced a significant number of novel results and believe that it will prepare the reader well for both applications and research.

We have benefited enormously by our on-going and historical collaborations with numerous gifted colleagues. We feel obliged to express our grateful appreciation for the many direct and indirect contributions of the following professional individuals:

B. Agrawal, K. Alfriend, H. Bang, M. Bell, H. Black, D. Bodden, J. Cochran, N. Creamer, A. Das, J. Godette, W. Haisler, S. Herrick, J. Hurtado, D. Hyland, G. James, J. Juang, S. Krishnan, A. Kurdila, B. Lawrence, S. Lee, P. Likins, K. Lim, R. Longman, A. Martins, L. Meirovitch, S. Morgan, H. Morton, R. Mukherjee, T. Papis, T. Pollock, J. Przemieniecki, Z. Rahman, D. Rew, T. Singh, M. Shuster, R. Skelton, R. Thompson, B. Tower, J. Turner, S. Vadali, V. Venkayya, L. Watson, B. Wie, S. Wie, L. Wood, and S. Wu.

Certainly our families have contributed both directly and indirectly to this text; their consistent encouragement and support are most appreciated and warmly acknowledged. The direct support of Elouise (Mrs. John) Junkins in the final editing of this manuscript contributed significantly to the timely completion.

We have been able to add a few significant terms to the infinite series in this field, but we remain grateful to those whose insights, previous work, and support have underwritten and stimulated our effort.

The support of the following organizations is also warmly acknowledged: Texas A&M University, Texas Engineering Experiment Station, Virginia Polytechnic Institute and State University, the Johns Hopkins Applied Physics Laboratory, U. S. Naval Postgraduate School, Royal Melbourne Institute of Technology, the Air Force Office of Scientific Research, the American Astronautical Society, and the American Institute of Aeronautics and Astronautics. The organizational support has been an important ingredient in the development of many excellent students who have collaborated in our research over the past decade. The support of basic research is central to the advancement of our individual and collective base of knowledge, and is an important catalyst enabling the advancement of our civilization. We express our appreciation to these individuals and institutions for their investments of time, confidence, and resources in our historical research.

Preface

Writing a book in a dynamic field can be frustrating in the sense that the field keeps re-inventing itself faster than one can understand and document its current state. The danger that a text can quickly become obsolete provides some damping of enthusiasm which encourages one to avoid heavy book-writing investments in "vogue" topics. On the other hand, one of our primary motivations in writing this book was the desire to integrate a significant subset of results from the research literature of the past decade into an up-to-date textbook, accessible by first year graduate students in engineering and applied science. Obviously, some judgement was exercised and some compromises were made, we believe that we have achieved a proper balance. We have strived to present a central subset of the most important material in dynamical and control system analysis which we consider illuminates key ideas in an interesting and general setting, with sufficient connectivity to classical concepts, and with sufficient illustration of practical implications through the use of examples. Having made a significant effort pursuing these objectives, we trust that this book will stand the test of time. Time will tell.

John L. Junkins
Youdan Kim
1992

The NOVA Spacecraft

The NOVA spacecraft depicted in the sketch of Figure 1.1 and on the book cover was reproduced with permission from the Johns Hopkins Applied Physics Laboratory (JHUAPL). The first author, along with several colleagues at JHUAPL, designed and implemented a near-minimum-time attitude maneuver control law for this spacecraft launched in May, 1981. We believe this to be the first on-orbit implementation of a Pontryagin's Principle-derived optimal control law.

TABLE OF CONTENTS

CHAPTER 1

Introduction

1.1 Some Biased Historical Notes

We begin with some historical observations. Modern dynamics has its roots in the classical mechanics of Newton, Euler, Lagrange, Jacobi, Gauss, Hamilton et al., during the eighteenth and nineteenth centuries. Classical mechanics, along with mathematics and astronomy, were the central intellectual disciplines of virtually all of the major European scholars; the physical problem that lay near their common focus was the dynamics of the solar system. These scholars were concerned with establishing the cause/effect relationships that govern nature (hence the British moniker "natural philosophy"); of specific concern was the motion of particles, rigid bodies, elastic bodies, and fluids under the influence of "natural forces" such as gravity. Fortunately (for our collective present livelihoods), they largely left for the twentieth century the consideration of dynamical behavior under the influence of active (man-made) forces, the field we now know as control of dynamical systems.

Classical mechanics, in terms of the rate of mathematical and physical innovation, reached its intellectual apogee in the late 1800s. While the field remained vital and there were some notable exceptions, the intellectual momentum and rate of innovation subsided by at least one order of magnitude during the first half of the twentieth century. Then came Sputnik and the many-faceted new wave of basic and applied research, which has aggressively extended the classical works and integrated these rapidly evolving ideas with computational, electronic, and allied physical process innovations to establish the underpinnings of modern theory and applications.

Control theory and control engineering developed in a somewhat uncoupled fashion (from the research of mechanics scholars) with the most obvious period of aggressive innovations initiating during the prewar effort of the late 1930s. From a historical viewpoint, this is somewhat surprising. Had control theory been developed during the 1800s, then it would almost certainly have been developed by Lagrange, Hamilton, Jacobi et al. as natural appendages of classical mechanics; the masters would have viewed optimal control theory (as we do) as simply a quest for the best forces and moments to use as inputs to their mathematical models, which drive a mechanical system from State A to State B. Of course, this mechanics egocentric view of control theory and control engineering oversimplifies and trivializes many important aspects of modern control methodology such as signal processing, which have very little direct relationship to classical mechanics. On the other

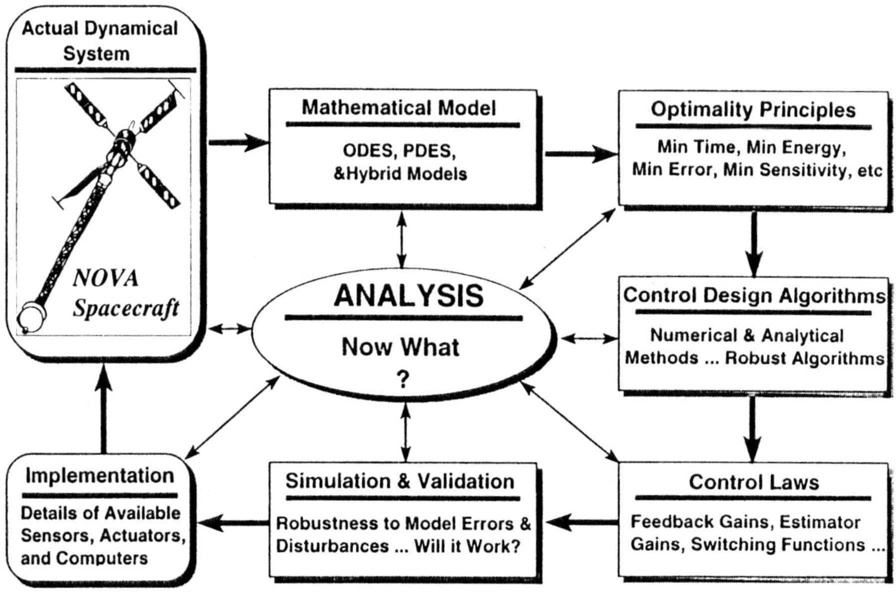

Figure 1.1 Mechanics and Control of Flexible Structures.

hand, virtually the entire field of signal processing has its roots in Gauss's "theory of errors" and the method of least squares which he digressed to invent while determining planetary orbits from telescope measurements of planetary motion! It is most curious, and should cause us mechanics types to pause for a humility lesson, that the fundamental control theory contributions made during the first half of this century were made primarily by mathematicians and electrical engineers with little formal training in classical mechanics. Where were the mechanics scholars?

During the first half of the twentieth century, there were in fact very few mechanics scholars and the most active subset of these (physicists) were focused mainly on quantum mechanics and were engaged in an important quest (which continues today) for the grand unification of theories for strong and weak forces, particles, and physical principles. Apparently classical mechanics and its natural appendage (control of dynamical systems) were felt by physicists to be too applied for their attention, and the level of mechanics competence among mechanical engineers was not yet sufficiently mature to undertake the needed developments. Thus the few available mechanics researchers were slow to recognize the importance of control theory; apparently servomechanisms had less esthetic appeal than quantum mechanics. This dearth of interest/competence in applied mechanics among the first half-century researchers and their relative lack of interest in the connections of control theory with mechanics, continued well into the 1960s. However, motivated by an unprecedented set of opportunities to apply dynamics and control methodology, during the past three decades, we have witnessed a tremendous acceleration in the number of engineering mechanics researchers and engineers who have acquired excellent breadth and depth of competence in both mechanics and controls. A very substantial mechanics/control research literature has evolved, especially since the mid 1980's, and in essence, this text is a direct effort to integrate some of the most important ideas into a coherent source. It is increasingly evident that many of the most important theoretical and practical problems cannot be satisfactorily addressed unless a unified view is taken. In control of high-dimensioned structural dynamic systems, for example, the issues of mathematical model errors, model simplification, order reduction and real time computation play pivotal roles in the design of practical control systems. Also of importance is the issue of physically motivated coordinate transformations, especially for nonlinear mechanical systems. It is easy to show examples in which the same physical motion can often be described by either quasilinear or highly nonlinear differential equations, depending on coordinate selection. For high-dimensioned linear models, it is extremely important to understand the extent to which computed high-frequency dynamics of the model are accurately converged; blind attempts to control the dynamics predicted by a many-degree-of-freedom finite element model can lead to absurd control laws and conclusions from simulations. The analyst who approaches the mechanics and control problem in a unified fashion has the best hope of obtaining satisfactory results. Also the electromechanical systems embodied as sensors and actuators often display subtle characteristics which require careful modeling. It is apparent to us that one not schooled in mechanics (who must be "given" the dynamical system model or obtain it through rote execution of a canned computer code) is at a serious disadvantage.

We believe that the trend toward unification of many aspects of dynamics and control is an extremely healthy development and should be fostered in our academic, governmental, and industrial institutions, most especially in the manner in which mechanics and control are taught. This text has grown out of our recent teaching, research, and applications, which are addressed to these issues. We anticipate that this text will provide a useful basis for teaching, learning, and applying this unified perspective.

1.2 Scope and Organization

With reference to Figure 1.1, we overview the problems associated with modeling and controlling a flexible structure such as the NOVA spacecraft shown. When designing control laws for a complicated structure, it is necessary to introduce idealizations of the actual structure to make analysis feasible. Regardless of how well this modeling process is done, we can anticipate the existence of model errors. Although model errors can usually be reduced through *system identification* based on experiments with the actual structure, the design of the control system should proceed with the explicit recognition that model errors are likely present and care should be taken to make the control laws developed to be tolerant of realistic levels of model and measurement errors.

This text strives for unification and balance in presenting some fundamental ideas and recent research on mechanics and control of structures. It is felt that this book will serve as a good text for a one-academic-year graduate course on structural dynamics and control. We present methodololgy for structural mathematical modeling, optimal control design and solution algorithms, and through case studies, we illustrate how these ideas are synthesized to resolve the modeling, control design, and implementation issues diagrammed in Figure 1.1. The organization of this text is unusual, as is the balance between theoretical, analytical, computational and experimental issues. There are only seven chapters, six dealing with methodololgy. The seven chapter format is deceptive, because each chapter excluding this first one is fairly long, initiates with basic ideas and theoretical discussions, develops a family of related ideas in detail with a number of examples, then ends with a significant case study to illustrate how the issues pertain to a prototype problem. While the treatment is not claimed to be comprehensive, we do feel we have developed a central core of ideas in a rigorous fashion and have efficiently presented selected applications-oriented issues to enliven the discussion. The computational exercises of this text were performed almost exclusively with MATLAB® (a trademark of MathWorks, Inc., of Natick, MA, USA), and using the operators discussed in Appendix A. The source code for these operators is a supplement for this text.

These operators are provided mainly for illustration and educational purposes; they are not intended to constitute a professional toolbox. However, they are likely to provide important starting points for many applications and will assist analysts in inventing new methodologies appropriate for their applications. We have achieved an unusual degree of completeness in this text in the sense that the basic theory, algorithms, and applications/experimental results are housed in a single text. While we have developed the analytical and theoretical discussions in a fashion to maximize their wide accessibility, we have attempted to retain a socially acceptable degree of rigor in their presentation. Some detailed proofs are not presented, but the key ideas are explained.

Requisite background material in linear algebra and matrix computations is presented in Chapter 2. Central to Chapter 2 is the detailed discussion of matrix decompositions, especially the singular value decomposition (SVD), and eigenvalue/sensitivity issues. In the authors' opinion, the SVD is the most important and widely useful computational linear algebra result of the twentieth century, and it is utilized throughout this text. A reasonable number of examples and illustrations are included in the text discussion, and a significant collection of exercises is provided for students to further test and develop their understanding. Chapter 2 concludes with a case study on parameterization of orthogonal matrices. This case study introduces some interesting classical material for parameterizing 3×3 orthogonal matrices and extends these results to the n-dimensional case through the use of the Cayley transform. Chapter 2 represents a unique synthesis and organization of a large volume of ideas from computational linear algebra that will be found useful across many engineering disciplines. For graduate teaching, depending on student preparation, much of Chapter 2 could be skipped and returned to as needed. For example, if the students have studied computational linear algebra from a text such as *Matrix Computations* by G.H. Golub and C.F. Van Loan, then Chapter 2 can be assigned for reading, with little formal lecturing. However, it is recommended that the students carry out a significant subset of the analytical and computational exercises prior to continuing proceeding in this text.

Chapter 3 provides a sharply focused course in Lyapunov stability theory and applications, including some recent extensions that are important when using Lyapunov functions to design control laws. The discussion begins with basic definitions and culminates in general methodolgy and a case study of a flexible structure maneuver control experiment. Methodology for both linear and nonlinear systems is developed, and both ordinary and partial differential equation models are considered. An important idea is presented in section 3.4: *the energy rate principle*, which vastly accelerates Lyapunov stability and control analysis for physical systems wherein a weighted "error energy" measure is used as the Lyapunov function. This principle allows one to bypass a large volume of algebra and calculus normally associated with Lyapunov analysis by making judicious use of the work/energy method from classical mechanics (here is a specific example where the unification of mechanics and control has proven highly advantageous; the sum is greater than the parts). The case study presented in section 3.5 makes use of this principle to design a globally stable tracking-type control law for a flexible body undergoing a near-minimum time, large angle maneuver with vibration suppression. Analytical, numerical, and experimental results are given for the tracking control law, as well as compared to experimental results using a simple constant gain output feedback control law.

Chapter 4 provides a terse but adequate development of the classical mechanics principles needed to derive mathematical models for the motion of structural dynamical systems. The generalized methods due to Lagrange et al. are adopted to formulate ordinary differential equation models for spatially discrete coordinate dynamical systems. Hamilton's principle and associated methods are presented for formulating the hybrid ordinary/partial differential equation systems and their natural boundary conditions, for describing the dynamics of distributed parameter systems. An explicit generalization of Lagrange's equations is presented for a significant family of distributed parameter systems. Using this formulation, analogous to the classical discrete coordinate Lagrange's equations, the governing hybrid system of partial differential equations is obtained by simply taking appropriate derivatives of energy functions, and the boundary conditions are derived from evaluation of associated transversality condition operators. We discuss the differential eigenvalue value problem and the properties of eigenfucntions associated with linear distributed parameter systems. The finite element method and the so-called *method of assumed modes* (a Ritz method) are introduced to discretize distributed parameter systems and obtain approximate finite-dimensional ordinary differential equation models. Several examples are given to illustrate these topics and a significant case study is discussed in section 4.4 to illuminate the potentially catastrophic consequences of neglecting the inertial coupling between vibration and rotation in a specific configuration. It is shown that a control law design that stabilizes the uncoupled model can destabilize the actual system, if the structure is sufficiently flexible.

The final three chapters deal with analysis and design of linear controllers for structural dynamic systems. In Chapter 5, we present two sets of topics on full state feedback control law design. We begin with a discussion of linear optimal control theory and develop the now classical results for the quadratic cost linear regulator (LQR). We present algorithms for solving the matrix Riccati equation whose solution is needed in the LQR feedback gain calculation. We also present a generalization of the quadratic performance measure that allows one to solve a simpler linear Lyapunov equation in lieu of the nonlinear Riccati equation. In section 5.2, we present a set of interesting analytical results and algorithms that permit, for controllable systems, arbitrary assignment of closed-loop eigenvalues and approximate assignment of eigenvectors. Several ideas are discussed regarding selection of target eigenvalues and eigenvectors, and simple examples are given to illustrate the ideas. A case study of eigenstructure assignment algorithms is given in section 5.3.

Chapter 6 presents a discussion of observability and controllability for linear dynamical systems. Several measures of controllability and observability are introduced, and relationships between them are discussed. These measures are mapped into modal coordinates, and concepts of *modal controllability* and *modal importance* are discussed. The consequences of actuator placement on the variation of these measures is considered, and a strategy for optimizing actuator placement is introduced. The validity of the actuator placement optimization is supported by the case study presented in section 6.4, where the ideas are applied to a simple grid structure. Laboratory experimental studies with the grid are summarized, and control of the system in the presence of unknown (to the control law) persistent disturbances is studied. The actuator locations are optimized, and control laws are designed by two methods (LQR and symmetric output feedback eigenstructure assignment).

Chapter 7 develops ideas and algorithms for output feedback of linear dynamical systems. The chapter opens with a fairly extensive literature survey. Section 7.2 provides some theoretical background for a guaranteed stabilizing symmetric parameterization of the feedback gain matrix. Stability and robustness characteristics, as well as the relation to the LQR parameterization of the gain matrix, are discussed. A nonlinear programming algorithm is introduced based on the symmetric gain matrix parameterization, and a homotopic continuation method is utilized to enhance convergence of a gain-tuning algorithm. Three case studies are given in section 7.5 to illustrate the utility of the methods in this chapter.

Nine appendices are given that are collectively the size of an eighth chapter for this book. We relegated this material to appendices with some trepidation. Use of extensive appendices is a dangerous practice in our experience, because there is a significant probability that some readers will place insufficient weight on the appendices relative to the main text, and not give serious attention to that material. On the other hand, the use of appendices is extremely important to make the text read smoothly, yet have adequate closure and good organization. For example, it is clearly preferable to discuss the MATLAB operators in a single appendix rather than sprinkling this material throughout the text. It is hoped that the reader will invest the attention required to digest the appendices; these details must be included in a serious study to obtain the full benefit of this text.

Finally some remarks are offered on what this text is not. A course based solely on this text may not be considered an adequate replacement for a traditional specialty course in either advanced analytical dynamics or control system analysis. For example there is no discussion herein of the traditional analytical dynamics topics: Hamilton-Jacobi theory, Gibbs-Appell equations, canonical transformation theory, phase plane analysis, or perturbation theory. Likewise, conspicuous by their absence here are the standard linear control topics on Laplace domain transfer function and frequency response analysis, and the classical frequency-domain control design methods for single input, single output controllers, as well as recently developed multiinput, multi-output control design methods based on H_∞ and other frequency-domain concepts. Given the diversity of approaches, it is not expected that our subjective decisions (on the scope and depth penetration in the text) will meet with universal concurrence. However, we feel that there is an emerging view that a large central core of the dynamics and control concepts can and should be unified as they have been in this text.

We anticipate that most readers will agree that the material included in this text certainly belongs to a set of important topics that lie near the centroid of the intersection of two traditionally disjoint disciplines of dynamics and control. In a unified approach, it is our contention that this material can be more efficiently taught and will allow the students to gain a unified perspective early in their graduate studies. In addition to the blending of mechanics and control, it is felt extremely important that we break down some of the traditional barriers between rigorous mechanics or controls analysis and the associated computational and experimental issues. While easy to state, treating such a broad scope of topics and methods in a coherent fashion is not an easy undertaking. We hope that this text is found attractive and useful in at least partially addressing this important need in a fashion consistent with the preparation of most first year graduate students in engineering. More specialized topics can and should be presented in sharply focused courses taken subsequently or in parallel with a course based on the present text.

Of course, with a modest amount of augmentation, experienced professors can supplement this text with material of their choosing in such a fashion to profitably customize or reinvent existing dynamics and control courses. This is expected and is consistent with our view of this dynamic subject matter.

CHAPTER 2

Mathematical Background: Matrix Analysis and Computation

The purpose of this chapter is to introduce a number of important results in matrix analysis, especially for matrix decomposition analysis. Although most of these results are well known, the conventional developments here and our extensions may provide some fresh insights. These developments are the algebraic foundation of the following chapters of this monograph. Several brief, low-dimensioned examples and computational exercises are presented. The computations were performed using MATLAB®, together with the associated Control System Toolbox (Ref. 1).

In section 2.1, basic notions are introduced, along with a family of linear system inverse problems:

(1) the determined case,
(2) the underdetermined case (minimum norm solution), and
(3) the overdetermined case (least squares solution).

Section 2.2 introduces some important matrix decompositions – *spectral decomposition, singular value decomposition (SVD), Cholesky decomposition,* and the *Schur decomposition,* – and discusses their usefulness. In section 2.3, sensitivities of eigenvalues, eigenvectors, and singular values are developed, and we discuss conditioning issues related to the sensitivities of linear systems. Section 2.4 presents a case study on parameterization of orthogonal matrices.

2.1 Introduction and Basic Notions

Most of the analysis of linear systems in modern control theory can be formulated in a finite dimensional linear space. This section introduces some basic notions related to the solutions of simultaneous linear algebraic equations. Tables 2.1 and 2.2 provides a summary of our notations for and properties of various types of matrices and, in particular, Table 2.2 summarizes the notations adopted throughout this text for elementary matrix operations.

2.1.1 Simultaneous Linear Algebraic Equations

Consider a set of simultaneous linear algebraic equations

$$\begin{aligned} \mathrm{y}_1 &= a_{11}\mathrm{x}_1 + a_{12}\mathrm{x}_2 + \ldots + a_{1n}\mathrm{x}_n \\ \mathrm{y}_2 &= a_{21}\mathrm{x}_1 + a_{22}\mathrm{x}_2 + \ldots + a_{2n}\mathrm{x}_n \\ &\vdots \\ \mathrm{y}_m &= a_{m1}\mathrm{x}_1 + a_{m2}\mathrm{x}_2 + \ldots + a_{mn}\mathrm{x}_n \end{aligned} \tag{2.1}$$

Table 2.1 Classification of Matrices $A=[a_{ij}]$

Type of Matrix	Defining Property	Necessary and Sufficient Condition
Symmetric	$A = A^T$	$a_{ij} = a_{ji}$
Hermitian	$A = A^H$	$a_{ij} = \text{conj}(a_{ji})$
Skew-Symmetric	$A = -A^T$	$a_{ij} = -a_{ji}$
Skew-Hermitian	$A = -A^H$	$a_{ij} = -\text{conj}(a_{ji})$
Orthogonal	$AA^T = A^TA = I$	
Unitary	$AA^H = A^HA = I$	
Idempotent	$A^2 = A$	
Nilpotent	$A^k = 0$ for some integer k	
Positive Definite	$\mathbf{x}^H A\mathbf{x} > 0 \ \forall\ \mathbf{x} \neq 0$	All eigenvalues of $A > 0$
Positive Semidefinite	$\mathbf{x}^H A\mathbf{x} \geq 0 \ \forall\ \mathbf{x} \neq 0$	All eigenvalues of $A \geq 0$

The above equation can be rewritten in matrix notation as follows:

$$\mathbf{y} = A\mathbf{x} \tag{2.2}$$

where

$$\mathbf{y} = \begin{Bmatrix} y_1 \\ y_2 \\ \vdots \\ y_m \end{Bmatrix}, \quad \mathbf{x} = \begin{Bmatrix} x_1 \\ x_2 \\ \vdots \\ x_n \end{Bmatrix}, \quad \text{and } A = \begin{bmatrix} a_{11} & a_{12} & \dots & a_{1n} \\ a_{21} & a_{22} & \dots & a_{2n} \\ \vdots & \vdots & \ddots & \vdots \\ a_{m1} & a_{m2} & \dots & a_{mn} \end{bmatrix}$$

The **standard inverse problem** is this: *"Find a solution vector* $\mathbf{x}$ *which satisfies the simultaneous linear algebraic equation [Eq. (2.2)] for given* $\mathbf{y}$ *and* A". Not every standard inverse problem has a solution; there are three possibilities:

(1) no exact solution,

(2) exactly one solution, or

(3) an infinity of exact solutions.

These three cases are further discussed and classified as follows.

Table 2.2 Notations for Elementary Matrix Operations $A=[a_{ij}]$

Function	Symbol	Description/Definition (computation)
Transpose	A^T	Interchange rows and columns (ij-*th* element of $A^T = a_{ji}$)
Conjugate Transpose	A^H	(ij-*th* element of $A^H = \text{conj}(a_{ji})$)
Rank	r(A)	The smaller of the number of lip† rows and the number of lip columns (compute via SVD)
Trace	tr(A)	Sum of the diagonal elements ($\sum a_{ii}$)
Column Space	$C(A)$	The vector space spanned by the columns of A (compute via SVD)
Null Space	$N(A)$	The vector space containing all $\mathbf{x} \neq 0$ satisfying $A\mathbf{x} = 0$ (determine via QR or SVD)

† Linearly independent (lip).

Overview of solutions of the inverse problem

(1) Determined case [$m = n$ =rank(A)]: one exact and unique solution exists.
(2) Underdetermined case [$n > m$ =rank(A)]: an infinity of solutions exist.
(3) Overdetermined case [$m > n$ =rank(A)]: usually no exact solution exists.

Note that if rank conditions are not satisfied, we can use the 'generalized inverse', as discussed below in section 2.1.2. The above three cases are now discussed in detail.

Determined Case

Since the matrix A is square and nonsingular, it has a unique inverse. Therefore, the unique and exact solution for $\mathbf{x}$ can be written in the symbolic form

$$\mathbf{x} = A^{-1}\mathbf{y} \tag{2.3}$$

In all cases where A^{-1} is written, it is understood that the solution for $\mathbf{x}$ can be obtained by many numerically more attractive elimination algorithms wherein it is not necessary to actually compute A^{-1}.

Underdetermined Case (Minimum Norm Solution)

When there are fewer independent equations than unknowns, the matrix A has dimensions $m < n$, and there is no possibility of a unique solution $\mathbf{x}$. Typically, we expect an infinity of exact solutions, and some criteria should be introduced to select a particular exact solution. One widely useful choice is the "minimum norm" solution, which is the solution $\hat{\mathbf{x}}$ that minimizes the norm of $\mathbf{x}$ subject to satisfying the given simultaneous equations, i.e.,

$$J = \frac{1}{2}\mathbf{x}^T\mathbf{x} \tag{2.4}$$

subject to

$$\mathbf{y} = A\mathbf{x} \tag{2.5}$$

The analytical solution can be obtained by applying the Lagrange multiplier rule. Straightforward minimization of the norm of $\mathbf{x}$ subject to the satisfaction of Eq. (2.5), is accomplished as follows.

Consider the augmented function

$$J_a = \frac{1}{2}\mathbf{x}^T\mathbf{x} + \underline{\lambda}^T(\mathbf{y} - A\mathbf{x}) \tag{2.6}$$

with $\underline{\lambda}$ denoting an $m \times 1$ vector of Lagrange multipliers.

To find the solution, we use the necessary conditions for a constrained minimum (Ref. 2) as follows*

$$\nabla_{\mathbf{x}} J_a = \hat{\mathbf{x}} - A^T\underline{\lambda} = 0 \tag{2.7}$$

$$\nabla_{\underline{\lambda}} J_a = \mathbf{y} - A\hat{\mathbf{x}} = 0 \tag{2.8}$$

Substituting Eq. (2.7) into Eq. (2.8) gives

$$(AA^T)\underline{\lambda} = \mathbf{y} \tag{2.9}$$

---

* The gradient of J_a can be obtained using the following properties (Ref. 2):
$\nabla_{\mathbf{x}}[\mathbf{x}^T M\mathbf{y}] = M\mathbf{y}$
$\nabla_{\mathbf{y}}[\mathbf{x}^T M\mathbf{y}] = M^T\mathbf{x}$
$\nabla_{\mathbf{x}}[\mathbf{x}^T M\mathbf{x}] = M\mathbf{x} + M^T\mathbf{x} = (M + M^T)\mathbf{x}$

(AA^T) is an $m \times m$ matrix with rank m and is therefore nonsingular. Solving for $\underline{\lambda}$ gives

$$\underline{\lambda} = \left(AA^T\right)^{-1}\mathbf{y} \tag{2.10}$$

and, therefore, the minimum norm solution is

$$\hat{\mathbf{x}} = A^T\left(AA^T\right)^{-1}\mathbf{y} \tag{2.11}$$

This result can also be derived by using an orthogonal projection method from a geometric point of view (Ref. 3).

A more general solution can be obtained by introducing a symmetric positive definite weight matrix W such that

$$J_a = \frac{1}{2}\mathbf{x}^T W\mathbf{x} + \underline{\lambda}^T(\mathbf{y} - A\mathbf{x}) \tag{2.12}$$

which leads to the weighted minimum norm solution

$$\hat{\mathbf{x}} = W^{-1}A^T\left(AW^{-1}A^T\right)^{-1}\mathbf{y} \tag{2.13}$$

The minimum norm solution is often adopted to solve constrained optimization problems, and this generalization has been shown (Ref. 4) to be mathematically equivalent to the gradient projection formulation.

Overdetermined Case (Least Squares Solution)

When there are more independent equations than unknowns, the $m \times n$ coefficient matrix A has $m > n$. Generally, no exact solution exists in this case, and an approximate solution vector $\mathbf{x}$ is desired. Since we expect that no one $\mathbf{x}$ can satisfy all of the simultaneous equations, it is inappropriate to write the equality $A\mathbf{x} = \mathbf{y}$. Rather, an $m \times 1$ error vector $\mathbf{e}$ is introduced

$$\mathbf{e} = \mathbf{y} - A\mathbf{x} \tag{2.14}$$

The least squares approach, due to Gauss, is used to find the approximate solution $\hat{\mathbf{x}}$ that minimizes the sum of the squares of the e_i components. That is, $\hat{\mathbf{x}}$ is determined to minimize

$$J = \frac{1}{2}\|\mathbf{e}\|^2 = \frac{1}{2}\mathbf{e}^T\mathbf{e} = \frac{1}{2}(\mathbf{y} - A\mathbf{x})^T(\mathbf{y} - A\mathbf{x}) \tag{2.15}$$

The necessary condition (for $\hat{\mathbf{x}}$ to yield the minimum of J) is that the gradient of J with respect to $\mathbf{x}$ must equal zero, i.e.,

$$\nabla_{\mathbf{x}} J = 0 = -A^T\mathbf{y} + A^T A\hat{\mathbf{x}}$$

or

$$(A^T A)\hat{\mathbf{x}} = A^T \mathbf{y} \tag{2.16}$$

If rank(A)=rank($A^T A$) is n, then the above "normal equations" [Eq. (2.16)] can be solved for

$$\hat{\mathbf{x}} = \left(A^T A\right)^{-1} A^T \mathbf{y} \tag{2.17}$$

and this is the classical least squares solution. In Gauss's original developments (not in matrix notation), Eq. (2.16) was solved by a process now known as "Gaussian elimination with double pivoting."

If some equations are more reliable than others, a weighted least squares approximation can be used. That is, $\hat{\mathbf{x}}$ is obtained by minimizing

$$J = \frac{1}{2} e^T W e = \frac{1}{2} (\mathbf{y} - A\mathbf{x})^T W (\mathbf{y} - A\mathbf{x}) \tag{2.18}$$

Using the same approach, the weighted least squares solution is obtained as follows:

$$\hat{\mathbf{x}} = \left(A^T W A\right)^{-1} A^T W \mathbf{y} \tag{2.19}$$

Note that if A is not full rank, the required inverse will not exist, and the classical least squares solution cannot be computed. In the following discussion, we show that the above explicit inversion results of Eqs. (2.11), (2.13), (2.17), and (2.19), while valid and widely used, are not the most efficient or numerically well conditioned methods for computing the minimum norm and least squares solutions.

Example 2.1 Polynomial Curve Fitting

Consider that we have m sets of measurement data (y_j, t_j) and that we want to obtain the coefficients of the best-fitting polynomial curve such that

$$y_j = x_1 + x_2 t_j + x_3 t_j^2 + \ldots + x_n t_j^{n-1}, \qquad j = 1, 2, \ldots, m$$

or

$$\begin{Bmatrix} y_1 \\ y_2 \\ \vdots \\ y_m \end{Bmatrix} = \begin{bmatrix} 1 & t_1 & t_1^2 & \ldots & t_1^{n-1} \\ 1 & t_2 & t_2^2 & \ldots & t_2^{n-1} \\ \vdots & \vdots & \vdots & \ddots & \vdots \\ 1 & t_m & t_m^2 & \ldots & t_m^{n-1} \end{bmatrix} \begin{Bmatrix} x_1 \\ x_2 \\ \vdots \\ x_n \end{Bmatrix}$$

or

$$\mathbf{y} = A\mathbf{x}$$

The above matrix is known as the *Vandermonde matrix*. As we will subsequently see, this family of matrices is very poorly conditioned for large m and especially for large n. Since the problem is overdetermined, usually no coefficient vector $\mathbf{x}$ will yield an exact solution, and so we let $\mathbf{e}$ denote the residual error vector associated with any particular choice $\hat{\mathbf{x}}$.

$$\mathbf{y} = A\hat{\mathbf{x}} + \mathbf{e}$$

Using Gauss's principle of least squares, we seek to find the solution $\hat{\mathbf{x}}$ that minimizes the norm of residual error $J = \mathbf{e}^T\mathbf{e}$. Therefore, the optimal coefficients of the curve-fitting polynomial can be obtained via the least squares method, for example, using Eq. (2.17) or (2.19).

■

Note that the general solution for the nonsingular case ($m = n$), the underdetermined minimum norm case ($m < n$), and the overdetermined least squares case ($m > n$) can *all* be expressed in terms of the *pseudoinverse* denoted as $\mathbf{x} = A^{\dagger}\mathbf{y}$ (Ref. 5), the details of the pseudoinverse will be discussed later in section 2.2.2. It is important note here that the pseudoinverse can also be computed for the case in which A does not have maximum rank, i.e., for the case in which $\text{rank}(A) < \min(m, n)$.

2.1.2 Numerical Methods to Solve the Least Squares Problem

There are many numerical methods available to solve least squares problems, and four of them will be discussed here, as a preview and as motivation for the subsequent developments in this chapter that covers these methods in detail.

(a) "Brute force" inversion

Direct inversion of (A^TA) is executed, i.e.,

$$\hat{\mathbf{x}}_1 = (A^TA)^{-1}A^T\mathbf{y} \tag{2.20}$$

The MATLAB operation to implement Eq. (2.20) is

$$> \mathbf{xhat1 = inv(A' * A) * A' * y}$$

Note that if the condition of (A^TA) is poor, it can be anticipated that $\hat{\mathbf{x}}_1$ will be inaccurate.

(b) Gaussian elimination solution

Equation (2.16) is used to solve for $\hat{\mathbf{x}}$ instead of Eq. (2.17)≡ (2.20); thus we use Gaussian elimination to solve the following linear equation for $\hat{\mathbf{x}}_2$:

$$(A^TA)\hat{\mathbf{x}}_2 = A^T\mathbf{y} \tag{2.21}$$

The solution is obtained by premultiplying this equation by a sequence of elementary row and column operator matrices that reduce A^TA to upper triangular form; $\hat{\mathbf{x}}_2$ can be computed by back-substitution. The advantage of doing this is that a matrix inversion is not directly needed; therefore, this approach is faster (requires fewer operations) than the brute force matrix inversion approach.

(c) Pseudoinversion using the singular value decomposition

The *Moore-Penrose generalized inverse* can be used to solve for $\mathbf{x}$ from

$$A\mathbf{x} = \mathbf{y} \tag{2.22}$$

for the determined, overdetermined, and underdetermined cases, *including all circumstances in which* A *is not of maximum rank.* This approach uses the singular value decomposition, which will be presented in section 2.2.2, therefore we will not discuss the details of singular value decomposition and the Moore-Penrose generalized inverse in the preview presented here. The solution using the Moore-Penrose generalized inverse (Ref. 5) is denoted as

$$\hat{\mathbf{x}}_3 = A^{\dagger}\mathbf{y} \tag{2.23}$$

The MATLAB operation for computing the pseudoinverse is

```
> xhat3 = pinv(A) * y
```

As is noted above, the Moore-Penrose generalized inverse is valid not only for least squares solution but also for minimum norm solution. This inverse is slow to compute, but it is numerically accurate.

(d) QR factorization

This approach uses the *QR decomposition*, as defined in the following theorem.

Theorem 2.1: QR Decomposition

Let $A \in R^{m\times n}$ have linearly independent columns. Then A can be written uniquely in the factored form

$$A = QR \tag{2.24}$$

where $Q \in R^{m\times m}$ has orthonormal columns and $R \in R^{m\times n}$ is upper triangular with positive diagonal elements, i.e.,

$$Q^TQ = QQ^T = I, \quad R = \begin{bmatrix} U \\ 0 \end{bmatrix} \tag{2.25}$$

■

For an $m \times n$ matrix A of rank r (with no restrictions on the relative size of m and n), the QR decomposition produces the matrix $Q \in R^{m \times r}$, which has orthonormal columns, and the matrix $R \in R^{r \times n}$, which is upper triangular in the sense that the (i,j) element r_{ij} of R equals zero if $i > j$.

Equation (2.22) can be rewritten via QR factorization as

$$QR\hat{\mathbf{x}}_4 = \mathbf{y}$$

Since Q is orthonormal, premultiplying both sides by Q^T, then the above equation becomes

$$R\hat{\mathbf{x}}_4 = Q^T\mathbf{y}$$

Using Eq. (2.25) in the above equation and partitioning the Q matrix yield

$$\begin{bmatrix} U \\ 0 \end{bmatrix} \hat{\mathbf{x}}_4 = \begin{bmatrix} Q_1^T \\ Q_2^T \end{bmatrix} \mathbf{y}$$

or

$$U\hat{\mathbf{x}}_4 = Q_1^T\mathbf{y} \tag{2.26}$$

Now, an approximate solution $\hat{\mathbf{x}}_4$ can be obtained by back-substitution. As proven in Ref. 5, a remarkable truth can be established: $\hat{\mathbf{x}}_4$ is the least squares solution that minimizes $\|\mathbf{y} - \mathbf{Ax}\|$.

Note that the Gaussian elimination approach and the QR factorization approach are faster $[o(n^2)]$ than the Moore-Penrose inversion approach and brute force approach $[o(n^3)]$; however, the QR factorization solution and Moore-Penrose inversion solution are more accurate and reliable than the Gaussian elimination solution and brute force inversion solution (Ref. 6). Because of the combination of efficiency and precision, the QR algorithm is probably the best approach to computing the overdetermined (least squares) solution.

Example 2.2

Consider the following polynomial curve-fitting problem

$$\begin{bmatrix} 1 & t_1 & t_1^2 & \cdots & t_1^{n-1} \\ 1 & t_2 & t_2^2 & \cdots & t_2^{n-1} \\ \vdots & \vdots & \vdots & \ddots & \vdots \\ 1 & t_m & t_m^2 & \cdots & t_m^{n-1} \end{bmatrix} \begin{Bmatrix} x_1 \\ x_2 \\ \vdots \\ x_n \end{Bmatrix} = \begin{Bmatrix} y_1 \\ y_2 \\ \vdots \\ y_m \end{Bmatrix}$$

Table 2.3 Accuracy of Least Squares Solution[a]

Dimension of $A(m, n)$	Brute Force Inversion	Gaussian LU Factorization and back-substitution[b]	SVD-based Moore-Penrose Pseudo Inverse
(20,2)	8×10^{-16}	5×10^{-16}	8×10^{-16}
(20,3)	9×10^{-14}	3×10^{-14}	4×10^{-14}
(20,4)	6×10^{-12}	2×10^{-12}	6×10^{-13}
(20,5)	6×10^{-10}	1×10^{-10}	7×10^{-12}
(20,6)	5×10^{-8}	8×10^{-9}	8×10^{-11}
(20,7)	7×10^{-6}	9×10^{-7}	4×10^{-9}
(20,8)	9×10^{-4}	6×10^{-4}	5×10^{-8}
(20,9)	1×10^{-1}	1×10^{-2}	3×10^{-6}
(20,10)	2×10^{1}	2	7×10^{-5}

[a] The numerical relative errors $\|\hat{\mathbf{x}} - \mathbf{x}\|/\|\mathbf{x}\|$ are shown. All computation was performed using MATLAB on an IBM AT math coprocessor.

[b] Solved by reducing $(A^T A)$ to upper triangular form and back substituting.

To generate the data for a known "true" $\mathbf{x}$ vector, take the values

$$\{t_1, t_2, \ldots, t_m\} = \{0, 1, 2, \ldots, m-1\}$$

$$\{x_1, x_2, \ldots, x_n\} = \{1, 1, 1, \ldots, 1\}$$

Then the data vector $\mathbf{y}$ is evaluated from the equation $\mathbf{y} = \mathbf{Ax}$; note that

$$y_j = \sum_{k=1}^{n} x_k t_j^{k-1}$$

The relative error between the true solution vector and the least squares solution is defined as

$$\text{error} \equiv \frac{\|\hat{\mathbf{x}} - \mathbf{x}\|}{\|\mathbf{x}\|} \tag{2.27}$$

Note that the negative exponent provides the approximate number of significant digits of the solution. For $m = 20$, $n = 5$, we find that the error using brute force inversion is 6×10^{-10}, that the error using Gaussian LU reduction and back-substitution is 1×10^{-10}, and that the error using Moore-Penrose pseudoinverse (via SVD computation) is 7×10^{-12}. The QR solution (not tabulated) was essentially the same as the SVD. These results were obtained using MATLAB with an 80-bit math co-processor. Increasing the dimensionality and the associated ill-conditioning of the Vandermonde matrix affects the above computation, and Table 2.3 shows the accuracy [Eq. (2.27)] of least squares solutions with different dimensions of the Vandermonde matrices; it is evident that the SVD and QR solutions are vastly superior in this application.

■

2.1.3 Partitioned Matrix Inversion Formulas and the Matrix Inversion Lemma

For the special case that $m = n$ in Eq. (2.2), $\mathbf{y} = \mathbf{A}\mathbf{x}$ can be written as

$$\mathbf{y} = \mathbf{A}\mathbf{x} \rightarrow \begin{Bmatrix} \mathbf{y}_1 \\ \mathbf{y}_2 \end{Bmatrix} = \begin{bmatrix} \mathcal{A} & \mathcal{B} \\ \mathcal{C} & \mathcal{D} \end{bmatrix} \begin{Bmatrix} \mathbf{x}_1 \\ \mathbf{x}_2 \end{Bmatrix} \tag{2.28}$$

or as the two equations

$$\mathbf{y}_1 = \mathcal{A}\mathbf{x}_1 + \mathcal{B}\mathbf{x}_2 \tag{2.29}$$

$$\mathbf{y}_2 = \mathcal{C}\mathbf{x}_1 + \mathcal{D}\mathbf{x}_2 \tag{2.30}$$

Assuming that the square diagonal submatrices $(\mathcal{A}, \mathcal{D})$ are nonsingular, we can solve Eqs. (2.29), (2.30) two different ways. In the process, some important matrix identities – which express $\mathbf{A}^{-1}$ as a function of its submatrices – will be established.

First form

(a) Solve Eq. (2.29)

$$\mathbf{x}_1 = \mathcal{A}^{-1}\mathbf{y}_1 - \mathcal{A}^{-1}\mathcal{B}\mathbf{x}_2 \tag{2.31}$$

(b) Substitute Eq. (2.31) into Eq. (2.30) and solve for $\mathbf{x}_2$:

$$\mathbf{x}_2 = -\Delta^{-1}\mathcal{C}\mathcal{A}^{-1}\mathbf{y}_1 + \Delta^{-1}\mathbf{y}_2 \tag{2.32}$$

with

$$\Delta \equiv \mathcal{D} - \mathcal{C}\mathcal{A}^{-1}\mathcal{B}$$

(c) Substitute Eq. (2.32) into Eq. (2.31) to obtain:

$$\mathbf{x}_1 = (\mathcal{A}^{-1} + \mathcal{A}^{-1}\mathcal{B}\Delta^{-1}\mathcal{C}\mathcal{A}^{-1})\mathbf{y}_1 - \mathcal{A}^{-1}\mathcal{B}\Delta^{-1}\mathbf{y}_2 \tag{2.33}$$

(d) It is evident from Eq. (2.28) that

$$\begin{Bmatrix} \mathbf{x}_1 \\ \mathbf{x}_2 \end{Bmatrix} = \begin{bmatrix} \mathcal{A} & \mathcal{B} \\ \mathcal{C} & \mathcal{D} \end{bmatrix}^{-1} \begin{Bmatrix} \mathbf{y}_1 \\ \mathbf{y}_2 \end{Bmatrix} \tag{2.34}$$

So, by comparing Eq. (2.34) with Eqs. (2.32) and (2.33)

$$A^{-1} = \begin{bmatrix} \mathcal{A} & \mathcal{B} \\ \mathcal{C} & \mathcal{D} \end{bmatrix}^{-1} = \begin{bmatrix} \mathcal{A}^{-1} + \mathcal{A}^{-1}\mathcal{B}\Delta^{-1}\mathcal{C}\mathcal{A}^{-1} & -\mathcal{A}^{-1}\mathcal{B}\Delta^{-1} \\ -\Delta^{-1}\mathcal{C}\mathcal{A}^{-1} & \Delta^{-1} \end{bmatrix} \tag{2.35}$$

where $\Delta = \mathcal{D} - \mathcal{C}\mathcal{A}^{-1}\mathcal{B}$, this is known as the *Schur complement.*

Second form

(a) Solve Eq. (2.30):

$$\mathbf{x}_2 = -\mathcal{D}^{-1}\mathcal{C}\mathbf{x}_1 + \mathcal{D}^{-1}\mathbf{y}_2 \tag{2.36}$$

(b) Substitute Eq. (2.36) into Eq. (2.29) and solve for $\mathbf{x}_1$:

$$\mathbf{x}_1 = \nabla^{-1}\mathbf{y}_1 - \nabla^{-1}\mathcal{B}\mathcal{D}^{-1}\mathbf{y}_2 \tag{2.37}$$

with the alternate Schur complement form

$$\nabla \equiv \mathcal{A} - \mathcal{B}\mathcal{D}^{-1}\mathcal{C}$$

(c) Substitute Eq. (2.37) into Eq. (2.36) to obtain

$$\mathbf{x}_2 = -\mathcal{D}^{-1}\mathcal{C}\nabla^{-1}\mathbf{y}_1 + (\mathcal{D}^{-1} + \mathcal{D}^{-1}\mathcal{C}\nabla^{-1}\mathcal{B}\mathcal{D}^{-1})\mathbf{y}_2 \tag{2.38}$$

(d) By comparing Eq. (2.34) with Eqs. (2.37) and (2.38)

$$A^{-1} = \begin{bmatrix} \mathcal{A} & \mathcal{B} \\ \mathcal{C} & \mathcal{D} \end{bmatrix}^{-1} = \begin{bmatrix} \nabla^{-1} & -\nabla^{-1}\mathcal{B}\mathcal{D}^{-1} \\ -\mathcal{D}^{-1}\mathcal{C}\nabla^{-1} & \mathcal{D}^{-1} + \mathcal{D}^{-1}\mathcal{C}\nabla^{-1}\mathcal{B}\mathcal{D}^{-1} \end{bmatrix} \tag{2.39}$$

and thus Eqs. (2.35), (2.39) provide alternate forms for A^{-1}.

Equations (2.35) and (2.39) suggest the possibility of finding the inverse of large matrices by computing inverses of smaller ones. Equating the (1,1) submatrices of Eqs. (2.35) and (2.39), we establish the following important identity:

$$\mathcal{A}^{-1} + \mathcal{A}^{-1}\mathcal{B}(\mathcal{D} - \mathcal{C}\mathcal{A}^{-1}\mathcal{B})^{-1}\mathcal{C}\mathcal{A}^{-1} = (\mathcal{A} - \mathcal{B}\mathcal{D}^{-1}\mathcal{C})^{-1} \tag{2.40}$$

Since the submatrices $\mathcal{A}$, $\mathcal{B}$, $\mathcal{C}$, $\mathcal{D}$ themselves represent rather general matrices, it follows from the above identity that the following *matrix inversion lemma* can be stated: For given nonsingular $(m \times m)$ A and $(n \times n)$ C matrices, and arbitrary $(m \times n)$ B and $(n \times m)$ D matrices,

$$(\mathrm{A} + \mathrm{BCD})^{-1} = \mathrm{A}^{-1} - \mathrm{A}^{-1}\mathrm{B}(\mathrm{DA}^{-1}\mathrm{B} + \mathrm{C}^{-1})^{-1}\mathrm{DA}^{-1} \tag{2.41}$$

While this procedure is not recommended as the basis for general-purpose matrix inversion algorithms, it finds many applications in iterative methods for solving nonlinear equations that require the inverse of a matrix that from step to step is modified by a matrix of low rank. This matrix inversion lemma has many other applications in system theory, for example, updating the covariance matrix in sequential estimation (Ref. 7), and evaluating certain transfer matrices that arise in state space realization theory (Ref. 8).

2.2 Matrix Decompositions (Ref. 9)

Matrix decomposition methods are playing an increasingly important role in the analytical and computational aspects of dynamics, estimation, and control of aerospace vehicles. In this section, several important matrix decompositions known as *spectral decomposition, singular value decomposition, Cholesky decomposition*, and *Schur decomposition* are introduced, and several illustrative examples are also given. Before concerning ourselves with the role that these ideas play in analysis of dynamical systems, it is instructive to first consider the "cleaner" applications of these ideas to solution of linear algebraic systems of equations.

2.2.1 Spectral Decomposition

The analysis of a system of algebraic or differential equations often becomes simplest and yields useful underlying properties if the coefficient matrices can be put into a diagonal form. Through solution of an eigenvalue problem, a square matrix A can usually be decomposed into the product of three matrices of the form.

$$\mathrm{A} = \Phi\Lambda\Phi^{-1} \tag{2.42}$$

where Λ is diagonal.

Before further discussion of the above *spectral decomposition*, let us introduce the following generalized eigenvalue problem (Ref. 9). Consider the "nondefective" system (the A, B pair have a set of n linearly independent eigenvectors) of differential equations

$$\mathrm{B}\dot{\mathbf{x}} = \mathrm{A}\mathbf{x} \tag{2.43}$$

Note that overdots denote time differentiation throughout this text. The *right and left eigenvalue problems* associated with $\mathbf{x} = \underline{\phi} e^{\lambda t}$ solutions of Eq. (2.43) are

$$\begin{aligned} &\text{right:} \qquad \lambda_i B \underline{\phi}_i = A \underline{\phi}_i, \quad i = 1, 2, \ldots, n \\ &\text{left:} \qquad \lambda_i B^T \underline{\psi}_i = A^T \underline{\psi}_i, \quad i = 1, 2, \ldots, n \end{aligned} \tag{2.44}$$

with the usual normalization of the *biorthogonality conditions*

$$\begin{aligned} \underline{\phi}_i^T B \underline{\phi}_i &= 1, \qquad i = 1, 2, \ldots, n \\ \underline{\psi}_j^T B \underline{\phi}_i &= \delta_{ij}, \qquad i, j = 1, 2, \ldots, n \end{aligned} \tag{2.45}$$

so that

$$\underline{\psi}_j^T A \underline{\phi}_i = \lambda_i \delta_{ij}, \qquad i, j = 1, 2, \ldots, n \tag{2.46}$$

The MATLAB statements to solve this generalized eigenvalue problem are

$$> \ [\mathbf{Phi}, \mathbf{Lambda}] = \mathbf{eig}(\mathbf{A}, \mathbf{B})$$

$$> \ [\mathbf{Psi}, \mathbf{Lambda}] = \mathbf{eig}(\mathbf{A}', \mathbf{B}')$$

where **Phi** and **Psi** are right and left *modal matrices*, and **Lambda** is a matrix with the eigenvalues on the diagonal.

Note that every eigenvalue solver (computer software package) uses its own special normalization equation for eigenvectors. In MATLAB, eigenvectors are normalized so that norm of each eigenvector becomes 1. In order to obtain eigenvectors that satisfy some other normalization conditions [e.g., Eqs. (2.45)], we need our own special normalization routine. The following code solves right and left eigenvalue problems and normalizes eigenvectors according to Eqs. (2.45). The MATLAB operator to compute this normalized eigensolution is

$$> \ [\mathbf{Lambda}, \mathbf{Phi}, \mathbf{Psi}] = \mathbf{eign}(\mathbf{A}, \mathbf{B})$$

Observe that the above code cannot be used for the repeated eigenvalue case since eigenvectors satisfying Eqs. (2.45) and (2.46) do not generally exist. The source code for **eign** is provided on the diskette and is discussed in Appendix A. There are many eigenvalue problem-related applications, such as system analysis and synthesis techniques that use modal frequencies and mode shapes, as well as stability analysis and controller design methods such as eigenstructure assignment techniques. We will consider these in subsequent chapters of this text.

Now let us introduce *spectral decomposition.* Defining the modal matrices of right and left eigenvectors as

$$\Phi = [\underline{\phi}_1, \dots, \underline{\phi}_n], \qquad \Psi = [\underline{\psi}_1, \dots, \underline{\psi}_n] \tag{2.47}$$

Then Eqs. (2.45) can be rewritten as follows

$$\begin{aligned} \Psi^{\mathrm{T}} \mathrm{B} \Phi &= \mathrm{I} \\ \Psi^{\mathrm{T}} \mathrm{A} \Phi &= \Lambda = \mathrm{diag}(\lambda_1, \dots, \lambda_n) \end{aligned} \tag{2.48}$$

Therefore, we can obtain spectral decomposition of A and B as follows

$$\begin{aligned} \mathrm{A} &= \Psi^{-\mathrm{T}} \Lambda \Phi^{-1} \\ \mathrm{B} &= \Psi^{-\mathrm{T}} \Phi^{-1} \end{aligned} \tag{2.49}$$

The following special spectral decompositions are frequently used.

Case I: B=I

Since $\Phi^{-1} = \Psi^{\mathrm{T}}$ from Eq. (2.45),

$$\mathrm{A} = \Phi \Lambda \Psi^{\mathrm{T}} = \Phi \Lambda \Phi^{-1} \tag{2.50}$$

Case II: B=I and A=A^{T}

Since $\Psi = \Phi$ from Eq. (2.44), and $\Phi^{-1} = \Psi^{\mathrm{T}}$ from Eq. (2.45), it is evident that $\Phi^{-1} = \Phi^{\mathrm{T}}$. Therefore Φ is orthogonal and we have the "self-adjoint" spectral decomposition

$$\mathrm{A} = \Phi \Lambda \Phi^{\mathrm{T}} \tag{2.51}$$

The diagonalization of A and existence of a nonsingular Φ depend upon matrix A having n linearly independent (but not necessarily unique) eigenvectors. If the eigenvalues of A are distinct, A can be diagonalized, otherwise we generally must solve a more complicated generalized eigenvalue problem for repeated eigenvalues for obtaining the modal matrix Φ, and Λ takes on a form that generally has entries only on the diagonal and the superdiagonal. For a repeated eigenvalue λ_i, one eigenvector can be obtained by Eq. (2.44). Additional generalized eigenvectors are obtained by the following linear operations on previous computed eigenvector corresponding to the same eigenvalues (Ref. 5), i.e.,

$$\mathrm{A} \underline{\phi}_{i+1} = \lambda_i \mathrm{B} \underline{\phi}_{i+1} + \underline{\phi}_i \tag{2.52}$$

If the resulting vector $\underline{\phi}_{i+1}$ is linearly independent of all vectors already found, it is a valid generalized eigenvector. If still more generalized eigenvectors are needed for λ_i, then solve

$$A\underline{\phi}_{i+2} = \lambda_i B\underline{\phi}_{i+2} + \underline{\phi}_{i+1} \tag{2.53}$$

and so on until the number of generalized eigenvectors computed equals the multiplicity of the repeated eigenvalue.

If A is symmetric and positive definite, then it can be shown that it can always be diagonalized by the spectral decomposition of Eq. (2.51), even for the repeated eigenvalue case.

Application 2.1: Solution of the Lyapunov Equation

Consider a matrix A such that all the eigenvalues of A have negative real parts. There exists a positive definite symmetric matrix P that satisfies the following Lyapunov equation:

$$AP + PA^T = -Q \tag{2.54}$$

if and only if the solution P of Eq. (2.54) satisfies $P = P^T > 0$ for any given matrix $Q = Q^T > 0$. This is known as the Lyapunov theorem.

The above Lyapunov equation can be solved via spectral decomposition, especially using special case I. Matrix A can be decomposed as Eq. (2.50),

$$A = \Phi\Lambda\Phi^{-1}$$

where Φ and Ψ $(=\Phi^{-T})$ are computed from the right and left eigenvalue problems, such as

$$A\Phi = \Phi\Lambda$$

$$A^T\Psi = \Psi\Lambda$$

with $\Psi^T\Phi = I$.

Using spectral decomposition, Eq. (2.54) can be rewritten as

$$(\Phi\Lambda\Phi^{-1})P + P(\Phi\Lambda\Phi^{-1})^T = -Q$$

Pre- and postmultiplication of the Lyapunov equation by Φ^{-1} and Φ^{-T}, respectively, gives

$$\Lambda\Phi^{-1}P\Phi^{-T} + \Phi^{-1}P\Phi^{-T}\Lambda = -\Phi^{-1}Q\Phi^{-T}$$

In order to avoid the inversion, we use the matrix Ψ^T instead of Φ^{-1}, so that the above equation becomes

$$\Lambda(\Psi^T P \Psi) + (\Psi^T P \Psi)\Lambda = -(\Psi^T Q \Psi)$$

Let us define T as

$$T = \Psi^T P \Psi \tag{2.55}$$

Rewrite the above equation by using index notation as

$$\sum_{k=1}^{n} [\Lambda]_{ik}[T]_{kj} + \sum_{k=1}^{n} [T]_{ik}[\Lambda]_{kj} = -[\Psi^T Q \Psi]_{ij}$$

Since Λ is a diagonal matrix, the above equation becomes

$$[\Lambda]_{ii}[T]_{ij} + [T]_{ij}[\Lambda]_{jj} = -[\Psi^T Q \Psi]_{ij}$$

Therefore,

$$[T]_{ij} = -\frac{[\Psi^T Q \Psi]_{ij}}{[\Lambda]_{ii} + [\Lambda]_{jj}} \tag{2.56}$$

Now, from the definition of T [Eq. (2.55)] with the relation $\Psi^T \Phi = I$, the Lyapunov solution P can be obtained as follows:

$$P = \Phi T \Phi^T \tag{2.57}$$

■

Note that the purpose of the above application is to show that spectral factorization may be used to solve the Lyapunov equation; however, one of the most efficient and robust algorithms to solve the Lyapunov equation utilizes QR factorization (Ref. 10).

2.2.2 Singular Value Decomposition

The singular values of a general matrix have many analogies with the eigenvalues of a Hermitian matrix; indeed, for a positive definite matrix, they are the same. Singular value decomposition has a number of practical applications in control theory, numerical analysis, and statistics. This is a most important result which is used extensively throughout the remainder of this text. In this section, we will establish the existence and fundamental properties of the singular value decomposition, and several elementary applications will be illustrated. The main features of singular value decomposition are captured in the following theorem.

Theorem 2.2: Singular Value Decomposition

Any rectangular complex-valued matrix $A \in C^{m \times n}$ that has rank r can be factored as

$$A = U\Sigma V^H = \begin{bmatrix} U_1 & U_2 \end{bmatrix} \begin{bmatrix} \Sigma_1 & 0 \\ 0 & 0 \end{bmatrix} \begin{bmatrix} V_1^H \\ V_2^H \end{bmatrix} \tag{2.58}$$

where $U \in C^{m \times m}$ and $V \in C^{n \times n}$ are unitary matrices normalized so that

$$\begin{aligned} U^H U &= I \\ V^H V &= I \end{aligned} \tag{2.59}$$

and the matrix $\Sigma_1 = \text{diag}(\sigma_1, \ldots, \sigma_r)$ contains as its elements the real *singular values* of A ordered as $\sigma_1 \geq \sigma_2 \geq \ldots \geq \sigma_r > 0$.

■

The MATLAB syntax for computing the singular value decomposition is

$$> \ \mathbf{[U, S, V] = svd(A)}$$

Note the notation that $S = \Sigma$ in Eq. (2.58).

In order to establish important relationships between the eigenvalue problem and $A \in C^{n \times n}$, matrix A is decomposed as follows:

$$A = U\Sigma V^H \tag{2.60}$$

$$U^H U = U U^H = I_n \tag{2.61}$$

$$V^H V = V V^H = I_n \tag{2.62}$$

with $\Sigma = \text{diag}(\sigma_1, \sigma_2, \ldots, \sigma_n)$.

Since U and V are both unitary matrices, we can use Eqs. (2.61) and (2.62) to rewrite Eq. (2.60) as

$$\Sigma = U^H A V \tag{2.63}$$

Equations (2.60)-(2.63) can be expressed as conditions on the $n \times 1$ complex column vectors of $U = [U_1 U_2 \ldots U_n]$ and $V = [V_1 V_2 \ldots V_n]$ as

$$AV_i = U_i \sigma_i \qquad i = 1, 2, \ldots, n \tag{2.64}$$

$$A^H U_i = V_i \sigma_i \qquad i = 1, 2, \ldots, n \tag{2.65}$$

$$U_i^H U_j = \delta_{ij} \qquad i, j = 1, 2, \ldots, n \tag{2.66}$$

$$V_i^H V_j = \delta_{ij} \qquad i, j = 1, 2, \ldots, n \tag{2.67}$$

$$\sigma_i \delta_{ij} = U_i^H A V_j \qquad i, j = 1, 2, \ldots, n \tag{2.68}$$

Multiplying Eqs. (2.64) and (2.65) by σ_i yields

$$AV_i\sigma_i = U_i{\sigma_i}^2 \tag{2.69}$$
$$A^H U_i\sigma_i = V_i{\sigma_i}^2 \tag{2.70}$$

Substituting Eq. (2.65) into Eq. (2.69), and Eq. (2.64) into Eq. (2.70), gives

$$\begin{aligned}(AA^H)U_i &= {\sigma_i}^2 U_i\\ (A^H A)V_i &= {\sigma_i}^2 V_i\end{aligned} \tag{2.71}$$

From the above equations, we can conclude that the singular values (σ_i^2) of A are the eigenvalues of (AA^H) and $(A^H A)$ and that the columns of U and V are the corresponding eigenvectors. From Eqs. (2.71), we can also notice that, if A is Hermitian, then the singular values of A are just the absolute values of the eigenvalues of A. While we are concerned with the case of a general A, certain of the details presented here apply only to the case of distinct singular values. The generalizations for the repeated singular value cases can probably be pursued in a fashion analogous to the developments in Ref. 11.

As we have mentioned in section 2.1.2, one important use of singular value decomposition is in the solution of the linear set of equations

$$A\mathbf{x} = \mathbf{y} \tag{2.72}$$

where $\mathbf{x} \in R^n$, $\mathbf{y} \in R^m$, and rank(A)=r.

The solution for $\mathbf{x}$ can be obtained using the Moore-Penrose generalized inverse, by considering the singular value decomposition [Eqs. (2.58) and (2.59)] of the A matrix.

Application 2.2: Moore-Penrose Generalized Inverse

Consider any $m \times n$ matrix A that has rank r, and A is decomposed as follows [Eqs. (2.58) and (2.59)]

$$A = U\Sigma V^H = [\,U_1 \quad U_2\,]\begin{bmatrix}\Sigma_1 & 0\\ 0 & 0\end{bmatrix}\begin{bmatrix}V_1^H\\ V_2^H\end{bmatrix}$$

The Moore-Penrose generalized inverse can be written as

$$A^\dagger = V\Sigma^\dagger U^H = V_1\Sigma^\dagger U_1^H \tag{2.73}$$

where

$$\Sigma^{\dagger} = \begin{bmatrix} \Sigma_1^{-1} & 0 \\ 0 & 0 \end{bmatrix}, \qquad \Sigma_1^{-1} = \text{diag}\left(\frac{1}{\sigma_1}, \ldots, \frac{1}{\sigma_r}\right)$$

■

Using the Moore-Penrose generalized inverse, the solution of Eq. (2.72) can be written as

$$\hat{\mathbf{x}} = \mathrm{A}^{\dagger}\mathbf{y} \tag{2.74}$$

We previously established for any real $m \times n$ matrix that [Eqs. (2.71)]

$$\mathrm{A}^{\mathrm{T}}\mathrm{A} = \mathrm{V}\Sigma_n\mathrm{V}^{\mathrm{T}}, \qquad \Sigma_n = \begin{bmatrix} \sigma^2 & 0 \\ 0 & 0 \end{bmatrix}_{n \times n}$$

$$\mathrm{A}\mathrm{A}^{\mathrm{T}} = \mathrm{U}\Sigma_m\mathrm{U}^{\mathrm{T}}, \qquad \Sigma_m = \begin{bmatrix} \sigma^2 & 0 \\ 0 & 0 \end{bmatrix}_{m \times m}$$

where $\sigma^2 = \text{diag}(\sigma_1^2, \sigma_2^2, \ldots, \sigma_r^2)$ since rank(A)=r.

From the above relations, the following three important algebraic equivalences can be derived:

(1) Least squares case: $m > n$ and r=rank(A)=n

$$\mathrm{A}^{\dagger} = (\mathrm{A}^{\mathrm{T}}\mathrm{A})^{-1}\mathrm{A}^{\mathrm{T}} \tag{2.75}$$

(2) Minimum norm case: $m < n$ and r=rank(A)=m

$$\mathrm{A}^{\dagger} = \mathrm{A}^{\mathrm{T}}(\mathrm{A}\mathrm{A}^{\mathrm{T}})^{-1} \tag{2.76}$$

(3) General or rank-deficient case: r=rank(A)$\leq \min(m, n)$

$$\mathrm{A}^{\dagger} = \mathrm{V}\Sigma^{\dagger}\mathrm{U}^{\mathrm{T}} \tag{2.77}$$

Note that the above algebraic equivalences [Eqs. (2.75)-(2.77)] do not imply the computational equivalences, as demonstrated earlier in section 2.1.2.

Example 2.3

Consider the equation

$$\mathbf{y} = \mathrm{A}\mathbf{x}$$

with

$$A = \begin{bmatrix} 1 & 2 & 3 & 0 & 0 \\ 0 & 0 & 4 & 5 & 6 \\ 0 & 0 & 0 & 0 & 7 \\ 1 & 2 & 7 & 5 & 6 \end{bmatrix}, \qquad \mathbf{x} = \begin{Bmatrix} 1 \\ 1 \\ 1 \\ 1 \\ 1 \end{Bmatrix}, \qquad \mathbf{y} = \begin{Bmatrix} 6 \\ 15 \\ 7 \\ 21 \end{Bmatrix}$$

The singular value decomposition of A is computed via the MATLAB operator

$$> \ [\mathbf{U}, \mathbf{S}, \mathbf{V}] = \mathbf{svd}(\mathbf{A})$$

For the given A matrix, this operator gives the following numerical solution for the singular vectors and singular values:

$$U = \begin{bmatrix} 0.1303 & -0.4318 & 0.6806 & 0.5774 \\ 0.5908 & 0.0686 & -0.5594 & 0.5574 \\ 0.3375 & 0.8227 & 0.4574 & 0.0000 \\ 0.7711 & -0.3632 & 0.1212 & -0.5774 \end{bmatrix}$$

$$V = \begin{bmatrix} 0.0584 & -0.1364 & 0.2708 & -0.9506 & 0.0322 \\ 0.1169 & -0.2729 & 0.5416 & 0.1746 & -0.7669 \\ 0.5355 & -0.6115 & 0.2205 & 0.2004 & 0.5005 \\ 0.4502 & -0.2528 & -0.7398 & 0.1604 & -0.4004 \\ 0.7202 & 0.6849 & 0.1935 & 0.0000 & 0.0000 \end{bmatrix}$$

$$S = \begin{bmatrix} \sigma_1 & 0 & 0 & 0 & 0 \\ 0 & \sigma_2 & 0 & 0 & 0 \\ 0 & 0 & \sigma_3 & 0 & 0 \\ 0 & 0 & 0 & 0 & 0 \end{bmatrix}$$

with $\sigma_1 = 14.5695$, $\sigma_2 = 5.8276$, and $\sigma_3 = 2.9611$.

Since A is rank-deficient [rank(A)=3 by inspection], the fourth row is the sum of rows 1 and 2 and therefore we cannot directly compute the minimum norm or least square solutions, since neither $(A^TA)^{-1}$ nor $(AA^T)^{-1}$ exists. We can, however, compute the Moore-Penrose solution as follows:

$$\hat{\mathbf{x}} = A^{\dagger}\mathbf{y} = \begin{Bmatrix} 0.3209 \\ 0.6419 \\ 1.4651 \\ 0.6279 \\ 1.0000 \end{Bmatrix}$$

with

$$A^{\dagger} = VS^{\dagger}U^{T}, \quad \text{and} \quad S^{\dagger} = \begin{bmatrix} \frac{1}{\sigma_1} & 0 & 0 & 0 \\ 0 & \frac{1}{\sigma_2} & 0 & 0 \\ 0 & 0 & \frac{1}{\sigma_3} & 0 \\ 0 & 0 & 0 & 0 \\ 0 & 0 & 0 & 0 \end{bmatrix}$$

Note that $\|\mathbf{x}\| = 2.236$, $\|\hat{\mathbf{x}}\| = 2.014$, and $\hat{\mathbf{x}}$ is the minimum norm solution.

Recall that the 4×5 A matrix is of rank three, therefore there are only three linearly independent rows and columns. The first three columns of V (associated with non-zero singular values) are an orthogonal set of vectors spanning the column space of A. The last two columns of V are orthogonal vectors spanning the (column) null space of A. Note that all vectors $\mathbf{x}$ lying in the null space of A satisfy $A\mathbf{x} = 0$; therefore, in this example,

$$AV_i = 0, \qquad i = 4, 5$$

i.e.,

$$\begin{bmatrix} 1 & 2 & 3 & 0 & 0 \\ 0 & 0 & 4 & 5 & 6 \\ 0 & 0 & 0 & 0 & 7 \\ 1 & 2 & 7 & 5 & 6 \end{bmatrix} \begin{Bmatrix} -0.9506 \\ 0.1746 \\ 0.2004 \\ 0.1604 \\ 0.0000 \end{Bmatrix} = \begin{Bmatrix} 0.0000 \\ 0.0000 \\ 0.0000 \\ 0.0000 \end{Bmatrix}$$

$$\begin{bmatrix} 1 & 2 & 3 & 0 & 0 \\ 0 & 0 & 4 & 5 & 6 \\ 0 & 0 & 0 & 0 & 7 \\ 1 & 2 & 7 & 5 & 6 \end{bmatrix} \begin{Bmatrix} 0.0322 \\ -0.7669 \\ 0.5005 \\ -0.4004 \\ 0.0000 \end{Bmatrix} = \begin{Bmatrix} 0.0000 \\ 0.0000 \\ 0.0000 \\ 0.0000 \end{Bmatrix}$$

In a similar fashion, we can show that the fourth column of U spans the (column) null space of A^T, or the (row) null space of A.

■

Application 2.3: Determination of Orthogonal Subspace (Ref. 12)

If the $m \times n$ matrix A has rank r and we partition the singular matrix U as

$$U = [\,U_1 \quad U_2\,] \tag{2.78}$$

where U_1 is $m \times r$, then the columns of U_1 correspond to the nonzero singular values and form an orthogonal basis for the column space of A. Consequently, the orthogonal projection onto the column space of A is given by

$$P_A = U_1 U_1^T \tag{2.79}$$

The projection onto the orthogonal complement is given by

$$\mathrm{P}_{\mathrm{A}}^{\perp} = \mathrm{U}_2\mathrm{U}_2^{\mathrm{T}} \tag{2.80}$$

Note that we can alternatively compute P_{A} and $\mathrm{P}_{\mathrm{A}}^{\perp}$ by using the right singular matrix V.

■

Application 2.4: Closest Unitary Matrix to a Given Matrix (Ref. 13)

If the $n \times n$ matrix A_o has rank n and is decomposed by

$$\mathrm{A}_o = \mathrm{U}\Sigma\mathrm{V}^{\mathrm{H}} \tag{2.81}$$

then the unitary matrix A_1, which minimizes $\|\mathrm{A} - \mathrm{A}_o\|_{\mathrm{F}}$ over all unitary matrices A, is given by

$$\mathrm{A}_1 = \mathrm{U}\mathrm{V}^{\mathrm{H}} \tag{2.82}$$

Note that the Frobenius norm $\|\cdot\|_{\mathrm{F}}$ of matrix A is just the Euclidean norm $\|\cdot\|_2$ of A considered as an element of C^{nxn}.

■

Another useful application of singular value decomposition is that it can be used to compute the condition number of a matrix (which is often used as a robustness index in computational linear algebra, and in control design). The most central matrix conditioning issues are discussed in section 2.4.

2.2.3 Cholesky Decomposition

We consider now the decomposition of a positive definite matrix $\mathrm{A} = [a_{\mathrm{ij}}]$.

Theorem 2.3

Every $n \times n$ real positive definite matrix A can be factored into a real lower triangular matrix (with positive diagonal terms) times its transpose such that

$$\mathrm{A} = \mathrm{L}\mathrm{L}^{\mathrm{T}} = \begin{bmatrix} l_{11} & 0 & 0 & \dots & 0 \\ l_{21} & l_{22} & 0 & \dots & 0 \\ l_{31} & l_{32} & l_{33} & \dots & 0 \\ \vdots & \vdots & \vdots & \ddots & \vdots \\ l_{n1} & l_{n2} & l_{n3} & \dots & l_{nn} \end{bmatrix} \begin{bmatrix} l_{11} & 0 & 0 & \dots & 0 \\ l_{21} & l_{22} & 0 & \dots & 0 \\ l_{31} & l_{32} & l_{33} & \dots & 0 \\ \vdots & \vdots & \vdots & \ddots & \vdots \\ l_{n1} & l_{n2} & l_{n3} & \dots & l_{nn} \end{bmatrix}^{\mathrm{T}} \tag{2.83}$$

■

The MATLAB syntax for computing the Cholesky decomposition is

```
> U = chol(A)
```

Note that the MATLAB operation gives us an upper triangular matrix U ($=L^T$). The Cholesky factor matrix (L) is widely known as the *matrix square root* of A, because of the obvious specialization to the square root if A is a scalar.

Carrying out the matrix multiplication in Eq. (2.83) and equating elements, we can establish the algorithm for computing the lower triangular Cholesky factor matrix (L) as follows:

Algorithm

For $i = 1, 2, \ldots, n$

For $j = 1, 2, \ldots, i-1$

$$l_{ij} = l_{jj}^{-1}\left(a_{ij} - \sum_{k=1}^{j-1} l_{jk} l_{ik}\right)$$

end

$$l_{ii} = \sqrt{a_{ii} - \sum_{k=1}^{i-1} l_{ik}^2}$$

end

■

If A is positive definite, the algorithm can always be carried to completion, and it requires about $n^3/6$ multiplications (Ref. 14). There are several useful applications of Cholesky decomposition such as:

(1) parameterization of any matrix for which we wish to guarantee positive definiteness (for example, weight matrices, and covariance matrices) ,

(2) solving linear systems with positive definite coefficient matrices, and

(3) the "square root form" of the Kalman filter covariance update algorithm.

Example 2.4

Consider the example A matrix for Cholesky decomposition:

$$A = \begin{bmatrix} 5 & 10 & 30 & 100 \\ 10 & 30 & 100 & 354 \\ 30 & 100 & 354 & 1300 \\ 100 & 354 & 1300 & 4890 \end{bmatrix}$$

MATLAB generates from > **U = chol(A)** the following results.

$$U = \begin{bmatrix} 2.2361 & 4.4721 & 13.4164 & 44.7214 \\ 0 & 3.1623 & 12.6491 & 48.6991 \\ 0 & 0 & 3.7417 & 22.4499 \\ 0 & 0 & 0 & 3.7942 \end{bmatrix}$$

It can be confirmed that $A = U^T U$. In most contexts, *matrix square root* is a qualitative description used to refer to the Cholesky decomposition.

■

2.2.4 Schur Decomposition

Schur decomposition is very closely related to the spectral decomposition. The next theorem shows how we may reduce a general matrix to upper triangular form by a unitary similarity transformation.

Theorem 2.4: Complex Schur Decomposition

Consider any $n \times n$ matrix A that has eigenvalues $\lambda_1, \lambda_2, \ldots, \lambda_n$ (not necessarily distinct). There exists a $n \times n$ unitary transformation matrix U such that S is upper triangular with the eigenvalues of the matrix on the diagonal.

$$S = U^H A U = \begin{bmatrix} \lambda_1 & * & * & \ldots & * \\ 0 & \lambda_2 & * & \ldots & * \\ 0 & 0 & \lambda_3 & \ldots & * \\ \vdots & \vdots & \vdots & \ddots & \vdots \\ 0 & 0 & 0 & \ldots & \lambda_n \end{bmatrix} \tag{2.84}$$

with $U^H U = U U^H = I$.

■

The *complex Schur form* of a matrix is upper triangular with the eigenvalues of the matrix on the diagonal as stated in the above theorem. The *real Schur form* has all real elements; the real eigenvalues lie on the diagonal and the complex eigenvalues in 2×2 blocks on the diagonal, such as

$$S = U^T A U = \begin{bmatrix} \lambda_1 & * & * & * & \ldots & * \\ 0 & \lambda_2 & * & * & \ldots & * \\ 0 & 0 & \lambda_{3r} & \lambda_{3i} & \ldots & * \\ 0 & 0 & -\lambda_{3i} & \lambda_{3r} & \ldots & * \\ \vdots & \vdots & \vdots & \vdots & \ddots & \vdots \\ 0 & 0 & 0 & 0 & \ldots & \lambda_n \end{bmatrix} \tag{2.85}$$

with $U^T U = U U^T = I$. In the most general case, the complex eigenvalues may not appear as shown, but rather the eigenvalues of each 2×2 block are equal to the corresponding eigenvalues of A. Notice that, if A is real symmetric, the Schur decomposition becomes the spectral decomposition. The MATLAB syntax for the Schur decomposition is

$$> \ [\mathrm{U},\mathrm{S}] = \mathbf{schur}(\mathbf{A})$$

The MATLAB operator **schur** computes the Schur form S and the associated orthogonal matrix U for the given matrix A. Note that, in MATLAB, if the matrix A is purely real, **schur** returns the real Schur form, if $imag(\mathrm{A})$ is not zero, the complex Schur form is returned. An important application of the Schur decomposition is to solve the algebraic Riccati equation (ARE), as illustrated in the following example.

Application 2.5: Solving the Algebraic Riccati Equation (Ref. 15)

Find the Riccati solution $\mathrm{P} \in \mathrm{R}^{n\times n}$ which satisfies the ARE

$$\mathrm{A^TP + PA - PBR^{-1}B^TP + Q = 0} \tag{2.86}$$

where Q is a positive semidefinite matrix, and R is a positive definite matrix. A solution for P can be constructed as follows:

(1) Construct the *Hamiltonian matrix*:

$$\mathcal{H} = \begin{bmatrix} \mathrm{A} & -\mathrm{BR^{-1}B^T} \\ -\mathrm{Q} & -\mathrm{A^T} \end{bmatrix} \tag{2.87}$$

(2) Compute the Schur decomposition of $\mathcal{H}$:

$$\mathcal{H} = \mathrm{USU^T} \tag{2.88}$$

(3) Partition S and U into $n \times n$ submatrices as

$$\mathrm{U} = \begin{bmatrix} \mathrm{U_{11}} & \mathrm{U_{12}} \\ \mathrm{U_{21}} & \mathrm{U_{22}} \end{bmatrix}, \qquad \mathrm{S} = \begin{bmatrix} \mathrm{S_{11}} & \mathrm{S_{12}} \\ 0 & \mathrm{S_{22}} \end{bmatrix} \tag{2.89}$$

with U ordered so that all real parts of $\mathrm{S_{11}}$ are negative and all real parts of $\mathrm{S_{22}}$ are positive.

(4) The solution P is computed by solving the linear equation

$$\mathrm{PU_{11} = U_{21}} \tag{2.90}$$

■

The Schur method for solving the ARE has been shown to be numerically stable, and it usually performs reliably on well conditioned systems with dense matrices of up to order 100 or so (Ref. 15).

In the above sections, some important matrix decomposition methods and their applications to several interesting problems of linear system theory have been reviewed. Another important decomposition – *QZ decomposition* (or generalized Schur decomposition) – is widely used to solve generalized eigenvalue problems, to find solutions of Riccati and Lyapunov equations, etc. For more information on matrix decompositions, readers may refer to Ref. 5. Table 2.4 summarizes matrix decomposition methods, and Table 2.5 shows most common applications of those matrix decompositions.

2.3 Sensitivity and Conditioning Issues

The usefulness of modal sensitivities for analysis and design of engineering systems is well known. Some specific applications include identification of dynamical systems (Refs. 16, 17), redesign of vibratory systems (Refs. 18-20), and design of control systems by pole placement (Refs. 21-25). In the above algorithms, eigenvalue and eigenvector derivatives with respect to design parameters are often useful.

The effect of uncertainties on the closed-loop performance of structural systems is an important and many-faceted issue in the design of control systems. Uncertainty in the system arises from dynamical modeling errors, parameter variations, and so on. The concept of "stability robustness" has been introduced in the recent literature to define the extent to which certain performance properties, especially stability of closed-loop systems, are maintained in the presence of system uncertainties.

The topic of robust control is an important issue in the research literature pertaining to design of control systems at present, and some key results are presented in the subsequent chapters. In this section we study eigenvalue sensitivity measures. The discussion serves to introduce some important concepts that will be used in this monograph (stability robustness theory); these ideas are directly related to the robustness of closed loop dynamical systems and other sensitivity measures subsequently studied for actuator placement.

In section 2.3.1, for the nondefective linear systems, analytic expressions for eigenvalue and eigenvector derivatives are derived using a modal expansion approach. In section 2.3.2, we review the concept of conditioning of eigenvalue problems. Following developments in the recent literature, we show that the eigenvector matrix condition number can be used to directly establish an upper bound on eigenvalue errors in the presence of variations in the matrix. In section 2.3.3, the robustness measure of Patel and Toda (Ref. 30) is introduced, and relationships between robustness and conditioning of eigenvalue problem are

Table 2.4 Matrix Decomposition: A Summary

Method	Matrix Condition of A	Form	Comments
LU	Square Nonsingular	A=LU	Gaussian elimination
Cholesky	Square Symmetric Positive definite	$A = LL^T$	Matrix square root L lower triangular
Spectral	Square Nondefective	$A = U\Lambda U^{-1}$	Eigenvalue problem U orthonormal Λ diagonal
Schur	Square	$A = USU^T$	U orthogonal S block upper triangular
QR	General matrix	A=QR	R upper triangular Q orthonormal
SVD	General matrix	$A = USV^H$	Singular value problem U,V unitary

discussed. It will be evident that that the Patel-Toda robustness measure, when applied to parameter variations in the system matrix reduces to the same result (the Bauer-Fike theorem) obtained from eigenvalue sensitivity analysis. Partial derivatives of the singular values are also introduced, which play an important role

Table 2.5 Matrix Decompositions' Most Common Applications

Method	Most Common Applications
LU	Solution of linear equations
Cholesky	Solution of linear equations with positive definite matrix
	Parameterization of positive definite matrices
Spectral	Diagonalization of matrices
	Solution of eigenvalue problems
Schur	Triangularization of matrices
	Orthogonal projection
	Solution of eigenvalue problems
	Solution of Riccati and Lyapunov equations
QR	Solution of least square problems[a]
	Determination of orthogonal vectors spanning the range space and null space of matrices
SVD	Solution of least square problems[a]
	Determination of orthogonal subspaces
	Determination of rank and condition numbers
	Determination of the Moore-Penrose pseudoinverse

[a] When using the above decompositions for the most fundamental problem of solving linear algebraic systems of equations, note that only QR and SVD can accommodate easily over- and underdetermined systems of equations, including rank-deficient and poorly conditioned cases. The QR and SV decompositions have been implemented and can often be carried through accurately for fully populated systems with dimensions ≈ 100.

in algorithms developed herein (Chapter 7) for optimizing the condition number, since indeed, the condition number is simply the ratio of the largest to the least of the singular values.

2.3.1 Eigenvalue/Eigenvector Sensitivities

Consider the nondefective system

$$\mathbf{B}\dot{\mathbf{x}} = \mathbf{A}\mathbf{x} \tag{2.91}$$

The right and left eigenvalue problems of the above system are represented as follows [Eqs. (2.44)-(2.46)]:

$$\lambda_i \mathrm{B}\underline{\phi}_i = \mathrm{A}\underline{\phi}_i, \qquad i = 1, 2, \ldots, n \tag{2.92}$$

$$\lambda_i \mathrm{B}^T\underline{\psi}_i = \mathrm{A}^T\underline{\psi}_i, \qquad i = 1, 2, \ldots, n \tag{2.93}$$

where the conventional normalization of the biorthogonality conditions for the eigenvectors are adopted as

$$\underline{\phi}_i^T \mathrm{B}\underline{\phi}_i = 1, \qquad i = 1, 2, \ldots, n \tag{2.94}$$

$$\underline{\psi}_j^T \mathrm{B}\underline{\phi}_i = \delta_{ij}, \qquad i, j = 1, 2, \ldots, n \tag{2.95}$$

so that

$$\underline{\psi}_j^T \mathrm{A}\underline{\phi}_i = \lambda_i \delta_{ij}, \qquad i, j = 1, 2, \ldots, n \tag{2.96}$$

Theorem 2.5: Eigenvector Sensitivities

Partial derivatives of eigenvalues and eigenvectors of the system (2.91) are given by the following equations:

$$\frac{\partial \lambda_i}{\partial \rho} = \underline{\psi}_i^T \left(\frac{\partial \mathrm{A}}{\partial \rho} - \lambda_i \frac{\partial \mathrm{B}}{\partial \rho}\right)\underline{\phi}_i, \qquad i = 1, \ldots, n \tag{2.97}$$

$$\frac{\partial \underline{\phi}_i}{\partial \rho} = \sum_{j=1}^{n} a_{ji}\underline{\phi}_j, \qquad \frac{\partial \underline{\psi}_i}{\partial \rho} = \sum_{j=1}^{n} b_{ji}\underline{\psi}_j \tag{2.98}$$

where a_{ji} and b_{ji} are the projections of the eigenvector gradients onto the eigenvectors (*not* the elements of A and B), and ρ is some arbitrary parameter upon which

A and B depend, and where the a_{ji} and b_{ji} are given by

$$
\begin{aligned}
a_{ji} &= \frac{1}{\lambda_i - \lambda_j} \underline{\psi}_j^T \left(\frac{\partial A}{\partial \rho} - \lambda_i \frac{\partial B}{\partial \rho} \right) \underline{\phi}_i, && j \neq i \\
&= -\frac{1}{2} \left[\underline{\phi}_i^T \frac{\partial B}{\partial \rho} \underline{\phi}_i + \sum_{\substack{k=1 \\ k \neq i}}^{n} a_{ki} \underline{\phi}_k^T (B + B^T) \underline{\phi}_i \right], && j = i \\
b_{ji} &= \frac{1}{\lambda_i - \lambda_j} \underline{\psi}_i^T \left(\frac{\partial A}{\partial \rho} - \lambda_i \frac{\partial B}{\partial \rho} \right) \underline{\phi}_j, && j \neq i \\
&= -\underline{\psi}_i^T \frac{\partial B}{\partial \rho} \underline{\phi}_i - a_{ii}, && j = i
\end{aligned}
\tag{2.99}
$$

■

Proof of Theorem 2.5

The derivation is an extension of the results in Ref. 31, where it is proven for the case $B = I$.

Eigenvalue Derivative

Let us derive the eigenvalue derivatives with respect to a scalar parameter ρ. First, we take the partial derivative of Eq. (2.92) to get

$$
\frac{\partial \lambda_i}{\partial \rho} B \underline{\phi}_i + \lambda_i \frac{\partial B}{\partial \rho} \underline{\phi}_i + \lambda_i B \frac{\partial \underline{\phi}_i}{\partial \rho} = \frac{\partial A}{\partial \rho} \underline{\phi}_i + A \frac{\partial \underline{\phi}_i}{\partial \rho} \tag{2.100}
$$

We then premultiply the above equation by $\underline{\psi}_j^T$ to obtain

$$
\frac{\partial \lambda_i}{\partial \rho} \underline{\psi}_j^T B \underline{\phi}_i + (\lambda_i \underline{\psi}_j^T B - \underline{\psi}_j^T A) \frac{\partial \underline{\phi}_i}{\partial \rho} = \underline{\psi}_j^T \left(\frac{\partial A}{\partial \rho} - \lambda_i \frac{\partial B}{\partial \rho} \right) \underline{\phi}_i
$$

Using Eqs. (2.93) and (2.95) in the above equation gives the analytical expressions of eigenvalue sensitivities as

$$
\frac{\partial \lambda_i}{\partial \rho} = \underline{\psi}_i^T \left(\frac{\partial A}{\partial \rho} - \lambda_i \frac{\partial B}{\partial \rho} \right) \underline{\phi}_i \tag{2.101}
$$

Eigenvector Derivatives

Here, we derive an expression for eigenvector derivatives paralleling and extending the developments of Ref. 32 by using $\{\underline{\phi}_i, i = 1, 2, ..., n\}$ as basis vectors for the eigenvector derivatives,

$$
\frac{\partial \underline{\phi}_i}{\partial \rho} = \sum_{j=1}^{n} a_{ji} \underline{\phi}_j \tag{2.102}
$$

Since the eigenvalues were assumed distinct, this always leads to a set of n eigenvectors that are linearly independent, and these may be used as a set of basis vectors for spanning the complex n-dimensional space.

To obtain the components $a_{ji}(i,j = 1,\ldots,n)$ of the eigenvector derivative projection of Eq. (2.102), we begin by substituting Eq. (2.102) into Eq. (2.100) and premultiply by $\underline{\psi}_k^T$ to get

$$\frac{\partial\lambda_i}{\partial\rho}\underline{\psi}_k^T B\underline{\phi}_i + \lambda_i\underline{\psi}_k^T\frac{\partial B}{\partial\rho}\underline{\phi}_i + \lambda_i\underline{\psi}_k^T B\sum_{j=1}^{n} a_{ji}\underline{\phi}_j = \underline{\psi}_k^T\frac{\partial A}{\partial\rho}\underline{\phi}_i + \underline{\psi}_k^T A\sum_{j=1}^{n} a_{ji}\underline{\phi}_j$$

Using Eqs. (2.95), (2.96), and (2.101) in the above equation gives

$$(\lambda_i - \lambda_k)a_{ki} = -\underline{\psi}_i^T(\frac{\partial A}{\partial\rho} - \lambda_i\frac{\partial B}{\partial\rho})\underline{\phi}_i\delta_{ik} + \underline{\psi}_k^T(\frac{\partial A}{\partial\rho} - \lambda_i\frac{\partial B}{\partial\rho})\underline{\phi}_i$$

Note that the above equation becomes trivial for the k=i case. Since A was assumed to have distinct eigenvalues, we can solve for a_{ki} $(k \neq i)$ as

$$a_{ki} = \frac{1}{\lambda_i - \lambda_k}\underline{\psi}_k^T(\frac{\partial A}{\partial\rho} - \lambda_i\frac{\partial B}{\partial\rho})\underline{\phi}_i, \qquad k \neq i \tag{2.103}$$

For the case of $i = k$, a_{ii} can be computed from the normalization condition of Eq. (2.94). By taking its partial derivatives and using Eq. (2.102), we find

$$(\sum_{k=1}^{n} a_{ki}\underline{\phi}_k)^T B\underline{\phi}_i + \underline{\phi}_i^T\frac{\partial B}{\partial\rho}\underline{\phi}_i + \underline{\phi}_i^T B(\sum_{k=1}^{n} a_{ki}\underline{\phi}_k) = 0$$

Separating the k=i case from the summation and using Eq. (2.94) leads to

$$a_{ii} = -\frac{1}{2}\Big[\underline{\phi}_i^T\frac{\partial B}{\partial\rho}\underline{\phi}_i + \sum_{\substack{k=1\\k\neq i}}^{n} a_{ki}\underline{\phi}_k^T(B + B^T)\underline{\phi}_i\Big] \tag{2.104}$$

To obtain left eigenvector derivatives, we similarly let

$$\frac{\partial\underline{\psi}_i}{\partial\rho} = \sum_{j=1}^{n} b_{ji}\underline{\psi}_j \tag{2.105}$$

By taking partial derivatives of Eq. (2.93), substituting Eq. (2.105), premultiplying by $\underline{\phi}_k^T$, and transposing the result, we obtain

$$\frac{\partial \lambda_i}{\partial \rho}\underline{\psi}_i^T B \underline{\phi}_k + \lambda_i \underline{\psi}_i^T \frac{\partial B}{\partial \rho}\underline{\phi}_k + \lambda_i \sum_{j=1}^{n} b_{ji}\underline{\psi}_j^T B \underline{\phi}_k = \underline{\psi}_i^T \frac{\partial A}{\partial \rho}\underline{\phi}_k + \sum_{j=1}^{n} b_{ji}\underline{\psi}_j^T A \underline{\phi}_k$$

By using Eqs. (2.95) and (2.96), we get

$$(\lambda_i - \lambda_k) b_{ki} = -\frac{\partial \lambda_i}{\partial \rho}\delta_{ki} + \underline{\psi}_i^T \left(\frac{\partial A}{\partial \rho} - \lambda_i \frac{\partial B}{\partial \rho}\right)\underline{\phi}_k$$

Therefore,

$$b_{ki} = \frac{1}{\lambda_i - \lambda_k}\underline{\psi}_i^T \left(\frac{\partial A}{\partial \rho} - \lambda_i \frac{\partial B}{\partial \rho}\right)\underline{\phi}_k, \qquad k \neq i \tag{2.106}$$

For k=i, b_{ii} and a_{ii} are related through the normalization relation of Eq. (2.95). Thus, by taking the partial derivative of Eq. (2.95) (for the i=j case), we get

$$\frac{\partial \underline{\psi}_i^T}{\partial \rho} B \underline{\phi}_i + \underline{\psi}_i^T \frac{\partial B}{\partial \rho}\underline{\phi}_i + \underline{\psi}_i^T B \frac{\partial \underline{\phi}_i}{\partial \rho} = 0$$

and, by using the expansions of Eqs. (2.102) and (2.105),

$$\sum_{j=1}^{n} b_{ji}\underline{\psi}_j^T B \underline{\phi}_i + \underline{\psi}_i^T \frac{\partial B}{\partial \rho}\underline{\phi}_i + \underline{\psi}_i^T B \sum_{j=1}^{n} a_{ji}\underline{\phi}_j = 0$$

Using Eq. (2.95) in the above equation gives

$$b_{ii} = -\underline{\psi}_i^T \frac{\partial B}{\partial \rho}\underline{\phi}_i - a_{ii} \tag{2.107}$$

Thus we obtain the complete expressions [Eqs. (2.101)-(2.107)] for the eigenvalue derivatives and the left and right eigenvector derivatives.

Q.E.D.

The MATLAB operator (**eigsen**) for computing the above eigenvalue/eigenvector sensitivities is explained in Appendix A.

It is apparent that we can also obtain expressions for the second partial derivatives of the right (or left) eigenvectors. The first equation of Eq. (2.98) can be rewritten as the matrix form

$$\frac{\partial \Phi}{\partial \rho} = \Phi[a_{ij}] \tag{2.108}$$

where $\Phi = [\underline{\phi}_1, \ldots, \underline{\phi}_n]$.

By differentiating Eq. (2.108) with respect to another parameter ν as

$$\frac{\partial^2 \Phi}{\partial \rho \partial \nu} = \frac{\partial \Phi}{\partial \nu}[\mathrm{a_{ij}}] + \Phi \frac{\partial [\mathrm{a_{ij}}]}{\partial \nu} \tag{2.109}$$

we can obtain explicit expressions for Eq. (2.109) by differentiating $[\mathrm{a_{ij}}]$ in Eq. (2.99), using the chain rule.

The importance of having rigorous analytical partial derivatives lies in the fact that we do not require finite difference approximation and the associated numerical pitfalls and experimentation. The above analytical expressions for computing eigenvector derivatives utilize the modal approach for computing eigenvector derivatives. For extremely high-dimensioned systems, this approach may not be desirable since this algebraic method requires the solution of auxiliary sets of linear equations that may be illconditioned. An improved method (Ref. 33) has been developed for the self-adjoint generalized eigenproblem in structural dynamics. Also of significance, only selected eigenvalue and eigenvector sensitivities may be of interest, and special algorithms (Ref. 31) are more efficient for this case. Note that the above expressions apply only when the eigenvalues and eigenvectors are smooth differentiable functions of the parameters, therefore for isolated events of multiple eigenvalues and root bifurcations, more generalized formulas (Ref. 34) should be used.

2.3.2 Conditioning of the Eigenvalue Problem

In numerical analysis, some linear systems are extremely sensitive to small changes in the system, and others are not. Of particular interest in the present discussion are the sensitivities of the eigenvalues to small variations in the matrix A. Qualitatively, sensitive systems are said to be "illconditioned," whereas insensitive systems are said to be "well conditioned". A condition number can be defined that represents a quantitative measure of ill-conditioning of the system and provides an upper bound on the perturbation of the eigenvalues due to a unit norm change in the system matrix (Refs. 13, 35, 36). We now review an important, well known result from matrix theory. Since the methodology of the proof is closely related to methods used in subsequent developments, we present the proof in detail.

Theorem 2.6 (Bauer and Fike, Ref. 35)

Let A have all distinct eigenvalues, i.e., A is guaranteed to have n linearly independent eigenvectors, and the eigenvalue and eigenvector matrices are denoted

$$\Lambda = \mathrm{diag}[\lambda_1, \ldots, \lambda_n]$$

and

$$\Phi = [\underline{\phi}_1, \ldots, \underline{\phi}_n]$$

Consider a perturbative matrix E such that μ and $\underline{\nu}$ (where $\|\underline{\nu}\| = 1$) is an eigenpair of a perturbed system matrix A + E.

Then, at least one eigenvalue of A satisfies

$$|\lambda_i - \mu| \leq \|E\| \; \|\Phi\| \; \|\Phi^{-1}\| \tag{2.110}$$

■

Proof of Theorem 2.6

Since A is assumed to have a linearly independent set of n eigenvectors, Φ, we can write

$$A = \Phi \Lambda \Phi^{-1}$$

From the definition of the $(\mu, \underline{\nu})$ eigenvalue/eigenvector pair

$$(A + E)\underline{\nu} = \mu \underline{\nu}$$

we see that,

$$-E\underline{\nu} = A\underline{\nu} - \mu\underline{\nu} = \Phi\Lambda\Phi^{-1}\underline{\nu} - \mu\Phi\Phi^{-1}\underline{\nu} = \Phi(\Lambda - \mu I)\Phi^{-1}\underline{\nu} \tag{2.111}$$

For the trivial case in which μ equals one of the eigenvalues of A, clearly Eq. (2.110) is automatically satisfied. If μ does not equal one of the eigenvalues of A, $(\Lambda - \mu I)$ is nonsingular and Eq. (2.111) can be inverted to get

$$\underline{\nu} = -\Phi(\Lambda - \mu I)^{-1}\Phi^{-1}E\underline{\nu}$$

By taking the operator norm of both sides,

$$\|\underline{\nu}\| \leq \|\Phi\| \; \|(\Lambda - \mu I)^{-1}\| \; \|\Phi^{-1}\| \; \|E\| \; \|\underline{\nu}\|$$

By using the normalization condition, $\|\underline{\nu}\| = 1$,

$$1 \leq \|\Phi\| \; \|(\Lambda - \mu I)^{-1}\| \; \|\Phi^{-1}\| \; \|E\| \tag{2.112}$$

Since $(\Lambda - \mu I)$ is a diagonal matrix,

$$(\Lambda - \mu I)^{-1} = \text{diag}\Big(\frac{1}{\lambda_1 - \mu}, \ldots, \frac{1}{\lambda_n - \mu}\Big)$$

And so, for any $(1,2,\infty)$ matrix operator norms,[a]

---

[a] Matrix operator norms (or induced norms) are defined as (Ref. 37)

$\|A\|_\infty \equiv \max_i \sum_j |a_{ij}|$ (row sum) ≡ "infinity" norm

$\|A\|_1 \equiv \max_j \sum_i |a_{ij}|$ (column sum) ≡ "one" norm

$\|A\|_2 \equiv \|A\|_s \equiv \sigma_{\max}(A)$ (maximum singular value) ≡ "two" or "spectral" norm

$$\|(\Lambda - \mu\mathrm{I})^{-1}\| = \max_{\mathrm{i}} \frac{1}{|\lambda_{\mathrm{i}} - \mu|} = \frac{1}{\min_{\mathrm{i}} |\lambda_{\mathrm{i}} - \mu|} \tag{2.113}$$

Using Eq. (2.113) in Eq. (2.112) gives

$$\min_{\mathrm{i}} |\lambda_{\mathrm{i}} - \mu| \leq \|\mathrm{E}\| \; \|\Phi\| \; \|\Phi^{-1}\|$$

thus verifying Eq. (2.110).

Q.E.D.

The condition number for the eigenvalue problem is defined as

$$\mathcal{K}(\Phi) \equiv \|\Phi\| \; \|\Phi^{-1}\| \tag{2.114}$$

Note that the condition number depends only on the eigenvectors of the unperturbed A matrix. Noting that the spectral norm of the matrix is by definition

$$\| \cdot \| = \text{maximum singular value of the matrix } [\cdot]$$

and because

$$\begin{aligned} \|\Phi^{-1}\| &= \text{maximum singular value of } \Phi^{-1} \\ &= \frac{1}{\text{minimum singular value of } \Phi} \end{aligned}$$

then, we can rewrite Eq. (2.114) as follows

$$\mathcal{K}(\Phi) = \frac{\sigma_1(\Phi)}{\sigma_n(\Phi)} \tag{2.115}$$

where σ_1 and σ_n denote the maximum and minimum singular values of the right modal matrix Φ.

It is easy to establish (Ref. 9) that the condition number satisfies the inequality

$$1 \leq \mathcal{K}(\Phi) < \infty \tag{2.116}$$

where the lower limit is achieved if Φ becomes any orthogonal (unitary) matrix, and the condition number goes to infinity for any/all rank-deficient (singular) Φ matrices. Thus, qualitatively, the more nearly orthogonal we can make the closed-loop eigenvectors, the less sensitive will be the closed-loop eigenvalues. One of the fundamental conceptual and practical difficulties (Refs. 34, 38, 39) associated with eigenstructure assignment (see Chapter 5) arises because *there exist an infinity of unitary matrices and some are more nearly achievable eigenvector sets than others.*

It is obvious from the above theorem that the condition number as defined in Eq. (2.114) is directly related to an upper bound on the perturbation of eigenvalues. As a consequence of Theorem 2.6, the following holds:

$$\min_{i} |\lambda_i - \mu_j| \leq \|E\| \; \mathcal{K}(\Phi), \qquad j = 1, \ldots, n \tag{2.117}$$

where λ_i and μ_i represent the eigenvalues of A and A+E, respectively.

The significance of the condition number is clear; it is directly related [through Eq. (2.117)] to the sensitivity of each eigenvalue with respect to perturbations in the system matrix elements.

Example 2.5

Consider the randomly chosen matrix

$$A_1 = A + E$$

where

$$A = \begin{bmatrix} 0 & 0 & 0 & 1 & 0 & 0 \\ 0 & 0 & 0 & 0 & 1 & 0 \\ 0 & 0 & 0 & 0 & 0 & 1 \\ -3 & -3 & -5 & -0.1 & 0 & 0 \\ -3 & -5 & -9 & 0 & -0.1 & 0 \\ -5 & -9 & -17 & 0 & 0 & -0.1 \end{bmatrix}, \qquad E = \begin{bmatrix} 0 & 0 & 0 & 0 & 0 & 0 \\ 0 & 0 & 0 & 0 & 0 & 0 \\ 0 & 0 & 0 & 0 & 0 & 0 \\ 0 & 0 & 0 & 0.1 & 0 & 0 \\ 0 & 0 & 0 & 0 & 0 & 0 \\ 0 & 0 & 0 & 0 & 0 & 0 \end{bmatrix}$$

The eigenvalues of A are

$$\lambda_{1,2} = -0.0500 \pm 0.3446i$$
$$\lambda_{3,4} = -0.0500 \pm 1.1843i$$
$$\lambda_{5,6} = -0.0500 \pm 4.8447i$$

The eigenvalues of the perturbed matrix A_1 are

$$\lambda_{1,2} = -0.0478 \pm 0.3450i$$
$$\lambda_{3,4} = -0.0060 \pm 1.1852i$$
$$\lambda_{5,6} = -0.0463 \pm 4.8447i$$

From the difference of these, the actual eigenvalues variations are

$$\Delta\lambda_{1,2} = 0.0022 \pm 0.0004i$$
$$\Delta\lambda_{3,4} = 0.0440 \pm 0.0009i$$
$$\Delta\lambda_{5,6} = 0.0037 \pm 0.0000i$$

The linearly predicted eigenvalue variations are [analytical partials, using Eq. (2.97)]:

$$\Delta\lambda_{1,2} = 0.0023 \pm 0.0003i$$
$$\Delta\lambda_{3,4} = 0.0440 \pm 0.0019i$$
$$\Delta\lambda_{5,6} = 0.0037 \pm 0.0000i$$

The bound on $\Delta\lambda$ based on the condition number of A's modal matrix(Φ) is

$$\mathcal{K}(\Phi) = \frac{\sigma_1(\Phi)}{\sigma_n(\Phi)} = 4.84 \implies |\Delta\lambda|_{\max} \leq \mathcal{K}(\Phi)\|\mathrm{E}\| = 0.484$$

Note that (1) the linear predictions are accurate in this case (but much more nonlinear behavior often occurs), and (2) the condition number bound is conservative by an order of magnitude (this is typical). However, the condition number remains a very useful robustness measure. Further note that the above condition number bound does not "know" that only the (4,4) element of A is perturbed, whereas the partial derivative prediction makes explicit use of the "structure" of the perturbation. Incorporating knowledge on the structure of perturbations generally decreases the degree of conservatism. Bounds on condition numbers and singular values arise in many contexts in modern system analysis, and the degree of convervatism has many roles to play in current robust control theory research.

■

2.3.3 Stability Robustness Criteria: Conditioning of the Eigenstructure

In this section, we study the usefulness of the condition number of the eigenvector (modal) matrix as a stability robustness criterion. The following theorem is related to the conditioning of the eigenstructure.

Theorem 2.7 (Patel and Toda Ref. 30)

Suppose that a general closed-loop linear system is asymptotically stable. If the perturbed closed-loop system matrix is A + E, it remains asymptotically stable for all perturbations E that are bounded by

$$\|\mathrm{E}\| < \frac{\min\ [-\Re\{\lambda_i(\mathrm{A})\}]}{\mathcal{K}(\Phi)} \tag{2.118}$$

where $\Re\{\cdot\}$ denotes the real part of $\{\cdot\}$.

■

The original proof utilizes a transient response formulation (Ref. 30), and also can be proven via a Lyapunov approach (Ref. 36). We include the complete proof via a Lyapunov approach in Appendix B. We discuss only the conceptual ideas of the proof of Theorem 2.7 here, since it is sufficient to understand the meaning of Theorem 2.7 in the present discussion. We note that the numerator on the right-handside of Eq. (2.118) is the stability margin, whereas the denominator is a measure of eigenvalue sensitivity. Thus the stability margin is divided by one if Φ is orthogonal and by infinity if Φ is rank-deficient. This indicates that a large stability margin is comforting only if the eigenstructure is well conditioned ($\mathcal{K}(\Phi)$ is small).

From Eq. (2.118) and Eq. (2.117) of Theorem 2.6,

$$\min_{\mathrm{i}} \; |\lambda_{\mathrm{i}} - \mu_{\mathrm{j}}| < \min_{\mathrm{i}} \; \left[\Re\{-\lambda_{\mathrm{i}}(\mathrm{A})\}\right], \qquad \mathrm{j} = 1, \ldots, n \tag{2.119}$$

The above condition states that the distance from every perturbed eigenvalue to the closest original eigenvalue is always less than the perpendicular distance of the "least damped" unperturbed eigenvalue to the imaginary axis. Hence, all the perturbed eigenvalues will remain in the left-half plane and, therefore the perturbed system A+E remains asymptotically stable [if the condition of Eq. (2.118) is satisfied].

Theorem 2.7 indicates clearly that stability robustness is directly related to the smallness of the condition number of the associated eigenvalue problem. From the above developments, it is evident that the maximum stability robustness condition, for a fixed stability margin, corresponds to the minimum condition number of the closed-loop eigenvalue problem. A significant consequence of the above is that a well conditioned (or robust) eigenstructure can be obtained by, for example, tuning feedback gains to minimize the condition number of the eigenvector matrix. Thus one can utilize the freedom beyond pole placement to achieve a less sensitive (i.e., more robust) control design. New algorithms for these purposes are developed in Chapter 7.

2.3.4 Partial Derivatives of the Singular Values

In the previous section, it was shown that the condition number of the closed-loop eigenvector matrix can be taken as the robustness measure of the system. By the definitions (2.114) and (2.115), a condition number is

$$\mathcal{K}(\Phi) \equiv \|\Phi\| \; \|\Phi^{-1}\| = \frac{\sigma_1(\Phi)}{\sigma_n(\Phi)} \tag{2.120}$$

The condition number of the eigenvector matrix is often adopted as the stability robustness measure (Refs. 24, 25). In order to apply gradient based algorithms to find the optimal solution, the partial derivatives of the condition number (and thus, the partial derivatives of singular values) are needed. This section is concerned with developing algebraic expressions for the first- and second-order sensitivities of the singular value decomposition of a general complex matrix. The singular value problem has already been reviewed in section 2.2.2. By considering Eqs. (2.60)-(2.68) for singular value decomposition, and using an approach similar to that for deriving partial derivatives of the eigenvalue and eigenvectors (section 2.3.1), the following analytical expressions for partial derivatives of singular values can be obtained.

Theorem 2.8: Partial Derivatives of Singular Values

Consider any complex-valued matrix Φ, factorized by the singular value decomposition

$$\Phi(\mathrm{p}) = \mathrm{U}\Sigma\mathrm{V}^{\mathrm{H}} \tag{2.121}$$

where U and V are $n \times n$ unitary matrices and Φ is a twice differentiable function of the $\mathrm{N} \times 1$ parameter vector $\mathbf{p}$ such as

$$\mathbf{p} = [\mathrm{p}_1 \quad \mathrm{p}_2 \quad \cdots \quad \mathrm{p}_\mathrm{N}] \tag{2.122}$$

The first and second partial derivatives of the singular values can be represented as follows:

$$\frac{\partial \sigma_i}{\partial \mathrm{p}_k} = \frac{1}{2}(\mathrm{U}_i^{\mathrm{H}}\frac{\partial \Phi}{\partial \mathrm{p}_k}\mathrm{V}_i + \mathrm{V}_i^{\mathrm{H}}\frac{\partial \Phi^{\mathrm{H}}}{\partial \mathrm{p}_k}\mathrm{U}_i) = \Re(\mathrm{U}_i^{\mathrm{H}}\frac{\partial \Phi}{\partial \mathrm{p}_k}\mathrm{V}_i) \tag{2.123}$$

$$\begin{aligned}\frac{\partial^2 \sigma_i}{\partial \mathrm{p}_k \partial \mathrm{p}_l} =& \Re[\sum_{\substack{j=1\\ j\neq i}}^{n} \frac{1}{\sigma_i^2 - \sigma_j^2}\{\sigma_i[(\mathrm{U}_j^{\mathrm{H}}\frac{\partial \Phi}{\partial \mathrm{p}_l}\mathrm{V}_i)^{\mathrm{H}}(\mathrm{U}_j^{\mathrm{H}}\frac{\partial \Phi}{\partial \mathrm{p}_k}\mathrm{V}_i) + (\mathrm{U}_i^{\mathrm{H}}\frac{\partial \Phi}{\partial \mathrm{p}_l}\mathrm{V}_j)^{\mathrm{H}}(\mathrm{U}_i^{\mathrm{H}}\frac{\partial \Phi}{\partial \mathrm{p}_k}\mathrm{V}_j)] \\ &+\sigma_j[(\mathrm{U}_i^{\mathrm{H}}\frac{\partial \Phi}{\partial \mathrm{p}_l}\mathrm{V}_j)(\mathrm{U}_j^{\mathrm{H}}\frac{\partial \Phi}{\partial \mathrm{p}_k}\mathrm{V}_i) + (\mathrm{U}_j^{\mathrm{H}}\frac{\partial \Phi}{\partial \mathrm{p}_l}\mathrm{V}_i)(\mathrm{U}_i^{\mathrm{H}}\frac{\partial \Phi}{\partial \mathrm{p}_k}\mathrm{V}_j)]\} + \mathrm{U}_i^{\mathrm{H}}\frac{\partial^2 \Phi}{\partial \mathrm{p}_k \partial \mathrm{p}_l}\mathrm{V}_i] \\ &+\frac{1}{\sigma_i}\Im(\mathrm{U}_i^{\mathrm{H}}\frac{\partial \Phi}{\partial \mathrm{p}_l}\mathrm{V}_i)\Im(\mathrm{U}_i^{\mathrm{H}}\frac{\partial \Phi}{\partial \mathrm{p}_k}\mathrm{V}_i)\end{aligned} \tag{2.124}$$

where $\Re(\cdot)$ and $\Im(\cdot)$ denote the real part and imaginary part of $(\cdot)$, respectively. ■

We include the complete proof in Appendix C, and the reader is referred to Ref. 24 for more information. Note that the first- and-second order sensitivities of the singular values of the eigenvector need the first and second partial derivatives of eigenvalues and eigenvectors that are reviewed in section 2.3.1. The description of the MATLAB operator **svsen** for computing partial derivatives of the singular values is given in Appendix A.

In this section, we have reviewed the basic principles underlying the stability robustness measures based on the conditioning of the eigenstructure. This type of robustness measure is believed to be simpler to implement and to be physically more interpretable than many competing robustness measures, such as those reviewed in Ref. 40. Note however, that, the condition number represents an upper bound only. Therefore, for some cases, this robustness measure may be too conservative and may not necessarily be a realistic estimate of actual errors. We also derived the first- and second-order sensitivities of the singular values, which can be used in optimization problems. It is seen that the sensitivities of the singular values are closely related to the sensitivities of the eigenvectors.

2.4 Case Study: Parameterization of Orthogonal Matrices

2.4.1 Some Geometrical and Kinematical Insights from $R^{3\times 3}$

Consider two right-handed orthogonal triads of unit vectors satisfying the orthogonality and right-handed conditions

$$\hat{\mathbf{a}}_i \cdot \hat{\mathbf{a}}_j = \delta_{ij}, \qquad \hat{\mathbf{b}}_i \cdot \hat{\mathbf{b}}_j = \delta_{ij}, \qquad \hat{\mathbf{a}}_i \cdot \hat{\mathbf{b}}_j = C_{ij} \tag{2.125}$$

$$\hat{\mathbf{a}}_i \times \hat{\mathbf{a}}_j = \epsilon_{ijk}\hat{\mathbf{a}}_k, \qquad \hat{\mathbf{b}}_i \times \hat{\mathbf{b}}_j = \epsilon_{ijk}\hat{\mathbf{b}}_k \tag{2.126}$$

where

$$\epsilon_{ijk} = \begin{cases} 1 & \text{if i,j,k is a cyclic permutation of 1,2,3} \\ -1 & \text{if i,j,k is a noncyclic permutation of 1,2,3} \\ 0 & \text{if i=j or i=k or j=k} \end{cases}$$

The C_{ij} are the direction cosine components of the projection of $\hat{\mathbf{b}}_i$ onto the $\hat{\mathbf{a}}_j$ in the sense

$$\hat{\mathbf{b}}_i = \sum_{j=1}^{3} C_{ij}\hat{\mathbf{a}}_j, \qquad i = 1, 2, 3 \tag{2.127}$$

where $C_{ij} = \cos(\angle \hat{\mathbf{b}}_i, \hat{\mathbf{a}}_j)$.

Note that the three projection equations (2.127) can be written as the following vector/matrix equation:

$$\begin{Bmatrix} \hat{\mathbf{b}}_1 \\ \hat{\mathbf{b}}_2 \\ \hat{\mathbf{b}}_3 \end{Bmatrix} = \begin{bmatrix} C_{11} & C_{12} & C_{13} \\ C_{21} & C_{22} & C_{23} \\ C_{31} & C_{32} & C_{33} \end{bmatrix} \begin{Bmatrix} \hat{\mathbf{a}}_1 \\ \hat{\mathbf{a}}_2 \\ \hat{\mathbf{a}}_3 \end{Bmatrix} \quad \Longrightarrow \quad \{\hat{\mathbf{b}}\} = C\{\hat{\mathbf{a}}\} \tag{2.128}$$

Substitution of Eqs. (2.127) into Eq. (2.125) leads immediately to the conclusion that the C matrix satisfies the orthogonality condition

$$C^T C = C C^T = I \tag{2.129}$$

Thus the inverse of the C matrix is simply its transpose. As a consequence of the requirement that both $\{\hat{\mathbf{a}}\}$ and $\{\hat{\mathbf{b}}\}$ satisfy the right-handed constraint of Eq. (2.126), it can be shown (Ref. 41) that C has one real positive eigenvalue (+1) and the determinant of C is +1. Examining the nine scalar equations from carrying out the matrix product in Eq. (2.129), there are only six independent conditions; thus three independent parameters are sufficient to parameterize the C matrix.

We now consider the common situation in which $\{\hat{\mathbf{b}}\}$ moves with respect to $\{\hat{\mathbf{a}}\}$; thus C is a time-varying matrix function of time. Introducing the angular velocity $\underline{\omega}$ of $\{\hat{\mathbf{b}}\}$ with respect to $\{\hat{\mathbf{a}}\}$

$$\underline{\omega} = \omega_1 \hat{\mathbf{b}}_1 + \omega_2 \hat{\mathbf{b}}_2 + \omega_3 \hat{\mathbf{b}}_3 \tag{2.130}$$

and from vector kinematics (Ref. 42),

$$\begin{aligned}
\frac{\mathrm{d}}{\mathrm{dt}}\left(\hat{\mathbf{b}}_1\right) &= \underline{\omega} \times \hat{\mathbf{b}}_1 = && -\omega_3 \hat{\mathbf{b}}_2 + \omega_2 \hat{\mathbf{b}}_3 \\
\frac{\mathrm{d}}{\mathrm{dt}}\left(\hat{\mathbf{b}}_2\right) &= \underline{\omega} \times \hat{\mathbf{b}}_2 = && \omega_3 \hat{\mathbf{b}}_1 - \omega_1 \hat{\mathbf{b}}_3 \\
\frac{\mathrm{d}}{\mathrm{dt}}\left(\hat{\mathbf{b}}_3\right) &= \underline{\omega} \times \hat{\mathbf{b}}_3 = && -\omega_2 \hat{\mathbf{b}}_1 + \omega_1 \hat{\mathbf{b}}_2
\end{aligned}$$

or

$$\frac{\mathrm{d}}{\mathrm{dt}}\{\hat{\mathbf{b}}\} = -\tilde{\omega}\{\hat{\mathbf{b}}\} \tag{2.131}$$

with

$$\tilde{\omega} \equiv \begin{bmatrix} 0 & -\omega_3 & \omega_2 \\ \omega_3 & 0 & -\omega_1 \\ -\omega_2 & \omega_1 & 0 \end{bmatrix} \tag{2.132}$$

Substitution of Eq. (2.128) into Eq. (2.131) yields the following linear kinematic differential equation for the direction cosine matrix

$$\dot{\mathrm{C}} = -\tilde{\omega}\mathrm{C} \tag{2.133}$$

It is easy to establish that the orthogonality condition of Eq. (2.129) is a rigorous integral of Eq. (2.133); this is accomplished by taking the time derivative of Eq. (2.129) and substituting Eq. (2.133).

The C matrix is most often parameterized as a function of sequential rotations (Euler angles θ_1, θ_2, θ_3) as

$$\mathrm{C} = \left[\mathrm{M}_\gamma(\theta_3)\right]\left[\mathrm{M}_\beta(\theta_2)\right]\left[\mathrm{M}_\alpha(\theta_1)\right] \tag{2.134}$$

where the rotation sequence (α, β, γ) is one of the following twelve sets of integers:

$$\{(1,2,3),(1,3,2),(1,2,1),(1,3,1),\\(2,1,3),(2,3,1),(2,1,2),(2,3,2),\\(3,2,1),(3,1,2),(3,2,3),(3,1,3)\}$$

and, using the abbreviations c=cos, s=sin, the $\mathrm{M}_\alpha(\theta)$ are the elementary orthogonal matrices

$$\mathrm{M}_1(\theta) = \begin{bmatrix} 1 & 0 & 0 \\ 0 & c\theta & s\theta \\ 0 & -s\theta & c\theta \end{bmatrix}, \quad \mathrm{M}_2(\theta) = \begin{bmatrix} c\theta & 0 & -s\theta \\ 0 & 1 & 0 \\ s\theta & 0 & c\theta \end{bmatrix}, \quad \mathrm{M}_3(\theta) = \begin{bmatrix} c\theta & s\theta & 0 \\ -s\theta & c\theta & 0 \\ 0 & 0 & 1 \end{bmatrix} \tag{2.135}$$

In 1776, Euler proved that the most general angular position of $\{\hat{\mathbf{b}}\}$ with respect to $\{\hat{\mathbf{a}}\}$ can be parameterized by a single rotation through an angle ϕ about the eigenvector $\hat{\underline{l}}$ of C corresponding to the +1 eigenvalue. The C matrix can be parameterized as a function of ϕ and the $\{\hat{\mathbf{b}}\}$ components of the unit vector $\hat{\underline{l}}$ as

$$\mathrm{C}(l_1,l_2,l_3,\phi) = \begin{bmatrix} l_1^2(1-c\phi)+c\phi & l_1 l_2(1-c\phi)+l_3 s\phi & l_1 l_3(1-c\phi)-l_2 s\phi \\ l_2 l_1(1-c\phi)-l_3 s\phi & l_2^2(1-c\phi)+c\phi & l_2 l_3(1-c\phi)+l_1 s\phi \\ l_3 l_1(1-c\phi)+l_2 s\phi & l_3 l_2(1-c\phi)-l_1 s\phi & l_3^2(1-c\phi)+c\phi \end{bmatrix} \tag{2.136}$$

Since $\hat{\underline{l}}$ is a unit vector, we know that $l_1^2 + l_2^2 + l_3^2 = 1$, so C has only three degrees of freedom, as expected. Note that the time evolution of C can also be described by the time evolution of (l_1, l_2, l_3, ϕ), by using Eq. (2.136) to generate the instantaneous C matrix.

There are two important parameterizations (Ref. 42) of C that are closely related to the principal rotation parameters (l_1, l_2, l_3, ϕ), these are respectively:

Euler parameters or *quaternions* $(\beta_0, \beta_1, \beta_2, \beta_3)$:

$$\begin{aligned} &\beta_0 \equiv \cos(\frac{\phi}{2}) \\ &\beta_\mathrm{i} \equiv l_\mathrm{i} \sin(\frac{\phi}{2}), \qquad \mathrm{i} = 1,2,3 \\ &\sum_{\mathrm{i}=0}^{3} \beta_\mathrm{i}^2 = 1 \end{aligned} \tag{2.137}$$

Euler-Rodrigues parameters or *Gibbs vector* (q_1, q_2, q_3):

$$q_i = \frac{\beta_i}{\beta_0} = l_i \tan(\frac{\phi}{2}) \tag{2.138}$$

Upon making use of Eqs. (2.137) and (2.138) in Eq. (2.136), we find that these parameterizations yield, respectively,

$$C(\beta_0, \beta_1, \beta_2, \beta_3) = \begin{bmatrix} \beta_0^2 + \beta_1^2 - \beta_2^2 - \beta_3^2 & 2(\beta_1\beta_2 + \beta_0\beta_3) & 2(\beta_1\beta_3 - \beta_0\beta_2) \\ 2(\beta_1\beta_2 - \beta_0\beta_3) & \beta_0^2 - \beta_1^2 + \beta_2^2 - \beta_3^2 & 2(\beta_2\beta_3 + \beta_0\beta_1) \\ 2(\beta_1\beta_3 + \beta_0\beta_2) & 2(\beta_2\beta_3 - \beta_0\beta_1) & \beta_0^2 - \beta_1^2 - \beta_2^2 + \beta_3^2 \end{bmatrix} \tag{2.139}$$

and

$$C(q_1, q_2, q_3) = \frac{1}{1 + q_1^2 + q_2^2 + q_3^2} \begin{bmatrix} 1 + q_1^2 - q_2^2 - q_3^2 & 2(q_1q_2 + q_3) & 2(q_1q_3 - q_2) \\ 2(q_1q_2 - q_3) & 1 - q_1^2 + q_2^2 - q_3^2 & 2(q_2q_3 + q_1) \\ 2(q_1q_3 + q_2) & 2(q_2q_3 - q_1) & 1 - q_1^2 - q_2^2 + q_3^2 \end{bmatrix} \tag{2.140}$$

The nine kinematical equations in Eq. (2.133) can be redeemed in favor of the above parameterizations by directly substituting the direction cosine functions into Eq. (2.133). In particular, for 3-1-3 Euler angles (ϕ, θ, ψ),

$$C(\phi, \theta, \psi) = [M_3(\psi)][M_1(\theta)][M_3(\phi)] \tag{2.141}$$

we find from Eq. (2.133) the kinematical differential equation

$$\begin{Bmatrix} \dot{\phi} \\ \dot{\theta} \\ \dot{\psi} \end{Bmatrix} = \frac{1}{\sin\theta} \begin{bmatrix} \sin\psi & \cos\psi & 0 \\ \cos\psi \sin\theta & -\sin\psi \sin\theta & 0 \\ -\sin\psi\cos\theta & -\cos\psi\cos\theta & \sin\theta \end{bmatrix} \begin{Bmatrix} \omega_1 \\ \omega_2 \\ \omega_3 \end{Bmatrix} \tag{2.142}$$

For the Euler parameters $(\beta_0, \beta_1, \beta_2, \beta_3)$ and the Gibbs vector (q_1, q_2, q_3), we find upon substituting Eqs. (2.139) and (2.140) into Eq. (2.133), that the corresponding kinematical differential equations are

$$\begin{Bmatrix} \dot{\beta}_0 \\ \dot{\beta}_1 \\ \dot{\beta}_2 \\ \dot{\beta}_3 \end{Bmatrix} = \frac{1}{2} \begin{bmatrix} 0 & -\omega_1 & -\omega_2 & -\omega_3 \\ \omega_1 & 0 & \omega_3 & -\omega_2 \\ \omega_2 & -\omega_3 & 0 & \omega_1 \\ \omega_3 & \omega_2 & -\omega_1 & 0 \end{bmatrix} \begin{Bmatrix} \beta_0 \\ \beta_1 \\ \beta_2 \\ \beta_3 \end{Bmatrix} \equiv \frac{1}{2} \begin{bmatrix} \beta_0 & -\beta_1 & -\beta_2 & -\beta_3 \\ \beta_1 & \beta_0 & -\beta_3 & \beta_2 \\ \beta_2 & \beta_3 & \beta_0 & -\beta_1 \\ \beta_3 & -\beta_2 & \beta_1 & \beta_0 \end{bmatrix} \begin{Bmatrix} 0 \\ \omega_1 \\ \omega_2 \\ \omega_3 \end{Bmatrix} \tag{2.143}$$

and

$$\begin{Bmatrix} \dot{q}_1 \\ \dot{q}_2 \\ \dot{q}_3 \end{Bmatrix} = \frac{1}{2} \begin{bmatrix} 1+q_1^2 & q_1q_2 - q_3 & q_1q_3 + q_2 \\ q_1q_2 + q_3 & 1+q_2^2 & q_2q_3 - q_1 \\ q_1q_3 - q_2 & q_2q_3 + q_1 & 1+q_3^2 \end{bmatrix} \begin{Bmatrix} \omega_1 \\ \omega_2 \\ \omega_3 \end{Bmatrix} \tag{2.144}$$

These classical parameterizations play an important role in many geometrical and kinematical descriptions of motion, especially in the dynamics of spacecraft and aircraft. The absence of trigonometric functions in Eqs. (2.143) and (2.144) make these differential equations more attractive than Eq. (2.142) for many applications. Also, notice that Eq. (2.143) are rigorously linear and universally nonsingular, in contrast to Eq. (2.142).

Orthogonal matrices of higher dimension arise often, and so it is natural to seek generalizations of the above results to parameterize and describe the evolution of orthogonal matrices for projections in R^n. As will be shown below, the *Cayley transform* and related results generalize completely Eqs. (2.133), (2.140), and (2.144). In fact, these R^3 results will be found to be special cases of the n-dimensional generalizations.

2.4.2 Parameterizations of Orthogonal Matrices in $\mathrm{R}^{n\times n}$

The key results that underlie the developments of this section are captured in the following theorem.

Theorem 2.9: Cayley Transform

If we let C denote an arbitrary $n \times n$ orthogonal matrix and $\mathrm{Q} = -\mathrm{Q}^\mathrm{T}$ denote a related $n \times n$ skew-symmetric matrix, the following transformations hold:

Forward Transformations

$$(1)\ \mathrm{C} = (\mathrm{I} - \mathrm{Q})(\mathrm{I} + \mathrm{Q})^{-1} \tag{2.145a}$$

$$(2)\ \mathrm{C} = (\mathrm{I} + \mathrm{Q})^{-1}(\mathrm{I} - \mathrm{Q}) \tag{2.145b}$$

Inverse Transformations

$$(3)\ \mathrm{Q} = (\mathrm{I} - \mathrm{C})(\mathrm{I} + \mathrm{C})^{-1} \tag{2.146a}$$

$$(4)\ \mathrm{Q} = (\mathrm{I} + \mathrm{C})^{-1}(\mathrm{I} - \mathrm{C}) \tag{2.146b}$$

Note the elegant identical form of the forward and inverse transformations, and the commutativity of the above matrix products. Since an arbitrary orthogonal matrix can be substituted into Eqs. (2.146), we can conclude that the Q matrix completely parameterizes all infinity of C matrices in $R^{n\times n}$.

Proof of Theorem 2.9

We first prove the equivalence of Eqs. (2.145a) and (2.145b). Consider Eq. (2.145a).

$$C = (I - Q)(I + Q)^{-1}$$

Premultiply by (I+Q):

$$\begin{aligned}(I + Q)C &= (I + Q)(I - Q)(I + Q)^{-1} \\ &= \Big[(I + Q) - (I + Q)Q\Big](I + Q)^{-1} \\ &= \Big[(I + Q) - Q(I + Q)\Big](I + Q)^{-1} \\ &= (I - Q)\end{aligned}$$

So,

$$C = (I + Q)^{-1}(I - Q)$$

Thus we have verified that Eq. (2.145b) follows directly from Eq. (2.145a). Similar manipulations verify that Eq. (2.146b) follows from Eq. (2.146a).

Now we prove that Eqs. (2.145a), (2.145b) parameterize an orthogonal matrix. Checking C^TC, we see

$$\begin{aligned}C^TC &= (I + Q^T)^{-1}(I - Q^T)(I + Q)^{-1}(I - Q) \\ &= (I - Q)^{-1}(I + Q)(I + Q)^{-1}(I - Q) \\ &= I\end{aligned}$$

So, we have established that an arbitrary skew-symmetric Q matrix substituted into Eqs. (2.145) indeed generates an orthogonal C matrix.

Now, we verify that Eqs. (2.146) are the inverse of Eqs. (2.145). This is done by the following manipulations: Consider Eq. (2.145a).

$$C = (I - Q)(I + Q)^{-1}$$

Postmultiply by (I+Q):

$$\begin{aligned}C(I + Q) &= I - Q \\ C + CQ &= I - Q \\ (I + C)Q &= I - C \\ Q &= (I + C)^{-1}(I - C)\end{aligned}$$

which verifies Eq. (2.146b). Since we already know that Eqs. (2.146a) and (2.146b) are equivalent, this completes the proof.

Q.E.D.

It can be immediately verified by substitution that Eqs. (2.145) and(2.146) specialize for n=3 to yield exactly Eq. (2.140) and the inverse solution for

$$\mathrm{Q} = \begin{bmatrix} 0 & -q_3 & q_2 \\ q_3 & 0 & -q_1 \\ -q_2 & q_1 & 0 \end{bmatrix} .$$

Note that the number of distinct elements in an $n \times n$ Q matrix is $(n^2 - n)/2$; this is the number of independent degrees of freedom of an $n \times n$ orthogonal matrix C.

Since the 3×3 Gibbs vector parameterization of C is seen to generalize fully to the $n \times n$ case, it is natural to be curious about whether the kinematical differential equations (2.133) and (2.144) also generalize. Indeed, these do generalize, as captured in the following theorem.

Theorem 2.10: Cayley Transform Kinematic Relationships

The following matrix $\tilde{\omega}$ is skew-symmetric:

$$\tilde{\omega} = 2(\mathrm{I} + \mathrm{Q})^{-1}\dot{\mathrm{Q}}(\mathrm{I} - \mathrm{Q})^{-1} \tag{2.147a}$$

and

$$\dot{\mathrm{C}} = -\tilde{\omega}\mathrm{C} \tag{2.147b}$$

$$\dot{\mathrm{Q}} = \frac{1}{2}(\mathrm{I} + \mathrm{Q})\tilde{\omega}(\mathrm{I} - \mathrm{Q}) \tag{2.147c}$$

Proof of Theorem 2.10

The skew-symmetry of $\tilde{\omega}$ in Eq. (2.147a) is established if $\tilde{\omega} = -\tilde{\omega}^{\mathrm{T}}$. Taking the transpose of Eq. (2.147a) immediately gives

$$\tilde{\omega}^{\mathrm{T}} = 2(\mathrm{I} - \mathrm{Q}^{\mathrm{T}})^{-1}\dot{\mathrm{Q}}^{\mathrm{T}}(\mathrm{I} + \mathrm{Q}^{\mathrm{T}})^{-1}$$

Since Q and $\dot{\mathrm{Q}}$ are skew symmetric, we immediately see that $\tilde{\omega} = -\tilde{\omega}^{\mathrm{T}}$; thus $\tilde{\omega}$ is skew-symmetric. Now we establish Eq. (2.147c). Premultiplying Eq. (2.147a) by (I+Q) and postmultiplying by (I-Q) immediately establishes Eq. (2,147c).

On account of the orthogonality of C, we know that $\mathrm{CC}^{\mathrm{T}} = \mathrm{I}$, so that from Eq. (2.147b), we can write Eq. (2.147b) in the alternate form

$$\tilde{\omega} = -\dot{\mathrm{C}}\mathrm{C}^{\mathrm{T}} \tag{2.148}$$

Substitution of Eqs. (2.145a) and (2.145b) into Eq. (2.148) yields

$$\begin{aligned}
\dot{C}C^T &= \frac{d}{dt}\left[(I-Q)(I+Q)^{-1}\right]\left[(I+Q)^{-1}(I-Q)\right]^T \\
&= -\left[\dot{Q}(I+Q)^{-1} + (I-Q)(I+Q)^{-1}\dot{Q}(I+Q)^{-1}\right]\left[(I+Q)(I-Q)^{-1}\right] \\
&= -\left[\dot{Q} + (I-Q)(I+Q)^{-1}\dot{Q}\right](I-Q)^{-1} \\
&= -\left[I + (I-Q)(I+Q)^{-1}\right]\dot{Q}(I-Q)^{-1} \\
&= -\left[(I+Q)(I+Q)^{-1} + (I-Q)(I+Q)^{-1}\right]\dot{Q}(I-Q)^{-1} \\
&= -\left[(I+Q) + (I-Q)\right](I+Q)^{-1}\dot{Q}(I-Q)^{-1} \\
&= -\left[2I + Q - Q\right](I+Q)^{-1}\dot{Q}(I-Q)^{-1}
\end{aligned}$$

So,

$$\dot{C}C^T = -2(I+Q)^{-1}\dot{Q}(I-Q)^{-1}$$

And, using Eq. (2.147a), we have proven

$$\dot{C}C^T = -\tilde{\omega}$$

or

$$\dot{C} = -\tilde{\omega}C$$

Q.E.D.

2.4.3 Applications of the Cayley Transform

Recall that the *spectral decomposition* of an $n \times n$ symmetric positive definite matrix P is given by

$$P = C\Lambda C^T \tag{2.149}$$

where $\Lambda = \text{diag}(\lambda_1, \ldots, \lambda_n)$.

Since C is an orthogonal matrix in Eq. (2.149), this is a redundant parameterization of P; a minimal parameterization is obtained by using the Cayley transform Eq. (2.145) to write

$$P(\lambda_1, \ldots, \lambda_n, q_1, \ldots, q_m) = (I-Q)(I+Q)^{-1}\Lambda(I-Q)^{-1}(I+Q) \tag{2.150}$$

where the qs are the distinct elements in Q and $m = (n^2 - n)/2$. Notice that the n λ_is and the $(n^2-n)/2$ q_is; total $(n^2+n)/2$ parameters, which is precisely equal to the number of distinct elements in $P = P^T$. We can show

$$\det(I+Q) = \det(I-Q) = \Sigma q_i^2 + 1$$

It is therefore evident that the transformation of Eq. (2.150) exists for all finite $Q = -Q^T$ matrices. Notice further that the Λ's control the "size" of P whereas the rotations implicit in the qs control the orientation. Note that, by constraining the λs to be positive, this parameterization forces P to be positive definite.

Symmetric positive definite matrices arise often in dynamics and control, for example:

Solution of matrix Lyapunov and Riccati differential equations:
covariance matrices in Kalman filter and estimation theory
gain matrices in optimal control theory

Parameterization of positive semidefinite physical property matrices in mechanics:
inertia and stress tensors
mass, stiffness and damping matrices

These examples and others will be encountered in this text. Obviously, Eqs. (2.149) and (2.150) represent a minimal parameter generator for symmetric positive definite matrices. The separation of rotational deformation $(q_1, \ldots, q_m)$ from the scale/size deformations $(\lambda_1, \ldots, \lambda_n)$ is physically and mathematically appealing.

In Refs. 43-44, this transformation has recently led to an attractive new algorithm for solving the matrix Riccati differential equation.

Problems

Problem 2-1

If A is a nonsingular square matrix, show that

$$\det \begin{bmatrix} A & B \\ C & D \end{bmatrix} = \det A \ \det(D - CA^{-1}B)$$

Note that the utility of this result depends on the existence of A^{-1}. What if we only know that B is a nonsingular square matrix?

Problem 2-2

A is an $n \times n$ square matrix, and λ is an eigenvalue of the A matrix, i.e., λ satisfies the *characteristic equation* obtained from $\det(\lambda I - A) = 0$; this equation can be written as the polynomial

$$\lambda^n + c_{n-1}\lambda^{n-1} + \cdots + c_0 = 0$$

Matrix A satisfies the following matrix equation (Cayley-Hamilton theorem)

$$A^n + c_{n-1}A^{n-1} + \cdots + c_0 I = 0$$

Thus a matrix satisfies its own characteristic equation. It is well-known that this theorem can be proven via the resolvent formula [adjoint of matrix $(\lambda I - A)$]. For the distinct eigenvalue case, prove the Cayley-Hamilton Theorem using spectral decomposition.

Problem 2-3

Let matrix A be of order n. Observe that λ is an eigenvalue of A if and only if $(\lambda I - A)$ is singular. Notice that $(\lambda I - A)$ is singular if and only if

$$\det(\lambda I - A) = 0 \tag{$*$}$$

The eigenvalues of A are precisely those values of λ that satisfy the equation $(*)$, which is called the *characteristic equation* of A, as mentioned in problem 2.2.

If matrix A and matrix B are *similar* (i.e., B is obtained from the *similarity transformation* $B = P^{-1}AP$), then show that A and B have the same characteristic equation.

Problem 2-4

Show that if a nonzero square matrix A is *nilpotent*, (i.e., $A^k = 0$, but $A^{k-1} \neq 0$ for some $k > 2$), then all eigenvalues of A are equal to zero.

Problem 2-5
Consider the linear system

$$\dot{\mathbf{x}}(t) = A\mathbf{x}(t) + B(t)\mathbf{u}(t)$$

where $\mathbf{x}(t)$ is an n-dimensional state vector, and $\mathbf{u}(t)$ is an m-dimensional control vector. The general solution of the above system is written by

$$\mathbf{x}(t) = e^{At}\mathbf{x}(0) + \int_0^t e^{A(t-\tau)}B\mathbf{u}(\tau)\, d\tau$$

where

$$e^{At} = I + tA + \frac{t^2}{2!}A^2 + \frac{t^3}{3!}A^3 + \cdots$$

Using the spectral decomposition of A, Eq. (2.50) shows that the response of linear dynamic systems can be represented using the eigenvalues and eigenvector matrices as follows:

$$\mathbf{x}(t) = \Phi e^{\Lambda t}\Psi^T\mathbf{x}(0) + \int_0^t \Phi e^{\Lambda(t-\tau)}\Psi^T B\mathbf{u}(\tau)\, d\tau$$

Problem 2-6
Consider a symmetric positive definite matrix A. Prove the following inequalities

$$\lambda_{min}(A) \le \left(\frac{\mathbf{x}^T A\mathbf{x}}{\mathbf{x}^T\mathbf{x}}\right) \le \lambda_{max}(A)$$

where $\lambda_{min}(A)$ and $\lambda_{max}(A)$ denote the smallest and the largest eigenvalues of matrix A, respectively. Note that the quantity $(\frac{\mathbf{x}^T A\mathbf{x}}{\mathbf{x}^T\mathbf{x}})$ is called a *Rayleigh quotient.*

Problem 2-7
Consider $n \times n$ real symmetric matrix A. Prove the following properties
(a) right eigenvector $\underline{\phi}_i$ = left eigenvector $\underline{\psi}_i$, i = 1, 2, ..., n
(b) right singular vector U_i = left singular vector V_i, i = 1, 2, ..., n
(c) For a positive semi-definite matrix A,
eigenvalue λ_i = singular value σ_i, and $U_i = V_i = \underline{\phi}_i = \underline{\psi}_i$, $i = 1, 2, ..., n$

Problem 2-8
Consider an $n \times n$ matrix A with eigenvalues $\lambda_1 \ge \lambda_2 \ge \ldots \ge \lambda_n$. Prove that the singular values of matrix A are $|\lambda_1|, \ |\lambda_2|, \ldots, \ |\lambda_n|$ for the following matrices:
(a) Hermitian matrix A
(b) Normal matrix A, (i.e., $A^H A = AA^H$)

Problem 2-9

A digital device is described by the following difference equation

$$\mathbf{x}(k+1) = \mathbf{A}\mathbf{x}(k) + \mathbf{B}\mathbf{u}(k)$$

where $\mathbf{x}(k)$ is the $n \times 1$ state vector, $\mathbf{u}(k)$ is the $m \times 1$ input vector at time $k > 0$, respectively. The initial state, $\mathbf{x}(0) = \mathbf{x}_0$, and the desired final state at the stage p, $\mathbf{x}(p) = \mathbf{x}_d$, are given.

(a) Describe how the pseudoinverse could be used to solve for the input sequence $\mathbf{u}(0), \mathbf{u}(1), \ldots, \mathbf{u}(p-1)$ causing the state to progress from $\mathbf{x}_0$ to $\mathbf{x}_d$. Discuss the practical implications and necessary conditions for the solutions when $n > mp$, $n = mp$, and $n < mp$.

(b) Illustrate your discussion by calculations for the system ($n = 2$, $m = 1$)

$$A = \begin{bmatrix} 1 & 0.5 \\ 0 & 0.5 \end{bmatrix}, \qquad B = \begin{Bmatrix} 1 \\ 1 \end{Bmatrix}$$

with $\mathbf{x}_d = \begin{Bmatrix} 0 \\ 0 \end{Bmatrix}$ and $\mathbf{x}_0 = \begin{Bmatrix} -40 \\ 50 \end{Bmatrix}$, for p=1,2,3, and 4.

Problem 2-10 (Geometric Interpretation of Minimum Norm Solution)

(a) Consider the line $2x_1 + x_2 = 2$ shown in Figure P2.1(a) in the two-dimensional space. Find the point (x_1^*, x_2^*) that is the shortest distance from the origin to the line.

(b) Consider the planes $x_1 + x_2 + x_3 = 1$ and $2x_1 + 3x_2 = 1$ shown in Figure P2.1(b) in the three-dimensional space. Intersection of these two planes is obviously a straight line. Find the point (x_1^*, x_2^*, x_3^*) that is the shortest distance from the origin to this line of intersection. This problem can be restated as a constrained optimization problem as follows:

$$\text{Minimize} \qquad x_1^2 + x_2^2 + x_3^2$$

$$\text{subject to} \qquad \begin{bmatrix} 1 & 1 & 1 \\ 2 & 3 & 0 \end{bmatrix} \begin{Bmatrix} x_1 \\ x_2 \\ x_3 \end{Bmatrix} = \begin{Bmatrix} 1 \\ 1 \end{Bmatrix}$$

(c) (Generalization to the n-dimensional case):
Consider the m linear equations ($\mathbf{Ax} = \mathbf{b}$) in the n-dimensional space ($m < n$). The intersection space of those m equations becomes $(n - m)$ dimensional hyperplane. Discuss the relation between this hyperplane and the solution of the following minimum norm problem.

$$\text{Minimize} \qquad \mathbf{x}^T\mathbf{x}$$
$$\text{subject to} \qquad \mathbf{Ax} = \mathbf{b}$$

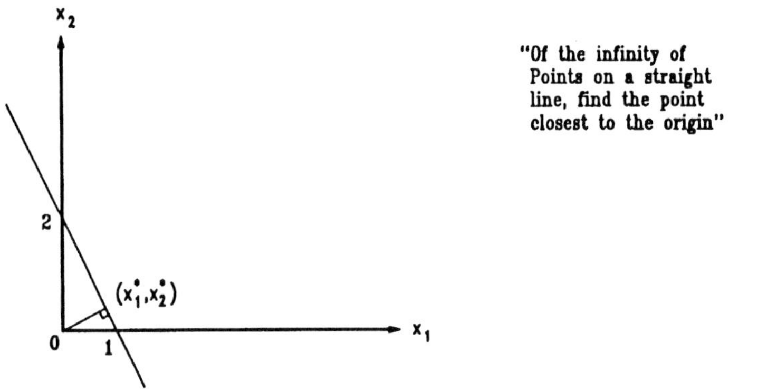

(a) Two Dimensional Space

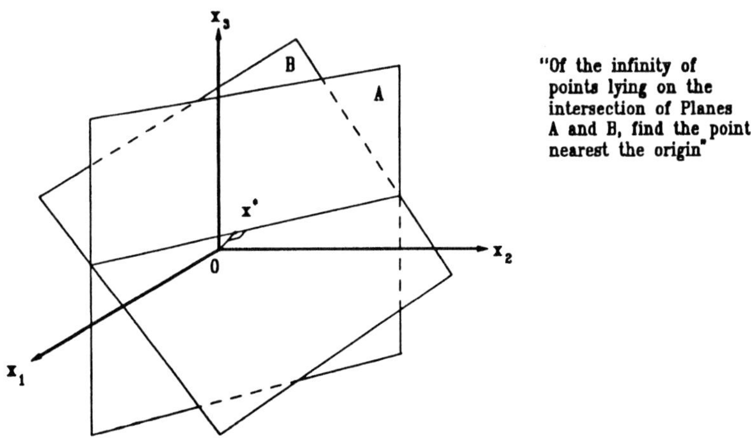

(b) Three Dimensional Space

Figure P2.1 Geometric Interpretation of Minimum Norm Solution

Specialize this general minimum norm problem and its solution to obtain the two special case solutions (a) and (b).

Problem 2-11 (Geometric Interpretation of the Least Squares Solution)

(a) Consider the following experimental data:

a	1	2	3	4	5
b	1.8	3.2	3.9	4.9	6.2

We want to fit this data to a straight line. The experimental data can be modeled as a straight line; we write the "observation equations," including the residual errors e_i as a function of the unknown slope (x_1) and intercept (x_2) of the best-fitting line.

$$b_i = x_1 a_i + x_2 + e_i, \qquad i = 1, 2, \ldots, 5$$

The least squares approach can be used to find the approximate solution $\hat{x}_1$, $\hat{x}_2$ that minimizes the sum of the squares $(\sum_i e_i^2)$ of the error components as shown in Figure P2.2. Find the least squares solution for the given data.

(b) Consider the m sets of experimental data $(a_1(i), a_2(i), b(i),\ i = 1, 2, \ldots, m)$. The data set can be mapped to the three-dimensional space by considering (a_1, a_2, b) as coordinates. Assume that we want to fit the data to an affine plane,

$$a_1 x_1 + a_2 x_2 = b \qquad (*)$$

A particular approximate solution can be obtained using the least squares approach that minimizes the squares of the errors between the data set and Eq. (∗). Discuss the relation of this least squares solution (for the best-fitting affine plane) to the data from a geometric point of view.

(c) (Generalization to the n-dimensional space):
Suppose we executed the same experiment m times, and obtained m sets of data with n variables and 1 output ($m > n$).

$$\text{input: } [a_1(i) \quad a_2(i) \quad \ldots \quad a_n(i)], \qquad \text{output: } b(i), \qquad i = 1, 2, \ldots, m$$

We want to fit these data to the linear equation

$$a_1 x_1 + a_2 x_2 + \cdots + a_n x_n = b$$

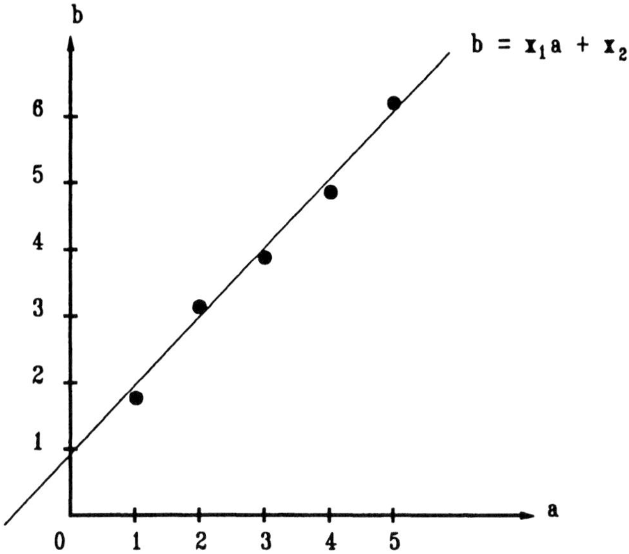

Figure P2.2 The Least Squares Solution.

Obtain the formula of the approximate solution for $\hat{\mathbf{x}} = [x_1 \ \ldots \ x_n]^T$ using the least squares approach, and discuss the best-fitting solution (a hyperplane) to the data from a geometric point of view.

Problem 2-12
Consider the orthogonal matrix

$$C = \begin{bmatrix} \cos\theta & \sin\theta & 0 & 0 \\ -\sin\theta & \cos\theta & 0 & 0 \\ 0 & 0 & \cos\phi & \sin\phi \\ 0 & 0 & -\sin\phi & \cos\phi \end{bmatrix}$$

Determine the distinct element of Q and $\tilde{\omega}$ in Eqs. (2.145), (2.146), and (2.147) as functions of the time-varying coordinates θ, $\dot{\theta}$, ϕ, $\dot{\phi}$.

Problem 2-13
Consider the skew-symetric matrix

$$Q = \begin{bmatrix} 0 & -1/3 & 1 & 1/4 \\ 1/3 & 0 & -1/2 & 1 \\ -1 & 1/2 & 0 & 1/3 \\ -1/4 & -1 & -1/3 & 0 \end{bmatrix}$$

Your tasks are to carry out the computations (e.g., using MATLAB) to establish the following:

(1) Determine the C matrix using the Cayley transform of Eq. (2.145a).
(2) Verify the inverse Cayley transform of Eq. (2.146a).
(3) Verify that $C^T C = I$.

References

1. Laub, A.J. and Little, J.N., *MATLAB User's Guide - Control System Toolbox*, The Mathworks, Inc., Sherborn, MA, 1986.

2. Kirk, D.E., *Optimal Control Theory, An Introduction*, Prentice-Hall, Englewood Cliffs, NJ, 1970.

3. Luenburger, D.G., *Optimization by Vector Space Method*, John Wiley, New York, NY, 1969.

4. Junkins, J.L., "Equivalence of the Minimum Norm and Gradient Projection Constrained Optimization Techniques," *AIAA Journal,* Vol.10, No.7, July 1972, pp. 927-929.

5. Stewart, G.W., *Introduction to Matrix Computations*, Academic Press, New York, NY, 1973.

6. Junkins, J.L., *Application of Modern State Space Analysis in Spacecraft Dynamics, Estimation and Control,* Tutorial Keynote Lecture, 1989 AAS Guidance and Control Conference, Keystone, CO., Feb. 5, 1989, also, *Advances in the Astronautical Sciences*,Vol. 68, pp. 555-678, Univelt, San Diego, CA. 1989.

7. Bryson, A.E. and Ho. Y-C., *Applied Optimal Control, Optimization, Estimation, and Control,* John Wiley, New York, NY, 1975.

8. Kailath, T., *Linear Systems*, Prentice-Hall, Englewood Cliffs, NJ, 1980.

9. Wilkinson, J.H., *The Algebraic Eigenvalue Problem*, Oxford University Press, New York, NY, 1965.

10. Bartels, R.H. and Stewart, G.W., "Solution of Matrix Equation AX+XB=C," *Communications of ACM*, Sept. 1972.

11. Mukhopadhyay, V. and Newsom, J.R., "Application of Matrix Singular Value Properties for Evaluating Gain and Phase Margins of Multiloop Systems," NASA TM-84524, July 1982.

12. Dongarra, J.J., Moler, C.B., Bunch, J.R., and Stewart, G.W., *LINPACK User's Guide*, SIAM Publication, Philadelphia, PA, 1979.

13. Noble, B. and Daniel, J.W., *Applied Linear Algebra*, Prentice Hall, Englewood Cliffs, NJ, 1977.

14. Stoer, J. and Bulirsch, R., *Introduction to Numerical Analysis*, Springer-Verlag, Berlin, 1980.

15. Laub, A.J., "A Schur Method for Solving Algebraic Riccati Equations," *IEEE Trans. on Automatic Control,* Vol. AC-24, No. 6, Dec. 1979.

16. Collins, J.D. et al., "Statistical Identification of Structures," *AIAA Journal,* Vol. 12, Feb. 1972, pp. 185-190.

17. Berman, A. and Flannelly, W.G., "Theory of Incomplete Models of Dynamic Structures," *AIAA Journal,* Vol. 9, Aug. 1971, pp. 1481-1487.

18. Haftka, R.T. et al., "Sensitivity of Optimized Control Systems to Minor Structural Modifications," AIAA Paper 85-0807, April 1985.

19. Lim, K.B. and Junkins, J.L., "Optimal Redesign of Dynamic Structures via Sequential Linear Programming," *Proceedings of 4th International Modal Analysis Conference,* Union College, New York, Feb. 1986, pp. 1615-1620.

20. Chou, Y-F. and Chen, J-S., "Structural Dynamics Modification via Sensitivity Analysis," *Proceedings of 3rd International Modal Analysis Conference,* Union College, New York, Jan. 1985, pp. 483-489.

21. Bodden, D.S. and Junkins, J.L., "Eigenvalue Optimization Algorithms for Structural/Controller Design Iterations," *Journal of Guidance, Control, and Dynamics,* Vol. 8, No. 6., Nov.-Dec. 1985.

22. Junkins, J.L., Bodden, D.S., and Turner, J.D., "A Unified Approach to Structure and Control System Design Iterations," *Fourth International Conference on Applied Numerical Modeling,* Tainan, Taiwan, Dec. 1984.

23. Lim, K.B. and Junkins, J.L., "Optimal Design of Dynamic Structures via Sequential Linear Programming," 4th International Modal Analysis Conference, Los Angeles, CA, Feb. 3-6, 1986.

24. Junkins, J.L. and Kim, Y., "First and Second Order Sensitivity of the Singular Value Decomposition," *Journal of the Astronautical Sciences,* Vol. 38, No. 1, Jan.-Mar., 1990, pp. 69-86.

25. Junkins, J.L. and Kim, Y., "A Minimum Sensitivity Design Method for Output Feedback Controllers," "Progress in AstronauticsAeronautics and Aeronautics Series" *Mechanics and Control of Large Space Structures*, Vol. 129, Chapter 15, published by AIAA, 1990.

26. Doyle, J.C. and Stein, G., "Multivariable Feedback Design: Concepts for a Classical/Modern Synthesis," *IEEE Transactions on Automatic Control,* Vol. AC-26, No. 1, Feb. 1981, pp. 4-16.

27. Francis, B.A., *A Course in* H_∞ *Control Theory,* Lecture Notes in Control and Information Sciences, Vol. 88, Springer-Verlag, Berlin, 1987.

28. McFarlane, D.C. and Glover, K., *Robust Controller Design Using Normalized Coprime Factor Plant Descriptions,* Lecture Notes in Control and Information Sciences, Vol. 138, Springer-Verlag, Berlin, 1989.

29. Maciejowski, J.M., *Multivariable Feedback Design,* Addison-Wesley, Reading, MA, 1989.

30. Patel, R.V. and Toda, M., "Quantitative Measures of Robustness for Multivariable Systems," TP-8A, *Proceedings of JACC,* San Francisco, CA, 1980.

31. Lim, K.B., Junkins, J.L., and Wang, B.P., "Re-examination of Eigenvector Derivatives," *Journal of Guidance, Control, and Dynamics,* Vol. 10, No. 6, 1987, pp. 581-587.

32. Plaut, R.H. and Huseyin, K., "Derivatives of Eigenvalues and Eigenvectors in Non-Self-Adjoint Systems," *AIAA Journal,* Vol.11, Feb. 1973, pp. 250-251.

33. Lim, K.B., Juang, J.N., and Ghaemmaghami, P., "Eigenvector Derivatives of Repeated Eigenvalue Using Singular Value Decompositions," *Journal of Guidance, Control, and Dynamics,* Vol. 12, No. 2, Mar.-Apr. 1989, pp. 282-283.

34. Juang, J.N., Ghaemmaghami, P., and Lim, K.B., "Eigenvalue and Eigenvector Derivatives of a Non-Defective Matrix," *Journal of Guidance, Control, and Dynamics,* Vol. 12, No. 4, July-Aug 1989, pp.480-486.

35. Bauer, F.L. and Fike, C.T., "Norms and Exclusion Theorems," *Numerische Mathematik,* Vol. 2, 1960, pp. 137-141.

36. Lim, K.B., and Junkins, J.L., "Robustness Optimization of Structural and Controller Parameters," AIAA Paper 87-0791-CP, AIAA 28th SDM Conference, Monterey, CA, 1987.

37. Vidyasagar, M., *Nonlinear System Analysis,* Prentice-Hall, Englewood Cliffs, NJ, 1978.

38. Rew, D.W. and Junkins, J.L., "Multi-Criterion Approaches to Optimization of Linear Regulators," AIAA Paper 86-2198-CP, *Proceedings of the AIAA Guidance, Navigation, and Control Conference,* Williamsburg, VA, Aug. 1986.

39. Wonham, W.M., "On Pole Assignment in Multi-Input Controllable Linear Systems," *IEEE Trans on Automatic Control,* Vol. AC-12, 1967, pp. 660-665.

40. Rew, D.W., *New Feedback Design Methodology for Large Space Structures: A Multi-Criterion Optimization Approach,"* Ph.D. Dissertation, Virginia Polytechnic Institute and State University, Blacksburg, VA, 1987.

41. Goldstein, H., *Classical Mechanics,* Addison-Wesley Press, Cambridge, 1950.

42. Junkins, J.L. and Turner, J.D., *Optimal Spacecraft Rotational Maneuvers,* Studies in Astronautics, Vol. 3, Chapter 2, Elsevier Scientific Publishing Company, New York, N.Y., 1985.

43. Bar-Itzhack, I.Y. and Markley, F.L., "Minimal Parameter Solution of the Orthogonal Matrix Differential Equation," NASA TM-4043, June, 1988.

44. Y. Oshman and Bar-Itzhack, I.Y., "Eigenfactor Solution of the Matrix Riccati Equation: A Continuous Square Root Algorithm," *IEEE Trans. on Automatic Control,* Vol. AC-30, No. 10, Oct. 1985, pp 971-978.

CHAPTER 3

Stability in the Sense of Lyapunov: Theory and Applications

In the present Chapter, we present some elegant concepts from stability theory. Although the spirit of our presentation is classical, we follow a Lyapunov approach and include some novel methodology for designing globally stable control laws for nonlinear and distributed parameter systems. A Lyapunov approach is attractive not only because it plays an important role in modern control theory but also because most of the classical and recent frequency-domain stability results have also been obtained or can be proven easily using Lyapunov methods. Especially significant is the fact that the Lyapunov approach leads to a unified stability perspective for both linear and nonlinear systems, as well as systems described by ordinary, partial, and hybrid systems of differential equations.

In section 3.1, several important definitions and concepts fundamental to stability analysis are stated. Section 3.2 reviews three basic theorems stemming from the so-called *direct method* of Lyapunov: (1) the stability theorem, (2) the asymptotic stability theorem, and (3) the instability theorem. In section 3.3, Lyapunov stability of systems described by linear differential equations is presented, and stability of second-order mechanical systems is also examined. Section 3.4 presents the *indirect method* of Lyapunov to characterize the local stability of motion of a nonlinear system around an equilibrium state. Also developed in section 3.4 is a method for generating globally stabilizing feedback control laws for maneuvers of distributed parameter structural systems by utilization of a Lyapunov control design approach. Section 3.5 summarizes a case study that includes a discussion of analytical, numerical, and laboratory experimental results on near-minimum-time maneuvers of a flexible structure.

3.1 Basic Definitions

A continuous, finite-dimensional dynamic system is described by a first-order nonlinear vector differential equation

$$\dot{\mathbf{x}} = \mathbf{f}(\mathbf{x}, t), \qquad \mathbf{x} \in \mathrm{R}^n \tag{3.1}$$

where $\mathbf{x}(t)$ is the $n \times 1$ state vector at time t, and the dot denotes time differentiation. The function $\mathbf{f}(\mathbf{x}, t)$ generates a vector field. In most of the developments herein, $\mathbf{f}(\mathbf{x}, t)$ is assumed to be a continuous function of all $n + 1$ arguments and at least piecewise differentiable one or more times with respect to all arguments. Clearly an infinite variety of motions can be generated by equations of the form of Eqs. (3.1), and discussions of stability properties must be posed carefully. In particular, we must address the fact that the stability is a descriptive property of *departure motion* from some particular reference state or motion. The most fundamental cases arise when the reference state is an *equilibrium state* of the system (3.1).

Definition 3.1: Equilibrium State

A vector $\mathbf{x_e} \in \mathrm{R}^n$ is said to be an *equilibrium state* of Eq. (3.1) at time t_0 if

$$\mathbf{f}(\mathbf{x_e}, t) = 0, \qquad \forall\, t \geq t_0 \tag{3.2}$$

■

If $\mathbf{x_e}$ is an equilibrium state of Eq. (3.1) at time t_0, then $\mathbf{x_e}$ is also an equilibrium state of Eq. (3.1) at all times $t_1 \geq t_0$. In other words, a solution that passes exactly through $\mathbf{x_e}$ at some time remains there for all time.

Definition 3.2: Stability of an Equilibrium State

The equilibrium state $\mathbf{x_e}$, or the equilibrium solution $\mathbf{x}(t) = \mathbf{x_e}$, is said to be stable if, for any given t_0 and positive ϵ, there exists a positive $\delta(\epsilon, t_0)$ such that *every* time-varying trajectory (or solution) $\mathbf{x}(t)$ initiating (time t_0) at a point $\mathbf{x_0}$ that lies within in a δ-neighborhood of $\mathbf{x_e}$

$$\|\mathbf{x_0} - \mathbf{x_e}\| < \delta, \qquad \mathbf{x_0} = \mathbf{x}(t_0)$$

remains for all time within an ϵ-neighborhood of $\mathbf{x_e}$:

$$\|\mathbf{x}(t) - \mathbf{x_e}\| < \epsilon, \qquad \forall\, t \geq t_0$$

■

The equilibrium state is said to be unstable if it is not stable.

Definition 3.3: Asymptotic Stability of an Equilibrium State

The equilibrium state $\mathbf{x_e}$ is said to be *asymptotically stable*

(1) if it is stable (Definition 3.2) and, in addition,

(2) if for any t_0, there exists a $\delta_1(t_0)$ such that

$$\|\mathbf{x_0} - \mathbf{x_e}\| < \delta_1$$

which implies that

$$\lim_{t\to\infty} \mathbf{x}(t) \to \mathbf{x_e}$$

■

If δ and δ_1 are not functions of t_0, then the equilibrium state is said to be *uniformly stable* and *uniformly asymptotically stable*, respectively. Definitions 3.2 and 3.3 constitute the two basic definitions of stability of an equilibrium state (a fixed point in the state space) for an unforced continuous time system. More generally, we need to consider the stability of a *trajectory* or a *motion*. Qualitatively, *stability of a trajectory* is concerned with whether the perturbed motion remains near the unperturbed trajectory or diverges from it. These isues are discussed further below. Stability of a motion is of central interest in many practical feedback control situations whereby a system is designed to execute a nominal state history, and control inputs must be developed not only to generate the nominal motion but also to stabilize perturbed neighboring motions so that the actual system will behave in a near-nominal fashion.

Definition 3.4: Stability of a Motion

The motion $\mathbf{x}(t)$ is said to be stable if, for all t_0 and positive ϵ there exists a positive $\delta(\epsilon, t_0)$, such that

$$\|\mathbf{x}(t) - \bar{\mathbf{x}}(t)\| < \epsilon, \quad \forall\, t \geq t_0, \text{ if } \quad \|\mathbf{x}_0 - \bar{\mathbf{x}}_0\| < \delta$$

where $\mathbf{x}(t)$ and $\bar{\mathbf{x}}(t)$ are neighboring trajectories with the given initial conditions $\mathbf{x}_0$ and $\bar{\mathbf{x}}_0$, respectively, at time t_0.

■

This bounded motion stability property is sometimes referred to as "path" stability. Qualitatively, path stability means that "if $\bar{\mathbf{x}}(t_0)$ is near $\mathbf{x}(t_0)$, then $\bar{\mathbf{x}}(t)$ will remain near $\mathbf{x}(t)$ for all t."

Definition 3.5: Asymptotic Stability of a Motion

The motion $\mathbf{x}(t)$ is said to be asymptotically stable if

(1) it is stable (Definition 3.4) and if, in addition

(2) for any t_0, there exists a positive $\delta_1(t_0)$, such that

$$\|\mathbf{x}_0 - \bar{\mathbf{x}}_0\| < \delta_1 \tag{3.4}$$

implies

$$\lim_{t \to \infty} \|\mathbf{x}(t) - \bar{\mathbf{x}}(t)\| = 0$$

Note that $\bar{\mathbf{x}}(t)$ is *any* member of the set of all neighboring (perturbed) trajectories satisfying Eq. (3.4), and any/all members of this set asymptotically approach $\mathbf{x}(t)$.

■

The above definitions are not directly concerned with the global properties of systems, but rather with properties of the motion in a finite local neighborhood of an equilibrium state (or reference trajectory, for Def. 3.4). If a system has a globally asymptotically stable equilibrium state, then it is obviously the *only* equilibrium state, and every motion converges to that unique equilibrium; therefore, the system can be said to be completely (or globally) stable. An analogous global stability property can be defined for the stability of a motion.

The simplest class of Lyapunov stability analysis methods arises in the context of systems described by linear unforced differential equations. We develop some of the central ideas below.

Consider the linear system

$$\dot{\mathbf{x}}(t) = A(t)\mathbf{x}(t)$$

which obviously has an equilibrium state at the origin. This linear system can be classified as stable, asymptotically stable, or unstable, depending on the stability of the origin (Refs. 1, 2).

Now, we introduce the concept of positive definite functions. These ideas are important to understand and apply Lyapunov stabilty theory.

Definition 3.6: Positive Definite Function

A singled-valued function $U(\mathbf{x})$, which is continuous and has continuous partial derivatives with respect to the components of the vector $\mathbf{x}$, is said to be *positive definite* in some region Ω about the origin if

(1) $U(0) = 0$

(2) $U(\mathbf{x}) > 0$ for all nonzero $\mathbf{x} \in \Omega$

■

If the positivity condition (2) is relaxed to the non-negative condition $U(\mathbf{x}) \geq 0$ for all $\mathbf{x} \in \Omega$, then $U(\mathbf{x})$ is said to be *positive semidefinite.* If the inequality sign in (2) is reversed, then the condition for a *negative definite* function is obtained. If a function is neither positive nor negative definite, then it is *indefinite.*

Definition 3.7: Positive Definite Quadratic Forms

In the study of dynamical systems, quadratic functions of the state vector arise often in the context of energy, stability, and control analyses. Especially important are *quadratic forms*: $U(\mathbf{x}) = \mathbf{x}^T Q\mathbf{x}$. $U(\mathbf{x})$ is said to be *positive definite* if

$$U(\mathbf{x}) = \mathbf{x}^T Q\mathbf{x} > 0, \qquad \text{for all nonzero } \mathbf{x} \in R^n$$

where Q is a real symmetric matrix.

■

Definition 3.7 is equivalent to the fact that all the eigenvalues of Q are strictly positive. Such a matrix is naturally called *positive definite.*

Note that the above concepts are central to the development of Lyapunov stability theory. Should the reader feel uncomfortable with the above discussion, it is important to first invest in a study of the underlying linear and functional analysis ideas before proceeding. For a detailed discussion of these issues, the reader is referred to Vidyasagar (Ref. 1) and Willems (Ref. 2). The following example is useful to illustrate the ideas underlying the above discussion. The reader is encouraged to invent several additional examples to illustrate the qualitative ideas underlying positive and negative definiteness in two- and three-dimensional state spaces.

Example 3.1

Consider the functions:

(1) $U_1(x) = x_1^2 + x_2^2$

(2) $U_2(x) = (x_1 + x_2)^2$

Clearly U_1 satisfies the condition of Definition 3.3. Therefore it is a positive definite function in a two-dimensional space, but U_1 is only positive semidefinite if the underlying space has more than two dimensions. U_2 is only positive semidefinite in a two-dimensional space since it is zero everywhere along the line $x_1 + x_2 = 0$.

■

3.2 Lyapunov Stability Theory (Direct Method)

The central ideas of the Lyapunov stability theorem are now introduced. Consider a system without any external forces on the system. Suppose that there exists only one equilibrium state $\mathbf{x_e}$ of the system. Suppose further that it is possible to find a positive definite function that is an exact integral of the system under some idealization. Such a function might represent a physical entity such as total system energy or the Hamiltonian. A candidate *Lyapunov function* must belong to a class of admissible functions that have as their most fundamental property that they are zero at the desired equilibrium state and positive everywhere else.

Now, suppose that the system, which is originally at the equilibrium state, is perturbed into a neighboring state where the energy level is positive by assumption, and we consider the time evolution of the distance to the equilibrium as measured by the energy function. Depending on the nature of the energy (Lyapunov) function, stability of the motion may be analyzed as follows:

(1) If the system dynamics are such that the initial energy of the system is not increasing with time for all starting points in a finite neighborhood, we can conclude that the equilibrium state is stable.

(2) If the system dynamics are such that the energy of the system is monotonically decreasing with time for all initial conditions in the neighborhood (and thus eventually approaches zero), the equilibrium state is asymptotically stable.

(3) If the energy of the system is increasing with time, for *any* initial condition in the neighborhood, then the equilibrium state is unstable.

(4) If the chosen energy measure is indefinite (i.e., it is neither strictly decreasing nor increasing), then no conclusion can be drawn about the stability of the system.

The following theorem, which is a rigorous statement of the above remarks, is the basic stability concept underlying Lyapunov's direct (second) method.

Theorem 3.1: Stability Theorem

The equilibrium state $\mathbf{x_e}$ is stable if there exists a continuously differentiable function $U(\mathbf{x})$ such that

(1) $U(\mathbf{x_e}) = 0$

(2) $U(\mathbf{x}) > 0$ for all $\mathbf{x} \neq \mathbf{x_e}$, $\mathbf{x} \in \Omega$

(3) $\dot{U}(\mathbf{x}) \leq 0$ for all $\mathbf{x} \neq \mathbf{x_e}$, $\mathbf{x} \in \Omega$

where $\dot{U}(\mathbf{x})$ denotes the time derivative of the function $U(\mathbf{x})$, and Ω is some region containing $\mathbf{x_e}$.

Notice that the "energy rate" $\dot{U}(\mathbf{x})$ is evaluated *along a typical trajectory* $\mathbf{x}(t)$, and the conditions (2) and (3) must hold along all members of *the infinite set of trajectories of the dynamical system*, that ensue from all initial states in Ω.

■

A modest perturbation of Theorem 3.1 (making the inequalities strict) results in the following theorem, which provides necessary and sufficient conditions for *asymptotic stability.*

Theorem 3.2: Asymptotic Stability Theorem

The equilibrium state $\mathbf{x_e}$ is asymptotically stable if there exists a continuously differentiable function U such that

(1) $U(\mathbf{x_e}) = 0$

(2) $U(\mathbf{x}) > 0$ for all $\mathbf{x} \neq \mathbf{x_e}$, $\mathbf{x} \in \Omega$

(3) $\dot{U}(\mathbf{x}) < 0$ for all $\mathbf{x} \neq \mathbf{x_e}$, $\mathbf{x} \in \Omega$

■

These theorems define local stability properties for motion in the vicinity of the equilibrium state. A system has global asymptotic stability with respect to a unique equilibrium point if the following theorem is satisfied.

Theorem 3.3: Global Asymptotic Stability Theorem

The equilibrium state $\mathbf{x_e}$ is globally asymptotically stable if there exists a continuously differentiable function U such that

(1) $U(\mathbf{x_e}) = 0$

(2) $U(\mathbf{x}) > 0 \quad$ for all $\mathbf{x} \neq \mathbf{x_e}$

(3) $\dot{U}(\mathbf{x}) < 0 \quad$ for all $\mathbf{x} \neq \mathbf{x_e}$

(4) $U(\mathbf{x}) \longrightarrow \infty$ as $\|\mathbf{x}\| \longrightarrow \infty$

■

Note that the stable region Ω extends to infinity in Theorem 3.3. Reference 1 provides further discussion, including the complete proofs of the above theorems. Also, observe that there is no one unique Lyapunov function for a given system; some may be better than others. This is especially important when we seek the "least conservative" stability information, when, for example, we seek to determine the size of the Ω region in which we have stability. If a poor choice of $U(\mathbf{x})$ results in a pessimistic conclusion that Ω is much smaller than it actually is, then this is an obvious (and very frequently occurring) concern. It should be also noted that if a Lyapunov function cannot be found, nothing can be concluded about the stability of the system, since the Lyapunov stability theorem provides only *sufficient* conditions for stability. Therefore, the conditions required to prove stability, based on a particular choice of Lyapunov function, may be very conservative.

Unfortunately, the Lyapunov theorems, in their classical setting, are not *constructive*; these stability theorems do not reveal a process to find a candidate Lyapunov function. It is often difficult to find a suitable Lyapunov function for a given nonlinear system. Although the physical and mathematical insights of the analyst has historically played an important role in most successful applications of this approach, more systematic methods have recently emerged (Refs. 3-5) for certain classes of control design problems. *In particular, when the stability analysis and the control design analysis are merged*, one is often able to exploit the additional freedom to simultaneously design control laws and select a Lyapunov function which guarantees stability of the closed-loop (controlled) system.

Example 3.2

Consider the system described by the nonlinear ordinary differential equation

$$\ddot{x}(t) - \epsilon x^2(t)\dot{x}(t) + x(t) = 0$$

The objective is to use Lyapunov analysis to investigate the stability of this system for motion near the origin.

Introducing the state variable representation of this system with the definitions $x_1 = x$, $x_2 = \dot{x}$, we write the equivalent first-order system

$$\dot{x}_1 = x_2$$
$$\dot{x}_2 = -x_1 + \epsilon x_1^2 x_2$$

It is easy to verify that the above "oscillator with nonlinear damping" has an equilibrium state at the origin $(x_1, x_2) = (0, 0)$. Our goal is to determine if this state is stable. For this purpose, let us choose the simplest candidate Lyapunov function,

$$U(x_1, x_2) = \frac{x_1^2 + x_2^2}{2}$$

We note a physical motivation for choosing this positive definite function as a candidate Lyapunov function: It is an exact (total mechanical energy) integral of the system, for $\epsilon = 0$. Clearly, $U(0,0) = 0$ and $U(x_1, x_2) > 0$ in any neighborhood of $(0,0)$ and, investigating the energy rate, we find

$$\begin{aligned} \dot{U}(x_1, x_2) &= x_1 \dot{x}_1 + x_2 \dot{x}_2 \\ &= x_1 x_2 + x_2(-x_1 + \epsilon x_1^2 x_2) \\ &= \epsilon x_1^2 x_2^2 \end{aligned}$$

Thus U *is a positive definite function that is strictly decreasing along all system trajectories* if $\epsilon < 0$. Therefore, by the above theorems, $(0,0)$ is a globally stable equilibrium point for $\epsilon = 0$, is globally asymptotically stable for $\epsilon < 0$, and is globally unstable for $\epsilon > 0$. Thus the Lyapunov analysis was completely successful in establishing the global stability characteristics of this system.

■

Example 3.3

Investigate the stability of the system of nonlinear differential equations:

$$\dot{x}_1 = x_1(x_1^2 + x_2^2 - 1) - x_2$$
$$\dot{x}_2 = x_1 + x_2(x_1^2 + x_2^2 - 1)$$

We try the candidate Lyapunov function

$$U(x_1, x_2) = x_1^2 + x_2^2$$

which is an exact integral of the simplified system $\dot{x}_1 = -x_2$, $\dot{x}_2 = x_1$.

U is obviously a positive definite function having its global minimun at the origin. It is also obvious by inspection that the origin is the only equilibrium point of the nonlinear system. Investigating the energy rate, we find

$$\dot{U}(x_1, x_2) = 2(x_1^2 + x_2^2)(x_1^2 + x_2^2 - 1)$$

Therefore, $\dot{U}$ is a negative definite function over the finite circular region

$$\{(x_1, x_2)|\ x_1^2 + x_2^2 < 1\}$$

which includes the equilibrium point at the origin. Hence, the origin $(0, 0)$ is an asymptotically stable equilibrium state of this system. Note that all points within the unit circle are asymptotically attracted to the origin. However, because $\dot{U}$ is not a negative definite function over all of R^n, we therefore cannot immediately conclude global asymptotic stability. While we are certain we have stability within the unit circle, this conclusion results from a particular choice of $U(x_1, x_2)$ and, without further analysis, we cannot conclude that the stable region is not actually larger than the unit circle. However, since $\dot{U}$ is positive *everywhere* outside the unit circle, we can finally conclude, using the following theorem 3.4, that we have instability for all trajectories that initiate outside the unit circle and asymptotic stability for all trajectories initiating inside the unit circle. Thus we are able to use the stability and instability insights simultaneously to "establish the complete story" vis-à-vis the global stability properties of this system, *since the stable and unstable regions have a mutual boundary; together, the stable and unstable regions span all of state space* R^2.

■

This final theorem is sometimes useful in avoiding a fruitless search for Lyapunov functions for systems that can be proven unstable. This theorem is also useful in obtaining *theoretical closure of the stability analysis*, in the sense that it is sometimes possible to simultaneously apply the instability theorem with the stability theorems to conclusively establish a particular system's global stability properties. In Ex. 3.3, for example, we concluded that our simple choice of U gave us all of the stability information (i.e., the system is stable only within the unit circle).

Theorem 3.4: Instability Theorem

The equilibrium state $\mathbf{x_e}$ is unstable if there exists a continuously differentiable function U such that

(1) $U(\mathbf{x_e}) = 0$ and $\dot{U}(\mathbf{x_e}) = 0$

(2) $\dot{U}(\mathbf{x}) > 0$ for all $\mathbf{x} \neq \mathbf{x_e}$, $\mathbf{x} \in \Omega$

(3) There exist points $\mathbf{x}$ arbitrarily close to $\mathbf{x_e}$ such that $U(\mathbf{x_e}) > 0$.

■

Some versions of the instability theorem require a function U such that both U and $\dot{U}$ are positive definite, and this unnecessarily more restrictive version is somewhat less efficient (Ref. 1) than the theorem stated above, which makes no requirement on the definiteness of U. Actually, if one can find *any* such function U satisfying the above conditions, then $\mathbf{x_e}$ is a completely unstable equilibrium point in Ω, and the quest for Lyapunov functions can be halted. In Example 3.3, the Ω for the instability theorem is clearly the complement of the Ω for the asymptotically stable region, and it is apparent that *the stable and unstable regions being complementary is the key to establishing global stability information.*

In this section, we have presented the most important theorems in Lyapunov stability theory. Taken collectively, the most positive features of these theorems are:

(1) We are able to draw rigorous conclusions about the stability/instability of the motion near an equilibrium state, without solving the system of generally nonlinear equations

(2) The Lyapunov function often has a physical interpretation as the energy of the system

(3) For particular nonlinear systems, we are sometimes able to rigorously establish global stability/instability knowledge

(4) This is one of the few available stability analysis approaches for establishing "stability in the large" characteristics for generally nonlinear systems whose differential equations cannot be solved analytically.

On the other hand, the most unfavorable aspects of the Lyapunov stability/instability theorems are:

(1) They provide only *sufficient conditions*, so that if a particular candidate Lyapunov function does not satisfy those conditions, no conclusions can be drawn; further conclusions may be possible, however, if we choose a different Lyapunov function candidate

(2) There is no systematic procedure (implicit or evident in the above theorems) for generating least conservative Lyapunov functions for arbitrary nonlinear systems.

However, it is usually easy to find Lyapunov functions for linear systems and/or weakly nonlinear systems.

There are many variations on Lyapunov's direct method of stability analysis such as the methods of Popov, Lure, etc. (Refs. 1, 2, 6, 7). In the developments below, we will develop and demonstrate more recent methods that combine the quest to define suitable Lyapunov functions with design of stabilizing control laws; these methods are especially promising (see section 3.4).

3.3 Stability of Linear Systems

3.3.1 Lyapunov Theorem for Linear Systems

Lyapunov's method is easily applied to determine the stability of a linear system. Consider an autonomous system described by the linear vector differential equation

$$\dot{\mathbf{x}}(t) = A\mathbf{x}(t) \tag{3.5}$$

The above system is said to be stable in the sense of Lyapunov, if the solution of Eq. (3.5) tends toward zero (which is obviously the *only* equilibrium state if A is of full rank) as $t \to \infty$ for an arbitrary initial condition.

If all eigenvalues of A are distinct, the response of system (3.5) ensuing from a given initial condition $\mathbf{x}_0$ can be written as

$$\mathbf{x}(t) = \sum_{i=1}^{n} (\underline{\psi}_i^T \mathbf{x}_0) e^{\lambda_i t} \underline{\phi}_i \tag{3.6}$$

where λ_i are the eigenvalues of A, $\underline{\phi}_i$ and $\underline{\psi}_i$ are, respectively, the right and left eigenvectors of A associated with λ_i.

For the repeated eigenvalue case, the situation is much more complicated (i.e., we should solve for the generalized eigenvectors of A). The generalization of Eq. (3.6) for the case of generalized eigenvectors has a similar form but will not be discussed here (Ref. 8).

From Eq. (3.6), we can see that the system will be asymptotically stable if and only if all the eigenvalues of A have negative real parts, i.e.,

$$\Re[\lambda_i(A)] < 0 \tag{3.7}$$

Thus, the stability of a linear system can be completely characterized by the eigenvalues of the system. This eigenvalue approach to stability yields both necessary and sufficient conditions. However, calculating all the eigenvalues of the system matrix is not always desirable. As will be evident below, other stability viewpoints lead to new insights and methods, especially vis-à-vis stability analysis time-varying, distributed parameter, and nonlinear systems.

For the linear dynamical system of Eq. (3.5), we choose a symmetric quadratic form as a candidate Lyapunov function

$$2\mathrm{U}(\mathbf{x}) = \mathbf{x}^{\mathrm{T}}\mathrm{P}\mathbf{x} \tag{3.8}$$

where P is a positive definite, real symmetric matrix. Thus U is positive definite with its global minimun at the origin, which is obviously an equilibrium state of Eq. (3.5).

Differentiating Eq. (3.8) and substituting Eq. (3.5) into the result gives the energy rate:

$$\dot{\mathrm{U}}(\mathbf{x}) = \mathbf{x}^{\mathrm{T}}(\mathrm{A}^{\mathrm{T}}\mathrm{P} + \mathrm{PA})\mathbf{x} \tag{3.9}$$

Using the Lyapunov stability theorem 3.2, we require $\dot{\mathrm{U}}(\mathbf{x})$ to be negative definite. We can rewrite the energy rate of Eq. (3.9) as

$$\dot{\mathrm{U}}(\mathbf{x}) = -\mathbf{x}^{\mathrm{T}}\mathrm{Q}\mathbf{x} \tag{3.10}$$

So, we see that, for asymptotic stability, P and Q must be positive definite matrices that satisfy the condition

$$\mathrm{A}^{\mathrm{T}}\mathrm{P} + \mathrm{PA} = -\mathrm{Q} \tag{3.11}$$

Equation (3.11) is commonly known as the *algebraic Lyapunov equation*, and it is possible to examine the stability of the given system via this equation. For example, if a pair of matrices (P,Q) satisfying Eq. (3.11) can be found such that both P and Q are positive definite, then U is positive definite [Eq. (3.8)] and $\dot{\mathrm{U}}$ is a negative definite function [Eq. (3.10)]. Hence, according to Theorem 3.2, global asymptotic stability of the linear system is assured, since initial conditions do not affect the stability of linear systems.

To examine the stability of a linear system via the above Lyapunov approach, we can proceed as follows: "Choose Q to be any positive definite matrix for a given A, and check the eigenvalues of the resulting P that we obtain by solving Eq. (3.11); if P is positive definite (*all* positive eigenvalues), the given system is asymptotically stable but, if P has *any* negative eigenvalues, the system is unstable." One of the potential difficulties with selecting Q and solving the Lyapunov equation (which, of course, depends on system matrix A) is the possible nonuniqueness of the resulting solution for P. The following theorem gives the necessary and sufficient conditions for the Lyapunov equation (3.11) to have a unique solution.

Theorem 3.5

If $\{\lambda_1, \ldots, \lambda_n\}$ are the eigenvalues of the system matrix A, then the Lyapunov equation (3.11) has a unique solution P if and only if

$$\lambda_{\mathrm{i}} + \lambda_{\mathrm{j}}^{\mathrm{H}} \neq 0, \qquad \mathrm{i,j} = 1, \ldots, n$$

where $(\ )^H$ denotes complex conjugate.

■

The proof of this theorem requires some lengthy linear algebra, and we will not cover it here. The reader is referred to Chen (Ref. 8) for the proof. Note from Theorem 3.5 that we cannot solve the Lyapunov equation for undamped second-order systems having pairs of eigenvalues on the imaginary axis (including rigid body modes, whose eigenvalues reside at the origin of the complex plane) and, therefore, stability analysis for systems having a neutrally stable subspace cannot be completed via solution of an algebraic Lyapunov equation. Now, we are ready to introduce the Lyapunov theorem for linear systems.

Theorem 3.6: Lyapunov Stability Theorem for Linear Systems

A linear system is asymptotically stable or, equivalently, all the eigenvalues of A have negative real parts, if and only if, for any given positive definite symmetric matrix Q, there exists a positive definite (symmetric) matrix P that satisfies the Lyapunov equation

$$A^TP + PA = -Q$$

■

Proof of Theorem 3.6

Necessity

The uniqueness condition of the solution (Theorem 3.5) is automatically met in Theorem 3.6, since $\Re[\lambda_i(A)] < 0$ for all i, for asymptotically stable systems. The unique solution P of the Lyapunov equation can be expressed as (Ref. 9)

$$P = \int_0^\infty e^{A^Tt}Qe^{At}dt \tag{3.12}$$

The integrand is an infinite sum of terms of the form $t^ke^{\lambda t}$, where λ is an eigenvalue of A and, since $\Re[\lambda_i(A)] < 0$, the integrand will exist for all values of t. Next, by direct substitution of Eq. (3.12) into the left hand side of Eq. (3.11), we find that the combined integrand is a perfect differential, permitting us to directly establish

$$\begin{aligned} A^TP + PA &= \int_0^\infty A^Te^{A^Tt}Qe^{At}dt + \int_0^\infty e^{A^Tt}Qe^{At}Adt \\ &= \int_0^\infty \frac{d}{dt}\left(e^{A^Tt}Qe^{At}\right)dt \\ &= e^{A^Tt}Qe^{At}\Big|_0^\infty \\ &= -Q \end{aligned}$$

Note that e^{At} (and e^{A^Tt}) $\rightarrow 0$ as $t \rightarrow \infty$, since all eigenvalues of A have negative real parts, and since e^{At} (and e^{A^Tt}) equal 1 at $t = 0$.

Sufficiency

Consider

$$U(\mathbf{x}) = \mathbf{x}^T P \mathbf{x}$$

From Eqs. (3.8)-(3.11),

$$\dot{U}(\mathbf{x}) = -\mathbf{x}^T Q \mathbf{x}$$

along all trajectories of $\dot{\mathbf{x}} = A\mathbf{x}$.

From the assumptions, P and Q are positive definite matrices; therefore, we see that, for any $\mathbf{x} \neq \mathbf{0}$,

$$U(\mathbf{x}) > 0$$

$$\dot{U}(\mathbf{x}) < 0$$

We can apply Theorem 3.3 and conclude that the equilibrium state [the state space origin $(\mathbf{x} = 0)$] is globally asymptotically stable and, of course all eigenvalues of A have negative real parts.

Q.E.D.

Note that the Lyapunov equation is equivalent to a set of $n(n+1)/2$ linear equations in $n(n+1)/2$ unknowns for an nth order system. As we mentioned in section 2.2, the Lyapunov equation can be solved by using numerical algorithms utilizing QR factorization; Schur decomposition, or spectral decomposition, however, we believe that the most efficient and robust algorithm utilizes the QR factorization. Note that the MATLAB operators for computing the solution of Lyapunov equations are

> $\mathbf{X} = \mathbf{lyap(A, C)}$, to solve the classical Lyapunov equation $\mathbf{AX} + \mathbf{XA}^T = -\mathbf{C}$
> $\mathbf{X} = \mathbf{lyap(A, B, C)}$, for the generalized Lyapunov equation $\mathbf{AX} + \mathbf{XB} = -\mathbf{C}$

Example 3.4

Consider the system matrix

$$A = \begin{bmatrix} -2 & 1 \\ -1 & 1 \end{bmatrix}$$

The simplest choice of Q is the identity matrix or some other diagonal matrix; we take $Q = I$ for this example and let

$$P = \begin{bmatrix} p_1 & p_2 \\ p_2 & p_3 \end{bmatrix}$$

be a symmetric matrix containing three unknowns.

Substituting this A and P into the Lyapunov equation (3.11) yields the following three linear algebraic equations:

$$-4p_1 - 2p_2 = -1$$
$$p_1 - p_2 - p_3 = 0$$
$$2p_2 + 2p_3 = -1$$

The solution of these three equations is straightforward,. We find

$$p_1 = -\frac{1}{2}, \qquad p_2 = \frac{3}{2}, \qquad p_3 = -2$$

Thus,

$$P = \begin{bmatrix} -\frac{1}{2} & \frac{3}{2} \\ \frac{3}{2} & -2 \end{bmatrix}$$

Even though we have a unique solution in this case, the resulting matrix P is not positive definite. Hence, we know that not all eigenvalues of A have negative real parts, and that the system is therefore unstable.

■

In the case of a linear time-varying system

$$\dot{\mathbf{x}}(t) = A(t)\mathbf{x}(t)$$

the sufficient conditions for the stability of the equilibrium state can be discussed based on the concept of matrix measure (Ref. 1) and, if the system is asymptotically stable, then a quadratic Lyapunov function exists for this system. Of course, eigenvalue analysis is not applicable to the time-varying case and, therefore, the more general Lyapunov approach provides one possible avenue to characterizing the stability of nonautonomous systems. The reader is referred to Ref. 1 for more information on this generalization of the above discussion.

3.3.2 Linear Dynamic Systems Subject to Arbitrary Disturbances

In section 2.3.3, we introduced stability robustness criteria that provide a bound on disturbances such that the system maintains its stability. To make the Lyapunov stability analysis in this section more complete, we will briefly discuss stability in the presence of disturbances.

Consider the class of systems described by the matrix differential equation

$$\dot{\mathbf{x}}(t) = A\mathbf{x}(t) + \mathbf{f}(t, \mathbf{x}(t)) \tag{3.13}$$

where the uncertainty and/or perturbations of the system are assumed representable by arbitrary nonlinear function $\mathbf{f}(\mathrm{t}, \mathbf{x}(\mathrm{t}))$ (except that we require $\mathbf{f}(t, 0) = 0$, so that the origin of the state space remains an equilibrium state for this class of model errors or disturbances). Furthermore, we assume that exact expressions for $\mathbf{f}(\mathrm{t}, \mathbf{x}(\mathrm{t}))$ are unknown and only bounds on $\mathbf{f}(\mathrm{t}, \mathbf{x}(\mathrm{t}))$ are known. A central question that we would like to answer is: Given that A is asymptotically stable, and without using specific knowledge of $\mathbf{f}(\mathrm{t}, \mathbf{x}(\mathrm{t}))$, is it possible to obtain a bound on all $\mathbf{f}(\mathrm{t}, \mathbf{x}(\mathrm{t}))$ such that the system maintains its stability? Put another way, can we determine how large $\mathbf{f}(\mathrm{t}, \mathbf{x}(\mathrm{t}))$ can be without destabilizing an otherwise stable system? These questions can be answered by applying the following theorem.

Theorem 3.7 (Patel and Toda Ref. 10)

Suppose that the system of Eq. (3.13) is asymptotically stable for $\mathbf{f}(\mathrm{t}, \mathbf{x}(\mathrm{t})) = 0$. The system remains asymptotically stable for all nonzero perturbations $\mathbf{f}(\mathrm{t}, \mathbf{x}(\mathrm{t}))$ that are sufficiently small that they satisfy the following inequality:

$$\frac{\|\mathbf{f}\|}{\|\mathbf{x}\|} \leq \frac{\min\ \lambda(\mathrm{Q})}{\max\ \lambda(\mathrm{P})} \equiv \mu_{\mathrm{PT}} \tag{3.14}$$

where P and Q satisfy the following Lyapunov equation

$$\mathrm{A}^{\mathrm{T}}\mathrm{P} + \mathrm{PA} = -2\mathrm{Q}$$

and where the otherwise arbitrary $\mathbf{f}(\mathrm{t}, \mathbf{x}(\mathrm{t}))$ vanishes at the origin $\mathbf{f}(\mathrm{t}, 0) = 0$.

■

One way to prove Theorem 3.7 is almost the same as (part of the) proof of Theorem 2.7 (this proof is included in Appendix B); the reader should be able to proceed analogously to prove the above theorem without any difficulties. Since P is a positive definite symmetric matrix, the maximum eigenvalue of P is the same as the largest singular value of P. It has also been shown in Appendix B (Lemma B.2) that when the identity matrix is chosen for Q, μ_{PT} in Eq. (3.14) is a maximum and, for this choice, μ_{PT} can be expressed as

$$\mu_{\mathrm{PT}} = \frac{1}{\max\ \lambda(\mathrm{P})} = \frac{1}{\sigma_{\max}(\mathrm{P})} \tag{3.15}$$

The above bound is often very conservative, since it is only a *sufficient* condition for the stability of the system, and the stringent bounds so determined are not usually *necessary*.

An important special case is for the class of perturbations having the linear structure

$$\mathbf{f}(\mathrm{t}, \mathbf{x}(\mathrm{t})) = \mathrm{E}\mathbf{x}(\mathrm{t}) \tag{3.16}$$

Clearly this corresponds to an additive error in the A matrix (i.e., A → A + E). We can easily apply Theorem 3.7 to arrive at the desired result (which is identical to Theorem 2.7). That is, the system remains stable if E is bounded by the following *modified stability margin*:

$$\|E\| \leq \frac{\min\left[-\Re\{\lambda_i(A)\}\right]}{\mathcal{K}(\Phi)} \tag{3.17}$$

where $\mathcal{K}(\Phi)$ is the *condition number* of Φ, and Φ is the normalized eigenvector (modal) matrix of A. The condition number, discussed in Chapter 2, is conveniently computed as the ratio of the largest and least singular values of Φ,

$$\mathcal{K}(\Phi) = \frac{\sigma_{\max}(\Phi)}{\sigma_{\min}(\Phi)}$$

As is evident in the above discussion, the stability margin is closely related to the Patel-Toda robustness margin; the more stable the nominal system is, the larger the bound on the allowable E becomes. However, the important ingredient evident in Eq. (3.17) is the fact that a large condition number (indicating high sensitivity of the eigenvalue problem to perturbations in the A matrix) degrades the effective stability margin. The intimate connection of the Patel-Toda robustness measure (for stability of linear dynamical systems in the presence of additive perturbations) to the Bauer-Fike Theorem 2.6 (for conditioning of the algebraic eigenvalue problem) is clear.

We can see that a large stability margin is comforting only if the condition number (of the modal matrix of eigenvectors) is small. Note that the condition number $\mathcal{K}(\Phi)$ approaches its smallest possible value of unity if Φ is *any* unitary matrix, and the upper bound on the condition number is infinity, which occurs if Φ is *any* singular matrix. Observe that an infinity of unitary matrices exists, some of which are "closer" to Φ than others. This fact has consequences when one has the freedom to modify A (and therefore Φ); a natural question arises: For a given class of A modifications, how can we make Φ as near as possible to a unitary matrix? Of course, one way to modify the A matrix is through design of a feedback controller, and one avenue toward designing gains in linear robust control laws is to maximize the right-hand side of Eq. (3.17) by minimizing $\mathcal{K}(\Phi)$. These ideas provide some of the motivation for the *robust eigenstructure algorithms* presented in Chapter 5.

3.3.3 Stability Analysis for Mechanical Second-Order Systems

The dynamics of mechanical systems, such as robot manipulators or space structures, are often directly approximated from physical principles (e.g., Newton's laws or Lagrange's equations) as second-order matrix systems of differential equations; therefore, it is useful to directly study stability of linear second-order systems, without converting to a first-order state space form. The motivation for this study is not purely academic, because the large dimension of M (resulting, for example, from application of a finite element analysis) often precludes accurate matrix inversion and, therefore, the obvious (pedestrian) transformation to the standard first-order form may encounter practical difficulties at step zero. There are other reasons that will become evident in the sequel. Stability analysis for open-loop systems (no feedback) has been examined in Ref. 11 for several families of second-order systems: conservative systems, viscously damped systems, gyroscopic systems, circulatory systems, etc. Since our objective is to stabilize the system response, we are naturally interested in feedback control and, therefore we shall study the closed-loop system stability herein.

A simple class of controlled mechanical systems are generally modeled by the equations of motion

$$\begin{aligned} \mathrm{M}\ddot{\mathbf{x}} + \mathrm{C}\dot{\mathbf{x}} + \mathrm{K}\mathbf{x} &= \mathbf{f} \\ \mathbf{f} &= -\mathrm{C_c}\dot{\mathbf{x}} - \mathrm{K_c}\mathbf{x} \end{aligned} \tag{3.18}$$

where $\mathbf{x} \in \mathrm{R}^n$ and $\mathbf{u} \in \mathrm{R}^m$ are configuration and control vectors, respectively, M is the $n \times n$ positive definite symmetric mass matrix, C is the $n \times n$ positive semidefinite symmetric structural damping matrix, K is the $n \times n$ positive semidefinite stiffness matrix, $\mathbf{f}$ is the external control force vector (which is more generally replaced by the sum of control forces, disturbances, and model errors), $\mathrm{C_c}$ and $\mathrm{K_c}$ are velocity and position feedback gain matrices, and $\dot{(\ )}$ represents differentiation with respect to time. Note that we consider here the case of full state feedback; in Chapter 7, we study the most important and practical case of output feedback.

Theorem 3.7

For the system [Eq. (3.18)], where M and $(\mathrm{K} + \mathrm{K_c})$ are symmetric and positive definite, when $(\mathrm{C} + \mathrm{C_c})$ is positive definite, the system is asymptotically stable and, when $(\mathrm{C} + \mathrm{C_c})$ is positive semidefinite, then the system is guaranteed stable in the sense of Lyapunov, but may (or may not) be asymptotically stable.

■

With the assumption that the sensors and actuators are perfect, i.e., linear and instantaneous, the closed-loop equations can be written as

$$\mathrm{M}\ddot{\mathbf{x}} + (\mathrm{C} + \mathrm{C_c})\dot{\mathbf{x}} + (\mathrm{K} + \mathrm{K_c})\mathbf{x} = 0 \tag{3.19}$$

Considering the first-order state space form of the system differential equations, which correspond to the second-order closed-loop system of Eq. (3.19), we can write

$$B\dot{\mathbf{z}} = A\mathbf{z} \tag{3.20}$$

where

$$\mathbf{z} = \begin{pmatrix} \mathbf{x} \\ \dot{\mathbf{x}} \end{pmatrix}, \quad A = \begin{bmatrix} 0 & M \\ -(K + K_c) & -(C + C_c) \end{bmatrix}, \quad B = \begin{bmatrix} M & 0 \\ 0 & M \end{bmatrix}$$

Consider the following candidate Lyapunov function (total mechanical energy of the closed-loop system):

$$U = \frac{1}{2}\dot{\mathbf{x}}^T M\dot{\mathbf{x}} + \frac{1}{2}\mathbf{x}^T(K + K_c)\mathbf{x} \tag{3.21}$$

By differentiating the above equation with respect to time and using the closed-loop system model of Eq. (3.19), $\dot{U}$ can be simplified to the quadratic form

$$\dot{U} = -\dot{\mathbf{x}}^T(C + C_c)\dot{\mathbf{x}} \tag{3.22}$$

We can rewrite the candidate Lyapunov function and its time derivative by using the $2n \times 1$ state vector $\mathbf{z}$ as follows:

$$U = \frac{1}{2}\mathbf{z}^T \begin{bmatrix} K + K_c & 0 \\ 0 & M \end{bmatrix} \mathbf{z} \tag{3.23}$$

$$\dot{U} = -\mathbf{z}^T \begin{bmatrix} 0 & 0 \\ 0 & C + C_c \end{bmatrix} \mathbf{z} \tag{3.24}$$

For the open-loop ($C_c = 0,\ K_c = 0$) system, the function U is positive semidefinite since M is positive definite and K is positive semidefinite. If there is no rigid body mode in the open-loop system, the function U becomes positive definite since K is positive definite. However, the rigid body modes are generally present in spacecraft systems, and K is singular for such cases. In order to control the rigid body modes successfully (for attitude control, station keeping control, or vibration suppression), the rigid body modes should be eliminated by judicious position ($\mathbf{x}$) feedback porportional to errors in rigid body position errors and, if the system is controllable, a suitable position feedback gain matrix K_c can always be chosen such that $(K + K_c)$ becomes positive definite. Therefore, whether or not the open-loop system has rigid body modes, we can arrange that controllable closed-loop systems do not have rigid body modes and that the Lyapunov function U of Eq. (3.21) is positive definite.

We direct attention now to $\dot{U}$. From Eq. (3.22) [or Eq. (3.24)], $\dot{U}$ is negative semidefinite. Several cases can be considered. First, if the open-loop system is undamped ($\mathrm{C} = 0$), $\dot{\mathrm{U}}$ is negative semidefinite and the closed-loop system is stable in the Lyapunov sense. Second, if the open-loop system is approximated as having structural damping (proportional to M and/or K) and rigid body modes are not present, then C is positive definite. In spite of a positive definite C, $\dot{\mathrm{U}}$ is negative semidefinite, since $\dot{\mathrm{U}}$ depends only on the velocity vector. However, it is sufficient for global asymptotic stability for this case, because nonzero position alone (identically zero velocities) is not a trajectory of the closed-loop system. We can establish that $\dot{\mathbf{x}}$ vanishes identically, for all future time, only if motion is initiated at the point $\mathbf{x} = 0$. Consider the conditions for an equilibrium point of the closed-loop system. The first part of Eq. (3.20) yields $\dot{\mathbf{x}} = 0$ or $\mathbf{x} = \mathbf{c}$. A nontrivial equilibrium ($\mathbf{x} = \mathbf{c} \neq 0$) exists only if $\dot{\mathbf{x}} = 0$ and $\ddot{\mathbf{x}} = 0$. Note that the second part of Eq. (3.20) yields

$$-(\mathrm{K} + \mathrm{K_c})\mathbf{x} = -(\mathrm{K} + \mathrm{K_c})\mathbf{c} = 0 \tag{3.25}$$

This is a contradiction since it requires $\mathbf{x} = \mathbf{c} = 0$ [remembering that $(\mathrm{K} + \mathrm{K_c})$ is positive definite and therefore does not have a null space]. Therefore, $\dot{\mathbf{x}}$ vanishes identically only at the origin, and Eq. (3.22) [or Eq. (3.24)] reveals that the system is asymptotically stable. So, as is evident in this simple example, $\dot{\mathrm{U}}$ being only negative semidefinite often corresponds to global asymptotic stability.

As will be evident in the sequel, care must be taken in more general nonlinear, time-varying, and distributed parameter systems (when $\dot{\mathrm{U}}$ is only negative semidefinite), especially for nonlinear dynamical systems having multiple equilibrium states, and for systems that have an *uncontrollable subspace.*

Now, consider the final case that the open-loop system has structural damping (C positive semidefinite) and there exist rigid body modes. In order to control the rigid body modes successfully, the zero eigenvalues should be eliminated, and those modes should have some artificial control-induced stiffness and damping via position and velocity feedback (proportional to errors in the postition and velocity coordinates). The conservative approach is to require $\mathrm{C_c}$ to be chosen such that $(\mathrm{C} + \mathrm{C_c})$ is positive definite, although coupling in M and K is almost always sufficient to permit $(\mathrm{C} + \mathrm{C_c})$ to be merely positive semidefinite, to guarantee that $\dot{\mathrm{U}}$ is negative. By the same reasoning as above, the closed-loop system for this case is asymptotically stable.

This closed-loop stability analysis leads us to the conclusion that if the system is controllable, then the closed-loop system is at least asymptotically stable if the gain matrices are chosen properly. Therefore, by imposing these requirements on the gain matrices, globally stabilizing controllers can be designed (Refs. 12, 13). For more information on the stability analysis for the closed-loop second order mechanical systems, readers may refer to Refs. 14-16.

3.4 Nonlinear, Time-Varying and Distributed Parameter Systems

In this section, we study several stability analysis methods for nonlinear systems. In section 3.4.1, we consider a method known as Lyapunov's indirect (or first) method, whereby we can determine partial stability information for nonlinear systems by examining the behavior of locally linearized systems. In section 3.4.2, we develop an important result that provides easy to test sufficient conditions to determine if we have asymptotic stability despite the common situation in which the energy function's time derivative is only a negative semidefinite function of the state variables. In addition to the classical stability analysis for which the Lyapunov methods were developed, these ideas can be used to motivate design methods that yield control laws for control of large maneuvers for distributed parameter systems.

This approach is used throughout the remainder of this Chapter. As an application of Lyapunov stability theory, a specific design method is presented in section 3.4.3 to establish globally stabilizing control laws for a particular distributed parameter system by judicious selection of Lyapunov functions and corresponding feedback laws. As will be evident, this method is rigorous and direct in the sense that it is not necessary to first approximate partial differential equation models by any finite-dimensional ordinary differential equation system derived, for example, via the finite element method.

While a particular system is studied, the methodololgy is developed in a fashion that should motivate a general approach to problems of similar structure. The avoidance of the usual spatial and time truncation approximations (in arriving at the feedback law and the associated gain space in which closed-loop stability is rigorously guaranteed for the hybrid ODE/PDE model) is a most important and elegant feature of this approach. Spatial and time discretizations are still useful, however to do numerical response studies and to approximately optimize the gains over the stable region. By determining the structure of a rigorously stabilizing feedback law, and the associated inequalities defining the stable gain space from the hybrid ODE/PDE model *before* introducing spatial approximations, we can be sure that these approximations do not result in instability due to truncation error effects.

3.4.1 Local Stability of Linearized Systems

Stability analysis of linear motion arises often in practical analysis of nonlinear systems when we are concerned with motion near an equilibrium state. The results presented in section 3.3.1 enable us to obtain necessary and sufficient conditions for the stability of linear systems but also provides us with a method for determining the local stability of a nonlinear system by linearization, which is called *Lyapunov's indirect method.*

Consider the autonomous system

$$\dot{\mathbf{x}}(t) = \mathbf{f}[\mathbf{x}(t)] \qquad \text{with} \quad \mathbf{f}(\mathbf{x_e}) = 0 \tag{3.26}$$

Let $\mathbf{z}(t)$ be the perturbation (departure motion) from the equilibrium state such that

$$\mathbf{x}(t) = \mathbf{x_e} + \mathbf{z}(t) \tag{3.27}$$

Using Taylor's series expansion of $\mathbf{f}(\cdot)$ around the equilibrium state $\mathbf{x_e}$, we can write

$$\mathbf{f}[\mathbf{z}(t) + \mathbf{x_e}] = \mathbf{f}(\mathbf{x_e}) + \left[\frac{\partial \mathbf{f}}{\partial \mathbf{x}}\right]_{\mathbf{x}=\mathbf{x_e}} \mathbf{z}(t) + O[\mathbf{z}(t)]^2 \tag{3.28}$$

where we use the order notation

$$\frac{\|O[\mathbf{z}(t)]^2\|}{\|\mathbf{z}\|} \to 0 \qquad \text{as} \quad \|\mathbf{z}\| \to 0$$

This gives the perturbation equation

$$\dot{\mathbf{z}}(t) = \mathrm{A}\mathbf{z}(t) + O[\mathbf{z}(t)]^2 \tag{3.39}$$

where A denotes the Jacobian matrix of $\mathbf{f}$ evaluated at $\mathbf{x} = \mathbf{x_e}$,

$$\mathrm{A} = \left[\frac{\partial \mathbf{f}}{\partial \mathbf{x}}\right]_{\mathbf{x}=\mathbf{x_e}}$$

and the linear, constant coefficient matrix differential equation

$$\dot{\mathbf{z}}(t) = \mathrm{A}\mathbf{z}(t) \tag{3.30}$$

The following theorem is given here (without proof); this is the main stability result of Lyapunov's *indirect method.*

Theorem 3.8: Lyapunov's Indirect Method

If the linearized system [Eq. (3.30)] is asymptotically stable, then the original nonlinear system [Eq. (3.26)] is also asymptotically stable given "sufficiently small" initial perturbations from the equilibrium state.

∎

The above theorem is useful since we can analyze the local stability of an equilibrium state of a given nonlinear system by examining a linear system. However, the conclusions based on linearizations are local and, therefore to study global stability, we should rely on Lyapunov's direct method. On the other hand, if one has found all equilibrium points and investigated their local stability, a fairly complete picture of the overall global stability characteristics can often be derived. Note that one key shortcoming (of the indirect approach) is the absence of information on the size or boundary of the "domain of attraction" of each locally stable equilibrium point; this is precisely the information that a completely successful application of the direct approach determines. Finally, we note the most important point: *If the linear motion is critical (e.g., zero damping), then the locally linearized analysis should be considered inconclusive, and nonlinear effects must be included to conclude local stability or instability.*

3.4.2 What To Do When $\dot{U}$ Is Only Negative Semidefinite

A variety of subtle possiblities arise if the function derived for $\dot{U}$ is not negative definite. For a significant fraction of the practical occurences of this condition, including several applications considered subsequently in this Chapter, *we can prove global asymptotic stability in spite of the fact that the function derived for* $\dot{U}$ *is negative semidefinite.* The main results from the traditional literature for dealing with this problem are embodied in a theorem due to LaSalle Ref. 29; this theorem sometimes allows us to conclude that we have local asymptotic stability for the case in which $U > 0$ and $\dot{U} \leq 0$, provided we can prove that the equilibrium point is contained in a region of state space LaSalle called the *maximum invariant subspace* M. It is usually easy to identify the subset Z of points in the state space for which $\dot{U} = 0$, but LaSalle's maximum invariant subspace M is, in general, a subset of Z. The main challenge in applying LaSalle's elegant theorem then reduces to a quest to identify or approximate M; this is indeed difficult when the differential equations are complicated nonlinear functions. While these ideas have significant esthetic appeal, we elect not to discuss this approach in detail, but rather we present a more widely useful, recently developed result that is easier to apply; this result was derived by Mukherjee, Chen and Junkins (Refs. 30, 31). The additonal results developed below were motivated by discussions with Mukherjee.

We need to introduce some elementary results from functional analysis, in order to state the main constructive result as a theorem. A *concave* function $f(t) \in C^2$ is one that satisfies $\ddot{f}(t) \leq 0 \; \forall \; t \in (a, b)$. If $f(t) \leq 0$ and $f(t_0) = 0$, then it can easily be shown from a Taylor series expansion that $\dot{f}(t_0) = 0$ and $\ddot{f}(t) < 0$ for some neighboring values of t from a set that includes t_0.

With these ideas in mind, the following theorem provides a sufficient condition for asymptotic stability when $\dot{U} \leq 0$.

Theorem 3.9

Prior to stating the theorem, we need some notations: Let $\mathbf{x} = \mathbf{0}$ be an equilibrium state of the nonlinear system

$$\dot{\mathbf{x}} = \mathbf{f}(t, \mathbf{x})$$

where **f** is a smooth, twice differentiable n-vector function of t and **x**. Note that the trajectories of the nonlinear differential equation $\dot{\mathbf{x}} = \mathbf{f}(t, \mathbf{x})$ generates a smooth vector field in the region Ω, which includes $\mathbf{x} = \mathbf{0}$. Let U(t, x) be a scalar analytic function in Ω, which is locally positive definite. Suppose $\dot{U}(t, \mathbf{x})$ is only negative semidefinite. Let Z denote the set of points for which $\dot{U}(t, \mathbf{x})$ vanishes. We will be concerned with the first k derivatives d^kU/dt^k, *evaluated on the set* Z. We are now prepared to state the theorem.

A sufficient condition for asymptotic stability, when $U > 0$ and $\dot{U} \leq 0$ for all $\mathbf{x} \in \Omega$ is that the first (k-1) derivatives of U vanish on Z, up through some *even* order (k-1) as

$$\frac{d^jU}{dt^j} = 0, \quad \forall \quad \mathbf{x} \in Z, \quad \text{for} \quad j = 1,\ 2,\ \dots,\ k-1$$

and *the first* (the kth) nonzero derivative of U (evaluated on Z) is of odd order and is negative definite for all points on Z:

$$\frac{d^kU}{dt^k} < 0, \quad \forall \quad \mathbf{x} \in Z, \quad \text{for k odd}$$

In the event that all infinity of U derivatives vanish on Z, sufficient conditions for stability are that U is positive definite and that $\mathbf{x} = \mathbf{0}$ is the only equilibrium point.

The proof of this theorem is given in Ref. 30 and further discussed in Ref. 31. In spite of the wordy structure (which is a modified version of the original statement of this result in Ref. 30), this theorem turns out to be fairly easy to apply and useful results can often be obtained, as will be evident in the several nonlinear and distributed parameter applications considered in the remainder of this Chapter. As is evident in the following example, this theorem is also useful for determining the stability of time-varying systems.

Example 3.5

Consider the damped Mathieu equation Ref. 30

$$\dot{x}_1 = x_2$$

$$\dot{x}_2 = x_2 - (2 + \sin t)x_1$$

We select the candidate Lyapunov function

$$U(t, x_1, x_2) = x_1^2 + \frac{x_2^2}{(2 + \sin t)}$$

which we observe is positive definite and analytic for all (t, x_1, x_2). Upon differentiation of U and substitution of the equations of motion, we find that

$$\dot{U}(\mathbf{x}) = -\mathbf{x}_2^2 g(t), \quad \text{where} \quad g(t) = \frac{4 + 2\sin t + \cos t}{(2 + \sin t)^2}$$

Since $\dot{U}(\mathbf{x})$ does not depend on $\mathbf{x}_1$, it is obviously not positive definite and without further analysis, we can only conclude mere *stability*; however, we'd like to make a stronger statement and conclude *asymptotic stability.* This can be done by considering the applicability of Theorem 3.9. Note that the set Z of points for which $\dot{U}(\mathbf{x})$ vanishes is the set of all real values for $\mathbf{x}_1$, and zero values for $\mathbf{x}_2$. Upon taking the second and third derivatives of U, and evaluating them on Z, we find that

$$\frac{d^2U}{dt^2} = 0, \quad \forall \quad \mathbf{x} \in Z$$

and

$$\frac{d^3U}{dt^3} = -2(2 + \sin t)^2 g(t) \mathbf{x}_1^2, \quad \forall \quad \mathbf{x} \in Z$$

Since the second derivative vanishes on Z and the third derivative is negative on Z, except at the origin, we conclude that all of the conditions of Theorem 3.9 are satisfied; indeed this system is globally asympotically stable.

3.4.3 Lyapunov Control Law Design Method

In this section, we present a method for generating globally stable feedback control laws for maneuvers of nonlinear systems and distributed parameter systems via a Lyapunov control design approach. The approach is to adopt a Lyapunov function that an energy integral of some idealized approximation of the uncontrolled system. When the control $\mathbf{u}(t)$ is considered, $\dot{U}(\mathbf{x})$ depends on the control $\mathbf{u}(t)$ through the equations of motion. One strategy is to select the control function $\mathbf{u}(t, x)$ (from a set of *admissible controls*) to make $\dot{U}(\mathbf{x})$ as negative as possible; this *Lyapunov optimal* strategy ensures that $U(\mathbf{x})$ will locally approach zero as fast as possible. On the other hand, any control law that makes $\dot{U}(\mathbf{x})$ negative is stablilizing and, in many instances, very simple, yet globally stable control laws can be determined that are attractive for applications.

We will use specific dynamical systems to introduce Lyapunov control design methods for nonlinear and distributed parameter systems using Lyapunov's direct method in this section. The broadest and most productive viewpoint is to simultaneously consider $U(\mathbf{x})$ and $\mathbf{u}(t, \mathbf{x})$ "up for grabs" in the design process; as will be seen, the class of problems for which globally stable feedback laws can be obtained is surprisingly large. As will be evident, we place the initial emphasis on using work/energy methods, together with stability theory, to determine the structure of a stabilizing feedback law and thereby parameterize an infinite family of stable controllers. Conventional nonlinear programming algorithms can then be invoked to optimize a specified closed-loop performance criterion over the stable set. Although we subsequently develop methods for controlling distributed parameter systems governed by hybrid coupled sets of ordinary and partial differential equations, we first consider a system described by a sixth-order set of nonlinear ordinary differential equations.

Motivation of the approach using rigid body maneuvers

The ideas are easily introduced by considering general three dimensional nonlinear maneuvers of a single rigid body. The equations governing large motion can be written as (Ref. 17)

$$
\begin{aligned}
I_1\dot{\omega}_1 &= (I_2 - I_3)\omega_2\omega_3 + u_1 \\
I_2\dot{\omega}_2 &= (I_3 - I_1)\omega_3\omega_1 + u_2 \\
I_3\dot{\omega}_3 &= (I_1 - I_2)\omega_1\omega_2 + u_3 \\
2\dot{q}_1 &= \omega_1 - \omega_2 q_3 + \omega_3 q_2 + q_1(q_1\omega_1 + q_2\omega_2 + q_3\omega_3) \\
2\dot{q}_2 &= \omega_2 - \omega_3 q_1 + \omega_1 q_3 + q_2(q_1\omega_1 + q_2\omega_2 + q_3\omega_3) \\
2\dot{q}_3 &= \omega_3 - \omega_1 q_2 + \omega_2 q_1 + q_3(q_1\omega_1 + q_2\omega_2 + q_3\omega_3)
\end{aligned}
\tag{3.31}
$$

where $(\omega_1, \omega_2, \omega_3)$ and (q_1, q_2, q_3) are the principal axis components of angular velocity and the Euler-Rodriguez parameters ("Gibbs vector"), respectively. The three differential equations above for (q_1, q_2, q_3) were encountered in Chapter 2 [see Eq. (2.144)], and represent but one of many alternative sets of coordinate and kinematic descriptions of general nonlinear rigid body motion. Note that (I_1, I_2, I_3) and (u_1, u_2, u_3) are the principal moments of inertia and the principal axis components of the external control torque, respectively.

For the case of zero control torque, it can be readily verified that total rotational kinetic energy is an exact integral of the motion described by differential Eq. (3.31), namely, $2T = (I_1\omega_1^2 + I_2\omega_2^2 + I_3\omega_3^2)$. Motivated by this total system energy integral, we investigate the trial Lyapunov function

$$U = \frac{1}{2}(I_1\omega_1^2 + I_2\omega_2^2 + I_3\omega_3^2) + k_0(q_1^2 + q_2^2 + q_3^2) \\ \equiv \text{kinetic energy } + k_0 \tan^2\left(\frac{\phi}{2}\right) \tag{3.32}$$

where ϕ is the instantaneous *principal rotation angle* (about the instantaneous Eulerian principal rotation axis, from the current angular position to the desired final angular position of the body [see Eqs. (2.136) - (2.138) and Ref. 17]. It is apparent that the additive term $k_0(q_1^2 + q_2^2 + q_3^2)$ can be viewed as the potential energy stored in a conservative spring. We can therefore anticipate that the system dynamics will evolve such that U is constant if the only external torque is the associated conservative moment. Of course, we are not interested in preserving U as a constant, but rather we seek to drive it to zero, because it measures the departure of the system from the desired equilibrium state at the origin. We therefore anticipate the necessity to determine an additional judicious control moment to guarantee that U is a decreasing function of time. It is obvious by inspection that U is positive definite and vanishes only at the desired state $q_i = \omega_i = 0$. Differentiation of Eq. (3.32) and substitution of Eqs. (3.31) lead directly to the following ("power") expression for $\dot{U}$:

$$\dot{U} = \sum_{i=1}^{3} \omega_i \left[u_i + k_0 q_i(1 + q_1^2 + q_2^2 + q_3^2)\right] \tag{3.33}$$

Of all of the infinity of possible control laws, we can see that *any control* u_i *that reduces the bracketed terms to a function whose sign is opposite to* ω_i *will guarantee that* $\dot{U}$ *is globally negative semidefinite.* The simplest choice consists of the following: Select u_i and positive constant k_i so that the ith bracketed term becomes $-k_i\omega_i$. This gives the control law

$$u_i = -\left[k_i\omega_i + k_0 q_i(1 + q_1^2 + q_2^2 + q_3^2)\right], \qquad i = 1, 2, 3 \tag{3.34}$$

The closed-loop equations of motion are obtained by substitution of the control law of Eq. (3.34) into the equations of motion of Eq. (3.31) to establish

$$I_1\dot{\omega}_1 = (I_2 - I_3)\omega_2\omega_3 - \left[k_1\omega_1 + k_0 q_1(1 + q_1^2 + q_2^2 + q_3^2)\right]$$

$$I_2\dot{\omega}_2 = (I_3 - I_1)\omega_3\omega_1 - \left[k_2\omega_2 + k_0 q_2(1 + q_1^2 + q_2^2 + q_3^2)\right]$$

$$I_3\dot{\omega}_3 = (I_1 - I_2)\omega_1\omega_2 - \left[k_3\omega_3 + k_0 q_3(1 + q_1^2 + q_2^2 + q_3^2)\right]$$

Since $\dot{U} = -(k_1\omega_1^2 + k_2\omega_2^2 + k_3\omega_3^2)$ does not depend on the qs, it is only a negative semidefinite function and, while we have *stability*, if we choose all $k_i > 0$, we cannot immediately conclude that we have *asymptotic stability*. We can prove that we do indeed have asymptotic stability. For illumination we establish this truth by two logical paths.

Path 1: This analysis is physically motivated; we try to see if there is some equilibrium point or trajectory other than the target state where the system can get "stuck" with $\dot{U}(\mathbf{x}) = 0$. We directly investigate the above three closed-loop equations of motion for the existence of equilibrium points in these nonlinear *closed-loop equations of motion*. It can be verified that $(q_1, q_2, q_3, \omega_1, \omega_2, \omega_3) = (0,0,0,0,0,0)$ is the *only* equilibrium state, where all velocity and acceleration coordinates vanish; in fact, imposing the conditions $(\dot{\omega}_1, \dot{\omega}_2, \dot{\omega}_3) = (0,0,0)$ and $(\omega_1, \omega_2, \omega_3) = (0,0,0)$ on the above closed-loop equations of motion immediately gives the requirement that the qs satisfy the equations

$$0 = -\left[k_0 q_1 (1 + q_1^2 + q_2^2 + q_3^2)\right]$$

$$0 = -\left[k_0 q_2 (1 + q_1^2 + q_2^2 + q_3^2)\right]$$

$$0 = -\left[k_0 q_3 (1 + q_1^2 + q_2^2 + q_3^2)\right]$$

and it is obvious by inspection that these equations are simultaneously satisfied only at the origin.

Since we have shown that $(\dot{\omega}_1 \neq 0, \dot{\omega}_2 \neq 0, \dot{\omega}_3 \neq 0)$, for $(q_1 \neq 0, q_2 \neq 0, q_3 \neq 0)$, everywhere except the origin $\mathbf{x} = (q_1, q_2, q_3, \omega_1, \omega_2, \omega_3) = (0,0,0,0,0,0)$, we conclude that $\dot{U}(\mathbf{x}) = 0$ can only be encountered for $(q_1 \neq 0, q_2 \neq 0, q_3 \neq 0)$ at (possibly) apogeelike points in the behavior of U ($\dot{U}$ instantaneously vanishes) but these points cannot be equilibrium states because $(\dot{\omega}_1 \neq 0, \dot{\omega}_2 \neq 0, \dot{\omega}_3 \neq 0)$. Therefore, we are guaranteed that $\dot{U}(\mathbf{x}) < 0$ *almost everywhere*. Thus we asymptotically approach the origin.

Path 2: This analysis is more formal (exactly analogous to Example 3.5), we simply apply Theorem 3.9. First, notice that the set Z where $\dot{U}(\mathbf{x})$ vanishes is the set of arbitrary real values for the qs and zero values for the ωs. It can be verified by direct differentiation of U that, for general motion,

$$\frac{d^2U}{dt^2} = -2\sum_{i=1}^{3} k_i \omega_i \dot{\omega}_i$$

and

$$\frac{d^3U}{dt^3} = -2\sum_{i=1}^{3} k_i (\dot{\omega}_i{}^2 + \omega_i \ddot{\omega}_i)$$

Upon evaluation of these derivatives for motion on Z where angular velocity vanishes $[(\omega_1, \omega_2, \omega_3) = (0, 0, 0)]$ but, from the closed-loop equations of motion, the nonzero acceleration components are $\dot{\omega}_i = -k_0(1 + q_1^2 + q_2^2 + q_3^2)(q_i/I_i)$, we find that

$$\frac{d^2U}{dt^2} = 0, \quad \forall \quad x \in Z$$

and

$$\frac{d^3U}{dt^3} = -k_0^2(1 + q_1^2 + q_2^2 + q_3^2)^2 \sum_{i=1}^{3} k_i \left(\frac{q_i}{I_i}\right)^2, \quad \forall \quad x \in Z$$

Since the second derivative of U vanishes everywhere on Z, the third derivative is negative definite everywhere on Z, the conditions of Theorem 3.9 are fully satisfied, and we again conclude that the nonlinear control law of Eq. (3.34) gives us globally asymptotically stable attitude control.

Since we have shown U to be a positive definite, decreasing function of time along all trajectories, and since it vanishes at the origin, then the necessary and sufficient conditions are satisfied for global Lyapunov stability. We have implicitly excluded the geometric singularity ($q_i \to \infty$) associated with this parameterization of rotational motion as $\phi \to n\pi$; we can use the quaternion or Euler parameter description of motion and avoid all geometric singularities as well. This more general path has been successfully pursued in Refs. 3 and 20.

The nonlinear feedback control law of Eq. (3.34) guarantees stability of the nonlinear closed-loop system under the assumption of zero model errors. In practice, of course, guaranteed stability in the presence of zero model error is not a sufficient condition to guarantee stability of the actual plant having arbitrary model errors and disturbances. On the other hand, rigorously defining a region in gain space guaranteeing global stability for our best model of the nonlinear system is an important step; it is reasonable to restrict the optimization of gains to this stable family of designs. The determination of the particular gain values, selected from the space of globally stabilizing gains, is usually based on performance optimization criteria specified in consideration of the disturbance environment, sensitivity to model errors, desired system time constants, actuator saturation, and sensor/actuator bandwidth limitations. For the purpose of gain optimization it is usually convenient to introduce spatial approximations and discretize partial differential equation systems to obtain a finite set of ordinary differential equations linearized about one or more points in the anticipated operation envelope.

Using a generalization of the preceding example, we can rigorously determine a region of admissible gains that stabilizes globally a general nonlinear, partial differential equation description of the system. Note that spatial discretization and order reduction approximations subsequently introduced in the gain design process cannot destabilize the zero model error system if the gains are optimized over this region that is known to be stable. In the preceding rigid body discussion, as long as we restrict $\{k_0, k_1, k_2, k_3\}$ to be positive, for example, approximations such as small angle linearizations used to simplify design calculations leading to particular positive gain values cannot destabilize the system. Of course, approximations introduced to facilitate the gain design process can indeed lead to undesired and unanticipated performance variations of the actual nonlinear distributed parameter system, as compared to predictions based on approximate linear models used to design the control gains.

Before generalizing the methodology to consider partial differential equation systems, it is important to reflect on the selection of the Lyapunov function previously given. Notice that, if a system has no inherent stiffness with respect to rigid body displacement, it is necessary to augment the open-loop energy integral by a pseudopotential energy term [such as $k_0(q_1^2 + q_2^2 + q_3^2)$ in the preceding example]; generally speaking, the pseudoenergy term should be defined, if possible, such that the resulting candidate Lyapunov function (U) is a positive definite measure of departure motion that has its global minimum at *the desired target state.* Then the still-to-be-determined controls are usually selected as simply as possible (from an implementation point of view) to force pervasive dissipation ($\dot{U} < 0$) of the modified energy (Lyapunov) function along all trajectories of the closed-loop system, and thereby guarantee closed-loop stability.

Although the above insights are useful, the Lyapunov function is generally not unique. However, these ideas lead to an attractive strategy that defines the Lyapunov function with relative weights on the portions of total mechanical energy associated with structural subsystems (Ref. 5), and leads to a systematic work/energy method to bypass much of the algebra and calculus leading to the power equations, analogous to Eq. (3.33), for each particular physical system (Ref. 3). The lack of uniqueness of the Lyapunov function is not necessarily a disadvantage in practice because it is a source of user flexibility, providing control design freedom qualitatively comparable to the freedom one has in selecting performance indices when applying optimal control theory. Indeed, formulating the Lyapunov function as a weighted error energy to be dissipated by the controller is qualitatively attractive for both linear and nonlinear systems, since this gives intuitive and physical meaning to the Lyapunov function and the control gains.

Maneuvers and vibration control for distributed parameter systems

In Figure 3.1 we consider control of a rigid hub with four cantilevered flexible appendages. We consider the appendages to be identical uniform flexible beams and make the Euler-Bernoulli assumptions of negligible shear deformation and distributed rotary inertia. Each beam is cantilevered rigidly to the hub and has

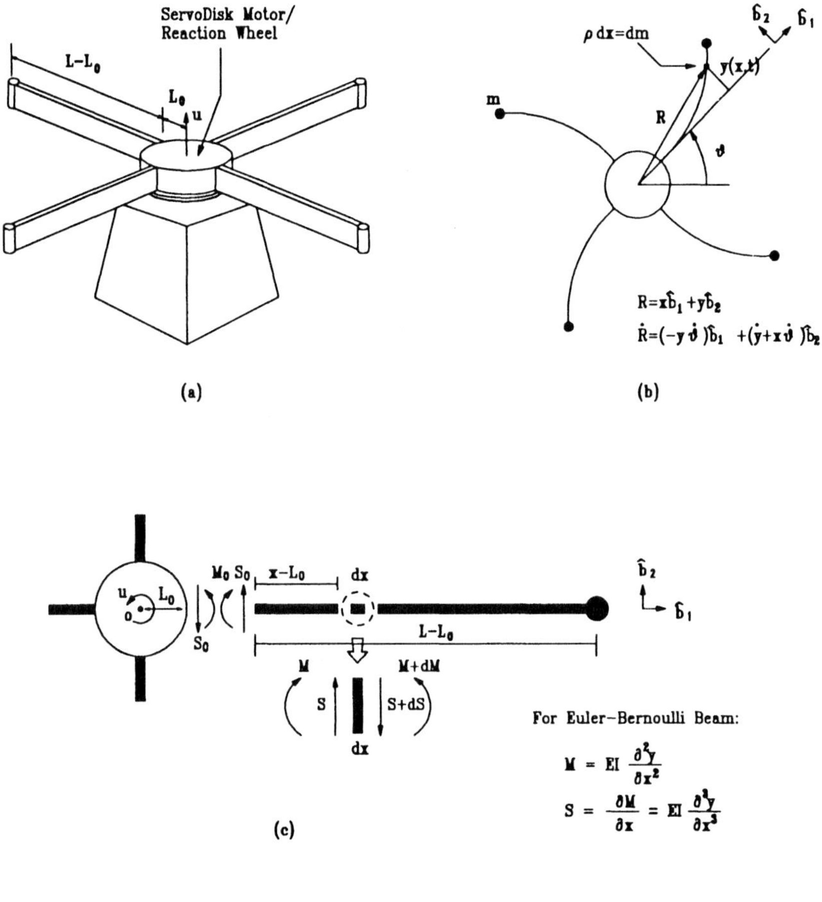

Figure 3.1 Texas A&M Maneuverable Space Structure Experiment:

(a) Configuration Sketch,

(b) Antisymmetric Deformation,

(c) Notation for Forces and Moments.

a finite tip mass. Motion is restricted to the horizontal plane and, control torque u(t) acting on the hub is the only external effect considered.

We are interested in a class of rest-to-rest maneuvers, and under the previously mentioned assumptions, we can show that the beams will deform in the antisymmetric fashion (Figure 3.1), with the configuration's instantaneous mass center remaining at the hub's geometric center. Also, because of the assumed antisymmetric deformation of the beams, in this section we need to concern ourselves only with the deformation $y(x,t)$ of a single beam. We subsequently relax this restriction, to permit more general kinematic assumptions and the analysis that flows form it. We adopt the continuum viewpoint and avoid introducing spatial approximations in the application of Lyapunov stability concepts; the resulting control law and stability arguments will therefore apply rigorously to the distributed parameter system. The hybrid system of ordinary and partial differential equations governing the dynamics of this system is readily obtained from Hamilton's principle to be (Ref. 5)

$$
\begin{aligned}
I_{hub}\frac{d^2\theta}{dt^2} &= u + 4(M_0 - S_0L_0) \\
-(M_0 - S_0L_0) &= \int_{L_0}^{L} \rho x\left(\frac{\partial^2 y}{\partial t^2} + x\frac{d^2\theta}{dt^2}\right)dx + m_t\left(L\frac{d^2\theta}{dt^2} + \frac{\partial^2 y}{\partial t^2}\bigg|_L\right) + \text{ HOT} \\
\rho\left(\frac{\partial^2 y}{\partial t^2} + x\frac{d^2\theta}{dt^2}\right) + EI\frac{\partial^4 y}{\partial x^4} &= 0 + \text{ HOT}
\end{aligned}
\tag{3.35}
$$

where

$$
\begin{aligned}
\rho &= \text{assumed constant mass/unit length of the beams} \\
EI &= \text{assumed constant bending stiffness of the beams} \\
(M_0, S_0) &= \text{bending moment and shear force at the root of the beam} \\
\theta &= \text{hub inertial rotation} \\
m_t &= \text{mass of the tip mass} \\
(L, L_0) &= \text{distance from the hub center to the beam tip and the hub radius}
\end{aligned}
$$

In Eq. (3.35), we denote higher-order terms by HOT to indicate other *known* linear and nonlinear effects (such as rotational stiffening, and shear deformation). The most fundamental developments do not consider these higher-order effects; however, we selectively discuss the generalizations that accommodate these effects as well. Of course, in general, there are *unknown* model errors and disturbances as well, and a practical control scheme must be stable in the presence of reasonable model errors.

The boundary conditions on Eqs. (3.35) are:

$$
\begin{aligned}
&\text{at} \quad x = L_0 : \ y(t, L_0) = \left.\frac{\partial y}{\partial x}\right|_{L_0} = 0 \quad \text{(clamped beam geometric B.C.s)} \\
&\text{at} \quad x = L : \ \left.\frac{\partial^2 y}{\partial x^2}\right|_{L} = 0 \quad \text{(moment)} \\
&\qquad \left.\frac{\partial^3 y}{\partial x^3}\right|_{L} = \frac{m_t}{EI}\left(L\frac{d^2\theta}{dt^2} + \left.\frac{\partial^2 y}{\partial t^2}\right|_{L}\right) \quad \text{(shear)}
\end{aligned}
\tag{3.36}
$$

The total energy of the system (constant in the absence of control or disturbances) is

$$
\begin{aligned}
2E = I_{hub}\left(\frac{d\theta}{dt}\right)^2 + 4\Bigg[\int_{L_0}^{L} \rho\left(\frac{\partial y}{\partial t} + x\frac{d\theta}{dt}\right)^2 dx \\
+ \int_{L_0}^{L} EI\left(\frac{\partial^2 y}{\partial x^2}\right)^2 dx + m_t\left(L\frac{d\theta}{dt} + \left.\frac{\partial y}{\partial t}\right|_{L}\right)^2\Bigg]
\end{aligned}
\tag{3.37}
$$

Motivated by results published in the recent literature (Refs. 3-5,19,21,22), we investigate the following weighted energy function as a candidate Lyapunov function:

$$
\begin{aligned}
2U &= a_1 I_{hub}\dot{\theta}^2 + a_2(\theta - \theta_f)^2 \\
&+ 4a_3\left[\int_{L_0}^{L} \rho\left(\frac{\partial y}{\partial t} + x\dot{\theta}\right)^2 dx + \int_{L_0}^{L} EI\left(\frac{\partial^2 y}{\partial x^2}\right)^2 dx + m_t\left(L\dot{\theta} + \left.\frac{\partial y}{\partial t}\right|_{L}\right)^2\right]
\end{aligned}
\tag{3.38}
$$

where the positive weighting coefficients a_i are included to allow relative emphasis on the three contributors to the "error energy" of the system. Note that this is one of many possible ways to weight the mechanical system error energy, and merely provides one illustration of an approach. It is physically reasonable to consider placing relative emphasis upon dissipating subsets of mechanical energy as a control strategy, because some energy subsets are obviously more degrading of system performance objectives than others in practical applications. Since θ does not appear in the total mechanical energy of Eq. (3.7), the total energy of Eq. (3.37) is only positive semidefinite. We have added the positive "torsional spring energy" term $a_2(\theta - \theta_f)^2$ as a pseudoenergy to make the target final state

$$
\left(\theta, \dot{\theta}, y(x,t), \frac{\partial y(x,t)}{\partial t}\right)_{desired} = (\theta_f, 0, 0, 0)
$$

be the global minimum of U. It is obvious by inspection that imposing $a_i > 0$ in Eq. (3.38) guarantees that $U \geq 0$ and that indeed the global minimum of $U = 0$ occurs only at the desired state (we wish to begin at rest and rotate to a new angular position $\theta = \theta_f$, suppressing vibration enroute and returning to zero flexural deformation in the final position). Differentiation of Eq. (3.38), substitution of the equations of motion [Eqs. (3.35) and (3.36)], and considerable calculus leads to the weighted power

$$\dot{U} = \frac{dU}{dt} = \dot{\theta}\Big[a_1 u + a_2(\theta - \theta_f) + 4(a_3 - a_1)(L_0 S_0 - M_0)\Big] \tag{3.39}$$

Since we require that $\dot{U} \leq 0$ to guarantee stability, we set the term in brackets to $-a_4\dot{\theta}$, and this leads to the control law

$$u = -\frac{1}{a_1}\Big[a_2(\theta - \theta_f) + a_4\dot{\theta} + 4(a_3 - a_1)(L_0 S_0 - M_0)\Big] \tag{3.40}$$

In Ref. 3, we developed a shortcut based upon the work/energy rate method that avoids most of the algebra and calculus required to establish the weighted power expressions like Eq. (3.39), we generalize this discussion in section 3.4.

From Eqs. (3.39) and (3.40), and considering all possible values for the a_i, we see that the following *linear, spatially discrete* output feedback law globally stabilizes this distributed parameter system:

$$\begin{aligned} u &= -\Big[g_1(\theta - \theta_f) + g_2\dot{\theta} + g_3(L_0 S_0 - M_0)\Big]; \\ g_1 &\geq 0, \quad g_2 \geq 0, \quad g_3 > -4 \qquad \text{for stability} \end{aligned} \tag{3.41}$$

This control law is elegant. Notice that the rigorous stability proof does not depend on introducing spatial discretization methods such as the finite element method. Furthermore, we have verified from root locus calculations that the gain stability boundaries are apparently exact in this case (to 10 digits for the first 10 eigenvalues). Of important practical consequence, notice that controllers based on this law of Eq. (3.41) are easy to implement since no state estimation is required. The root shear and bending moment can be measured by using conventional strain gauges. The value and sign of the shear/moment feedback gain $g_3 = 4(a_3 - a_1)/a_1$ depends on whether we wish to emphasize dissipation of the beam vibration energy (for $a_3 > a_1$) or the energy of hub motion (for $a_3 < a_1$), as is evident from Eq. (3.38).

Since $\dot{U} = -a_4\dot{\theta}^2$ is not an explicit, negative definite function of the subset of state variables

$$\left[\theta, y(x,t), \frac{\partial y(x,t)}{\partial t}\right]$$

the stability arguments implicitly depend on the truth *that all infinity of the antisymmetric modes of motion of this structure*, have generally nonzero hub angular velocity ($\dot{\theta}$). Note under the kinematic assumptions leading to Eqs. (3.35), only antisymmetric modes are present, and no non-trivial motion can exist while the hub angular velocity vanishes identically for finite time intervals. A more elegant proof of global asymptotic stability using the feedback law of Eq. (3.41) can be done by applying Theorem 3.9. This has been carried to completion in Ref. 30, including consideration of the cases in which we relax the antisymmetric deformation assumption applied in deriving Eqs. (3.35), thereby admit a richer and more general set of motions (the four beams are described by four distinct functions of space and time, and there are now four PDEs and one hybrid differential/integral equation). For this more general configuration, it can be shown that a single hub actuator cannot provide rigorous asymptotic stability, because only an antisymmetric subset of the modes are controllable by a hub actuator (the uncontrollable modes have identical adjacent beams moving in opposition, which results in equal and opposite root moments and, because of this cancellation, zero hub motion). For rest-to-rest maneuvers, however, only the antisymmetric modes considered here are disturbable (by a hub torque actuator), and they are also controllable. Thus, for the model and differential equations developed above, the control law of Eq. (3.41) is globally stabilizing.

It is significant that this same linear feedback law of Eq. (3.41) maintains its globally stabilizing character even when the Euler-Bernoulli assumptions are relaxed to include the most common additional linear and nonlinear effects. In particular, we have verified that closed-loop stability is maintained when we include the following: rotational stiffening, Coriolis kinematic coupling terms, aerodynamic drag, shear deformation, beam rotary inertia, and finite inertia of the tip mass. The verification of these truths requires appropriate modifications of the kinetic and potential energy functions and, of course, the differential equations of motion must be generalized consistently. In particular, $\dot{U} = -a_4\dot{\theta}^2$ can vanish only if the conditions $\dot{\theta} = 0$, $\ddot{\theta} = 0$ can be encountered at some point other than U=0 (the target state), so the nonlinear proof proceeds directly from the closed-loop system differential equations by showing that the condition $\dot{\theta} = \ddot{\theta} = 0$ occurs only at the desired equilibrium:

$$\left(\theta, \dot{\theta}, y(x,t), \frac{\partial y(x,t)}{\partial t}\right)_{\text{desired}} = (\theta_f, 0, 0, 0)$$

In short, global stability of the system using the simple linear output feedback control law of Eq. (3.41) has been found to be very forgiving of the usual variations in modeling assumptions and therefore modeling errors.

In this section, an indirect method of Lyapunov for analyzing the motion of a nonlinear system near the equilibrium state has been presented, and also a method for generating globally stabilizing feedback control law for distributed parameter structural systems has been discussed as an important application of Lyapunov direct method.

3.4.4 Work Energy Rate Principle and Lyapunov Stable Control Laws

Consider a mechanical system that consists of N multiply connected substructures. The *Work Energy Rate Principle* for the kth substructure is expressed as (see Appendix D and Refs. 3, 26)

$$\frac{dE_k}{dt} = \frac{d}{dt}(T_k + V_k) = \sum_i \mathbf{F}_{ik} \cdot \dot{\mathbf{R}}_{ik} + \sum_j \mathbf{M}_{jk} \cdot \underline{\omega}_{jk}, \qquad k = 1, 2, \ldots, N \quad (3.42)$$

where E_k denotes the total mechanical energy of the kth substructure, T_k is the total kinetic energy of the kth substructure, V_k is the potential energy of the kth substructure (all conservative forces are absorbed into V_k), $\mathbf{F}_{ik}$ and $\mathbf{M}_{jk}$ include all working forces and moments other than those absorbed into V_k, therefore include (1) control and other external forces, (2) dissipative forces, and (3) substructure boundary forces and moments, and - $\dot{\mathbf{R}}_{ik}$ and $\underline{\omega}_{jk}$ are linear and angular velocities represented in the generalized coordinates. Implicit in the present discussion is a target state of zero inertial motion, we will subsequently generalize this discussion so that the target state can be an arbitrary smooth reference trajectory of the system.

A simultaneously obvious and subtle feature of Eq. (3.42) is that it is essentially a *kinematic* equation that applies regardless of the dynamical and constituitive models of the various substructures. Thus we could be modeling a given substructure by any of a variety of constituitive laws, and could be using finite-dimensional (e.g., obtained using the finite element method) models, or partial differential equation models. Notice that the rate of change of energy can be written down without knowing the energy function itself! The kinematic rate of change of energy due the work of boundary forces can be computed from Eq. (3.42), regardless of how we have modeled the dynamical cause/effect mechanics of the substructure's internal dynamics. On the other hand, we can differentiate a given energy function (however complicated), substitute the *corresponding* equations of motion, carry out the ensuing algebra and calculus and, if we are persistent, can ultimately verify that a specialized form of the general energy rate equation holds for our special case. More expediently, we can specialize Eq. (3.42) directly instead of re-deriving a special case of it every time we need it. When using the energy rate principle in the design of Lyapunov stable controllers, however, we will need to investigate the particular energy functions corresponding to our model of the system to confirm that the Lyapunov function (constructed from substructure energy functions) has its global minimum at a desired target state. One subtle advantage is implicit in this approach, *an infinity of admissible system models often have distinct energy functions, all of whose global minima occur at the same target state.* For example all the classical elastic beam models (Euler-Bernoulli, Timenshenko, etc.) have the common feature that the corresponding potential energy functions are positive for all nonzero deformations and vanish identically at a global minimum for zero deformation. These developments are considered below.

Motivated by the analysis of the previous section, a candidate Lyapunov function U can be defined for the purpose of designing a control law of the form

$$\mathrm{U} = \sum_{k} a_k E_k + \sum_{k} b_k W_k \tag{3.43}$$

where positive coefficients a_k are energy weights, W_k denotes the pseudoenergies (non-negative functions of selected coordinates), and b_k are non-negative pseudoenergy weights. The pseudoenergy functions W_k are included for generality and to permit sufficient flexibility to guarantee that U is a positive definite function which assumes its global minimum value of zero at the target state. As is evident in the sequel, the error energy parameterization of Eq. (3.43) leads to control laws whose gains are functions of the weights a_k and b_k.

Note that these weights a_k allow relative emphasis to be placed on each subset of the error energy of the system and, by choosing $a_k > 0$, we guarantee that $U \geq 0$. A candidate Lyapunov function including only energy of the substructures may vanish for an infinity of not necessarily desired rigid body states of the given system. This occurs, for example, when one or more of the rigid body coordinates do not appear in *any* of the E_k. Therefore, by adding the pseudoenergy functions (usually quadratic type) W_k usually with $b_k \geq 0$, the Lyapunov function U can be made to vanish only at the desired equilibrium state; i.e., the global minimum of U=0 occurs at the desired state.

Differentiation of Eq. (3.43) and using the work /energy rate principle, Eq. (3.42), lead directly to the following expression for $\dot{U}$

$$\begin{aligned}\dot{U} &= \sum_{k} a_k \Big[\sum_{i} \mathbf{F}_{ik} \cdot \dot{\mathbf{R}}_{ik} + \sum_{j} \mathbf{M}_{jk} \cdot \underline{\omega}_{jk}\Big] + \sum_{k} b_k \dot{W}_k \\ &= \sum_{k} a_k \left[\sum_{i} \left\{\mathbf{F}_{ik} + \frac{b_k}{a_k}\left(\frac{\partial W_k}{\partial \mathbf{R}_{ik}}\right)^T\right\} \cdot \dot{\mathbf{R}}_{ik} + \sum_{j} \left\{\mathbf{M}_{jk} + \frac{b_k}{a_k}\left(\frac{\partial W_k}{\partial \underline{\omega}_{jk}}\right)^T\right\} \cdot \underline{\omega}_{jk}\right]\end{aligned} \tag{3.44}$$

$\dot{U} \leq 0$ is required to guarantee stability. Therefore, if admissible control forces and moments in $\mathbf{F}_{ik}$ and $\mathbf{M}_{ik}$ can be chosen to ensure $\dot{U} \leq 0$, then this provides a Lyapunov stable control law. Of course, the question of whether or not the stability claim can be upgraded to asymptotic stability requires further investigation for each application of this approach, and optimization of the gains over the stable region requires appropriate numerical methods to be introduced.

Note that, in obtaining a Lyapunov control law, we use the work/energy rate principle instead of direct differentiation of Eq. (3.43) and substitution of the corresponding governing equations of motion. This avoids some often tedious differentiation and integration by parts for obtaining Eq. (3.44). However, it is dangerous to jump directly Eq. (3.44) to obtain the control law, since one may not have sufficient insight to choose the W_k. Selection of the W_k should be based on answering the question: What must be added to the total weighted energy $\Sigma a_k E_k$ to guarantee that U of Eq. (3.43) is positive definite and that the global minimum of U occurs at the target state? In particular, if certain *cyclic* coordinates are missing from the mechanical energy functional, then an obvious candidate for the W_k would be positive definite functions of all cyclic coordinates' departure from the target state.

Now, we summarize a method for obtaining Lyapunov control laws using the work/energy rate principle. First, the candidate Lyapunov function is constructed as a linear combination of the error energy (departure from a desired stable equilibrium point in the state space) contribution of all substructures. The division of the overall system into substructures is not unique, and some physical insight into the objectives of the particular system is required to make a meaningful error energy parameterization. Second, by examining the positive definiteness of the candidate Lyapunov function at the target equilibrium state, energy-type (positive) quantities W_k are added, in particular, to make a positive definite dependence on the errors in all cyclic coordinates. Third, by considering all working forces and moments (other than those absorbed into the potential energy), and directly differentiating the W_k with respect to the generalized velocities, we obtain the weighted energy rate $\dot{U}$. Finally, if possible, we choose admissible control forces and moments to make $\dot{U}$ negative semidefinite. If the admissible controls are selected to make $\dot{U}$ as negative as possible, then we establish a *Lyapunov optimal* policy for feedback control.

It is important to analyze the particular dynamical system to establish whether or not it is controllable; if the system is controllable, then we know that there is a family of choices for feedback controls for which $\dot{U} \leq 0$, guaranteeing asymptotic stability, otherwise the system is merely stable (residual energy may be "trapped" in the uncontrollable subspace). In some problems, we find this fortuitious circumstance: the uncontrollable subspace is essentially undisturbable (i.e., there is no mechanism for either disturbing or controlling the motion in this subspace). For such fortuitious problems, we may be able to achieve practical asymptotic stability, although it is not possible to rigorously claim asymptotic stability. There is practical insight here: *we wish to avoid disturbing dynamics over which we have poor controllability.* This insight affects the qualitative strategy underlying the design of target reference motions for the flexible body maneuvers generally, and those discussed in the following section. A corollary to this practical insight, especially for active control of flexible structures, is to "let sleeping dogs lie," and avoid disturbing the dynamical behavior of the system in a motion regime where the system is almost certainly poorly modeled (e.g., for flexible structures, seek to avoid disturbing the initially quiescent very high frequency modes). These qualitative ideas can be combined with the elegant Lyapunov analysis to establish some very attractive and practical control laws.

The work/energy rate principle can be rigorously generalized for distributed parameter systems and, therefore the associated Lyapunov control law design approach can also be generalized immediately for distributed parameter systems. These ideas can also be extended so that the target state is a time-varying trajectory rather than a fixed point in state space; the result is a Lyapunov stable tracking law design methodology, as developed in the following section.

3.4.5 Globally Stable Tracking Control: A Lyapunov Approach

Although the constant gain feedback law designed in the previous section can be tuned to work well for terminal pointing and vibration suppression, it is a poor law for carrying out both a large-angle maneuver and the ensuing pointing/vibration suppression "end game" to null small errors near the final target state. In fact, it is evident from both analytical/numerical studies and hardware experimental studies that sophisticated gain schedules are required if we wish to use the simplest linear feedback law to control both the large maneuvering motions and the small terminal motions. To obtain a control law more appropriate for large angle maneuvers with vibration suppression, the preceding developments (Lyapunov stable control laws) can be modified to obtain stable tracking-type feedback control laws.

Before studying the tracking problems, let us introduce the following three conceptual systems:

(1) The actual physical system
(2) A best theoretical model of the system
(3) A practical (reduced-order) model of the system

Computed motion for idealized open-loop control inputs can be determined from the models (2) and (3) and used as target trajectories for tracking-type feedback control laws. While model (3) is compatible with real time computation constraints, model (2) can be used for off-line evaluation studies.

For the sake of the current discussion, we assume that the motion of the actual system in the absence of disturbances is best modeled by a coupled hybrid set of ordinary differential equations and partial differential equations

$$\mathbf{f}(\mathbf{y}, \mathbf{u}, t) = 0 \tag{3.45}$$

where $\mathbf{f}(\mathbf{y}, \mathbf{u}, t)$ denotes the vector of differential operators that produces the hybrid ordinary differential equations and partial differential equations, $\mathbf{y}(P, t)$ denotes the generalized coordinate vector describing both the overall rigid body motions and the superimposed elastic motions of the hybrid system, and $\mathbf{u}(t)$ denotes the control vector. P denotes a spatial point and t denotes time.

Since Eq. (3.45) is a mathematical model, in general, we must anticipate that there exist modeling errors and unmodeled disturbances, when we concern ourselves with the dynamics of the actual physical system. However, we assume that the motion of the theoretical reference (target) motion is governed by the same hybrid system of ordinary and partial differential equations, Eq. (3.45), with the motion driven by a judicious open-loop control $\mathbf{u}_{ref}(t)$ that we have determined such that the ensuing motion achieves the desired mission; that is, we have a specific target reference trajectory satisfying

$$\mathbf{f}(\mathbf{y}_r, \mathbf{u}_{ref}, t) = 0 \tag{3.46}$$

where $\mathbf{y}_r(P, t)$ denotes the (target) theoretical open-loop reference trajectory for the system.

Since the system (3.46) is the assumed correct model, at this stage of the discussion, we do not include any modeling errors or disturbances. Once we have a reference control $\mathbf{u}_{ref}(t)$ and the theoretical reference trajectory $\mathbf{y}_r(P, t)$, then the object is to find the actual control $\mathbf{u}(t)$ in Eq. (3.45) so that the actual (initially disturbed) state vector $\mathbf{y}(P,t)$ will approach the reference trajectory $\mathbf{y}_r(P, t)$. However, it is usually difficult to compute the optimal reference control $\mathbf{u}_{ref}(t)$ that exactly satisfies Eq. (3.46) and the desired boundary conditions for a specific mission. The theoretical reference trajectory $\mathbf{y}_r(P, t)$ is not generally computable for distributed parameter systems, (even) for the bang-bang-type (minimum time) control law because of practical difficulties with infinite-dimensions and solving nonlinear partial differential equations. Therefore, for distributed parameter systems, a discretization approximation process is usually introduced to obtain a sufficiently good finite-dimensional solution that can be assumed to be an accurate approximation of the solution of the full-order system.

Since the theoretical reference dynamical system is often of high order, a model reduction process is typically required to obtain a practical reference system consistent, for example, with real time computing constraints. The reduced-order system is usually governed by a relatively low-dimensioned system of linear ordinary differential equations

$$\hat{\mathbf{f}}(\mathbf{y}_{ref}, \mathbf{u}_{ref}, t) = 0 \tag{3.47}$$

where $\hat{\mathbf{f}}(\mathbf{y}_{ref}, \mathbf{u}_{ref}, t)$ denotes the vector of operators that produce the reduced-order model, and $\mathbf{y}_{ref}(P,t)$ is the corresponding approximate reference trajectory obtained using open-loop reference control $\mathbf{u}_{ref}(t)$. The order reduction is motivated by the need to compute $\mathbf{y}_{ref}(P,t)$ using a tolerable computational budget; it is often necessary to determine $\mathbf{y}_{ref}(P,t)$ in near real time.

Now, the "path tracking" control objective is to *find control law* $\mathbf{u}(t)$ *so that the actual system state* $\mathbf{y}(P,t)$ *makes an asymptotic approach to the real time computable reference trajectory* $\mathbf{y}_{ref}(P,t)$.

Using the above three motions, three *departure motions* can be defined as:

$$\Delta\mathbf{y}(P,t) = \mathbf{y}_r(P,t) - \mathbf{y}_{ref}(P,t) \tag{3.48}$$

$$\delta\mathbf{y}(P,t) = \mathbf{y}(P,t) - \mathbf{y}_r(P,t) \tag{3.49}$$

$$\hat{\delta}\mathbf{y}(P,t) = \mathbf{y}(P,t) - \mathbf{y}_{ref}(P,t) \tag{3.50}$$

Note that $\Delta\mathbf{y}(P,t)$ is the open-loop residual tracking error arising only from model reduction, discretization, etc., $\delta\mathbf{y}(P,t)$ is a real time departure motion error of theoretical interest, and $\hat{\delta}\mathbf{y}(P,t)$ is the departure motion of practical interest. Using the above definitions, the Lyapunov stable controller introduced in the previous section can be rigorously generalized to obtain globally stable tracking controllers. In some cases, we will be able to establish global asymptotic stability. For nonlinear systems, however, we find that global stability of the departure motion is not generally guaranteed when we use the reduced-order reference trajectory for the target motion. We initially deal with only linear distributed parameter systems to illustrate the method for designing a globally stable tracking controller.

Consider the candidate Lyapunov function for the actual system of Eq. (3.45) by summing the energy of the kth substructures [see Eq. (3.43)]

$$U_a = \sum_k a_k E_k + \sum_k b_k W_k \tag{3.51}$$

Note that the total energy E_k and pseudoenergy E_k of the kth substructure consists of the generalized variable vector $\mathbf{y}$ and partial derivatives of $\mathbf{y}$ with respect to the time and spatial variables, i.e.,

$$E_k = E_k(\mathbf{y}, \dot{\mathbf{y}}, \mathbf{y}', \mathbf{y}'', \ldots) \tag{3.52}$$

$$W_k = W_k(\mathbf{y}, \dot{\mathbf{y}}) \tag{3.53}$$

Considering Eqs. (3.52) and (3.53) for the actual system, the candidate error energy Lyapunov function is introduced as follows:

$$U = \sum_k a_k \delta E_k + \sum_k b_k \delta W_k \tag{3.54}$$

where δE_k and δW_k are defined, using Eqs. (3.52), (3.53) and the departure motion of theoretical interest, as follows:

$$\delta E_k = E_k(\delta \mathbf{y}, \delta \dot{\mathbf{y}}, \delta \mathbf{y}', \delta \mathbf{y}'', \ldots) \tag{3.55}$$

$$\delta W_k = W_k(\delta \mathbf{y}, \delta \dot{\mathbf{y}}) \tag{3.56}$$

with $\delta(\ ) = (\ ) - (\ )_r$.

It is guaranteed that δE_k and δW_k are positive, since they are obtained by substituting all departure motion variables into Eqs. (3.52), (3.53). Note that the candidate error energy Lyapunov function U in Eq. (3.54) becomes zero only along the desired trajectory, and if the Lyapunov function is zero, we can conclude that the system is following the reference trajectory exactly.

Now, imagine a *virtual system* (some aspects of this discussion must be modified if the hybrid ordinary differential equations and partial differential equations are nonlinear) governed by the following equations:

$$\mathbf{f}(\delta \mathbf{y}, \hat{\delta}\mathbf{u}, t) = 0 \tag{3.57}$$

where $\hat{\delta}\mathbf{u} = \mathbf{u} - \mathbf{u}_{ref}$; note that $\hat{\delta}\mathbf{u}(t, \delta \mathbf{y}, \delta \dot{\mathbf{y}}, \delta \mathbf{y}', \delta \mathbf{y}'', \ldots)$ is the desired feedback control law.

Since E_k of Eq. (3.52) is the total energy of the actual system of Eq. (3.45), it is easy to understand that δE_k of Eq. (3.55) is the total energy of the virtual system of Eq. (3.57). Therefore, for this virtual system, the work/energy rate principle can be applied, i.e.,

$$\frac{d}{dt}\left(\delta E_k\right) = \sum_i \delta \mathbf{F}_{ik} \cdot \delta \dot{\mathbf{R}}_{ik} \tag{3.58}$$

where $\delta \mathbf{F}_{ik}$ denotes all working forces and moments, including control and other external forces and moments and substructure boundary forces and moments, and where $\delta \dot{\mathbf{R}}_{ik}$ denotes linear and angular velocities. Note that, in the generalization of the above developments for nonlinear systems, Eq. (3.58) has more terms (including cross product terms of the actual trajectory and theoretical time-varying reference trajectory) resulting from the nonlinearities in addition to $\delta \mathbf{F}_{ik}$. To obtain correct expressions, including these time-varying and nonlinear terms in the virtual energy rate $d(\delta E_k)/dt$, direct differentiation of δE_k and substitution of equations of motions into the resulting equations constitute the only known way to develop the corresponding energy rate equation.

Differentiating Eq. (3.54) and using Eq. (3.58) leads to the following expression for the energy rate (power) function $\dot{U}$:

$$\dot{U} = \sum_k a_k \left[\sum_i \delta \mathbf{F}_{ik} \cdot \delta \dot{\mathbf{R}}_{ik}\right] + \sum_k b_k \delta \dot{W}_k \tag{3.59}$$

Equation (3.59) can be rewritten by separation of all control forces and moments $\hat{\delta}\mathbf{u}$ ($= \mathbf{u} - \mathbf{u}_{ref}$) and all boundary forces and moments $\delta\mathbf{F}_{ik}^{B}$ from all working forces and moments $\delta\mathbf{F}_{ik}$ as follows:

$$\dot{U} = \sum_{k} a_k \sum_{i} \left[\hat{\delta}\mathbf{u}_{ik} + \delta\mathbf{F}_{ik}^{B} + \delta\mathbf{W}_{ik}\right] \cdot \delta\dot{\mathbf{R}}_{ik} \tag{3.60}$$

where $\mathbf{u}_{ik}$ denotes the ith control force vector or moment vector acting on the kth substructure and

$$\delta\mathbf{W}_{ik} = \frac{b_k}{a_k}\left[\frac{\partial(\delta\dot{\mathbf{W}}_k)}{\partial(\delta\dot{\mathbf{R}}_{ik})}\right]$$

If we choose control $\hat{\delta}\mathbf{u}$ such that $\dot{U} \leq 0$, then global stability of the virtual system is guaranteed. In other words, when displaced initially, the actual trajectory approaches the theoretical reference trajectory asymptotically. A key necessary condition, that underlies the existence of globally stabilizing control for the present family of problems, and envisioned extensions to more general distributed parameter systems, is that the system must be controllable. Thus, all infinity of the modes of the system must be controllable by the control input as a necessary condition to establish a globally stabilizing control (in this case, a stabilizing tracking law). For example, for a flexible body system to be controllable, it is necessary that the number of actuators be at least equal to the number of rigid body freedoms of the system. This necessary condition is most usually a sufficient condition in practice; however, the degree of controllability may be very low for some modes, depending on the location of the actuators and the mode shapes. Note that all of this discussion must be modified to discuss stability in the presence of model errors or unmodeled disturbances.

From Eq. (3.60), it is clear that stability is obtained by choosing the control such that

$$\hat{\delta}\mathbf{u}_{ik} + \delta\mathbf{F}_{ik}^{B} + \delta\mathbf{W}_{ik} = -c_{ik}\delta\dot{\mathbf{R}}_{ik} \tag{3.61}$$

where coefficients c_{ik} are positive.

Dependence on the velocity $\delta\dot{\mathbf{R}}$ of a given actuator location is often included in the terms associated with several substructures' kinematics. For example, the angular velocity vector components of the central rigid body rotation can be found in the brackets corresponding to the error energy in all substructures' contributions in Eq. (3.60). Also, the boundary forces and moments resulting at the joints of the substructures have equal magnitudes and opposite direction and, of course, they share the same generalized velocities. In these cases, redundancies in the coordinates of Eq. (3.60) are found. In order to deal with this problem, before determining the tracking law, Eq. (3.60) should be manipulated (i.e., all forcing terms related to the same generalized velocity should be collected in the same bracket, and Newton's third law then imposed as appropriate) so that the remaining vectors $\delta\dot{\mathbf{R}}_{ik}$ are independent of each other.

Equation (3.60) can be rearranged by collecting on distinct terms $\delta\mathbf{F}_{ik}^{B}$ and $\delta\mathbf{W}_{ik}$ as

$$\dot{U} = \sum_{k} a_k \sum_{i} \left\{ \hat{\delta}\mathbf{u}_{ik} + g_{ik}^{B}\delta\mathbf{F}_{ik}^{B} + g_{ik}^{W}\delta\mathbf{W}_{ik} \right\} \cdot \delta\dot{\mathbf{R}}_{ik} \tag{3.62}$$

where constants g_{ik}^{B} and g_{ik}^{W} are functions of energy weights a_k and pseudoenergy weights b_k. Algebraic conditions defining the stable region of these gain constants $g_{ik}(a_1, a_2, ..., b_1, b_2, ...)$ can be determined by examining those functions, requiring Eq. (3.62) to be strictly negative and the constants a_k and b_k to be strictly positive.

The global tracking control law can be obtained by setting

$$\hat{\delta}\mathbf{u}_{ik} + g_{ik}^{B}\delta\mathbf{F}_{ik}^{B} + g_{ik}^{W}\delta\mathbf{W}_{ik} = -g_{ik}^{V}\delta\dot{\mathbf{R}}_{ik} \tag{3.63}$$

with appropriate selection of positive constants g_{ik}^{V} for the velocity feedback. Finally, the stable tracking control law $\mathbf{u}(t)$ for the actual system can be obtained from Eq. (3.63) as

$$\mathbf{u}_{ik}(t) = \Big[\mathbf{u}_{ik}(t)\Big]_{ref} - \Big[g_{ik}^{B}\delta\mathbf{F}_{ik}^{B} + g_{ik}^{W}\delta\mathbf{W}_{ik} + g_{ik}^{V}\delta\dot{\mathbf{R}}_{ik}\Big] \tag{3.64}$$

Note that $\mathbf{u}_{ik}$ denotes the ith control force vector or moment vector acting on the kth substructure. Obviously, the globally stabilizing control law of Eq. (3.64) requires presolution for the open-loop solutions for the theoretical reference system which is most generally a nonlinear system of hybrid ordinary and partial differential equations. Note again that for nonlinear systems, Eq. (3.64) also includes nonlinear cross product functions of the actual trajectory and theoretical reference trajectory, and the global stability of the tracking law can still be guaranteed by including those effects in determining a feedback law that ensures $\dot{U} < 0$. It is also implicit that the system must be fully controllable; otherwise, one cannot claim that requiring $\dot{U}$ in Eq. (3.62) to be nonpositive will actually drive U to zero. This is physically equivalent to requiring that *the only equilibrium state of the closed-loop system is the desired target state in which U vanishes.* This condition can often be checked by setting all departure velocities and accelerations to zero in the closed-loop equations of motion and checking to see if there is only one equilibrium state.

As discussed above, it is typically difficult (and often impossible) to obtain the open-loop solutions for the theoretical reference system of differential equations.

Therefore, from a practical point of view, it is more attractive to design control laws that make the actual system follow an easy-to-compute motion satisfying a reduced-order reference system of differential equations instead of attempting to compute the open-loop solution satisfying the infinite-dimensional theoretical reference system. Motivated by the structure of the globally stabilizing control law of Eq. (3.64), the following form for the tracking control law is hypothesized:

$$\mathbf{u}_{ik}(t) = \left[\mathbf{u}_{ik}(t)\right]_{ref} - \left[g_{ik}^{B}\hat{\delta}\mathbf{F}_{ik}^{B} + g_{ik}^{W}\hat{\delta}\mathbf{W}_{ik} + g_{ik}^{V}\hat{\delta}\dot{\mathbf{R}}_{ik}\right] \tag{3.65}$$

where $\hat{\delta}(\ ) \equiv (\ ) - (\ )_{ref}$, and $(\ )_{ref}$ quantities are evaluated along the open-loop solution of the reduced-order reference system of Eq. (3.47) with $\mathbf{u} = \mathbf{u}_{ref}(t)$.

Now, the following question naturally arises: What are the global stability characteristics of the closed-loop system using the simpler law of Eq. (3.65)? We can use Eq. (3.62) to evaluate the stability of motion resulting from using the feedback control $\mathbf{u}(t)$ of Eq. (3.65). The region possessing Lyapunov stability can be found directly by substituting Eq. (3.65) into Eq. (3.62) and finding the region for which $\dot{U}$ is guaranteed to be negative. Substitution of Eq. (3.65) into Eq. (3.62) leads to

$$\dot{U} = -\sum_{k} a_k \sum_{i} \left\{ g_{ik}^{V}\delta\dot{\mathbf{R}}_{ik} + \left[g_{ik}^{B}\Delta\mathbf{F}_{ik}^{B} + g_{ik}^{W}\Delta\mathbf{W}_{ik}\right]\right\} \cdot \delta\dot{\mathbf{R}}_{ik} \tag{3.66}$$

where $\Delta(\ ) = (\ )_r - (\ )_{ref}$. Notice that a sufficient condition characterizing the region where $\dot{U} \leq 0$ is the dominance of the first term in each of the braces of Eq. (3.66) for each controller $\mathbf{u}_{ik}$ acting on the kth substructure; this gives the Lyapunov stability condition

$$\delta\dot{\mathbf{R}}_{ik} \cdot \delta\dot{\mathbf{R}}_{ik} > \mu_{ik} \equiv \frac{1}{g_{ik}^{V}} \left\| \left\{g_{ik}^{B}\Delta\mathbf{F}_{ik}^{B} + g_{ik}^{W}\Delta\mathbf{W}_{ik}\right\} \cdot \delta\dot{\mathbf{R}}_{ik} \right\| \tag{3.67}$$

If the norm of all velocity variable tracking errors, $\delta\dot{\mathbf{R}}_{ik} \cdot \delta\dot{\mathbf{R}}_{ik}$, exceeds μ_{ik} (for all i and k), then $\dot{U}$ is negative and apparently U decreases until the boundary layer region bounded by Eq. (3.67) is approached. The Δ quantities on the right side of Eq. (3.67) are open-loop residuals between the open-loop solution of the practical reference system and that of the theoretical reference system. Frequently, especially for systems described by hybrid differential equations, it is not possible to obtain the exact solution $(\ )_r$ resulting from $\mathbf{u}_{ref}$.

As discussed before, a sufficiently accurate approximate (full-order) system can usually be obtained via a discretization process; the order of this highly accurate system is typically much greater than the reduced-order reference system since the model reduction process has been invoked to obtain a model compatible with near real time computing constraints. Therefore, considering the full-order approximate system as an acceptable approximation of the theoretical reference system, we can use simulations to numerically estimate the upper bounds μ_{ik} directly by *a priori* computation of the two open-loop motions (one from the reduced-order reference system and the other from the full-order system) and use of a particular (candidate design) set of feedback gains.

Equation (3.67) thus determines a generalized velocity inequality with zero defining the edge of a boundary layer region Γ near the $(\ )_{ref}$ motion. Note that *large motions are globally attracted to the region* Γ *near the target motion because the error energy rate satisfies* $\dot{U} \leq 0$ *everywhere outside of this region.* Thus, the control law of Eq. (3.65) is almost globally stabilizing, and the only region where asymptotic stability is not guaranteed is in Γ near the target trajectory (the reduced-order model reference trajectory). Furthermore, the right side of Eq. (3.65) is essentially a measure of *how nearly the target trajectory satisfies the hybrid partial and ordinary differential equations of the system* (theoretical exact reference system); a judicious choice of the shaping parameters (in $\mathbf{u}_{ref}$) and the reduced-order model that defines the target trajectory can usually be made to result in μ being sufficiently small. Again, this can be determined a priori based on a one-time open-loop simulation study. If μ and therefore Γ are made sufficiently small through suitable selection of the reference model, shaping of the reference input, and choice of feedback gains, then, in many applications (for all practical purposes), one may achieve globally stable tracking control. While the analysis is tedious for some problems, as is evident in section 3.5 below, a significant class of problems submits fully to this approach.

Even though globally stabilizing controllers can be designed using any smooth target trajectory, we have found that it is especially attractive to use a closed-form solution for the reference trajectory. For nonlinear systems, we find that a theoretically globally stabilizing controller that is similar to Eq. (3.64) can usually be designed; however, these laws are typically not practical because the needed reference motion is difficult to obtain with suitable efficiency/precision for most applications. When we pursue a simpler tracking law that makes use of a finite dimensional approximate system model to generate a target reference trajectory, stability must be established on a case-by-case basis. Also it is often difficult to algebraically determine the stability boundaries in the gain space because of the complexity of the energy rate equation induced by the nonlinearities of the particular problem. As is evident from the case study below, in spite of these difficulties, this line of thinking can lead to very attractive and practical control laws in many hardware implementations.

3.5 Case Study: Application for Near-Minimum-Time Large Angle Maneuvers of Distributed Parameter Systems

We have discussed the vibration suppression problem of the hub-appendage configuration in the previous sections. As discussed before, the constant gain linear feedback control law works poorly if we try to use the same constant gains for the large angular motions and for the small terminal motions. This is because the large gains required for effective vibration suppression near the target state are typically several orders of magnitude too large for near-minimum-time large-angle maneuvers (i.e., the large gains appropriate for vibration suppression, when used during a large-angle maneuver, typically result in significant θ overshoots and, often, actuator saturation). Also, the large initial torque command typically introduces a large vibratory transient into the structure. From a qualitative point of view, if we wish to maneuver a highly flexible structure while suppressing vibration, then it is unlikely that we should initiate this process by hitting the structure with a large hammer! To obtain a control law more appropriate for near-minimum-time large-angle maneuvers with vibration suppression, stable tracking-type feedback control laws discussed in this section can be applied.

Consider briefly the near-minimum-time maneuver of a rigid body. We know that the strict minimum-time control is a bang-bang law which, for the rest-to-rest maneuver-to-the origin case, saturates negatively during the first half of the maneuver and positively during the last half of the maneuver (Refs. 17, 23-27). From an implementation point of view, the instantaneous switches of the bang-bang law are sometimes troublesome because (1) no torque-generating device exists that can switch instantaneously; (2) when generalized and applied to a flexible structure, the bang-bang class of controls excite poorly modeled higher modes; and (3) the switch times (and, therefore, the dynamics of the actual system) are usually very sensitive to modeling errors.

Recently, an attractive family of *parameterized sharpness* approximations of the switch function has been introduced to modify the admissible controls in near-minimum-time control formulations. The approximation presented in Refs. 21, 28 involves transcendental functions, but recent experimental work (Refs. 4, 5) indicates that a much simpler piecewise continuous spline approximation of the sign function is attractive from an implementation point of view. Using this approach, a typical near-minimum-time control law (for single axis, rest-to-rest maneuver of a rigid body) has the form

$$I\ddot{\theta} = u = \pm u_{max} f(\Delta t, t_f, t) \tag{3.68}$$

where t_f is the maneuver time and $\alpha = \Delta t/t_f$. We choose the (+) sign if $\theta_f > \theta_o$. As a torque shaping function, we adopt the smooth sign function approximation $f(\Delta t, t_f, t)$:

Table 3.1 Texas A&M Maneuverable Flexible Structure: Configuration Parameters

Total undeformed system inertia, I	2128, oz-s^2-in.
Hub radius, L_0	5.5470, in.
Hub center to tip mass, L	51.07, in.
Tip mass, m_t	0.15627, oz-s^2/in.
Appendage modules of elasticity, E	161.6×10^6, oz/in.^{2}
Inertia of bending section, I	0.000813, in.^{4}
Mass density of appendage/length, ρ	0.003007, oz-s^2/in.^{2}

$$f(\Delta t, t_f, t) = \begin{cases} \left(\frac{t}{\Delta t}\right)^2\left[3 - 2\left(\frac{t}{\Delta t}\right)\right], & \text{for } 0 \le t \le \Delta t \\ 1, & \text{for } \Delta t \le t \le \frac{t_f}{2} - \Delta t \equiv t_1 \\ 1 - 2\left(\frac{t - t_1}{2\Delta t}\right)^2\left[3 - 2\left(\frac{t - t_1}{2\Delta t}\right)\right], & \text{for } t_1 \le t \le \frac{t_f}{2} + \Delta t \equiv t_2 \\ -1, & \text{for } t_2 \le t \le t_f - \Delta t \equiv t_3 \\ -1 + \left(\frac{t - t_3}{\Delta t}\right)^2\left[3 - 2\left(\frac{t - t_3}{\Delta t}\right)\right], & \text{for } t_3 \le t \le t_f \end{cases}$$

Adopting the positive sign, Eq. (3.68) integrates to yield

$$\dot{\theta}(t) = \dot{\theta}_0 + \frac{u_{max}}{I}\int_{t_0}^{t} f(\Delta t, t_f, \tau)d\tau \tag{3.69a}$$

$$\theta(t) = \theta_0 + (t - t_0)\dot{\theta}_0 + \frac{u_{max}}{I}\int_{t_0}^{t}\int_{t_0}^{\tau_1} f(\Delta t, t_f, \tau_2)d\tau_2 d\tau_1 \tag{3.69b}$$

The integrations in Eqs. (3.69) can be carried out in terms of elementary functions, which are not presented here for the sake of brevity; the results of these integrations give Eqs. (3.71), (3.72) below. Figure 3.2 shows a maneuver resulting from these integrations for a typical selection of parameters ($\alpha = 0.25$, $u_{max} = 400$ oz-in.), and a 40 deg rest-to-rest maneuver of a rigid approximation of the structure in Fig. 3.1 and Table 3.1. For rest-to-rest maneuvers, we impose the boundary conditions:

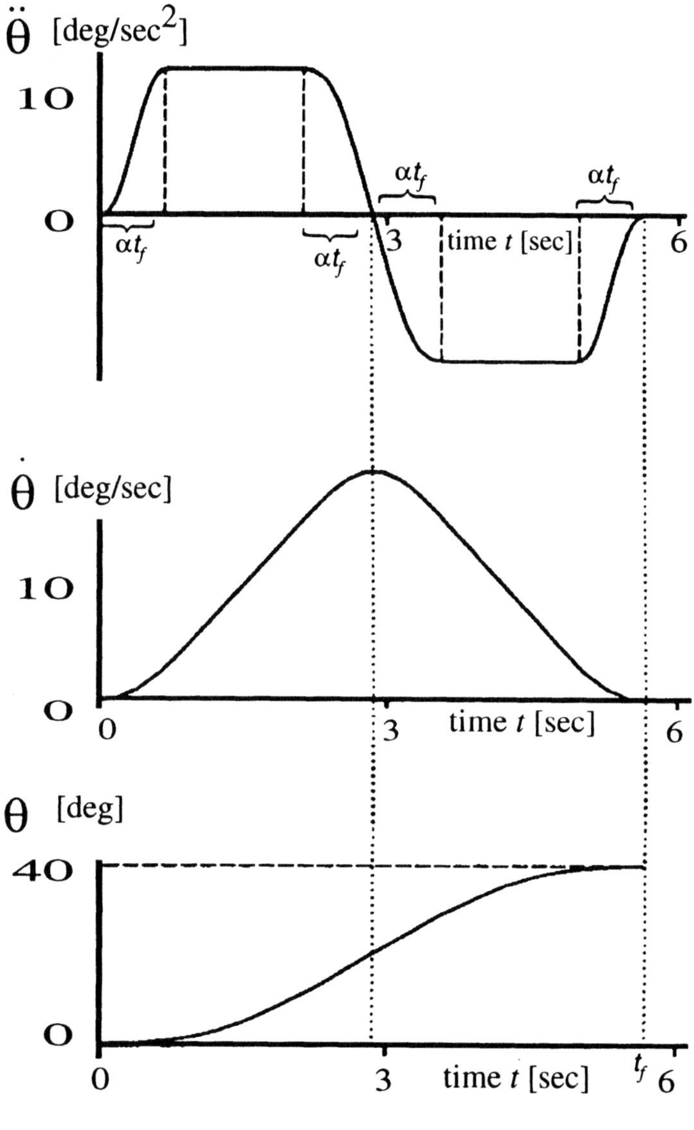

Figure 3.2 Torque-Shaped Rigid Body Maneuver.

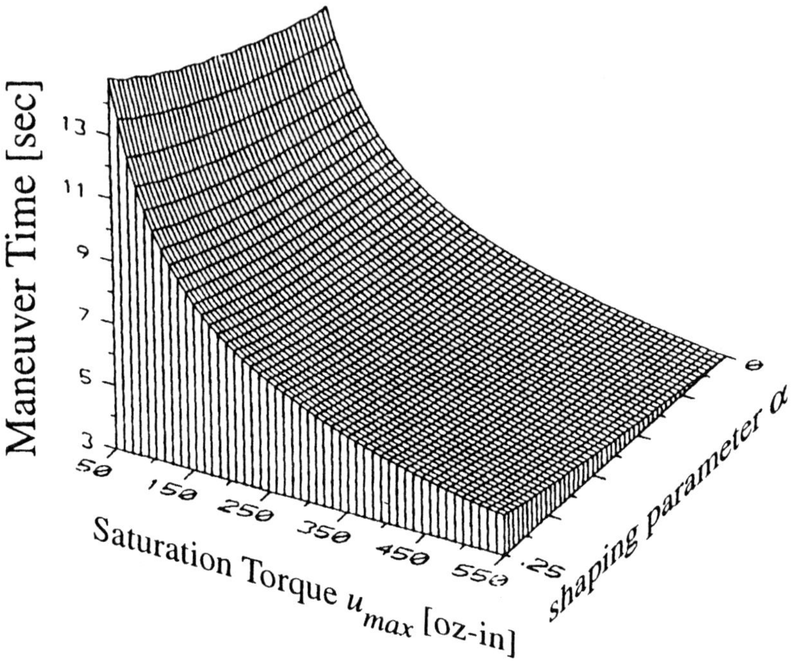

Figure 3.3 Rigid Body Maneuver Time vs Saturation Torque and Torque-Shaped Parameter.

$$
\begin{aligned}
&\text{at } t_0 = 0: \qquad && \theta(0) = \theta_0, \qquad && \dot{\theta}(0) = 0 \\
&\text{at time } t_f: \qquad && \theta(t_f) = \theta_f, \qquad && \dot{\theta}(t_f) = 0
\end{aligned}
\tag{3.70}
$$

and upon carrying out the integrations implied in Eq. (3.69), we obtain the useful relationship

$$
\theta_f - \theta_0 = \left(\frac{u_{max}}{I}\right)\left[\frac{1}{4} - \frac{1}{2}\alpha - \frac{1}{10}\alpha^2\right]t_f^{\,2}, \qquad \Delta t = \alpha t_f, \qquad 0 < \alpha < \frac{1}{4} \tag{3.71}
$$

or

$$
t_f = \left\{ \frac{I(\theta_f - \theta_0)}{u_{max}[(1/4) - (1/2)\alpha + (1/10)\alpha^2]} \right\}^{\frac{1}{2}} \tag{3.72}
$$

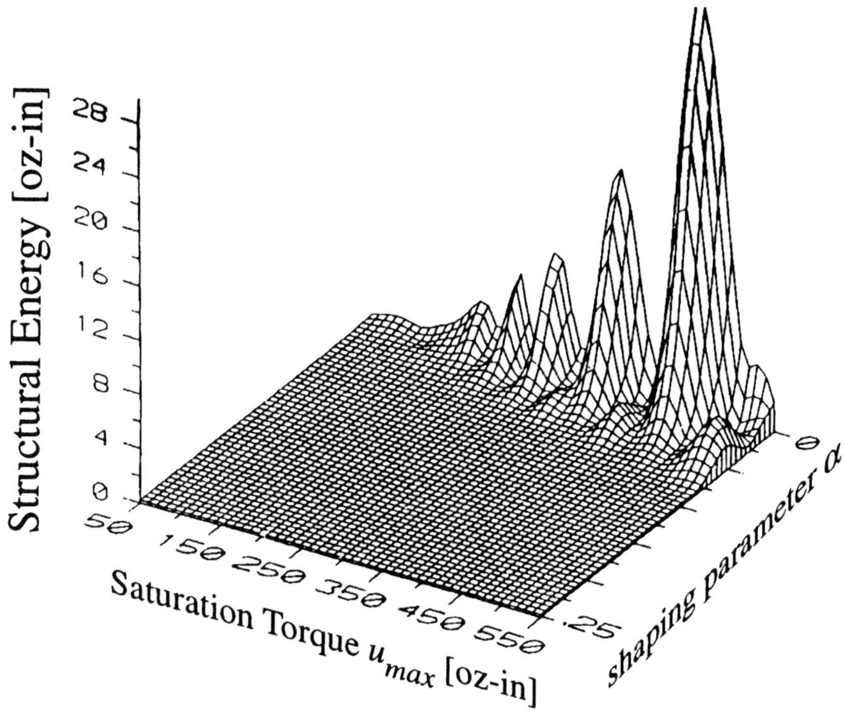

Figure 3.4 Flexible Body Open-Loop Residual Vibration Energy vs Saturation Torque and Torque-Shaped Parameter.

In Eq. (3.72), we see the explicit tradeoff between torque shaping α, target maneuver time t_f, maneuver angle $\theta_f - \theta_0$, and maximum angular acceleration u_{max}/I. Obviously, Eq. (3.71) can be inverted for any of these as a function of the remaining parameters. If we set $\alpha = \Delta t/t_f = 0$, of course, we obtain the well-known special case result expressing the relationship between the minimum time, maneuver angle, inertia, and saturation torque for bang-bang control. It is obvious that selection of α controls the sharpness of the switches, with $\alpha = 0$ corresponding to bang-bang control (instantaneous switches) and $\alpha = 0.25$ being the smoothest member of this family of torque-shaped maneuvers. Figure 3.3 shows the rigid body maneuver time t_f vs α, from Eq. (3.72), whereas Figure 3.4 shows the residual total energy (at time t_f) when the torque-command $u_{ref} = u_{max} f(\alpha t_f, t_f, t)$ is applied to simulate the flexible body response [first six modes from a discrete assumed mode model (Chapter 4 and Ref. 5) of order 20]. Notice (Figure 3.4) that open-loop torque shaping reduces residual vibration at time t_f by three orders of magnitude ($\alpha = 0.1$) with only a modest ten percent increase over the theoretical minimum time rigid body maneuver ($\alpha = 0$).

The preceding results and Refs. 4, 5, 21, 22, and 27 support the intuitively obvious truth that applying judiciously smoothed bang-bang controls such as Eq. (3.68) to generate an open-loop maneuver of a flexible body can result in near negligible structural vibration for sufficiently slow maneuvers (small u_{max} and large α) and neglecting disturbance torques. Of course, unmodeled disturbances, control implementation errors, and model errors will always negate some of these apparent gains. However, sharper control switches obviously increase the probability that higher frequency, less well-modeled modes will be excited and, therefore, robustness with respect to model errors is generally more of an issue for bang-bang control than for smoother torque profiles. Even for relatively small departures (smoothed switches) from bang-bang control, torque-shaped maneuvers of highly flexible structures typically enjoy a reduction of several orders of magnitude in residual vibration. Thus, the overall maneuver time (including terminal vibration suppression) can be reduced significantly by torque shaping.

These observations suggest the following strategy: Use an optimized shaped-input profile to establish a "trackable" a priori reference rigid (or reduced-order flexible) body maneuver; then, based on real-time measurements of the actual flexible body's departure from this smooth reference motion, superimpose a perturbation feedback control on the reference shaped-torque history that stabilizes the departure motion from the reference motion. Also of significance, it is usually desirable to select the reference torque profile parameters (e.g., u_{max}, α, etc.) to consider the available sensor and actuator dynamics and thereby make the commanded torque history more nearly achievable physically.

Pursuing this logic judiciously, attractive tracking-type feedback control laws can be established for near-minimum-time, large angle maneuvers. Since bang-bang flexible body controllers are sensitive to modeling and control implementation errors, we seek control laws that are a smooth torque-shaped compromise between the competing objectives of minimizing: (1) maneuver time, (2) residual vibration, and (3) sensitivity of closed-loop performance measures with respect to model and control implementation errors.

Suppose we adopt a reference rigid body maneuver

$$\{\theta_{ref}(t), \dot{\theta}_{ref}(t), \ddot{\theta}_{ref}(t) = u_{ref}/I\}$$

satisfying Eqs. (3.68)-(3.72), where I is the undeformed moment of inertia, and we have implicitly selected α, u_{max} and computed the corresponding t_f from Eq. (3.72) for specified initial and final angles. For designing a globally stable tracking controller, the candidate error energy Lyapunov function can be established by considering Eq. (3.38) and (3.54) as

$$2U = a_1 I_{hub}\delta\dot{\theta}^2 + a_2\delta\theta^2 + 4a_3\left\{\int_{L_0}^{L}\rho\left[\delta\frac{\partial y}{\partial t} + x\delta\dot{\theta}\right]^2 dx + \int_{L_0}^{L} EI\left(\delta\frac{\partial^2 y}{\partial x^2}\right)^2 dx + m_t\left[L\delta\dot{\theta} + \delta\frac{\partial y}{\partial t}\bigg|_L\right]^2\right\} \tag{3.73}$$

where $\delta(\ ) \equiv (\ ) - (\ )_r$ and the $(\ )_r$ quantities are evaluated along the open-loop flexible body solution of Eqs. (3.35) with $u(t) = u_{ref}(t)$. Considering Eqs. (3.39) and (3.60), the time derivative of U is given by

$$\dot{U} = (\dot{\theta} - \dot{\theta}_r)\left\{a_1 u - a_1 u_{ref} + a_2(\theta - \theta_r) + 4(a_3 - a_1)[(L_0S_0 - M_0) - (L_0S_0 - M_0)_r]\right\} \tag{3.74}$$

Pursuing the objective of globally stable control, it is clear that setting the [ ] term equal to $-a_4(\dot{\theta} - \dot{\theta}_r)$ leads to the following globally stabilizing [with $\dot{U} = -a_4(\dot{\theta} - \dot{\theta}_r)^2$] control law [see Eq. (3.64)]:

$$u = u_{ref}(t) - \left\{g_1(\theta - \theta_r) + g_2(\dot{\theta} - \dot{\theta}_r) + g_3[(L_0S_0 - M_0) - (L_0S_0 - M_0)_r]\right\} \tag{3.75}$$

For a class of practical applications, the following structure for the tracking control law [Eq. (3.65)] can be hypothesized:

$$u = u_{ref}(t) - \left\{g_1(\theta - \theta_{ref}) + g_2(\dot{\theta} - \dot{\theta}_{ref}) + g_3[(L_0S_0 - M_0) - (L_0S_0 - M_0)_{ref}]\right\} \tag{3.76}$$

where it is easy to show that the root moment for the special case of a reference (rigid body) motion is proportional to the angular acceleration:

$$(L_0S_0 - M_0)_{ref} = [\rho(L^3 - L_0^3)/3 + m_t L^2]\ddot{\theta}_{ref}(t) \tag{3.77}$$

Obviously, the globally stabilizing control law of Eq. (3.75) is similar to the conjectured law (for practical implementation) of Eq. (3.76), the difference being that Eq. (3.76) requires presolution for the open-loop rigid body $(\ )_{ref}$ quantities, whereas the globally stabilizing control law of Eq. (3.75) requires solution for the open-loop flexible body $(\ )_r$ quantities from the partial differential equations. Since near-minimum-time control implies a certain urgency(!), it is obvious that the negligible computational overhead of Eq. (3.76) is more attractive than Eq. (3.75) from the point of view of real time implementations. For the purpose of finding the region possessing Lyapunov stability, substitute Eq. (3.76) into Eq. (3.74) [see Eq. (3.66)]

$$\dot{U} = -a_1(\dot{\theta} - \dot{\theta}_r)\Big\{g_2(\dot{\theta} - \dot{\theta}_r) + \big[g_1\Delta\theta + g_2\Delta\dot{\theta} + g_3\Delta(L_0S_0 - M_0)\big]\Big\} \tag{3.78}$$

The Lyapunov stability condition comes from requiring $\dot{U}$ of Eq. (3.78) to be negative; a sufficient condition is

$$|\dot{\theta} - \dot{\theta}_r| > \mu \equiv \frac{1}{g_2}\big|g_1\Delta\theta + g_2\Delta\dot{\theta} + g_3\Delta(L_0S_0 - M_0)\big| \tag{3.79}$$

If the angular velocity tracking error $|\dot{\theta} - \dot{\theta}_r|$ exceeds μ, then $\dot{U}$ is negative and apparently U decreases until encountering the region bounded by Eq. (3.79). It is further apparent that the Δ quantities on the right side of Eq. (3.79) are finite and (pre-)computable differences between open-loop flexible $(\ )_r$ and rigid body $(\ )_{ref}$ motions. Thus, an upper bound μ can be established directly by precomputation of a family of two open-loop motions and the use of a particular set of feedback gains. Equation (3.79) thus determines an angular velocity variable boundary defining a region Γ near the $(\ )_{ref}$ motion. Note that large motions are globally attracted to Γ because $\dot{U} \leq 0$ everywhere outside of this region. Thus, the control law of Eq. (3.76) is almost globally stabilizing, and the only region where asymptotic stability is not guaranteed is the small Γ boundary layer region near the target trajectory. Furthermore, the right side of Eq. (3.79) is essentially a measure of how nearly the reference target trajectory satisfies the flexible body equations of motion; a judicious choice of the shaping parameters defining the target trajectory and the associated reference control input can usually be made to result in μ, and therefore Γ, being sufficiently small.

A bounded-input/bounded-output (BIBO) viewpoint of stability can be used to establish some insight into the motion in the Γ region. Departure motion differential equations for $\delta(\ ) = (\ ) - (\ )_r$ quantities can be obtained by differencing Eqs. (3.35), driven by the control law of Eq. (3.76), from the rigid body equations of motion, driven by u_{ref}. Upon formulating these equations, one can verify that the departure motion is governed by a linear, otherwise asymptotically stable system of differential equations, forced by the known Δ terms that appear in Eq. (3.79). The $\delta(\ )$ motion in the Γ region is thus bounded because the Δ forcing terms are bounded; the finite maxima of these terms can be found by direct calculation. The resulting departure motion is therefore bounded everywhere in the Γ region, which was already known to have a (typically small) finite dimension μ. Since the actual numerical bounds on the Δ and μ quantities can be made arbitrarily small (depending on how nearly the user-defined reference trajectory is made to satisfy the open-loop equations of motion), we have a very elegant theoretical and practical situation vis-à-vis stability of the closed-loop tracking motion. We see that the closed-loop motion is globally attracted to the controllably small Γ region near the target trajectory and, considering the motions within Γ, we have BIBO stability.

In this application, we use a torque-shaped rigid body reference trajectory, which is very attractive since the reference maneuver can be calculated in closed form [such as the family of Eqs. (3.68)-(3.72)] and since the ensuing tracking law performs extremely well. Note that Eqs. (3.68)-(3.72) have a C^1 continuous transition to the final fixed state:

$$\left\{\mathrm{u}_{\mathrm{ref}}(\mathrm{t}), \theta_{\mathrm{ref}}(\mathrm{t}), \dot{\theta}_{\mathrm{ref}}(\mathrm{t}), [\mathrm{M}_0(\mathrm{t})]_{\mathrm{ref}}, [\mathrm{S}_0(\mathrm{t})]_{\mathrm{ref}}\right\} = \left\{0, \theta_f, 0, 0, 0\right\}, \qquad \text{as } \mathrm{t} \to \mathrm{t_f}$$

so that, for $\mathrm{t} > \mathrm{t_f}$, only the three feedback terms of Eqs. (3.76) are contributing to the terminal fine-pointing/vibration arrest control. Thus, the controls blend continuously from the large-angle tracking law of Eq. (3.76) into a constant gain controller (for $\mathrm{t} > \mathrm{t_f}$) identical to the globally stable fixed point output feedback case of Eq. (3.40). Thus we have unqualified global stability for $\mathrm{t} > \mathrm{t_f}$.

3.5.1 Simulated Results for Large Angle Maneuvers

Returning to the family of 40-deg open-loop maneuvers used to generate the energy surface of Figure 3.4, we computed the velocity tracking bound μ for Lyapunov stability [as given by Eq. (3.79)] and found the maximum value ($\mu_{\max}$) of $\mu(\mathrm{t})$ along each trajectory. Figure 3.5 displays this worst-case tracking bound (maximum value of μ) surface $\mu_{\max}(\alpha, \mathrm{u}_{\max})$ region used to generate Figures 3.3 and 3.4. The closed-loop tracking error bound has a roughly analogous behavior to the open-loop residual vibration energy surface of Figure 3.4. Recall that, outside the region bounded by the inequality of Eq. (3.79), we have guaranteed Lyapunov stability, using the control law of Eq. (3.76) and the reference rigid body torque given by Eqs. (3.68)-(3.72). From Figure 3.4, it is clear that sufficiently small $\mu_{\max}$ and large α result in arbitrarily small tracking errors, but the (small α, large $\mathrm{u}_{\max}$) near-bang reference maneuvers cannot be tracked as precisely. It is easy to see how a subset of the candidate (α, $\mathrm{u}_{\max}$) designs can be found that satisfy specified inequalities on maneuver times, tracking errors, and residual vibration energy by direct examination of the surfaces of Figures 3.3-3.6.

The results obtained from the simulations (and in the actual hardware implementations presented later and in Refs. 4 and 5) support the conclusion that these surfaces can be used to establish a large region of feasible designs for near-minimum-time controls in the space of torque-shaped parameters and control gains.

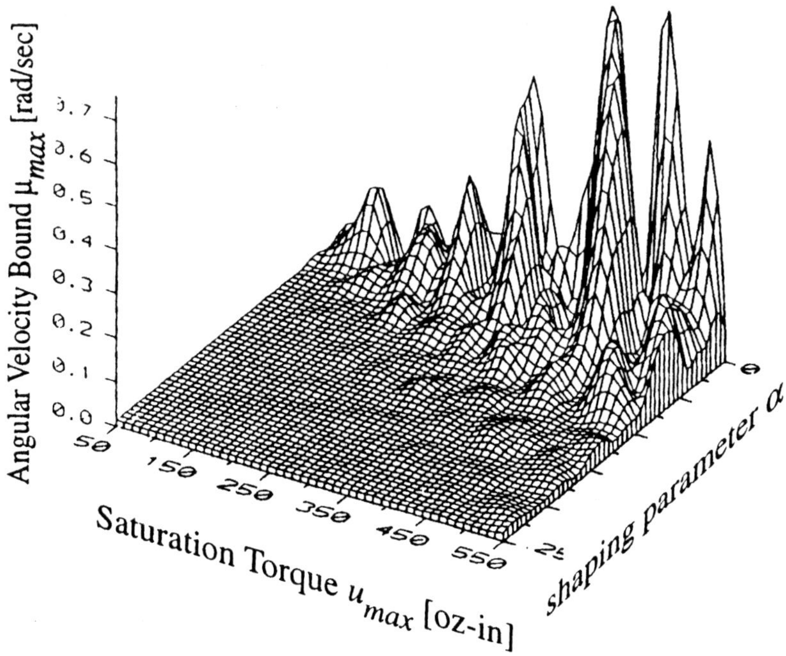

Figure 3.5 Boundary of the Lyapunov-Stable Tracking Region vs Saturation Torque and Torque-Shaped Parameter.

Optimization over the set of feasible designs should, in general, include consideration of the nature of expected disturbances to be rejected. One detailed simulation is now considered to show state and control variable histories along a typical trajectory of the family of trajectories underlying the above surfaces. In these simulations the effects of worst case disturbance torques are included in order to illustrate the effectiveness of controls in the presence of unmodeled effects. For simplicity, only the case of 40-deg rest-to-rest maneuver is considered here, along with setting $u_{max}=$ 400 oz-in. for all cases.

For the computational studies, two control laws are considered: namely, the output feedback law (control law I) of Eq. (3.40), and the tracking-type feedback control law (control law II) of Eq. (3.76). Although control law II could be used with an arbitrary reference trajectory, the torque-shaped rigid body trajectories of Eqs. (3.68)-(3.72) are specifically selected for investigation. The torque-shaped open-loop control history u_{ref} can be precomputed (in a fraction of a second!) from Eqs. (3.68)-(3.72) and stored, whereas the instantaneous trajectory variables $\{\theta_{ref}, \dot{\theta}_{ref}, [L_0 S_0(t) - M_0(t)]_{ref}\}$ are integrated easily in real time. Note that the boundary conditions of Eqs. (3.70) are enforced by using Eq. (3.72) to compute the trajectory maneuver time as a function of the maneuver angle, saturation torque, and torque-shaped parameter.

We now discuss the simulation results using control law II, which obviously blends into control law I in the end game (for $t \geq t_f$). In the experimental results in the subsequent discussion, maneuvers carried out by both control laws are reported. Both open-loop (all $g_i = 0$) and closed-loop time histories of selected state variables are shown in Figures 3.6 and 3.7.

Figures 3.6(a) and (b) show the hub angle and angular velocity for the case of an open-loop control and in the presence of substantial impulsive and quasi-random (5 oz-in., 1σ) disturbance torques. It is evident that the disturbance torque history is very significant vis-à-vis disturbing flexible dynamics in our experimental hardware; however, certain nonrandom, nonlinear effects associated with the bearing friction cause disturbances that are highly correlated in time and are not well represented by the present white noise model of the disturbance torques. In spite of the substantial disturbance torques (Figure 3.6), however, it is evident that the simulations indicate that the closed-loop flexible body dynamics in fact follows the near-minimum-time rigid body motion closely while effectively suppressing vibration as shown Figure 3.7. In addition to the variables graphed in Figures 3.6 and 3.7, we confirmed that the energy of the first 10 modes was effectively suppressed. These simulated results are very consistent with the experimental results discussed in the following section and those presented in Ref. 5.

3.5.2 Experimental Results

In all of the experiments in the following discussion, the target final angle is set to 40-deg and $u_{max} = 400$ oz-in. A detailed description of the hardware is given in Appendix I. We overview the system as follows. The configuration (Figure 3.1, Table 3.1) has a span of approximately 9 ft and has six natural frequencies below 20 Hz. The system is accurately balanced, and the four aluminum appendages' geometric, mass, and stiffness parameters are matched to high precision; the first three measured cantilevered natural frequencies of the four individual beams were found to be identical to within 0.05 Hz.

With this design the appendages vibrate almost exclusively in the horizontal plane; the hub is balanced on a custom-designed needle-jewel bearing that constrains the hub to rotate about the vertical axis. Our measurements confirm that negligible out-of-plane motion occurs in our experiments, although there is occasional evidence

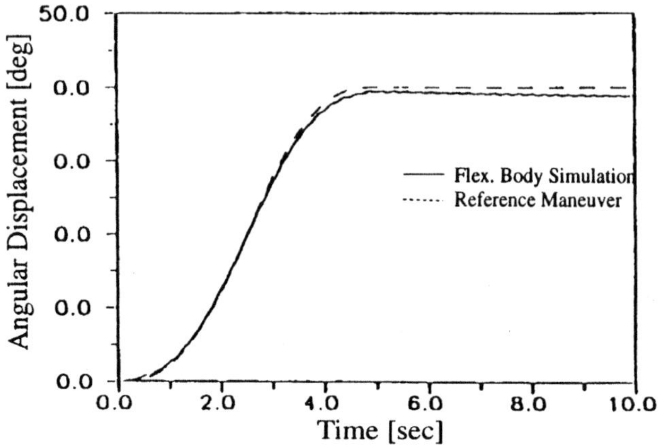

(a) Open-loop hub angle

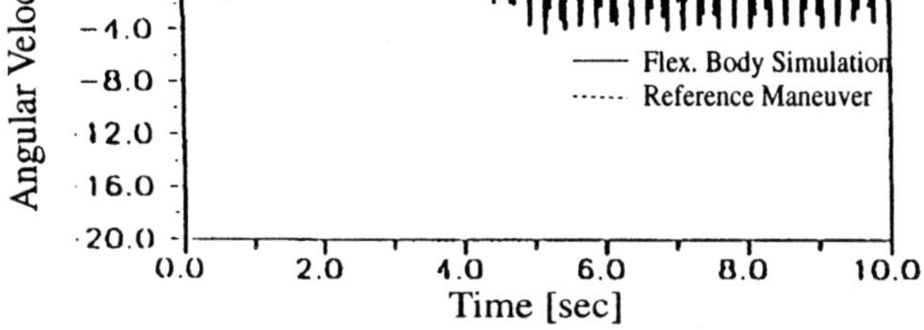

(b) Open-loop hub angular velocity

Figure 3.6 Open-Loop 40-Deg Maneuver with Random Disturbances.

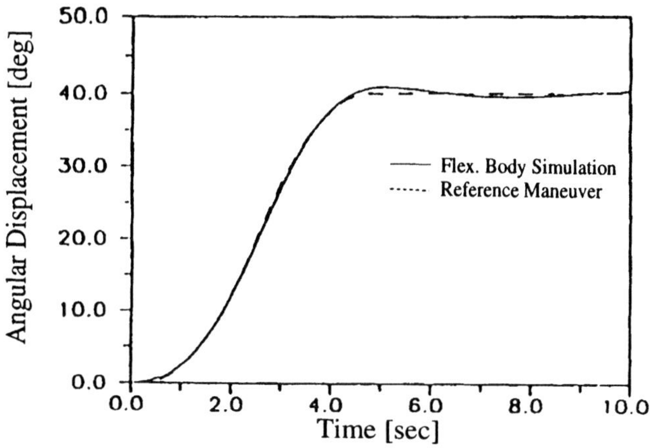

(a) Closed-loop hub angle

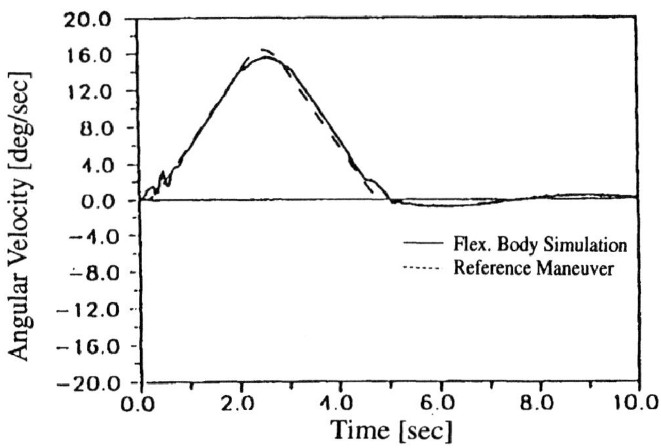

(b) Closed-loop hub angular velocity

Figure 3.7 Closed-Loop 40-Deg Maneuver with Random Disturbances.

of small beam torsional vibrations. Also, to very high accuracy, we can state that our experimental results confirmed that only the antisymmetric in-plane modes [implicit in the derivation of Eqs. (3.35)] were excited during rest-to-rest maneuvers using the hub torque actuator. The bearing stiction/friction torque is significant ($\sim$ 20 oz-in.), but is sufficiently small and predictable to permit meaningful experiments. Aerodynamic damping is important only during the most rapid slew maneuvers; in most cases, it represents a small perturbation as compared to the larger active vibration damping introduced by the feedback controller. The control torque is achieved by means of a reaction wheel mounted to the shaft of a DC motor [Figure 3.1(c)], which is, in turn, mounted to the hub. The commanded motor torque is achieved by precision current control using power amplifiers, as described in Appendix I. The angular rotation of the hub is measured using a Teledyne-Gurley angle encoder, accurate to about 0.01 deg, whereas the root bending moment and shear force estimates are derived from conventional full bridge strain gauge measurements. The derived estimates of the angular velocity history have a variance of approximately 1 deg/s and a time lag of 0.01 s. The noise and phase lag in the angular velocity estimates and the strain-gauge-derived root shear force and bending moment estimates limit the bandwidth of the closed-loop system to the range from approximately 0 to 10 Hz. The errors (noise and phase lag) in the derived hub angular velocity estimates represent the main source of the precision and bandwidth constraints of the experimental implementations. The control loops were closed, for all experiments discussed later, at 75 Hz; the angle encoder was also sampled at 75 Hz, whereas the strain gauges were sampled an order of magnitude faster and filtered to reduce the effects of sensor noise and higher-frequency modes outside the bandwidth of our controller.

Figure 3.8 shows the experimental system response for a maneuver using control law I [the constant gain control law of Eq. (3.40)] with $g_1 = 600$ oz-in./rad, $g_2 =$ 800 oz-in./rad/s, and $g_3 = 0$. Even though control law I [Eq. (3.40)] is anticipated to be poorly suited for large-angle maneuvers, we nonetheless apply this law to carry out 40 deg maneuvers to provide a reference for the subsequent discussion. Since the initial position error is large, the maneuvers start from zero with a large initial discontinuity to a large torque. For this gain selection, we see a large hub angle overshoot ($\sim$10 deg) and significant structural vibration that was effectively suppressed by around 12 s; the control was terminated at 16 s. These results were repeatable; however, the residual angle was typically $\sim$ 0.25 deg because the constant gain g_1 could not be set sufficiently large to overcome terminal bearing stiction without causing initial actuator saturation and large overshoots, and a compromise value was adopted for the sake of illustration. As is demonstrated in Ref. 5, the overall maneuver shape and settling time is sensitive to the gains selected; however, less than 10% reductions in the 12 s settling time can be achieved without initially saturating the actuator.

Control law II, on the other hand, leads to very attractive near-minimum-

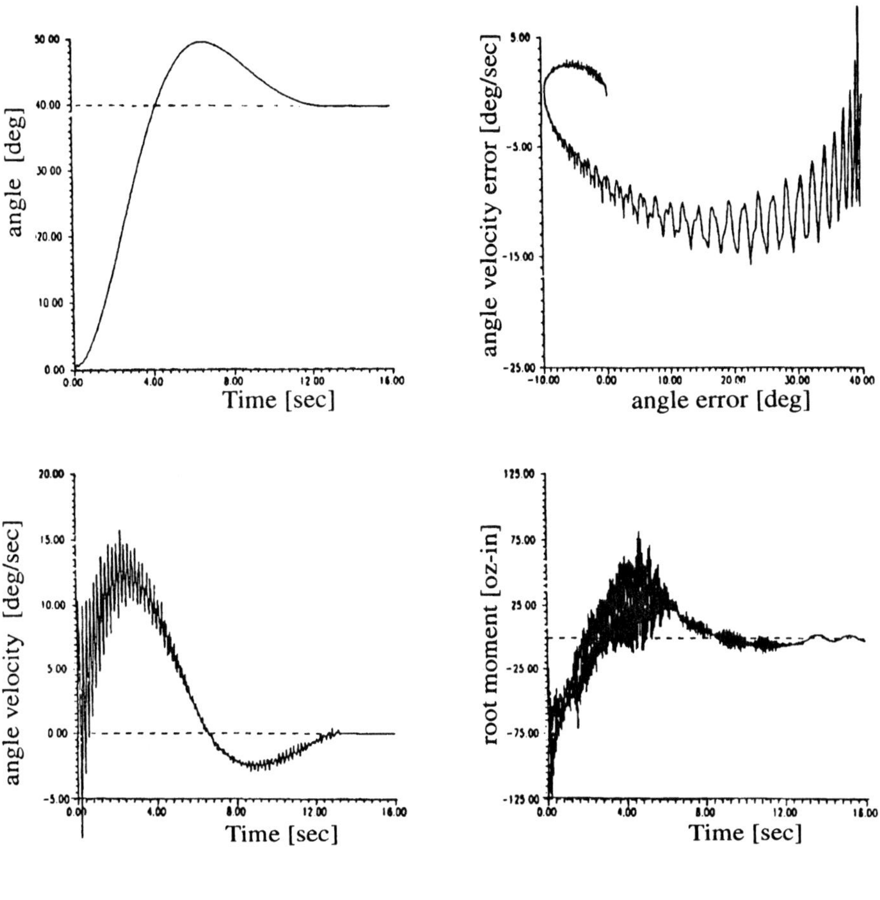

Figure 3.8 Experimental Results:

A 40-Deg Maneuver Using Control Law I

$(g_1 = 600, g_2 = 800, g_3 = 0.0)$.

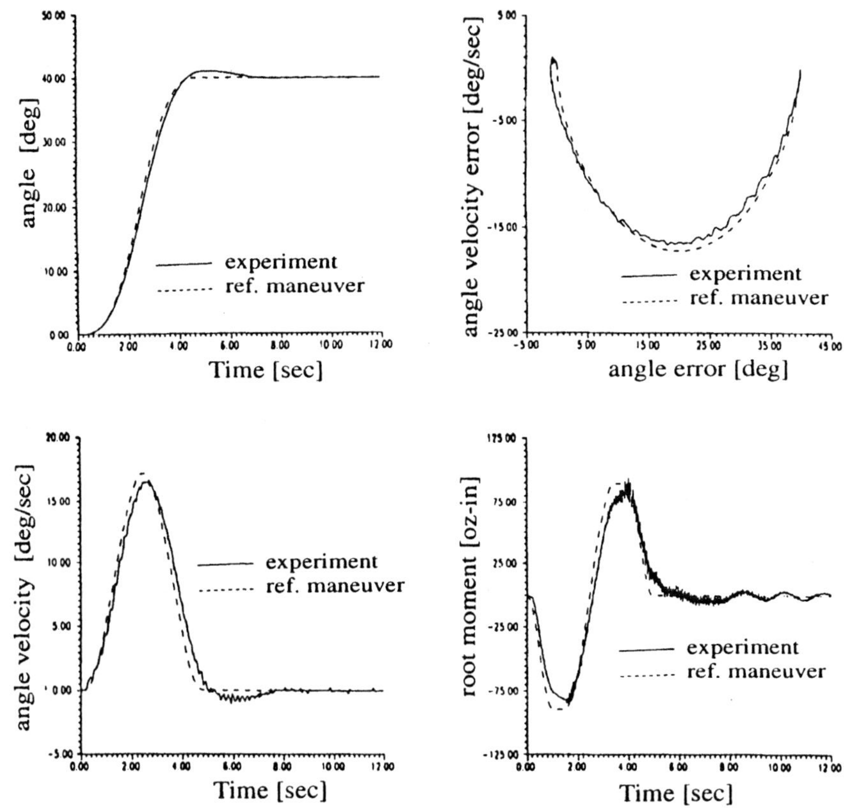

Figure 3.9 Experimental Results:

A 40-Deg Maneuver Using Control Law II

$(g_1 = 3000, g_2 = 800, g_3 = 0.0, \alpha = 0.2, \mathrm{u}_{\mathbf{max}} = 400)$.

time maneuvers. One feasible set of gain settings and torque shaped parameters leads to the experimental results shown in Figure 3.9. The effect of using a smooth, judiciously shaped reference torque history is evident if one compares the output and control variable histories in Figure 3.9 with those of Figure 3.8. This implementation of control law II produced much smaller overshoot ($\approx$ 1.5 deg vs $\sim$ 10 deg) and shorter maneuver time (6 s vs 12 s), and greatly reduced the severity of peak vibration, compared to control law I. These results, especially when considered in conjunction with numerous other cases, are reported in Refs. 5 and 21, provide convincing evidence that control law II is a versatile and highly effective way to incorporate open-loop torque-shaped optimization with en route and terminal vibration suppression. The fact that a globally continuous control structure is implicit in this approach leads to minimal difficulties in realizing robust control laws.

We encountered several practical difficulties in our experimental work, but these difficulties are not central to our control law design approach. First, the root shear force and bending moment apprximations obtained using strain gauge measurements resulted in sufficiently noisy and nonlinear measurements that, using this feedback ($g_3 \neq 0$), only marginally improved the controlled response over, for example, the results in Figure 3.9. These anomalies resulted, we hypothesize, from the non-ideal beam clamp effects near the station where the strain measurements were being made. Any slight play in the clamp due to large root moment variations would manifest itself in spurious strain measurements. Also, deriving the angular velocity estimate from the noisy angle encoder readout was difficult to accomplish with high precision and, as a consequence, we constructed a digital filter to process the angle encoder data and roll off the frequency content in the rate estimates above 10 Hz. We found this was useful to avoid erroneous, phase-lagged high-frequency components of the feedback that disturbed the higher frequency modes. These problems can be essentially eliminated, of course, by investing in a more precise sensor to measure angular displacement and/or angular velocity, as well as a load cell to measure the root shear and bending moments. Finally, our bearing presented us with another set of practical difficulties. Based on analysis of our bearing hardware, it became evident that interaction of the structure with the bearing accounts for the overwhelming source of disturbance torques. The bearing friction/stiction model developed from our analysis (Ref. 5) has the form

$$\tau_{\text{bearing}} = -c_1 \text{sign}(\dot{\theta}) - c_2\dot{\theta} + \text{HOT} \tag{3.80}$$

where we find $c_1 \sim 20$ oz-in. and $c_2 \sim 0.001$ oz-in./rad/s.

Thus, the first (stiction) term of Eq. (3.80) dominates the bearing torque for moderate $\dot{\theta}$ and is about 5% of the peak commanded torque of 400 oz-in. Although we believe that Eq. (3.80) models the bearing friction well, we found that it is difficult to use this model to compensate for bearing friction in real time because angle encoder noise results in uncertainty in the estimated instants that $\dot{\theta}$ switches sign. This difficulty has significant practical consequences. If we modify our control to compensate for bearing disturbance torques (essentially, attempt to cancel it) using Eq. (3.80), the commanded discontinuity (at the estimated time that $\dot{\theta}$ changes sign) will not coincide exactly with the actual stiction discontinuity; even slightly mistimed compensation torque discontinuities can actually worsen the disturbance! Although we experimented with several bearing torque compensation schemes, we ultimately decided to simply consider bearing torque an anticipated and well-modeled disturbance. Our simulations (such as the results shown in Figure 3.7) indicated that our control approach could easily tolerate disturbances of this magnitude, and our successful experiments in Figures 3.8 and 3.9 and Ref. 5 certainly confirm that our implemented control laws are robust in the presence of the actual disturbances from all sources.

This case study provides a good illustration of the mix of theoretical analysis, numerical computation, and engineering judgment required to carry out successful applications. The ultimate objective, of course, is to obtain perfect closure between theory and experiment. However, it is not realistic to expect the high degree of closure obtained above, when faced with more complicated dynamical systems. Note that excellent results were obtained, in spite of modest investments in sensors and actuators, however, for systems requiring high precision and wide control bandwith, it would be necessary to have corresponding improvements in the precision and bandwith of the sensors. In the context ot the above numerical and experimental results, however, we observe that a large degree of model error robustness implicit in our approach stems from our theoretical verification that the control of Eq. (3.40) remains stabilizing for most of the usual variations in modeling assumptions, and we used judicious sensor filtering to roll off the effects of the system dynamics outside the sensors' bandwidth. In conclusion, the excellent agreement between theory and experiment evident in Figures 3.7, 3.9 represents prototypical (rather than usual) results.

Problems

Problem 3-1

(a) Consider a nonlinear system

$$\dot{x}_1 = x_2 - a x_1 (x_1^2 + x_2^2)$$
$$\dot{x}_2 = -x_1 - a x_2 (x_1^2 + x_2^2)$$

where a is a positive constant.

Show that the function $U(\mathbf{x}) = x_1^2 + x_2^2$ is a Lyapunov function.

(b) Consider the second-order nonlinear system

$$\dot{x}_1 = x_2$$
$$\dot{x}_2 = -a_1 x_2 - a_2 x_1 - (b_1 x_2 + b_2 x_1)^2 x_2$$

where a_1 and a_2 are positive constants.

Show that the equilibrium state $(x_1, x_2) = (0, 0)$ is asymptotically stable in the large by means of the Lyapunov function

$$U(\mathbf{x}) = a_2 x_1^2 + x_2^2$$

Problem 3-2

Consider symmetric positive definite matrices P and Q satisfying

$$A^T P + PA - 2kP = -Q$$

where k is any real number.

Prove that all eigenvalues of A have real parts smaller than k.

Problem 3-3

Consider the following second-order system

$$\ddot{\eta} + 2\zeta\omega\dot{\eta} + \omega^2\eta = 0$$

Find the necessary condition for the above system to be asymptotically stable via

(a) an eigenvalue approach

(b) Lyapunov's first approach

Problem 3-4 (Ref. 12, 13)

Consider the sensor-actuator collocated second-order system

$$M\ddot{\mathbf{x}} + C\dot{\mathbf{x}} + K\mathbf{x} = D\mathbf{u}$$
$$\mathbf{y} = D^T\mathbf{x}, \qquad \dot{\mathbf{y}} = D^T\dot{\mathbf{x}}$$

where M is a symmetric positive definite matrix, C and K are positive semidefinite matrices, $\mathbf{x}$ is the $n \times 1$ generalized coordinate vector, $\mathbf{u}$ is the $m \times 1$ control vector, and $\mathbf{y}$, $\dot{\mathbf{y}}$ are $m \times 1$ measurement vectors.

Show that the following controllers guarantee the stability of the closed-loop system in the Lyapunov sense:

(a) $\mathbf{u} = -[G_1G_1^T \quad G_2G_2^T]\begin{Bmatrix} \mathbf{y} \\ \dot{\mathbf{y}} \end{Bmatrix}$

where G_i are lower triangular matrices with positive diagonal terms.

(b) $\mathbf{u} = -[G \quad H]\begin{Bmatrix} \mathbf{y} \\ \dot{\mathbf{y}} \end{Bmatrix}$

where

$$G = \sum_{k=1}^{l} g_i G_i, \qquad H = \sum_{k=1}^{l} h_i H_i$$

G_i and H_i are prescribed positive semidefinite gain submatrices, g_i and h_i are nonzero scalar gains, and l $(= m(m+1)/2)$ is the number of parameters.

Problem 3-5

Consider the nonlinear system with aerodynamic damping

$$\ddot{x} + a|\dot{x}|\dot{x} + x + \frac{1}{3}x^3 = 0, \qquad a > 0$$

(a) Apply Lyapunov's indirect (or first) method for the trivial equilibrium point ($\mathbf{x} = \mathbf{0}$) of the system. What can you conclude regarding the stability of the system?

(b) Apply Lyapunov's direct (or second) method by selecting an appropriate Lyapunov function. Is the system stable, asymptotically stable, or unstable?

Problem 3-6 (Ref. 22)

The tether dynamics of in-plane deployment and retrieval (of mass m on a weightless tether from, for example, the space shuttle) can be written as follows:

$$\ddot{l} - l\Big[(\dot{\theta} + \Omega)^2 - \Omega^2 + 3\Omega^2 \cos^2\theta\Big] = -\frac{\mathrm{T}}{\mathrm{m}}$$

$$\ddot{\theta} + 2\Big(\frac{\dot{l}}{l}\Big)(\dot{\theta} + \Omega) + 3\Omega^2 \cos\theta \sin\theta = 0$$

where l indicates the instantaneous tether length, θ is the in-plane pitch angle, Ω is the orbital rate, T is the tension, and m is the mass of the subsatellite.

Find a stabilizing nonlinear control law (tension T is a control variable) by using the following Lyapunov function:

$$\mathrm{U} = \frac{1}{2}\left[\Big(\frac{\dot{l}}{\Omega}\Big)^2 + \mathrm{k}_1(l - l_f)^2 + (\mathrm{k}_2 + 3l^2)\Big\{\frac{1}{3}\Big(\frac{\dot{\theta}}{\Omega}\Big)^2 + \sin^2\theta\Big\}\right]$$

where l_f is the desired final length of the tether, variable k_1 is a positive constant, and k_2 can be either positive or zero.

References

1. Vidyasagar, M., *Nonlinear System Analysis*, Prentice-Hall, Englewood Cliffs, NJ, 1978.
2. Willems, J.L., *Stability Theory of Dynamical Systems,* John Wiley & Sons, New York, NY, 1970.
3. Oh, H.S., Vadali, S.R., and Junkins, J.L., "On the Use of the Work-Energy Rate Principle for Designing Feedback Control Laws," *AIAA Journal of Guidance, Control, and Dynamics*, Vol 15, No. 1, Jan-Feb 1992, pp 275-277.
4. Junkins, J.L., Rahman, Z.H., and Bang, H., "Near-Minimum-Time Control of Distributed Parameter Systems: Analytical and Experimental Results," *Journal of Guidance, Control, and Dynamics,* Vol. 14, No. 2, Mar.-Apr. 1991, pp. 406-415.
5. Junkins, J.L., Rahman, Z.H., and Bang, H., "Near-Minimum Time Maneuvers of Flexible Vehicles: A Lyapunov Control Law Design Method," *Mechanics and Control of Large Flexible Structures,* Chapter 22, Progress in Astronautics and Aeronautics, Vol. 129, edited by J.L. Junkins, AIAA, New York, NY, 1990.
6. Kalman, R.E., "Lyapunov Function for the problems of Lure in Automatic Control," *Proceedings National Academy of Science,* Vol. 49, Feb. 1963, pp. 201-205.
7. Popov, V.M., "Hyperstability and Optimality of Automatic Systems with Several Control Functions," *Review of Roumanian Science and Technology,* Vol. 9, 1964, pp. 629-690.
8. Chen, C.T., *Linear System Theory and Design*, Holt, Rinehart and Winston, New York, NY, 1984.
9. Kalman, R.E. and Bertram, J.E., "Control System Analysis and Design via the Second Method of Lyapunov," Transactions of the ASME, *Journal of Basic Engineering*, Vol. 82, 1960, pp. 371-392.
10. Patel, R.V. and Toda, "Quantitative Measures of Robustness for Multivariable Systems," TP-8A, *Proceedings of JACC,* San Francisco, CA, 1980.
11. Inman, D.J., *Vibration with Control, Measurement, and Stability,* Prentice Hall, Englewood Cliffs, NJ, 1989.
12. Creamer, N. G., and Junkins, J. L., "Low-Authority Eigenvalue Placement for Second Order Systems," *Journal of Guidance, Control, and Dynamics,* Vol 14, No 3, May-June 1991, pp. 698-701.
13. Junkins, J.L. and Kim, Y., "A Minimum Sensitivity Design Method for Output Feedback Controllers," *Mechanics and Control of Large Flexible Structures,* Chapter 15, Progress in Astronautics and Aeronautics, Vol. 129, edited by J.L. Junkins, AIAA, New York, NY, 1990.

14. Canavin, J.R., "The Control of Spacecraft Vibrations Using Multivariable Output Feedback," AIAA/AAS Astrodynamics Conference, Palo Alto, CA, Aug. 7-9, 1978.

15. Goh, C.J. and Caughey, T.K., "On the Stability Problem Caused by Finite Actuator Dynamics in the Collocated Control of Large Space Structures," *International Journal of Control,* Vol. 41, No. 3, 1985, pp. 787-802.

16. Joshi, S.M., *Control of Large Flexible Space Structures,* Lecture Notes in Control and Information Sciences, Vol. 131, Springer-Verlag, New York, NY, 1989.

17. Junkins, J.L. and Turner, J.D., *Optimal Spacecraft Rotational Maneuvers,* Elsevier, Amsterdam, 1986.

18. Juang, J.N., Horta, L.G., and Robertshaw, N.H., "A Slewing Control Experiment for Flexible Structures," *Proceedings of the 5th VPI & SU Symposium on Dynamics and Control of Large Structures,* Virginia Polytechnic Institute and State University, Blacksburg, VA, 1985, pp. 547-551.

19. Fujii, H., Ohtsuka, T., and Udou, S., "Mission Function Control for Slew Maneuver Experiment," *Journal of Guidance, Control, and Dynamics,* Vol. 12, No. 6, Nov.-Dec. 1989, pp. 858-865.

20. Wie. B., Weiss, H., and Araposthathis, A., "Quaternion Feedback for Spacecraft Eigenaxis Rotations," *Journal of Guidance, Control, and Dynamics,* Vol. 12, No. 3, May-June 1989, pp. 375-380.

21. Thompson, R.C., Junkins, J.L., and Vadali, S.R., "Near-Minimum Time Open-Loop Slewing of Flexible Vehicles," *Journal of Guidance, Control, and Dynamics,* Vol. 12, No. 1, Jan.-Feb. 1989, pp. 82-88.

22. Vadali, S.R., "Feedback Control of Space Structures: A Lyapunov Approach," *Mechanics and Control of Large Flexible Structures,* Chapter 24, Progress in Astronautics and Aeronautics, Vol. 129, edited by J.L. Junkins, AIAA, New York, NY, 1990.

23. Meirovitch, L. and Quinn, R., "Maneuvering and Vibration Control of Flexible Spacecraft," *Journal of Astronautical Sciences,* Vol. 35, No. 3, 1987, pp.301-328.

24. Singh, G., Kabamba, P., and McClamroch, N., "Planar Time Optimal Slewing Maneuvers of Flexible Spacecraft," *Journal of Guidance, Control, and Dynamics,* Vol. 12, No. 1, Jan.-Feb. 1989, pp.71-81.

25. Breakwell, J.A., "Optimal Feedback Control for Flexible Spacecraft," *Journal of Guidance, Control, and Dynamics,* Vol. 4, No. 5, 1981, pp. 427-479.

26. Slotine, J. E. and Weiping, L., *Applied Nonlinear Control,* Prentice-Hall, 1991, New York, NY.

27. VanderVelde, W. and He, J., "Design of Space Structure Control Systems Using On-Off Thrusters," *Journal of Guidance, Control, and Dynamics,* Vol. 6, No 1, 1983, pp. 759-775.

28. Byers, R. M., Vadali, S. R., and Junkins, J. L., "Near-Minimum-Time Closed-Loop Slewing of Flexible Spacecraft," *Journal of Guidance, Control, and Dynamics,* Vol. 13, No 1, 1990, pp. 57-65.

29. LaSalle, J. and Lefschetz, S., *Stability by Lyapunov's Direct Method with Applications*, Academic Press, New York, NY, 1961.

30. Mukherjee, R., and Chen, D., "An Asymptotic Stability Theorem for Autonomous Systems," preprint 1992, to appear, AIAA *Journal of Guidance, Control, and Dynamics.*

31. Mukherjee, R., and Junkins, J.L., "An Invariant Set Analysis of the Hub-Appendage Problem," preprint 1992, to appear, AIAA *Journal of Guidance, Control, and Dynamics.*

CHAPTER 4

Mathematical Models of Flexible Structures

The purpose of this chapter is to develop a number of the most fundamental methods for formulating the equations of motion for vehicles and structures which are modeled as an elastic continuum, or a general collection of deformable bodies, rigid bodies, and particles. We develop mathematical models for these configurations in the form of differential equation systems; these differential equations may be ordinary, partial, or hybrid, and they are, in general, nonlinear as is evident in the example applications throughout this chapter. Broadly speaking, there are two issues: How to efficiently formulate valid differential equations, and how to solve these differential equations. Both issues are addressed in this chapter.

In section 4.1, the Lagrangian approach to formulating the differential equations of motion is introduced. This approach is widely used to derive the differential equations that govern the motion of discrete dynamical systems. Section 4.2 presents some classical and novel developments associated with the *extended Hamilton's principle* for deriving the equations of motion of distributed parameter systems. Of special note, section 4.2.2 presents an *explicit generalization of Lagrange's equations*, which generate the hybrid systems of ordinary and partial differential equations governing the dynamics of a family of flexible multibody configurations. This formulation provides an efficient alternative path (to the tedious integration-by-parts that ensue from conventional application of Hamilton's principle, for example); only differentiation of energy functions is required to generate the hybrid ordinary/partial/integral equations of motion and boundary conditions. The analytical developments are supported by several examples having varying degrees of generality, illustrating the utility and ease-of-application of these generalizations of Lagrange's equations.

The *differential eigenvalue problem* associated with systems described by partial differential equations is discussed in section 4.2.3. To help illuminate the properties of the differential eigenvalue problem, several classical structures with simple geometry and boundary conditions are considered, so that we can determine analytically the eigenfunctions that satisfy the governing partial equations and associated boundary conditions. Also presented in this section are some insights useful in numerically solving oscillatory/divergent characteristic equations for the higher eigenvalues.

Two spatial discretization methods (the so-called *assumed modes method*, and the *finite element method*) for obtaining approximate finite-dimensional ordinary differential equation models for flexible structure dynamics are discussed and compared in section 4.3. Finally, in section 4.4, some approximation and convergence issues for modeling dynamics of large flexible space structures are discussed. Special attention is focused on the consequences of neglecting coupling effects between the overall rigid body motion and the vibration of flexible appendages, for flexible structures having various degrees of flexibility. It is shown in these examples that closed-loop instability can result directly as a consequence of such modeling errors, and that poor choice of spatial basis functions in the "assumed modes" method can result in apparent "false convergence" of the structural model.

4.1 Lagrangian Approach to Equation of Motion Formulation

It is well known (Refs. 1-6) that Hamilton's principle is an attractive starting point to derive equations of motion for both lumped and distributed parameter systems. Hamilton's principle is perhaps the most famous variational principle of mechanics, and is conceptually easy to apply since it employs variations of kinetic energy and potential energy that are scalar functions of the system's state variables. While the idea is conceptually straightforward, the algebra and calculus are tedious for many applications. In particular, the integration by parts required in the taking of the variations for distributed parameter systems, especially for multibody problems, presents a practical difficulty that limits the utility of this approach. However, the development of the explicit generalization of Lagrange's equations, in section 4.2.2, circumvents these difficulties for a significant family of problems. We will begin this chapter by reviewing the generalized Hamilton's principle which, in its most general form, is given by the variational statement

$$\int_{t_1}^{t_2} \delta(T - V)dt + \int_{t_1}^{t_2} \delta W_{nc} dt = 0 \tag{4.1}$$

where T is the total kinetic energy of the system, V is the potential energy of the system (including the strain energy and the potential energy of conservative external forces), $\delta(\;)$ represents the virtual displacement (or variational) operator, and δW_{nc} is the virtual work done by nonconservative forces (including damping forces and external forces not accounted for in V).

Hamilton's principle is essentially an integrated form of d'Alembert's principle (Refs. 2, 7), and is applicable to the same wide variety of mechanical systems. However, the most common form of Hamilton's principle applies to a more restricted class of system (most mechanical and structural systems have belong to this class). Specifically, the kinetic energy is expressed in terms of the generalized coordinates and their first time derivatives, the potential energy is expressed in terms of generalized coordinates alone, and the virtual work done by nonconservative forces is expressed as a linear function of virtual displacements in generalized coordinates, as follows:

$$\begin{aligned} T &= T(q_1, q_2, \ldots, q_N, \dot{q}_1, \dot{q}_2, \ldots, \dot{q}_N, t) \\ V &= V(q_1, q_2, \ldots, q_N, t) \\ \delta W_{nc} &= Q_1 \delta q_1 + Q_2 \delta q_2 + \cdots + Q_N \delta q_N \end{aligned} \tag{4.2}$$

where Q_i denote generalized forces, and q_i are generalized coordinates defined as any set of N independent variables (that are sufficient to completely specify the inertial position of every element of mass in the system). For discrete systems, the q_i are functions of time only whereas, for distributed parameter systems, some or all of these variables are functions of both space and time. Also, for distributed parameter systems, the above assumed T and V forms must be generalized to include spatial derivatives and integrals over the elastic domains, as we shall see in subsequent developments. For the present, we restrict the discussion to the case of discrete coordinate systems and consider the independent generalized coordinates $q_i = q_i(t)$ to be functions of time only.

Note that the coordinate variations δq_i are defined as *virtual displacements*, that are independent, infinitesimal, arbitrary changes in the coordinates of a system, but which should be compatible with the constraints on the system. The admissible variations are also constrained such that the initial and final variations vanish as $\delta q_i(t_1) = \delta q_i(t_2) = 0$; this requirement is imposed to be consistent with the two-point boundary value version of Hamilton's principle.

Substituting Eqs. (4.2) into Eq. (4.1), taking variations using the chain rule, and integrating by parts, yield Lagrange's equations (Refs. 1-6); these are widely used to derive the equations of motion for holonomic discrete coordinate systems:

$$\frac{d}{dt}\left(\frac{\partial \mathcal{L}}{\partial \dot{q}_i}\right) - \frac{\partial \mathcal{L}}{\partial q_i} = Q_i, \qquad i = 1, 2, \ldots, N \tag{4.3}$$

where the *Lagrangian* $\mathcal{L}$ is defined in the classical form $\mathcal{L} = T - V$. Restrictions imposed in deriving Eqs. (4.3) are such that the coordinates q_i are independent functions of time only and that T, V, and δW_{nc} have the forms in Eqs. (4.2). Thus, Eqs. (4.3) are valid for nonlinear systems as well as linear systems.

A modest generalization allows Eqs. (4.3) to be applied to significant class of constrained systems (i.e., the coordinates q_i are not independent); for kinematic constraints which depend on the qs and their time derivatives, Lagrange multipliers can be introduced to generate additive generalized constraint forces on the right-hand side of Eqs. (4.3) (Refs. 2-5). In particular, for m Pfaffian (linear in the generalized velocities) constraints of the matrix form $A\dot{\mathbf{q}} + \mathbf{a}_o = 0$, the generalized constraint force that needs to be added to the right-hand side of Eqs. (4.3) is the ith element of the vector $A^T\lambda$, where $\mathbf{q}$ is an $N \times 1$ vector containing the generalized coordinates, $A = A(\mathbf{q})$ is an $m \times n$ continuous, differentiable matrix function, $\mathbf{a}_o(\mathbf{q})$ is an $m \times 1$ vector function, and λ is an $m \times 1$ vector of Lagrange multipliers. One standard solution process is to differentiate $A\dot{\mathbf{q}} + \mathbf{a}_o = 0$ to obtain $A\ddot{\mathbf{q}} + \dot{A}\dot{\mathbf{q}} + \dot{\mathbf{a}}_o = 0$; this latter equation is solved simultaneously with Eqs. (4.3) for $\ddot{\mathbf{q}}$ and λ, to determine the coordinate accelerations and constraint forces as a function of the qs and their time derivatives. Note that the N differential equations of Eqs. (4.3) must be solved simultaneously with the m algebraic equations $A\dot{\mathbf{q}} + \mathbf{a}_o = 0$ in order to determine the N+m unknowns in the vectors $\mathbf{q}(t)$ and $\lambda(t)$.

One of the most important nonconservative forces is *viscous damping*, which is proportional to the generalized velocities $\dot{q}_i$. It is possible to represent the entire set of viscous damping forces as a single scalar using the *Rayleigh dissipation function* (Ref. 4), in the form

$$\mathcal{F} = \frac{1}{2}\sum_{i=1}^{n}\sum_{j=1}^{n} c_{ij}\dot{q}_i\dot{q}_j \tag{4.4}$$

where the constant coefficients c_{ij} are symmetric in i and j. In a manner analogous to the "automatic generation" of conservative forces via taking the gradient of the scalar potential energy function, viscous damping forces can be obtained from the gradient of the Rayleigh dissipation function as $Q_j = -\frac{\partial \mathcal{F}}{\partial \dot{q}_j}$. Assuming that the nonconservative forces in Eqs. (4.3) have been divided into the viscous dissipative forces (derivable from the gradient of $\mathcal{F}$) and all other forces acting on the system (absorbed into the redefined generalized forces Q_j), a special form of Lagrange's equations can be obtained as follows:

$$\frac{d}{dt}\left(\frac{\partial \mathcal{L}}{\partial \dot{q}_i}\right) - \frac{\partial \mathcal{L}}{\partial q_i} + \frac{\partial \mathcal{F}}{\partial \dot{q}_i} = Q_i, \qquad i = 1, 2, \ldots, N \tag{4.5}$$

We emphasize that the generalized forces Q_i in Eqs. (4.5) include only the external forces not accounted for in $\mathcal{L}$ and $\mathcal{F}$.

Example 4.1

Consider the spring-mass-dashpot system shown in Fig. 4.1. The inertial displacements of the masses are taken as generalized coordinates. The kinetic energy, potential energy, and Raleigh dissipation function are therefore as follows:

$$T = \frac{1}{2}\left(m_1\dot{y}_1^2 + m_2\dot{y}_2^2 + m_3\dot{y}_3^2\right)$$

$$V = \frac{1}{2}\left\{k_1 y_1^2 + k_2(y_2 - y_1)^2 + k_3(y_3 - y_2)^2\right\}$$

$$\mathcal{F} = \frac{1}{2}\left\{c_1\dot{y}_1^2 + c_2(\dot{y}_2 - \dot{y}_1)^2 + c_3(\dot{y}_3 - \dot{y}_2)^2\right\}$$

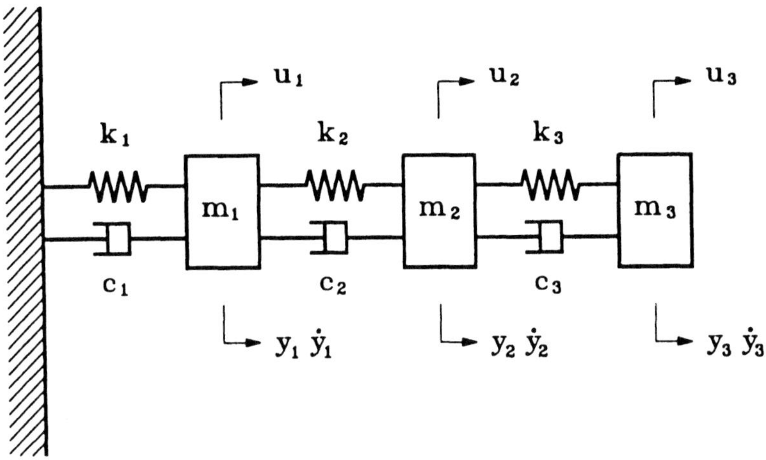

Figure 4.1 Lumped Parameter System.

The virtual work done by external forces u_i is $\delta W_{nc} = u_1\delta y_1 + u_2\delta y_2 + u_3\delta y_3$. Using the appropriate special case of Lagrange's equations, the differential equations of motion can be obtained immediately from Eq. (4.5) with $\{q_1, q_2, q_3\} = \{y_1, y_2, y_3\}$:

$$m_1\ddot{y}_1 + (c_1 + c_2)\dot{y}_1 - c_2\dot{y}_2 + (k_1 + k_2)y_1 - k_2 y_2 = u_1$$

$$m_2\ddot{y}_2 - c_2\dot{y}_1 + (c_2 + c_3)\dot{y}_2 - c_3\dot{y}_3 - k_2 y_1 + (k_2 + k_3)y_2 - k_3 y_3 = u_2$$

$$m_3\ddot{y}_3 - c_3\dot{y}_2 + c_3\dot{y}_3 - k_3 y_2 + k_3 y_3 = u_3$$

These equations of motion can be written compactly in matrix second order form:

$$M\ddot{y} + C\dot{y} + Ky = Du$$

The mass (M), damping (C), stiffness (K), and control influence (D) matrices are

$$M = \begin{bmatrix} m_1 & 0 & 0 \\ 0 & m_2 & 0 \\ 0 & 0 & m_3 \end{bmatrix}, \qquad C = \begin{bmatrix} c_1 + c_2 & -c_2 & 0 \\ -c_2 & c_2 + c_3 & -c_3 \\ 0 & -c_3 & c_3 \end{bmatrix}$$

$$K = \begin{bmatrix} k_1 + k_2 & -k_2 & 0 \\ -k_2 & k_2 + k_3 & -k_3 \\ 0 & -k_3 & k_3 \end{bmatrix}, \qquad D = \begin{bmatrix} 1 & 0 & 0 \\ 0 & 1 & 0 \\ 0 & 0 & 1 \end{bmatrix}$$

and $\mathbf{y}$ and $\mathbf{u}$ denote vectors of configuration coordinates (y_1, y_2, y_3) and control inputs (u_1, u_2, u_3), respectively.

■

For a significant class of systems, the algebra and calculus required in a straightforward application of Lagrange's equations can be dramatically reduced. For the the most common case of *natural* systems that the kinetic energy is a symmetric quadratic form in the generalized coordinate time derivatives, one finds:

$$T = \frac{1}{2}\sum_{i=1}^{N}\sum_{j=1}^{N} m_{ij}(\mathbf{q})\,\dot{q}_i\dot{q}_j = \frac{1}{2}\dot{\mathbf{q}}^T M\dot{\mathbf{q}} \tag{4.6}$$

Note that $\mathbf{q}$ is a $N \times 1$ configuration vector of generalized coordinates. It is convenient to collect the mass matrix $M = M(\mathbf{q})$ *before* the differentiations implied by Lagrange's equations are carried out. Including the possiblity of Pfaffian nonholonomic constraints, the equations of motion follow from Eqs. (4.3) as the following N+m system of differential and algebraic equations:

$$M\ddot{\mathbf{q}} + \mathbf{G} + \frac{\partial V}{\partial \mathbf{q}} = \mathbf{Q} + A^T\lambda, \quad A\dot{\mathbf{q}} + \mathbf{a}_o = 0 \tag{4.7}$$

where $\frac{\partial V}{\partial \mathbf{q}}$ is the $N \times 1$ vector gradient of the potential energy function, and $\mathbf{G} = \mathbf{G}(\mathbf{q}, \dot{\mathbf{q}})$ is the $N \times 1$ vector:

$$\mathbf{G} = [\dot{\mathbf{q}}^T C^{(1)}\dot{\mathbf{q}} \ \cdots \ \dot{\mathbf{q}}^T C^{(N)}\dot{\mathbf{q}}]^T, \quad c_{jk}^{(i)} = \frac{1}{2}\left(\frac{\partial m_{ij}}{\partial q_k} + \frac{\partial m_{ik}}{\partial q_j} - \frac{\partial m_{jk}}{\partial q_i}\right)$$

and where the last equation that generates the typical element $c_{jk}^{(i)}$ of the $N \times N$ symmetric matrix $C^{(i)} = C^{(i)}(\mathbf{q})$ is the *Christoffel* operator. It apparent that deriving the equations of motion, for natural systems subject to Pfaffian nonholonomic constraints, has been reduced to formation of the kinetic energy to identify the mass matrix, then carrying out the indicated gradient operations on the mass matrix elements and the potential energy to form the vectors $\mathbf{G} = \mathbf{G}(\mathbf{q}, \dot{\mathbf{q}})$ and $\frac{\partial V}{\partial \mathbf{q}}$.

4.2 Infinite-Dimensional Models of Distributed Parameter Systems

4.2.1 Classical Application of Hamilton's Principle

From the most general continuum mechanics viewpoint, most flexible structures can be regarded as distributed parameter systems (continuous systems). If we apply the principles of classical mechanics to distributed parameter systems (DPS), we are most rigorously led to systems of partial differential equations that govern the motion of such systems. The motion of DPS thus rigorously evolves such that it satisfies the underlying systems of partial differential equations and the associated boundary conditions and, obviously, these systems have an infinite number of degrees of freedom (Refs. 2-6).

As an aside, the continuum viewpoint breaks down at very high mode numbers, clearly, when natural frequencies multiplied times any finite vibration amplitude exceeds the speed of light, we know that we've gone too far! Also, when computed modal vibration amplitudes are less than the mean free molecular path of the material from which the structure is manufactured, we have clearly carried the infinite-dimensional viewpoint to an absurd contusion. For all engineering structural dynamics and control applications, it is obviously justifiable to introduce finite-dimensional approximations, the only issue is justifying the process by which we arrive at the finite-dimensional models. However, since low-dimensioned engineering approximations are often introduced to idealize a system with many more disturbable degrees of freedom, it remains useful to explore the continuum-based formulations and to normalize the finite-dimensional approximations.

For simple continuous systems, the extended Hamilton's principle is widely used to derive the governing equations of motion, and the use of Hamilton's principle offers an advantage over, for example, Newton's laws, in that the spatial boundary conditions are obtained automatically from the resulting *transversality conditions.* As the classical example, let us consider the transverse vibration of an Euler-Bernoulli (fixed-free) beam as shown in Figure 4.2; for simplicity, shear deformation and rotary inertia are ignored to derive the equations of motion. For generality, however, we retain variable mass density and bending stiffness EI(x). Consistent with these constitutive and kinematic assumptions, the kinetic energy is

$$T = \frac{1}{2}\int_0^L \rho(x)\left(\frac{\partial y(x,t)}{\partial t}\right)^2 dx \tag{4.8}$$

where y(x,t) denotes the transverse displacement from the neutral axis, $\rho(x)$ is mass per unit length, and the potential energy (simple beam constitutive model) is

$$V = \frac{1}{2}\int_0^L EI(x)\left(\frac{\partial^2 y(x,t)}{\partial x^2}\right)^2 dx \tag{4.9}$$

where E is the (Young's) modulus of elasticity and I(x) is the cross-sectional area moment of inertia of the variable geometry beam about the neutral axis. The virtual work done by the nonconservative distributed force has the form

$$\delta W_{nc} = \int_0^L p(x,t)\delta y(x,t)dx \tag{4.10}$$

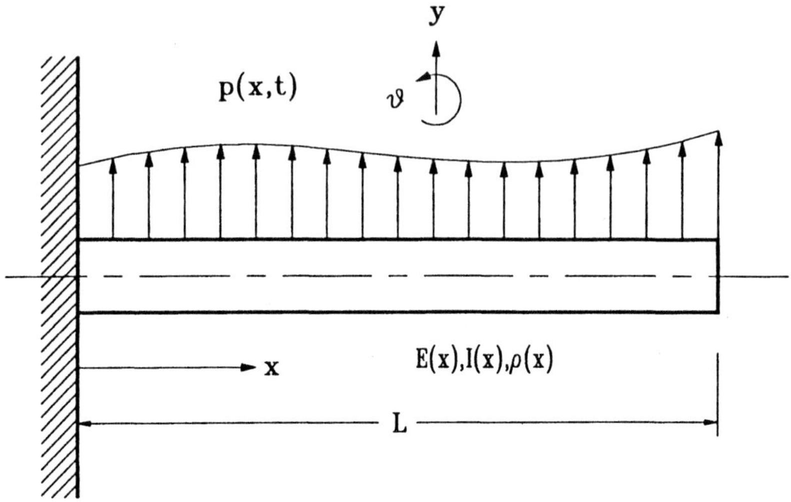

Figure 4.2 Transverse Vibration of Euler-Bernoulli Beam.

where p(x,t) is an arbitrary distributed external force and δy(x,t) is an arbitrary virtual displacement.

Substituting Eqs. (4.8)-(4.10) into the extended Hamilton's principle [Eq. (4.1)], taking the variation, integrating by parts with respect to t and x (to eliminate *all time and spatial derivatives* of δq and $\delta y(x,t)$, in favor of δq and $\delta y(x,t)$), and restricting attention to the class of virtual displacements that vanish at t_1, t_2, we obtain (after some algebra and calculus, note that the boundary terms arise due to the integration by parts)

$$\int_{t_1}^{t_2}\left[\int_0^L\left\{-\frac{\partial^2}{\partial x^2}\left(EI\frac{\partial^2 y}{\partial x^2}\right)-\rho\frac{\partial^2 y}{\partial t^2}+p\right\}\delta y\,dx + \frac{\partial}{\partial x}\left[EI\frac{\partial^2 y}{\partial x^2}\right]\delta y\Big|_0^L - EI\frac{\partial^2 y}{\partial x^2}\delta\frac{\partial y}{\partial x}\Big|_0^L\right]dt = 0 \tag{4.11}$$

Because of the arbitrariness of δy, Eq. (4.11) is satisfied for all possible δy only if the integrand vanishes identically for all s and t, we arrive at the partial differential the equation of motion:

$$\frac{\partial^2}{\partial x^2}\left(EI\frac{\partial^2 y}{\partial x^2}\right)+\rho\frac{\partial^2 y}{\partial t^2}=p, \qquad \text{for all t, and for} \qquad 0<x<L \tag{4.12}$$

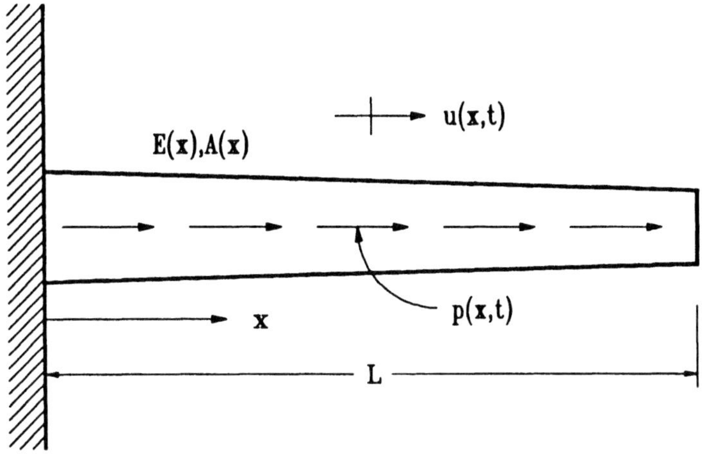

Figure 4.3 Axial Vibration of Rod.

The geometric constraints of a clamped-free beam require that y and $\partial y/\partial x$ be zero at the left end (x=0); moreover, the variations in the deflection and slope, δy and $\delta\left(\partial y/\partial x\right)$ are generally unknown and not zero at the free right end (x=L). Hence, the *geometric* and *natural* spatial boundary conditions are

$$\begin{aligned}&\text{For all t, at } x=0: \quad y(0,t)=0, \quad \frac{\partial y}{\partial x}|_0 = 0\\ &\text{For all t, at } x=L: \quad \frac{\partial}{\partial x}\left(EI\frac{\partial^2 y}{\partial x^2}\right)|_L = 0, \quad EI\frac{\partial^2 y}{\partial x^2}|_L = 0\end{aligned} \tag{4.13}$$

Example 4.2

Let us consider an elastic rod in axial vibration with distributed axial load shown in Fig. 4.3. The kinetic and potential energies are given by

$$T = \frac{1}{2}\int_0^L \rho(x)\left(\frac{\partial u(x,t)}{\partial t}\right)^2 dx, \qquad V = \frac{1}{2}\int_0^L EA(x)\left(\frac{\partial u(x,t)}{\partial x}\right)^2 dx$$

where u(x,t) is the displacement of the cross section along the axial direction, and where A(x) is the cross-sectional area. The virtual work done by a generally nonconservative distributed force $p(x,t)$ is given by

$$\delta W_{nc} = \int_0^L p(x,t)\delta u \, dx$$

Observe that $p(x,t)$ is the externally applied axial force per unit length, and $\delta u(x,t)$ is the axial virtual displacement.

The equations of motion and boundary conditions are obtained by means of the extended Hamilton's principle as follows:

$$\frac{\partial}{\partial x}\left(EA\frac{\partial u}{\partial x}\right) + p = \rho\frac{\partial^2 u}{\partial t^2}, \qquad 0 < x < L$$

$$u = 0 \quad \text{at } x{=}0, \qquad EA\frac{\partial u}{\partial x} = 0 \quad \text{at } x{=}L$$

■

4.2.2 Explicit Generalizations of Lagrange's Equations for Hybrid Coordinate Systems

Motivation

The integration by parts and the associated algebra implicit in applying Hamilton's principle [for example, the process required to obtain Eqs. (4.11)-(4.13) from Eq. (4.1), for the specific form of the energy functions of Eqs. (4.8)-(4.10)] are rather tedious, and the analogous developments for more complicated problems are a serious practical obstacle that limits applications of this approach to a rather small class of the systems encountered in applications. It would be attractive, indeed, if one had to carry out only *explicitly implied derivative operations* on energy functions (instead of integration by parts) to obtain the hybrid system of ordinary and partial differential equations of motion. Essentially, we seek to generalize the classical Lagrange's equations [e.g., Eqs. (4.3)] for discrete coordinate systems to obtain analogous equations that apply to distributed parameter systems. [This would be especially attractive for more mature analysts because, on the average, integration by parts ceases to be fun sometime around age forty-five. We have arrived at this qualitative "folk theorem" based on some of our own empirical experience coupled with the fact that we have found (inconclusive) evidence that Lagrange stopped doing integration by parts between age forty-five and fifty. We feel that it would be presumptous for anyone to (*even pretend to*) enjoy integration by parts longer than Lagrange did!]

Of course, completely generalizing the Lagrangian approach to automatically generate the differential equations and boundary conditions for every conceivable multiflexible body system is indeed difficult and, as a practical matter, impossible. However, based on recent developments (Refs. 22, 23), we have been able to accomplish a significant generalization of Lagrange's equations for a large family of flexible body configurations. In the presentation below, we introduce a mathematical structure and define an argument list for the energy functions, such that they embody a family of flexible multibody problems. We carry out the integrations by parts symbolically [using the specified structure and arguments for the Lagrangian function] once and for all. The result is an explicit generalization of Eqs. (4.3) such that the hybrid partial differential equation version of Lagrange's equations is generated, and associated transversality condition variational statements are determined that specialize to give the natural boundary conditions.

We consider a family of multibody hybrid discrete/distributed parameter systems that can be regarded as consisting of a collection of interconnected rigid and elastic bodies, i.e., a chain configuration. For the present discussion, we assume that the geometry/mechanics of the joints permits us to treat them as point connections; this includes body-to-body connections such as pinned joints, clamped joints, or connections by any point-to-point mechanism. We admit discrete elements such as particles or rigid bodies, and we admit any constitutive laws for the deformable structures that depend continuously on distributed space-time deformation states, and the space-time derivatives thereof. At this point, no restrictions are introduced on body geometry, size of deformations, spatial loading, or constitutive models. Thus the resulting equations will be a hybrid coordinate system of ordinary/partial/integral differential equations. Such models are useful for dynamics and control analysis of flexible spacecraft and multilink flexible robotic systems as they undergo (for example) large-angle three-dimensional maneuvers with general reconfiguration and vibratory deformations. However they are generated, the equations of motion are usually hybrid, in the sense that the overall rigid body motions and any other discrete degrees of freedom are described by time-varying coordinates and the elastic motions are described by time- and space-varying coordinates; the resulting hybrid system of ordinary and partial integro-differential equations usually embodies significant coupling between the rigid body and elastic motions.

Basic Developments

We consider a hybrid coordinate dynamical system and assume that the Lagrangian can be written in the general form $\mathcal{L} = \mathcal{L}(t, P, q_i, \dot{q}_i, w_j, \dot{w}_j, w_j', w_j'')$, where $\{q_i = q_i(t),\ i = 1, 2, \ldots, m\}$ are generalized coordinates describing rigid body motions of the hybrid system and $\{w_j = w_j(P, t),\ j = 1, 2, \ldots, n\}$ are distributed coordinates describing elastic motions relative to the rigid body motions of an undeformed body-fixed spatial position P. We define $\mathbf{q}$, $\mathbf{w}$ generalized coordinate vectors $\mathbf{q} = [q_1, q_2, \ldots, q_m]^T$ and $\mathbf{w} = [w_1, w_2, \ldots, w_n]^T$. The overdots designate derivatives with respect to time and the primes designate derivatives with respect to spatial position.

First, we consider the case in which there is only one elastic domain Ω. For convenience, we assume that the Lagrangian consists of three terms such as

$$\mathcal{L} = \mathrm{L_D} + \int_\Omega \hat{\mathrm{L}}\, d\Omega + \mathrm{L_B},$$

where Ω is the spatial domain occupied by the undeformed flexible body. $\mathrm{L_D}$ is the discrete portion of $\mathcal{L}$, and is a function of $t, \mathbf{q}$ and $\dot{\mathbf{q}}$; $\hat{\mathrm{L}}$ is the Lagrangian density which is a function of $\mathbf{q}, \dot{\mathbf{q}}, \dot{\mathbf{w}}, \mathbf{w}, \mathbf{w}', \mathbf{w}''$ and (P, t); $\mathrm{L_B}$, which is a function of $\mathbf{w}(l), \dot{\mathbf{w}}(l), \mathbf{w}'(l), \dot{\mathbf{w}}'(l), \mathbf{q}$ and $\dot{\mathbf{q}}$, is the "boundary term" portion of the Lagrangian energy functional, which in turn depends on the boundary motions. Observe for notational compaction, $\mathbf{w}(l)$ denotes $\mathbf{w}(l, t)$, above and throughout this discussion.

Next, we consider the nonconservative virtual work of hybrid systems. The virtual work can be written in the form

$$\delta \mathrm{W_{nc}} = \mathbf{Q}^{\mathrm{T}} \delta \mathbf{q} + \int_\Omega \hat{\mathbf{f}}^{\mathrm{T}} \delta \mathbf{w}\, d\Omega + \mathbf{f}_1{}^{\mathrm{T}} \delta \mathbf{w}(l) + \mathbf{f}_2{}^{\mathrm{T}} \delta \mathbf{w}'(l)$$

Here, $\mathbf{Q}$ is the nonconservative generalized force vector associated with $\mathbf{q}$, $\hat{\mathbf{f}}$ is the nonconservative generalized force density vector associated with $\mathbf{w}$, $\delta\mathbf{q}$ and $\delta\mathbf{w}$ are associated virtual displacements and $\mathbf{f}_1{}^{\mathrm{T}}\delta\mathbf{w}(l)$ and $\mathbf{f}_2{}^{\mathrm{T}}\delta\mathbf{w}'(l)$ are the nonconservative virtual work which depend on the boundary forces and associated boundary virtual displacements. If $\mathbf{w}$ is a linear displacement vector, then $\mathbf{f}_1$ is the nonconservative boundary force vector associated with $\delta\mathbf{w}(l)$ and $\mathbf{f}_2$ is the nonconservative boundary torque vector associated with $\delta\mathbf{w}'(l)$.

Hybrid Coordinate Systems Having One Spatial Independent Variable

Now we consider the simplest case of a dynamical system with discrete bodies (particles and rigid bodies) and a deformable member whose deformation state can be described using only one spatial variable. We derive here an explicit version of Lagrange's equations for this class of hybrid systems.

Usually, one spatial independent variable is used to identify the undeformed flexible body position using one deformation dependent variable and only one associated deformable structural member. For certain symmetric configurations, for example, (see Ref. 28), one spatial variable and the associated deformation coordinates can represent multiple domains if symmetric or antisymmetric deformation is assumed. Also, considering beam and rodlike bodies, we can introduce several dependent time- and space-varying generalized coordinates to describe in-plane bending, out-of-plane bending, shear deformation, torsion, and axial deformation, and so on, even though we employ a single spatial independent variable.

We assume that the kinetic energy, potential energy, and Lagrangian are expressed by Eqs.(4.14), (4.15), and (4.16), respectively. To simplify expressions, we do not show independent variable arguments, thus $\mathbf{q} = \mathbf{q}(t)$, $\mathbf{w} = \mathbf{w}(x,t)$ and $\mathbf{w}(l)$ represents $\mathbf{w}(x,t)|_l$. Also, we suppress the appearance of t everywhere for notational compaction. Adopting these conventions, the kinetic and potential energy functions are given by

$$T = T_D(\mathbf{q}, \dot{\mathbf{q}}) + \int_{l_0}^{l} \hat{T}(\text{arg})dx + T_B(\mathbf{w}(l), \dot{\mathbf{w}}(l), \mathbf{w}'(l), \dot{\mathbf{w}}'(l), \mathbf{q}, \dot{\mathbf{q}}) \tag{4.14}$$

$$V = V_D(\mathbf{q}, \dot{\mathbf{q}}) + \int_{l_0}^{l} \hat{V}(\text{arg})dx + V_B(\mathbf{w}(l), \dot{\mathbf{w}}(l), \mathbf{w}'(l), \dot{\mathbf{w}}'(l), \mathbf{q}, \dot{\mathbf{q}}) \tag{4.15}$$

where $\text{arg} \equiv \{\mathbf{q}, \dot{\mathbf{q}}, \dot{\mathbf{w}}, \mathbf{w}, \mathbf{w}', \mathbf{w}''\}$, so that

$$\mathcal{L} = T - V = L_D(\mathbf{q}, \dot{\mathbf{q}}) + \int_{l_0}^{l} \hat{L}(\text{arg})dx + L_B(\mathbf{w}(l), \dot{\mathbf{w}}(l), \mathbf{w}'(l), \dot{\mathbf{w}}'(l), \mathbf{q}, \dot{\mathbf{q}}) \tag{4.16}$$

where $L_D = T_D - V_D$, $\hat{L} = \hat{T} - \hat{V}$, $L_B = T_B - V_B$.

The nonconservative virtual work of the hybrid system is given by

$$\delta W_{nc} = \mathbf{Q}^T \delta\mathbf{q} + \int_{l_0}^{l} \hat{\mathbf{f}}^T(x)\delta\mathbf{w}(x)dx + \mathbf{f}_1{}^T\delta\mathbf{w}(l) + \mathbf{f}_2{}^T\delta\mathbf{w}'(l) \tag{4.17}$$

where $\mathbf{f}_1$ is the generally nonconservative force vector applied at the boundary and $\mathbf{f}_2$ is the nonconservative torque vector applied at the boundary. Note that it is easy to admit other point forces and moments. However, we have considered here the most common case of boundary loading, and admitted the possibility of coupling with discrete coordinate subsystems on the boundaries; we have not considered other possibilities to simplify notation.

The extended Hamilton's principle can be stated as

$$\int_{t_1}^{t_2} (\delta\mathcal{L} + \delta W_{nc})dt = 0, \quad \text{with } \delta\mathbf{q} = \delta\mathbf{w} = 0 \text{ at arbitrary times } t = t_1, t_2 \tag{4.18}$$

Using Eq. (4.16), carrying out the variation of the Lagrangian $\mathcal{L}$ yields

$$\int_{t_1}^{t_2} \delta\mathcal{L}dt = \int_{t_1}^{t_2} \left[\frac{\partial L_D}{\partial \mathbf{q}}\delta\mathbf{q} + \frac{\partial L_D}{\partial \dot{\mathbf{q}}}\delta\dot{\mathbf{q}} \right.$$

$$+\int_{l_0}^{l}\left(\frac{\partial \hat{L}}{\partial \mathbf{q}}\delta\mathbf{q}+\frac{\partial \hat{L}}{\partial \dot{\mathbf{q}}}\delta\dot{\mathbf{q}}+\frac{\partial \hat{L}}{\partial \dot{\mathbf{w}}}\delta\dot{\mathbf{w}}+\frac{\partial \hat{L}}{\partial \mathbf{w}}\delta\mathbf{w}+\frac{\partial \hat{L}}{\partial \mathbf{w}'}\delta\mathbf{w}'+\frac{\partial \hat{L}}{\partial \mathbf{w}''}\delta\mathbf{w}''\right)\mathrm{dx}$$
$$+\frac{\partial L_B}{\partial \mathbf{q}}\delta\mathbf{q}+\frac{\partial L_B}{\partial \dot{\mathbf{q}}}\delta\dot{\mathbf{q}}+\frac{\partial L_B}{\partial \mathbf{w}(l)}\delta\mathbf{w}(l)+\frac{\partial L_B}{\partial \dot{\mathbf{w}}(l)}\delta\dot{\mathbf{w}}(l)$$
$$\left.+\frac{\partial L_B}{\partial \mathbf{w}'(l)}\delta\mathbf{w}'(l)+\frac{\partial L_B}{\partial \dot{\mathbf{w}}'(l)}\delta\dot{\mathbf{w}}'(l)\right]\mathrm{dt} \tag{4.19}$$

The symbolic integration by parts is tedious (see Refs. 22-23) but can be carried to completion to obtain the following results. Equation (4.18) can be written as

$$\int_{t_1}^{t_2}(\delta\mathcal{L}+\delta \mathrm{W_{nc}})\mathrm{dt}$$
$$=\int_{t_1}^{t_2}\left[\frac{\partial L_D}{\partial \mathbf{q}}-\frac{\mathrm{d}}{\mathrm{dt}}\left(\frac{\partial L_D}{\partial \dot{\mathbf{q}}}\right)+\int_{l_0}^{l}\frac{\partial \hat{L}}{\partial \mathbf{q}}\mathrm{dx}-\int_{l_0}^{l}\frac{\mathrm{d}}{\mathrm{dt}}\left(\frac{\partial \hat{L}}{\partial \dot{\mathbf{q}}}\right)\mathrm{dx}\right.$$
$$\left.+\frac{\partial L_B}{\partial \mathbf{q}}-\frac{\mathrm{d}}{\mathrm{dt}}\left(\frac{\partial L_B}{\partial \dot{\mathbf{q}}}\right)+\mathbf{Q}^{\mathrm{T}}\right]\delta\mathbf{q}\,\mathrm{dt}$$
$$+\int_{t_1}^{t_2}\int_{l_0}^{l}\left[-\frac{\mathrm{d}}{\mathrm{dt}}\left(\frac{\partial \hat{L}}{\partial \dot{\mathbf{w}}}\right)+\frac{\partial \hat{L}}{\partial \mathbf{w}}-\frac{\partial}{\partial \mathrm{x}}\left(\frac{\partial \hat{L}}{\partial \mathbf{w}'}\right)+\frac{\partial^2}{\partial \mathrm{x}^2}\left(\frac{\partial \hat{L}}{\partial \mathbf{w}''}\right)+\hat{\mathbf{f}}^{\mathrm{T}}\right]\delta\mathbf{w}\,\mathrm{dx}\,\mathrm{dt}$$
$$+\int_{t_1}^{t_2}\left[\left\{\frac{\partial \hat{L}}{\partial \mathbf{w}'}-\frac{\partial}{\partial \mathrm{x}}\left(\frac{\partial \hat{L}}{\partial \mathbf{w}''}\right)\right\}\delta\mathbf{w}\Big|_{l_0}^{l}+\left\{\frac{\partial L_B}{\partial \mathbf{w}(l)}-\frac{\mathrm{d}}{\mathrm{dt}}\left(\frac{\partial L_B}{\partial \dot{\mathbf{w}}(l)}\right)\right\}\delta\mathbf{w}(l)+\mathbf{f}_1{}^{\mathrm{T}}\delta\mathbf{w}(l)\right]\mathrm{dt}$$
$$+\int_{t_1}^{t_2}\left[\frac{\partial \hat{L}}{\partial \mathbf{w}''}\delta\mathbf{w}'\Big|_{l_0}^{l}+\left\{\frac{\partial L_B}{\partial \mathbf{w}'(l)}-\frac{\mathrm{d}}{\mathrm{dt}}\left(\frac{\partial L_B}{\partial \dot{\mathbf{w}}'(l)}\right)\right\}\delta\mathbf{w}'(l)+\mathbf{f}_2{}^{\mathrm{T}}\delta\mathbf{w}'(l)\right]\mathrm{dt}$$
$$=0 \tag{4.20}$$

In the above equation and elsewhere in the present chapter, if F is a function of t, x, $\mathbf{q}(\mathrm{t})$, $\dot{\mathbf{q}}(\mathrm{t})$, $\mathbf{w}(\mathrm{x,t})$, $\dot{\mathbf{w}}(\mathrm{x,t})$, $\mathbf{w}'(\mathrm{x,t})$, $\mathbf{w}''(\mathrm{x,t})$, $\mathbf{w}(1)$, $\dot{\mathbf{w}}(1)$, $\mathbf{w}'(1)$, and $\dot{\mathbf{w}}'(l)$, then the derivative dF/dt is defined using the extended chain rule as

$$\frac{\mathrm{dF}}{\mathrm{dt}}\equiv\frac{\partial \mathrm{F}}{\partial \mathrm{t}}+\frac{\partial \mathrm{F}}{\partial \mathbf{q}}\frac{\mathrm{d}\mathbf{q}}{\mathrm{dt}}+\frac{\partial \mathrm{F}}{\partial \dot{\mathbf{q}}}\frac{\mathrm{d}\dot{\mathbf{q}}}{\mathrm{dt}}+\frac{\partial \mathrm{F}}{\partial \mathbf{w}}\frac{\partial \mathbf{w}}{\partial \mathrm{t}}+\frac{\partial \mathrm{F}}{\partial \dot{\mathbf{w}}}\frac{\partial \dot{\mathbf{w}}}{\partial \mathrm{t}}+\frac{\partial \mathrm{F}}{\partial \mathbf{w}'}\frac{\partial \mathbf{w}'}{\partial \mathrm{t}}+\frac{\partial \mathrm{F}}{\partial \mathbf{w}''}\frac{\partial \mathbf{w}''}{\partial \mathrm{t}}$$
$$+\frac{\partial \mathrm{F}}{\partial \mathbf{w}(l)}\frac{\mathrm{d}\mathbf{w}(l)}{\mathrm{dt}}+\frac{\partial \mathrm{F}}{\partial \dot{\mathbf{w}}(l)}\frac{\mathrm{d}\dot{\mathbf{w}}(1)}{\mathrm{dt}}+\frac{\partial \mathrm{F}}{\partial \mathbf{w}'(l)}\frac{\mathrm{d}\mathbf{w}'(1)}{\mathrm{dt}}+\frac{\partial \mathrm{F}}{\partial \dot{\mathbf{w}}'(l)}\frac{\mathrm{d}\dot{\mathbf{w}}'(1)}{\mathrm{dt}}$$

In a strict sense, dF/dt should be considered a partial derivative (in our original developments, we denoted dF/dt by $\delta\mathrm{F}/\delta\mathrm{t}$), since both t and x are independent variables. If t does not appear explicitly, then of course $\partial F/\partial t = 0$. If F does not depend on $\mathbf{w}$ (and derivatives thereof), then obviously dF/dt reduces the first three terms which is our usual interpretation of dF/dt as the total derivative of $F(t, \mathbf{q}, \dot{\mathbf{q}})$.

Using the usual arguments on the arbitrariness and independence of the variations $\delta\mathbf{q}(t)$, $dw(x,t)$ and the boundary variations, the above equation gives the governing differential equations and the boundary conditions. First, we consider the integrand associated with $\delta\mathbf{q}$. Since L is expressed by Eq. (4.16), the first term of Eq. (4.20) must vanish at every instant in time for all $\delta\mathbf{q}$. This gives the classical form of Lagrange's equations as

$$\frac{d}{dt}\left(\frac{\partial \mathcal{L}}{\partial \dot{\mathbf{q}}}\right) - \frac{\partial \mathcal{L}}{\partial \mathbf{q}} = \mathbf{Q}^T \tag{4.21}$$

While we see that the discrete coordinates satisfy the usual form of Lagrange's equations, however, because of the presence of the elastic coordinates, their partial derivatives, and the integration of Eq. (4.16) over the elastic domain, the Lagrangian $\mathcal{L}$ and therefore the resulting differential equations must be considered to be *hybrid* (ordinary/partial/integral) differential equations rather than the usual ordinary differential equations generated by the classical Lagrange's equations.

To obtain the associated partial differential equations governing $\mathbf{w}(x,t)$, we consider the integrand associated with $\delta\mathbf{w}$. The second term on the right side of Eq. (4.20) must vanish for all $\delta\mathbf{w}$ at every x and t, and therefore we obtain the following partial differential equations

$$\frac{d}{dt}\left(\frac{\partial \hat{L}}{\partial \dot{\mathbf{w}}}\right) - \frac{\partial \hat{L}}{\partial \mathbf{w}} + \frac{\partial}{\partial x}\left(\frac{\partial \hat{L}}{\partial \mathbf{w}'}\right) - \frac{\partial^2}{\partial x^2}\left(\frac{\partial \hat{L}}{\partial \mathbf{w}''}\right) = \hat{\mathbf{f}}^T \tag{4.22}$$

Equations (4.21) and (4.22) can be found in Ref. 22, 23, and 20, and a related formulation is derived in Ref. 24.

Next, we consider the boundary conditions. Since the remaining terms are zero in Eq. (4.20), we conclude that in order that the left side vanish, the boundary terms must also vanish independently. We therefore obtain the following symbolic variational statements from which the spatial boundary conditions can be obtained:

$$\left\{\frac{\partial \hat{L}}{\partial \mathbf{w}'} - \frac{\partial}{\partial x}\left(\frac{\partial \hat{L}}{\partial \mathbf{w}''}\right)\right\}\delta\mathbf{w}\Big|_{l_0}^{l} + \left\{\frac{\partial L_B}{\partial \mathbf{w}(l)} - \frac{d}{dt}\left(\frac{\partial L_B}{\partial \dot{\mathbf{w}}(l)}\right)\right\}\delta\mathbf{w}(l) + \mathbf{f}_1{}^T\delta\mathbf{w}(l) = 0 \tag{4.23}$$

$$\frac{\partial \hat{L}}{\partial \mathbf{w}''}\delta\mathbf{w}'\Big|_{l_0}^{l} + \left\{\frac{\partial L_B}{\partial \mathbf{w}'(l)} - \frac{d}{dt}\left(\frac{\partial L_B}{\partial \dot{\mathbf{w}}'(l)}\right)\right\}\delta\mathbf{w}'(l) + \mathbf{f}_2{}^T\delta\mathbf{w}'(l) = 0 \tag{4.24}$$

Equations (4.21) and (4.22) generate directly the coupled hybrid system of ordinary and partial differential equations, and the variational statements of Eqs. (4.23) and (4.24) directly generate the associated boundary conditions. Thus, we have established an explicit generalization of Lagrange's equations for nonconservative distributed parameter systems that have one elastic domain.

Generalizations for a Class of MultiBody Hybrid Coordinate Systems

Now we consider a more general case which has more than one elastic domain, i.e., we consider a point-connected system of n flexible bodies. We admit chain and tree topologies of bodies connected to each other with joints which can be idealized as point connections. For simplicity, we consider each (i.e., the ith) elastic body to be beamlike, with only one spatial variable (x_i). Boundary elements are admitted so that rigid bodies or masses can be concentrated, if desired, at points on the mutual boundaries of the elastic domains. Let us assume that the kinetic energy, potential energy, and Lagrangian are expressed by Eqs. (4.25), (4.26), and (4.27) respectively. We introduce notation for the n one-dimensional elastic domains Ω_i $(i = 1, \ldots, n)$, $\mathrm{x}_i \in \Omega_i$, where $l_{0_i} < \mathrm{x}_i \leq l_i$, and where $\hat{\mathrm{T}}^i$, $\hat{\mathrm{V}}^i$, and $\mathbf{w}_i$ are defined in domain D_i. For notational compaction, we define $\underline{\mathbf{w}}(\underline{l})$ as $\mathbf{w}_1(l_1), \mathbf{w}_2(l_2), \ldots, \mathbf{w}_n(l_n)$, and $\underline{\dot{\mathbf{w}}}(\underline{l})$, $\underline{\mathbf{w}}'(\underline{l})$. Note that $\underline{\dot{\mathbf{w}}}'(\underline{l})$ are defined in a similar manner. In the general case, $\hat{\mathrm{T}}^i$ and $\hat{\mathrm{V}}^i$ are functions of $\mathbf{q}$, $\dot{\mathbf{q}}$, $\mathbf{w}_i$, $\dot{\mathbf{w}}_i$, $\mathbf{w}_i'$, $\mathbf{w}_i''$, $\underline{\mathbf{w}}(\underline{l})$, $\underline{\dot{\mathbf{w}}}(\underline{l})$, $\underline{\mathbf{w}}'(\underline{l})$ and $\underline{\dot{\mathbf{w}}}'(\underline{l})$, and $\mathrm{T_B}$ and $\mathrm{V_B}$ are functions of $\underline{\mathbf{w}}(\underline{l})$, $\underline{\dot{\mathbf{w}}}(\underline{l})$, $\underline{\mathbf{w}}'(\underline{l})$, $\underline{\dot{\mathbf{w}}}'(\underline{l})$, $\mathbf{q}$, and $\dot{\mathbf{q}}$. We assume the following structure for T and V:

$$\mathrm{T} = \mathrm{T_D}(\mathbf{q}, \dot{\mathbf{q}}) + \sum_{i=1}^{n} \int_{l_{0_i}}^{l_i} \hat{\mathrm{T}}^i(\mathbf{q}, \dot{\mathbf{q}}, \dot{\mathbf{w}}_i, \mathbf{w}_i, \mathbf{w}_i', \mathbf{w}_i'', \underline{\mathbf{w}}(\underline{l}), \underline{\dot{\mathbf{w}}}(\underline{l}), \underline{\mathbf{w}}'(\underline{l}), \underline{\dot{\mathbf{w}}}'(\underline{l}))\mathrm{dx}_i$$
$$+\mathrm{T_B}(\underline{\mathbf{w}}(\underline{l}), \underline{\dot{\mathbf{w}}}(\underline{l}), \underline{\mathbf{w}}'(\underline{l}), \underline{\dot{\mathbf{w}}}'(\underline{l}), \mathbf{q}, \dot{\mathbf{q}}) \tag{4.25}$$

$$\mathrm{V} = \mathrm{V_D}(\mathbf{q}, \dot{\mathbf{q}}) + \sum_{i=1}^{n} \int_{l_{0_i}}^{l_i} \hat{\mathrm{V}}^i(\mathbf{q}, \dot{\mathbf{q}}, \dot{\mathbf{w}}_i, \mathbf{w}_i, \mathbf{w}_i', \mathbf{w}_i'', \underline{\mathbf{w}}(\underline{l}), \underline{\dot{\mathbf{w}}}(\underline{l}), \underline{\mathbf{w}}'(\underline{l}), \underline{\dot{\mathbf{w}}}'(\underline{l}))\mathrm{dx}_i$$
$$+\mathrm{V_B}(\underline{\mathbf{w}}(\underline{l}), \underline{\dot{\mathbf{w}}}(\underline{l}), \underline{\mathbf{w}}'(\underline{l}), \underline{\dot{\mathbf{w}}}'(\underline{l}), \mathbf{q}, \dot{\mathbf{q}}) \tag{4.26}$$

The Lagrangian is therefore given by

$$\begin{aligned} \mathcal{L} &= \mathrm{T} - \mathrm{V} \\ &= \mathrm{L_D}(\mathbf{q}, \dot{\mathbf{q}}) + \sum_{i=1}^{n} \int_{l_{0_i}}^{l_i} \hat{\mathcal{L}}^i(\mathbf{q}, \dot{\mathbf{q}}, \dot{\mathbf{w}}_i, \mathbf{w}_i, \mathbf{w}_i', \mathbf{w}_i'', \underline{\mathbf{w}}(\underline{l}), \underline{\dot{\mathbf{w}}}(\underline{l}), \underline{\mathbf{w}}'(\underline{l}), \underline{\dot{\mathbf{w}}}'(\underline{l}))\mathrm{dx}_i \\ &\quad +\mathrm{L_B}(\underline{\mathbf{w}}(\underline{l}), \underline{\dot{\mathbf{w}}}(\underline{l}), \underline{\mathbf{w}}'(\underline{l}), \underline{\dot{\mathbf{w}}}'(\underline{l}), \mathbf{q}, \dot{\mathbf{q}}) \end{aligned} \tag{4.27}$$

where $L_D = T_D - V_D$, $\hat{\mathrm{L}}^i = \hat{\mathrm{T}}^i - \hat{\mathrm{V}}^i$, and $\mathrm{L_B} = \mathrm{T_B} - \mathrm{V_B}$.

The nonconservative virtual work of the hybrid system is given by

$$\delta \mathrm{W_{nc}} = \mathbf{Q}^{\mathrm{T}}\delta\mathbf{q} + \sum_{i=1}^{n}\left(\int_{l_{0_i}}^{l_i} \hat{\mathbf{f}}^{i\,\mathrm{T}}(\mathrm{x}_i)\delta\mathbf{w}_i \mathrm{dx}_i + \mathbf{f}_1^{i\,\mathrm{T}}\delta\mathbf{w}_i(l_i) + \mathbf{f}_2^{i\,\mathrm{T}}\delta\mathbf{w}_i{}'(l_i)\right) \tag{4.28}$$

where $\mathbf{f}_1^i$ is the nonconservative force vector applied at the boundary (at $\mathrm{x}_i = l_i$) of domain Ω_i and $\mathbf{f}_2^i$ is the nonconservative torque vector applied at the boundary of domain Ω_i.

We use the extended Hamilton's principle which can be stated as

$$\int_{t_1}^{t_2} (\delta\mathcal{L} + \delta \mathrm{W_{nc}})\mathrm{dt} = 0, \qquad \delta\mathbf{q} = \delta\mathbf{w}_i = 0 \quad \text{at} \quad \mathrm{t} = \mathrm{t}_1, \mathrm{t}_2 \tag{4.29}$$

After carrying out the variation of L and symbolic integration by parts, see Ref. 23, then Eq. (4.29) can be written in the following compact form

$$\begin{aligned}
\int_{t_1}^{t_2} (\delta\mathcal{L} + \delta \mathrm{W_{nc}})\mathrm{dt} &= \int_{t_1}^{t_2} \left(\frac{\partial \mathcal{L}}{\partial \mathbf{q}} - \frac{\mathrm{d}}{\mathrm{dt}}\left(\frac{\partial \mathcal{L}}{\partial \dot{\mathbf{q}}}\right) + \mathbf{Q}^{\mathrm{T}}\right)\delta\mathbf{q}\mathrm{dt} \\
&+ \sum_{i=1}^{n}\int_{t_1}^{t_2}\int_{l_{0_i}}^{l_i}\left\{-\frac{\mathrm{d}}{\mathrm{dt}}\left(\frac{\partial \hat{\mathrm{L}}^i}{\partial \dot{\mathbf{w}}_i}\right) + \frac{\partial \hat{\mathrm{L}}^i}{\partial \mathbf{w}_i}\right. \\
&\qquad \left. - \frac{\partial}{\partial \mathrm{x_i}}\left(\frac{\partial \hat{\mathrm{L}}^i}{\partial \mathbf{w}_i{}'}\right) + \frac{\partial^2}{\partial \mathrm{x_i}^2}\left(\frac{\partial \hat{\mathrm{L}}^i}{\partial \mathbf{w}_i{}''}\right) + \hat{\mathbf{f}}^{i\,\mathrm{T}}\right\}\delta\mathbf{w}_i\mathrm{dx}_i\mathrm{dt} \\
&+ \sum_{i=1}^{n}\int_{t_1}^{t_2}\left[\left\{\frac{\partial \hat{\mathrm{L}}^i}{\partial \mathbf{w}_i{}'} - \frac{\partial}{\partial \mathrm{x_i}}\left(\frac{\partial \hat{\mathrm{L}}^i}{\partial \mathbf{w}_i{}''}\right)\right\}\delta\mathbf{w}_i\Big|_{l_{0_i}}^{l_i}\right. \\
&\qquad \left. + \left\{\frac{\partial \mathcal{L}_{\mathrm{B}}}{\partial \mathbf{w}_i(l_i)} - \frac{\mathrm{d}}{\mathrm{dt}}\left(\frac{\partial \mathcal{L}_{\mathrm{B}}}{\partial \dot{\mathbf{w}}_i(l_i)}\right)\right\}\delta\mathbf{w}_i(l_i) + \mathbf{f}_1^{i\mathrm{T}}\delta\mathbf{w}_i(l_i)\right]\mathrm{dt} \\
&+ \sum_{i=1}^{n}\int_{t_1}^{t_2}\left[\frac{\partial \hat{\mathrm{L}}^i}{\partial \mathbf{w}_i{}''}\delta\mathbf{w}_i{}'\Big|_{l_{0_i}}^{l_i}\right. \\
&\qquad \left. + \left\{\frac{\partial \mathcal{L}_B}{\partial \mathbf{w}_i{}'(l_i)} - \frac{\mathrm{d}}{\mathrm{dt}}\left(\frac{\partial \mathcal{L}_B}{\partial \dot{\mathbf{w}}_i'(l_i)}\right)\right\}\delta\mathbf{w}_i{}'(l_i) + \mathbf{f}_2^{i\mathrm{T}}\delta\mathbf{w}_i{}'(l_i)\right]\mathrm{dt} \\
&= 0
\end{aligned} \tag{4.30}$$

where

$$\mathcal{L}_{\mathrm{B}} \equiv \mathrm{L_B} + \sum_{i=1}^{n}\int_{l_{0_i}}^{l_i} \hat{\mathrm{L}}^i\mathrm{dx}_i \tag{4.31}$$

$$\mathcal{L} = \mathrm{L_D} + \sum_{i=1}^{n}\int_{l_{0_i}}^{l_i} \hat{\mathrm{L}}^i\mathrm{dx}_i + \mathrm{L_B} \equiv \mathrm{L_D} + \mathcal{L}_{\mathrm{B}} \tag{4.27}$$

Then by using the above definitions and the usual arguments on the arbitrariness and independence of the variations, we can obtain the following relatively compact hybrid coordinate differential equations and boundary condition operators. Note that these are the generalizations of the classical discrete coordinate versions of Lagrange's equations [viz., Eqs. (4.21)-(4.24)] which apply to a family of multiple flexible body distributed coordinate systems:

$$\frac{d}{dt}\left(\frac{\partial \mathcal{L}}{\partial \dot{\mathbf{q}}}\right) - \frac{\partial \mathcal{L}}{\partial \mathbf{q}} = \mathbf{Q}^{T} \tag{4.32}$$

$$\frac{d}{dt}\left(\frac{\partial \hat{L}^{i}}{\partial \dot{\mathbf{w}}_{i}}\right) - \frac{\partial \hat{L}^{i}}{\partial \mathbf{w}_{i}} + \frac{\partial}{\partial x_{i}}\left(\frac{\partial \hat{L}^{i}}{\partial \mathbf{w}_{i}'}\right) - \frac{\partial^{2}}{\partial x_{i}^{2}}\left(\frac{\partial \hat{L}^{i}}{\partial \mathbf{w}_{i}''}\right) = \hat{\mathbf{f}}^{iT} \tag{4.33}$$

$$\left\{\frac{\partial \hat{L}^{i}}{\partial \mathbf{w}_{i}'} - \frac{\partial}{\partial x_{i}}\left(\frac{\partial \hat{L}^{i}}{\partial \mathbf{w}_{i}''}\right)\right\}\delta \mathbf{w}_{i}\Bigg|_{l_{0_i}}^{l_i} + \left\{\frac{\partial \mathcal{L}_{B}}{\partial \mathbf{w}_{i}(l_i)} - \frac{d}{dt}\left(\frac{\partial \mathcal{L}_{B}}{\partial \dot{\mathbf{w}}_{i}(l_i)}\right)\right\}\delta \mathbf{w}_{i}(l_i) + \mathbf{f}_{1}^{iT}\delta \mathbf{w}_{i}(l_i) = 0 \tag{4.34}$$

$$\frac{\partial \hat{L}^{i}}{\partial \mathbf{w}_{i}''}\delta \mathbf{w}_{i}'\Bigg|_{l_{0_i}}^{l_i} + \left\{\frac{\partial \mathcal{L}_{B}}{\partial \mathbf{w}_{i}'(l_i)} - \frac{d}{dt}\left(\frac{\partial \mathcal{L}_{B}}{\partial \dot{\mathbf{w}}_{i}'(l_i)}\right)\right\}\delta \mathbf{w}_{i}'(l_i) + \mathbf{f}_{2}^{iT}\delta \mathbf{w}_{i}'(l_i) = 0 \tag{4.35}$$

Equations (4.32) and (4.33) generate directly the coupled hybrid system of ordinary and partial differential equations, and the variational statements of Eqs. (4.34) and (4.35) directly generate the associated boundary conditions. Actually, Eqs. (4.21)-(4.24) are the special case ($n = 1$) of the above Eqs. (4.32)-(4.34). Thus, we have an explicit generalization of Lagrange's equations, for a large family of hybrid systems that consist of interconnected rigid and elastic bodies. Each elastic body is to be beamlike and can have several dependent distributed variables, but only one independent spatial variable.

In essence, we have done the integrations by parts once and for all, for a large class of systems. Thus, the governing equations become quite analogous to the discrete version of Lagrange's equations; i.e., the equations of motion are generated through simply taking appropriate derivatives of energy functions. The utility of these equations [Eqs. (4.32)-(4.35)] can be appreciated by considering several examples.

Explicit Generalizations of Lagrange's Equations: Illustrative Examples

When there is only one domain for the elastic motion and there are no boundary-dependent terms in the Lagrangian, then the above developments are especially simple. The system Lagrangian $\mathcal{L}$ is expressed as $\mathcal{L} = L_D + \int_{l_0}^{l} \hat{L}dx$, and then Eqs. (4.21)-(4.24) or, more generally, Eqs. (4.32)-(4.35), are simplified to the following form:

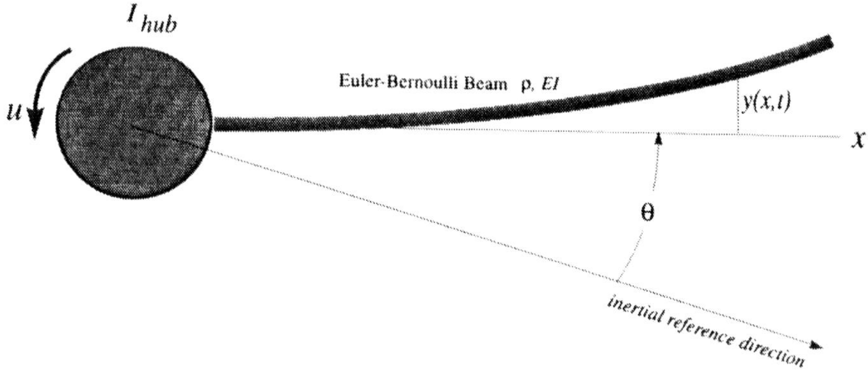

Figure 4.4 Rigid Hub with a Cantilevered Timoshenko Beam.

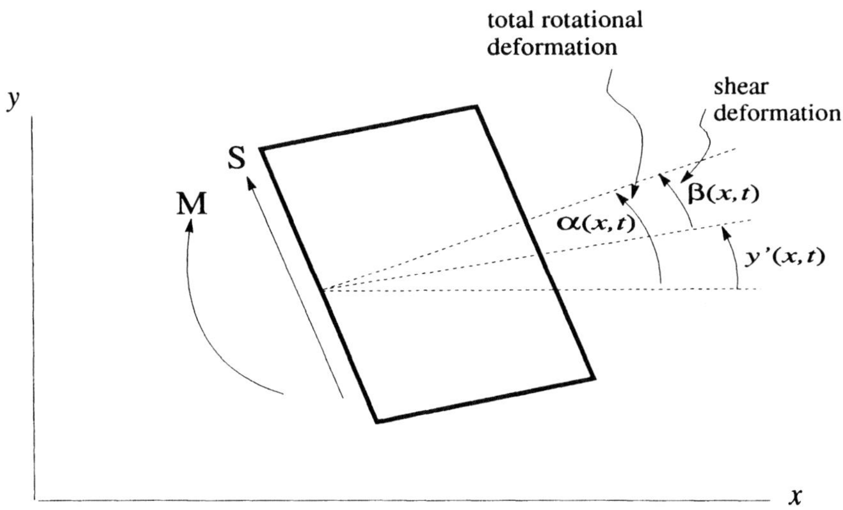

Figure 4.5 Kinematics of Deformation of a Timoshenko Beam.

$$\frac{d}{dt}\left(\frac{\partial \mathcal{L}}{\partial \dot{\mathbf{q}}}\right) - \frac{\partial \mathcal{L}}{\partial \mathbf{q}} = \mathbf{Q}^T \tag{4.21}$$

$$\frac{d}{dt}\left(\frac{\partial \hat{L}}{\partial \dot{\mathbf{w}}}\right) - \frac{\partial \hat{L}}{\partial \mathbf{w}} + \frac{\partial}{\partial x}\left(\frac{\partial \hat{L}}{\partial \mathbf{w}'}\right) - \frac{\partial^2}{\partial x^2}\left(\frac{\partial \hat{L}}{\partial \mathbf{w}''}\right) = \hat{\mathbf{f}}^T \tag{4.22}$$

$$\left\{\frac{\partial \hat{L}}{\partial \mathbf{w}'} - \frac{\partial}{\partial x}\left(\frac{\partial \hat{L}}{\partial \mathbf{w}''}\right)\right\}\delta \mathbf{w}\Bigg|_{l_0}^{l} = 0 \tag{4.23a}$$

$$\frac{\partial \hat{L}}{\partial \mathbf{w}''}\delta \mathbf{w}'\Bigg|_{l_0}^{l} = 0 \tag{4.24a}$$

For this simplest class of problems, it is apparent that the boundary conditions do not make allowance for lumped masses, springs, and similar forces at the boundaries and, obviously, the above equations do not apply to multilink flexible body chains.

Example 4.3 Rigid Hub with a Cantilevered Timoshenko Beam

With reference to Figures 4.4 and 4.5, we consider a rigid hub with a cantilevered flexible appendage. The appendage is considered to be a uniform flexible beam, and we make the Timoshenko assumptions. The beam is cantilevered rigidly to the hub. Motion is restricted to the horizontal plane, and a control torque u(t) acting on the hub (normal to the plane of motion) is the only external effect. Figure 4.5 shows the kinematics of deformation of a beam that undergoes shear deformation in addition to pure bending. In this example, we neglect the velocity component $-y\dot{\theta}$, which is perpendicular to the y direction.

Under these assumptions, the kinetic and potential energies of this hybrid system are as follows:

$$T = \frac{1}{2}I_{hub}\dot{\theta}^2 + \frac{1}{2}\int_{l_0}^{l}\left\{\rho(\dot{y} + x\dot{\theta})^2 + \left(\frac{\rho I}{A}\right)(\dot{\alpha} + \dot{\theta})^2\right\}dx$$

$$V = \frac{1}{2}\int_{l_0}^{l}\{EI(\alpha')^2 + kGA(\alpha - y')^2\}\,dx$$

where
$$\begin{cases} E & \text{Young's modulus of the beam} \\ I & \text{moment of inertia of the cross section about centroidal axis} \\ \rho & \text{constant mass per unit length of the beam} \\ k & \text{shear coefficient} \\ G & \text{modulus of rigidity} \\ A & \text{area of cross section on which shear force acts} \\ I_{hub} & \text{moment of inertia of the rigid hub} \\ \theta & \text{denotes the hub inertial rotation} \\ y & \text{deformation} \\ \alpha & \text{rotation of the cross section due to shear deformation} \end{cases}$$

Therefore, the Lagrangian is expressed by the following equation:

$$\begin{aligned} \mathcal{L} &= L_D + \int_{l_0}^{l} \hat{L} dx \\ &= \frac{1}{2} I_{hub} \dot{\theta}^2 + \frac{1}{2} \int_{l_0}^{l} \left\{ \rho(\dot{y} + x\dot{\theta})^2 + \left(\frac{\rho I}{A}\right)(\dot{\alpha} + \dot{\theta})^2 - EI(\alpha')^2 - kGA(\alpha - y')^2 \right\} dx \end{aligned}$$

The discrete and distributed coordinates for this case are

$$\mathbf{q}(t) = \theta \ , \quad \mathbf{w}(x,t) = [y \ \alpha]^T$$

and the only external force is u(t), and so $\mathbf{Q} = u$ and $\hat{\mathbf{f}} = 0$.

From Eqs. (4.21) and (4.22) and the Lagrangian, we determine the governing differential equations for this hybrid system.

$$\begin{aligned} & I_{hub} \frac{d^2\theta}{dt^2} + \int_{l_0}^{l} \left\{ \rho x \left(\frac{\partial^2 y}{\partial t^2} + x \frac{d^2\theta}{dt^2} \right) + \left(\frac{\rho I}{A}\right) \left(\frac{\partial^2 \alpha}{\partial t^2} + \frac{d^2\theta}{dt^2} \right) \right\} dx = u \\ & \rho \left(\frac{\partial^2 y}{\partial t^2} + x \frac{d^2\theta}{dt^2} \right) + kGA \left(\frac{\partial \alpha}{\partial x} - \frac{\partial^2 y}{\partial x^2} \right) = 0 \\ & \frac{\rho I}{A} \left(\frac{\partial^2 \alpha}{\partial t^2} + \frac{d^2\theta}{dt^2} \right) + kGA \left(\alpha - \frac{\partial y}{\partial x} \right) - EI \frac{\partial^2 \alpha}{\partial x^2} = 0 \end{aligned}$$

The boundary conditions are obtained from specialization of Eqs.(4.23a) and (4.24a) as follows:

$$kGA\left(\alpha - \frac{\partial y}{\partial x}\right)\delta y \Big|_{l_0}^{l} = 0 \ , \quad -EI\left(\frac{\partial \alpha}{\partial x}\right)\delta\alpha \Big|_{l_0}^{l} = 0$$

Since $y(l_0) = \alpha(l_0) = 0$, and $\delta y(l)$ and $\delta\alpha(l)$ are free, the boundary conditions are:

at x= l_0:

$$y = 0, \text{ and } \alpha = 0$$

at x = l:

$$EI\frac{\partial \alpha}{\partial x}\Big|_l = 0, \text{ and } kGA\left(\alpha - \frac{\partial y}{\partial x}\right)\Big|_l = 0$$

■

Example 4.4 Flexible Three-Body Problem (Hub, Beam, and Tip Mass)

In Figure 4.6, we show a rigid hub which has a cantilevered flexible appendage, and which has a finite tip mass. The appendage is considered to be a uniform flexible beam and we make the Euler-Bernoulli assumptions of negligible shear deformation and negligible distributed rotatory inertia. The other assumptions are identical to Example 4.3, except in this example, we account for the finite tip mass and the rotatory inertia of the tip mass. Under these kinematic and constitutive assumptions, the kinetic and potential energies of this hybrid coordinate system are as follows:

$$T = \frac{1}{2}I_{hub}\dot{\theta}^2 + \frac{1}{2}\int_{l_0}^{l}\{\rho(\dot{y} + x\dot{\theta})^2\}\,dx + \frac{1}{2}m(l\dot{\theta} + \dot{y}(l))^2 + \frac{1}{2}I_{tip}(\dot{\theta} + \dot{y}'(l))^2$$

$$V = \frac{1}{2}\int_{l_0}^{l}\{EI(y'')^2\}\,dx$$

where $\begin{cases} m & \text{is the mass of the tip mass} \\ I_{tip} & \text{is the rotatory inertia of the tip mass} \end{cases}$

Therefore, the Lagrangian may be expressed as

$$\mathcal{L} = L_D + \int_{l_0}^{l}\hat{L}dx + L_B$$

$$= \frac{1}{2}I_{hub}\dot{\theta}^2 + \frac{1}{2}\int_{l_0}^{l}\{\rho(\dot{y}+x\dot{\theta})^2 - EI(y'')^2\}\,dx + \frac{1}{2}m(l\dot{\theta}+\dot{y}(l))^2 + \frac{1}{2}I_{tip}(\dot{\theta}+\dot{y}'(l))^2$$

The discrete and distributed coordinates for this case are $\mathbf{q}(t) = \theta, \mathbf{w}(x,t) = y$, and the only external force is $u(t)$, so $\mathbf{Q} = u$, $\hat{\mathbf{f}} = 0$, $\mathbf{f}_1 = 0$, and $\mathbf{f}_2 = 0$.

From the Eqs.(4.21) and (4.22), the governing equations for this hybrid system are

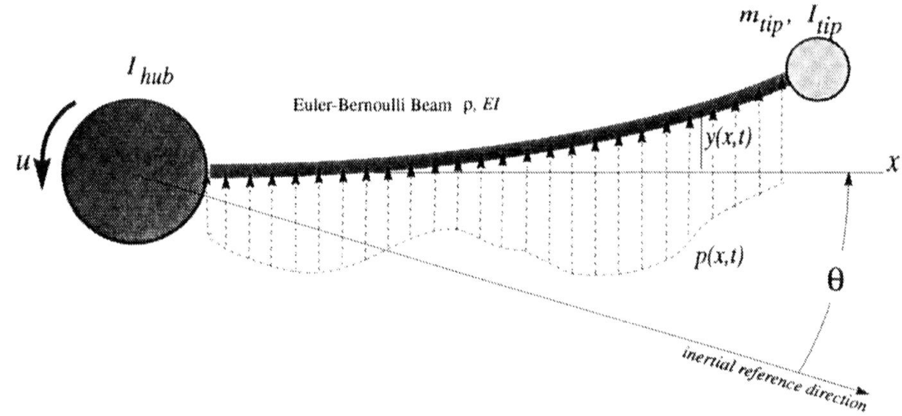

Figure 4.6 Flexible Three-Body Problem (Hub, Beam & Tip Mass).

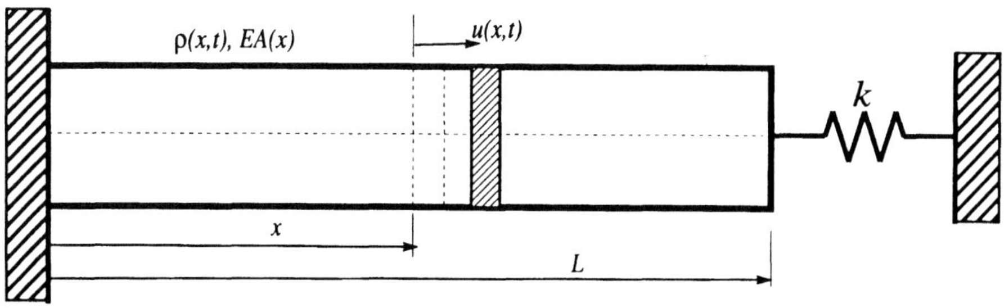

Figure 4.7 Rod in Axial Vibration.

$$I_{hub}\frac{d^2\theta}{dt^2}+\int_{l_0}^{l}\rho x\left(\frac{\partial^2 y}{\partial t^2}+x\frac{d^2\theta}{dt^2}\right)dx+ml\left(l\frac{d^2\theta}{dt^2}+\left.\frac{\partial^2 y}{\partial t^2}\right|_l\right)$$

$$+I_{tip}\left(\frac{d^2\theta}{dt^2}+\frac{d^2}{dt^2}\left(\left.\frac{\partial y}{\partial x}\right|_l\right)\right) = u$$

$$\rho\left(\frac{\partial^2 y}{\partial t^2}+x\frac{d^2\theta}{dt^2}\right)+EI\frac{\partial^4 y}{\partial x^4}=0$$

Boundary conditions are obtained from Eqs. (4.23) and (4.24) by inserting the descriptive variables $\mathbf{q}(t)=\theta, \mathbf{w}(x,t)=y$:

$$EI\frac{\partial^3 y}{\partial x^3}\delta y\Big|_{l_0}^{l}-m(l\ddot{\theta}+\ddot{y}(l))\delta y(l)=0,\quad -EI\frac{\partial^2 y}{\partial x^2}\delta\left(\frac{\partial y}{\partial x}\right)\Big|_{l_0}^{l}-I_{tip}(\ddot{\theta}+\ddot{y}'(l))\delta y'(l)=0$$

Since $y(l_0)=y'(l_0)=0$, and $\delta y(l)$ and $\delta y'(l)$ are free, the boundary conditions are:

at $x=l_0$: $y=0$, and $y'=0$

at $x=l$: $EI\left.\frac{\partial^3 y}{\partial x^3}\right|_l=m(l\ddot{\theta}+\ddot{y}(l))$, $EI\left.\frac{\partial^2 y}{\partial x^2}\right|_l=-I_{tip}(\ddot{\theta}+\ddot{y}'(l))$

(shear force), (bending moment)

■

Example 4.5 Axial Vibration of a Rod

With reference to Figure 4.7, we consider the rod in axial vibration. For simplicity, we assume that the rod has uniform properties along the axial coordinates. In this case, there are no discrete coordinates. Then, the kinetic and potential energies of this system are as follows:

$$T=\frac{1}{2}\int_0^L \rho\dot{u}^2\,dx,\qquad V=\frac{1}{2}\int_0^L EA(u')^2\,dx+\frac{1}{2}Ku^2(L)$$

where
- E is Young's modulus of the rod
- A is the area of cross section
- K is the spring constant
- L is the length of the rod
- ρ is the constant mass per unit length of the rod
- u is the axial displacement $u=u(x,t)$

Therefore, the Lagrangian may be expressed as

$$\mathcal{L} = \int_0^L \hat{\mathrm{L}}\,\mathrm{dx} + \mathrm{L_B} = \frac{1}{2}\int_0^L \{\rho\dot{\mathrm{u}}^2 - \mathrm{EA}(\mathrm{u}')^2\}\,\mathrm{dx} - \frac{1}{2}\mathrm{Ku}^2(L)$$

and there is no external force, so $\mathbf{Q} = 0$, $\hat{\mathbf{f}} = 0$, $\mathbf{f}_1 = 0$, $\mathbf{f}_2 = 0$, and $\mathbf{w}(\mathrm{x,t}) = \mathrm{u}(\mathrm{x,t})$.

The governing equations for this system follow immediately from Eq. (4.22) as

$$\rho\ddot{\mathrm{u}} - \mathrm{EA}\frac{\partial^2 \mathrm{u}}{\partial \mathrm{x}^2} = 0$$

and the boundary condition variational statement is obtained by Eq. (4.23).

$$-\mathrm{EAu}'\delta\mathrm{u}\,\Big|_0^L + (-\mathrm{Ku}(L))\delta\mathrm{u}(L) = 0$$

Since $\mathrm{u}(0) = 0$, and $\delta\mathrm{u}(L)$ is free, the boundary conditions are as follows:

$$\text{at } \mathrm{x} = 0:\ \mathrm{u} = 0, \qquad \text{at } \mathrm{x} = L:\ \mathrm{EAu}'\big|_L + \mathrm{Ku}\big|_L = 0.$$

■

Example 4.6 Two-Link Flexible Manipulator Model

We consider a planar two-link flexible manipulator as shown in Figure 4.8. Each link is modeled as a uniform flexible beam, and we make the Euler-Bernoulli assumptions. This system has two spatial variables x_1 and x_2, and elastic motions relative to the rigid body motions are described by $\mathrm{y}_1(\mathrm{x}_1,\mathrm{t})$ and $\mathrm{y}_2(\mathrm{x}_2,\mathrm{t})$, respectively. In this example, we neglect the velocity components that are perpendicular to the y_1 and y_2 directions. This system is controlled by the torque inputs u_1 and u_2 as shown in Figure 4.8.

The kinetic and potential energies of this hybrid system are as follows:

$$\mathrm{T} = \frac{1}{2}\int_0^{l_1} \rho_1(\mathrm{x}_1\dot{\theta}_1 + \dot{\mathrm{y}}_1)^2\,\mathrm{dx}_1 + \frac{1}{2}\mathrm{m}_2(l_1\dot{\theta}_1 + \dot{\mathrm{y}}_1(l_1,\mathrm{t}))^2 + \frac{1}{2}\int_0^{l_2} \rho_2(\mathrm{x}_2\dot{\theta}_2 + \dot{\mathrm{y}}_2)^2\,\mathrm{dx}_2$$
$$+ (l_1\dot{\theta}_1 + \dot{\mathrm{y}}_1(l_1,\mathrm{t}))\cos(\theta_2 - \theta_1)\int_0^{l_2} \rho_2(\mathrm{x}_2\dot{\theta}_2 + \dot{\mathrm{y}}_2)\,\mathrm{dx}_2$$

$$\mathrm{V} = \frac{1}{2}\int_0^{l_1} \mathrm{EI}_1\left(\frac{\partial^2 \mathrm{y}_1}{\partial {\mathrm{x}_1}^2}\right)^2 \mathrm{dx}_1 + \frac{1}{2}\int_0^{l_2} \mathrm{EI}_2\left(\frac{\partial^2 \mathrm{y}_2}{\partial {\mathrm{x}_2}^2}\right)^2 \mathrm{dx}_2$$

where
- ρ_i is the assumed constant mass/unit length of the ith beam
- EI_i is the assumed constant bending stiffness of the ith beam
- l_i is the length of the ith beam
- y_i is the elastic deformation of the ith beam
- m_2 is the mass of the second beam

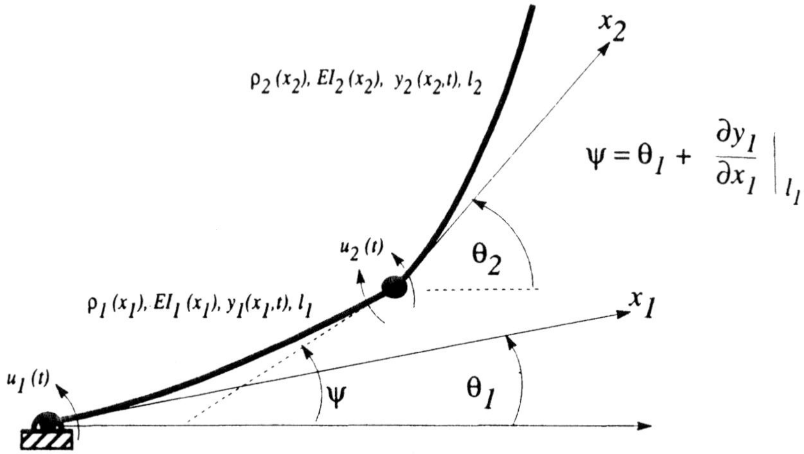

Figure 4.8 Two-Link Flexible Manipulator.

$$
\begin{aligned}
\mathcal{L} &= \int_0^{l_1} \hat{\mathrm{L}}^1 \, \mathrm{dx}_1 + \int_0^{l_2} \hat{\mathrm{L}}^2 \, \mathrm{dx}_2 + \mathrm{L_B} \\
&= \int_0^{l_1} \left\{ \frac{1}{2}\rho_1(\mathrm{x}_1\dot{\theta}_1 + \dot{\mathrm{y}}_1)^2 - \frac{1}{2}\mathrm{EI}_1\left(\frac{\partial^2 \mathrm{y}_1}{\partial \mathrm{x}_1{}^2}\right)^2 \right\} \mathrm{dx}_1 \\
&+ \int_0^{l_2} \left\{ \frac{1}{2}\rho_2(\mathrm{x}_2\dot{\theta}_2 + \dot{\mathrm{y}}_2)^2 + (l_1\dot{\theta}_1 + \dot{\mathrm{y}}_1(l_1,t))\cos(\theta_2 - \theta_1)\rho_2(x_2\dot{\theta}_2 + \dot{\mathrm{y}}_2) \right. \\
&\left. -\frac{1}{2}\mathrm{EI}_2\left(\frac{\partial^2 \mathrm{y}_2}{\partial \mathrm{x}_2{}^2}\right)^2 \right\} \mathrm{dx}_2 + \frac{1}{2}\mathrm{m}_2(l_1\dot{\theta}_1 + \dot{\mathrm{y}}_1(l_1,\mathrm{t}))^2
\end{aligned}
$$

The discrete and distributed generalized coordinates for this case are

$$
\mathrm{q}_1 = \theta_1(\mathrm{t}), \ \mathrm{q}_2 = \theta_2(\mathrm{t}), \ \mathrm{w}_1 = \mathrm{y}_1(\mathrm{x}_1,\mathrm{t}), \ \mathrm{w}_2 = \mathrm{y}_2(\mathrm{x}_2,\mathrm{t})
$$

The nonconservative virtual work is expressed by the following equation:

$$
\delta \mathrm{W_{nc}} = \mathrm{u}_1\delta\theta_1 + \mathrm{u}_2\delta\Big\{\theta_2 - \big(\theta_1 + \frac{\partial \mathrm{y}_1}{\partial \mathrm{x}_1}\Big|_{l_1}\big)\Big\} = (\mathrm{u}_1 - \mathrm{u}_2)\delta\theta_1 + \mathrm{u}_2\delta\theta_2 - \mathrm{u}_2\delta\big(\frac{\partial \mathrm{y}_1}{\partial \mathrm{x}_1}\Big|_{l_1}\big)
$$

Therefore, the generalized forces are

$$
\mathbf{Q} = [\,\mathrm{u}_1 - \mathrm{u}_2 \quad \mathrm{u}_2\,]^{\mathrm{T}}, \ \hat{\mathbf{f}}^1 = 0, \ \hat{\mathbf{f}}^2 = 0 \ \mathbf{f}_1^1 = 0, \ \mathbf{f}_1^2 = 0, \ \mathbf{f}_2^1 = -\mathrm{u}_2, \text{ and } \mathbf{f}_2^2 = 0
$$

In order to apply Eqs. (4.32)-(4.35), first we record $\hat{L}^1$, $\hat{L}^2$, and $\mathcal{L}_B$. In this example, $\mathcal{L}_B$ is identical to $\mathcal{L}$ since L_D is zero.

$$\hat{L}^1 = \frac{1}{2}\rho_1(x_1\dot{\theta}_1 + \dot{y}_1)^2 - \frac{1}{2}EI_1\left(\frac{\partial^2 y_1}{\partial {x_1}^2}\right)^2$$

$$\hat{L}^2 = \frac{1}{2}\rho_2(x_2\dot{\theta}_2+\dot{y}_2)^2 - \frac{1}{2}EI_2\left(\frac{\partial^2 y_2}{\partial {x_2}^2}\right)^2 + (l_1\dot{\theta}_1+\dot{y}_1(l_1,t))\cos(\theta_2-\theta_1)\rho_2(x_2\dot{\theta}_2+\dot{y}_2)$$

$$\mathcal{L}_B = \frac{1}{2}m_2(l_1\dot{\theta}_1 + \dot{y}_1(l_1,t))^2 + \sum_{i=1}^{2}\int_0^{l_i}\hat{L}^i dx_i$$

The resulting coupled system of nonlinear hybrid differential equations and the boundary conditions are obtained from Eqs. (4.32)-(4.35) as the following eight equations:

θ_1 Equation

$$\int_0^{l_1}\rho_1 x_1(x_1\ddot{\theta}_1+\ddot{y}_1)\,dx_1 + m_2 l_1(l_1\ddot{\theta}_1+\ddot{y}_1(l_1,t)) + l_1\cos(\theta_2-\theta_1)\int_0^{l_2}\rho_2(x_2\ddot{\theta}_2+\ddot{y}_2)\,dx_2$$
$$- \sin(\theta_2-\theta_1)(l_1\dot{\theta}_2 + \dot{y}_1(l_1,t))\int_0^{l_2}\rho_2(x_2\dot{\theta}_2+\dot{y}_2)\,dx_2 = u_1 - u_2$$

θ_2 Equation

$$\int_0^{l_2}\rho_2 x_2(x_2\ddot{\theta}_2+\ddot{y}_2)\,dx_2 + \frac{1}{2}m_2 l_2(l_1\ddot{\theta}_1 + \ddot{y}_1(l_1,t))\cos(\theta_2-\theta_1)$$
$$+ \sin(\theta_2-\theta_1)(l_1\dot{\theta}_1+\dot{y}_1(l_1,t))\left\{\int_0^{l_2}\rho_2(x_2\dot{\theta}_2+\dot{y}_2)\,dx_2 - \frac{1}{2}m_2 l_2(\dot{\theta}_2-\dot{\theta}_1)\right\} = u_2$$

y_1 & y_2 Equations

$$\rho_1(x_1\ddot{\theta}_1 + \ddot{y}_1) + EI_1\frac{\partial^4 y_1}{\partial {x_1}^4} = 0, \quad 0 < y_1 < l_1$$

$$\rho_2(x_2\ddot{\theta}_2+\ddot{y}_2) + EI_2\frac{\partial^4 y_2}{\partial {x_2}^4} + \rho_2[l_1\ddot{\theta}_1 + \ddot{y}_1(l_1,t)]\cos(\theta_2-\theta_1)$$
$$-\rho_2[l_1\dot{\theta}_1+\dot{y}_1(l_1,t)](\dot{\theta}_2-\dot{\theta}_1)\sin(\theta_2-\theta_1) = 0, \quad 0 < y_2 < l_2$$

Boundary Conditions of Beam 1

at $x_1 = 0$: $\quad y_1 = 0, \quad y_1' = 0$

at $x_1 = l_1$: $\quad EI_1\frac{\partial^2 y_1}{\partial {x_1}^2}\Big|_{l_1} = -u_2$

$$\mathrm{EI}_1 \frac{\partial^3 y_1}{\partial x_1{}^3}\bigg|_{l_1} = m_2[l_1\ddot{\theta}_1 + \ddot{y}_1(l_1,t)] + \cos(\theta_2 - \theta_1)\int_0^{l_2} \rho_2(x_2\ddot{\theta}_2 + \ddot{y}_2)\,dx_2$$

$$- \sin(\theta_2 - \theta_1)\,(\dot{\theta}_2 - \dot{\theta}_1)\int_0^{l_2} \rho_2(x_2\dot{\theta}_2 + \dot{y}_2)\,dx_2$$

Boundary Conditions of Beam 2

at $x_2 = 0$: $\quad y_2 = 0, \qquad y_2' = 0$

at $x_2 = l_2$: $\quad \mathrm{EI}_2 \frac{\partial^2 y_2}{\partial x_2{}^2}\bigg|_{l_2} = 0, \qquad \mathrm{EI}_2 \frac{\partial^3 y_2}{\partial x_2{}^3}\bigg|_{l_2} = 0$

The volume of algebra and calculus associated with carrying through the general developments leading to Eqs. (4.32)-(4.35) is comparable to doing one special case application of Hamilton's principle (for example, the above flexible manipulator)! However, having derived Eqs. (4.32)-(4.35), we can very efficiently obtain the system equations and boundary conditions for any system which belongs to the family to which Eqs. (4.32)-(4.35) apply.

■

Example 4.7 Multibody Flexible Space Structure Model

We consider now a more complicated space structure model that consists of four elastic domains (two of them curved to represent cross sections of a parabolic antenna), three boundary elements and, as shown in Figures 4.9 and 4.10. This example has sufficient complexity to illustrate the utility of the generalized Lagrange's equations, Eqs. (4.32)-(4.35); in fact, it is unlikely that one would attempt to directly apply Hamilton's principle in this situation due to the extensive calculus and algebra involved.

We now discuss the geometric and modeling assumptions in detail so that the kinematics leading to the system Lagrangian can be carried to completion. The rigid hub's center is pinned, and the rotational inertia of the boundary elements m_2 and m_3 is neglected. This structure is qualitatively similar to the SCOLE(Spacecraft Control Laboratory Experiment) experimental model (Ref. 29), and corresponds to a recently developed laboratory experiment at the Naval Postgraduate School (Ref. 31). Each elastic domain is idealized as a uniform beam clamped to two discrete mass elements. The only rigid body degree of freedom is the rotation θ of the hub, and this rigid body motion has highly nonlinear coupling effects with the other six substructures. In carrying out the further developments, we simplify the kinematics: We restrict all motion to occur in a horizontal plane, however, we include the velocity components in the x direction which result in centrifugal effect (rotational stiffening) and we also include the coriolis terms.

Three control torque actuators are assumed to be acting as shown in Figure 4.10; the first actuator torque (u_1) is applied to the rigid hub and the other two (u_2, u_3) are applied to the discrete mass elements m_2 and m_3. In recent developments, Junkins and Bang (Ref. 30) have used this model to establish a globally stabilizing control law for large rapid maneuvers of this structure. Consistent with the geometric and kinematic assumptions, the position vectors for the four flexible appendages and two discrete mass elements can be represented as in Figure 4.10, with components of all vectors taken along the body fixed axes $\hat{\mathbf{b}}_1, \hat{\mathbf{b}}_2$ as follows:

$$
\begin{aligned}
\mathbf{R}_1 &= x_1\hat{\mathbf{b}}_1 + w_1\hat{\mathbf{b}}_2 \\
\mathbf{R}_2 &= [l_1 - x_2\sin\alpha - w_2\cos\alpha]\hat{\mathbf{b}}_1 + [w_1(l_1) + x_2\cos\alpha - w_2\sin\alpha]\hat{\mathbf{b}}_2 \\
\mathbf{R}_3 &= [l_1 - l_2\sin\alpha - w_2(l_2)\cos\alpha + x_3\sin(\alpha+\beta) - w_3\cos(\alpha+\beta)]\hat{\mathbf{b}}_1 \\
&\quad + [w_1(l_1) + l_2\cos\alpha - w_2(l_2)\sin\alpha - x_3\cos(\alpha+\beta) - w_3\sin(\alpha+\beta)]\hat{\mathbf{b}}_2 \\
\mathbf{R}_4 &= [l_1 - l_2\sin\alpha - w_2(l_2)\cos\alpha - x_4\sin(\alpha+\beta) - w_4\cos(\alpha+\beta)]\hat{\mathbf{b}}_1 \\
&\quad + [w_1(l_1) + l_2\cos\alpha - w_2(l_2)\sin\alpha + x_4\cos(\alpha+\beta) - w_4\sin(\alpha+\beta)]\hat{\mathbf{b}}_2 \\
\mathbf{R}_5 &= l_1\hat{b}_1 + w_1(l_1)\hat{\mathbf{b}}_2 \\
\mathbf{R}_6 &= [l_1 - l_2\sin\alpha - w_2(l_2)\cos\alpha]\hat{\mathbf{b}}_1 + [w_1(l_1) + l_2\cos\alpha - w_2(l_2)\sin\alpha]\hat{\mathbf{b}}_2
\end{aligned}
$$

where α and β denote slopes of deflections at the tip of first and second domains

$$
\alpha = \left.\frac{\partial w_1}{\partial x_1}\right|_{l_1}, \qquad \beta = \left.\frac{\partial w_2}{\partial x_2}\right|_{l_2}
$$

on differentiation, we find the inertial velocity vectors are given by

$$
\begin{aligned}
\dot{\mathbf{R}}_1 &= (-w_1\dot{\theta})\hat{b}_1 + (\dot{w}_1 + x_1\dot{\theta})\hat{\mathbf{b}}_2 \\
\dot{\mathbf{R}}_2 &= \left[-x_2\dot{\alpha}\cos\alpha - \dot{w}_2\cos\alpha + w_2\dot{\alpha}\sin\alpha - \dot{\theta}(w_1(l_1) + x_2\cos\alpha - w_2\sin\alpha)\right]\hat{\mathbf{b}}_1 \\
&\quad + [\dot{w}_1(l_1) - x_2\dot{\alpha}\sin\alpha - \dot{w}_2\sin\alpha - w_2\dot{\alpha}\cos\alpha + \dot{\theta}(l_1 - x_2\sin\alpha - w_2\cos\alpha)]\hat{\mathbf{b}}_2 \\
\dot{\mathbf{R}}_3 &= [-l_2\dot{\alpha}\cos\alpha - \dot{w}_2(l_2)\cos\alpha + w_2(l_2)\dot{\alpha}\sin\alpha + x_3(\dot{\alpha}+\dot{\beta})\cos(\alpha+\beta) - \dot{w}_3\cos(\alpha+\beta) \\
&\quad + w_3(\dot{\alpha}+\dot{\beta})\sin(\alpha+\beta) - \dot{\theta}\big(w_1(l_1) + l_2\cos\alpha - w_2(l_2)\sin\alpha - x_3\cos(\alpha+\beta) \\
&\quad - w_3\sin(\alpha+\beta)\big)]\hat{\mathbf{b}}_1 \\
&\quad + [\dot{w}_1(l_1) - l_2\dot{\alpha}\sin\alpha - \dot{w}_2(l_2)\sin\alpha - w_2(l_2)\dot{\alpha}\cos\alpha + x_3(\dot{\alpha}+\dot{\beta})\sin(\alpha+\beta) \\
&\quad - \dot{w}_3\sin(\alpha+\beta) - w_3(\dot{\alpha}+\dot{\beta})\cos(\alpha+\beta) + \dot{\theta}\big(l_1 - l_2\sin\alpha - w_2(l_2)\cos\alpha \\
&\quad + x_3\sin(\alpha+\beta) - w_3\cos(\alpha+\beta)\big)]\hat{\mathbf{b}}_2
\end{aligned}
$$

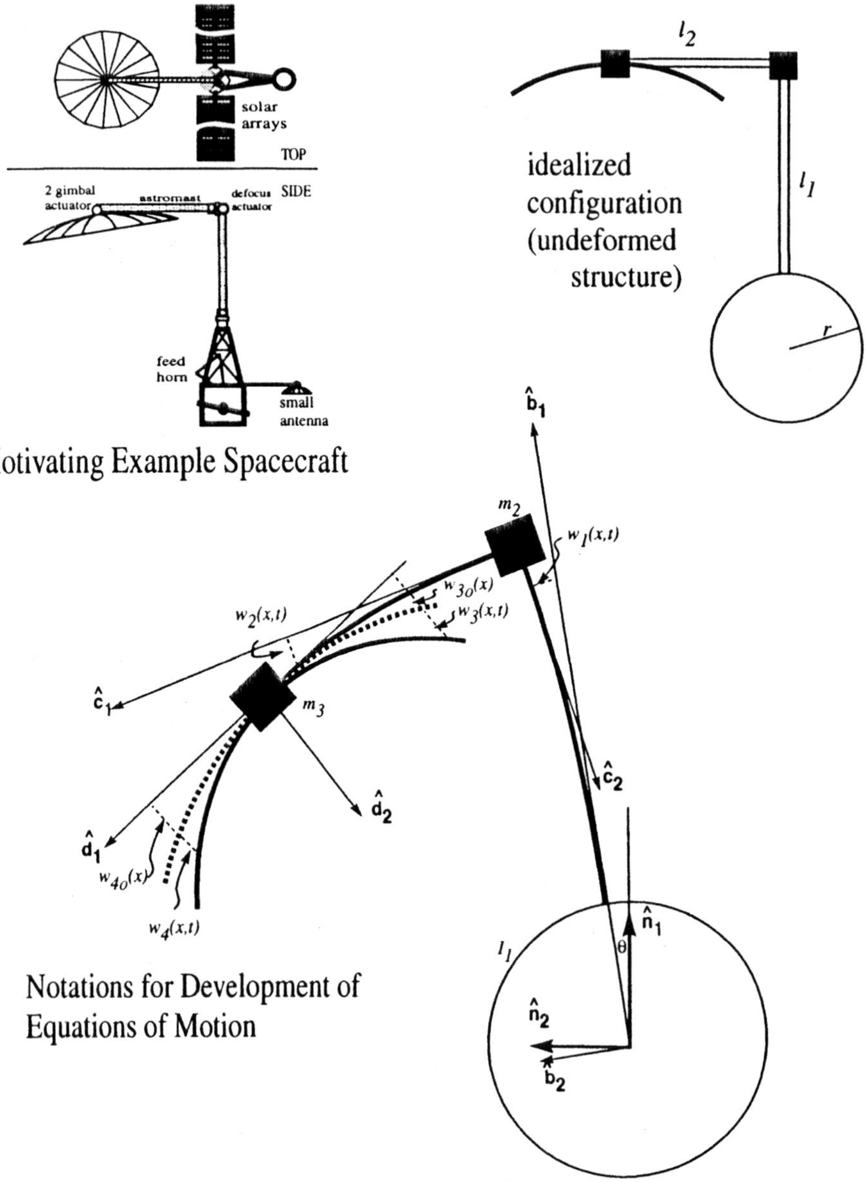

Figure 4.9 Multibody Flexible Spacecraft Configuration.

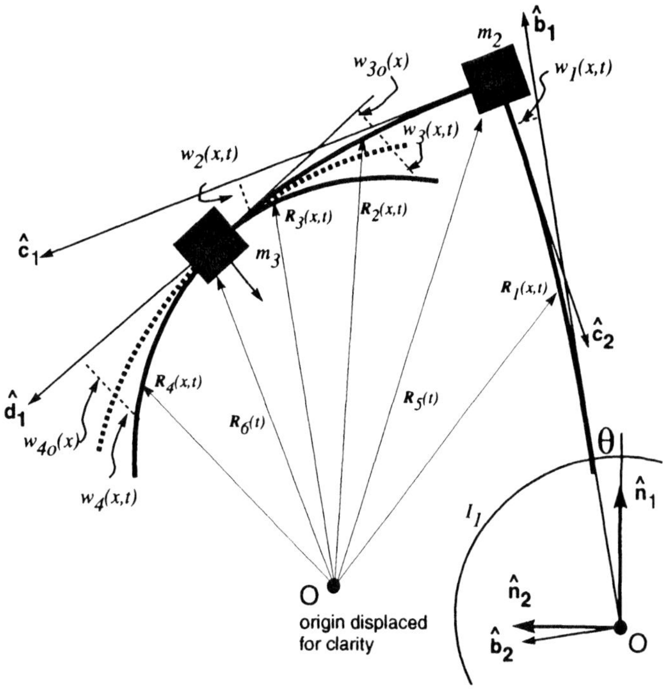

Figure 4.10 Notation for Equation of Motion Development.

$$\begin{aligned}
\dot{\mathbf{R}}_4 = {} & [-l_2\dot{\alpha}\cos\alpha - \dot{\mathrm{w}}_2(l_2)\cos\alpha + \mathrm{w}_2(l_2)\dot{\alpha}\sin\alpha - \mathrm{x}_4(\dot{\alpha}+\dot{\beta})\cos(\alpha+\beta) - \dot{\mathrm{w}}_4\cos(\alpha+\beta) \\
& + \mathrm{w}_4(\dot{\alpha}+\dot{\beta})\sin(\alpha+\beta) - \dot{\theta}\big(\mathrm{w}_1(l_1) + l_2\cos\alpha - \mathrm{w}_2(l_2)\sin\alpha + \mathrm{x}_4\cos(\alpha+\beta) \\
& - \mathrm{w}_4\sin(\alpha+\beta)\big)]\hat{\mathbf{b}}_1 \\
& + [\dot{\mathrm{w}}_1(l_1) - l_2\dot{\alpha}\sin\alpha - \dot{\mathrm{w}}_2(l_2)\sin\alpha - \mathrm{w}_2(l_2)\dot{\alpha}\cos\alpha - \mathrm{x}_4(\dot{\alpha}+\dot{\beta})\sin(\alpha+\beta) \\
& - \dot{\mathrm{w}}_4\sin(\alpha+\beta) - \mathrm{w}_4(\dot{\alpha}+\dot{\beta})\cos(\alpha+\beta) + \dot{\theta}\big(l_1 - l_2\sin\alpha - \mathrm{w}_2(l_2)\cos\alpha \\
& - \mathrm{x}_4\sin(\alpha+\beta) - \mathrm{w}_4\cos(\alpha+\beta)\big)]\hat{\mathbf{b}}_2 \\
\dot{\mathbf{R}}_5 = {} & -\mathrm{w}_1(l_1)\dot{\theta}\hat{\mathbf{b}}_1 + (\dot{\mathrm{w}}_1(l_1) + l_1\dot{\theta})\hat{\mathbf{b}}_2 \\
\dot{\mathbf{R}}_6 = {} & [-l_2\dot{\alpha}\cos\alpha - \dot{\mathrm{w}}_2(l_2)\cos\alpha + \mathrm{w}_2(l_2)\dot{\alpha}\sin\alpha \\
& - \dot{\theta}\big(\mathrm{w}_1(l_1) + l_2\cos\alpha - \mathrm{w}_2(l_2)\sin\alpha\big)]\hat{\mathbf{b}}_1 \quad + \quad [\dot{\mathrm{w}}_1(l_1) - l_2\dot{\alpha}\sin\alpha \\
& - \dot{\mathrm{w}}_2(l_2)\sin\alpha - \mathrm{w}_2(l_2)\dot{\alpha}\cos\alpha + \dot{\theta}(l_1 - l_2\sin\alpha - \mathrm{w}_2(l_2)\cos\alpha)]\hat{\mathbf{b}}_2
\end{aligned}$$

For convenience and to compact the subsequent expressions, we introduce the notation

$$\dot{\mathbf{R}}_i = v_i^1 \hat{\mathbf{b}}_1 + v_i^2 \hat{\mathbf{b}}_2$$

The definition of the $(v_i^1,\ v_i^2)$ components of the velocity vectors are obvious from inspection of the above six equations.

The kinetic and potential functions for this system are expressed as a function of the position and velocity vectors as follows:

$$2T = \sum_{i=1}^{4} \int_{l_{0_i}}^{l_i} \rho_i \dot{\mathbf{R}}_i \cdot \dot{\mathbf{R}}_i dx_i + I_h \dot{\theta}^2 + m_2 \dot{\mathbf{R}}_5 \cdot \dot{\mathbf{R}}_5 + m_3 \dot{\mathbf{R}}_6 \cdot \dot{\mathbf{R}}_6$$

$$2V = \sum_{i=1}^{4} \int_{l_{0_i}}^{l_i} EI_i \left(\frac{\partial^2 \mathbf{R}_i}{\partial {x_i}^2} \right)^2 dx_i$$

The nonconservative forces are obtained from determination of the virtual work as follows:

$$\begin{aligned} \delta W &= u_1 \delta\theta + u_2 \delta(\theta + \alpha) + u_3 \delta(\alpha + \beta + \theta) \\ &= (u_1 + u_2 + u_3)\delta\theta + (u_2 + u_3)\delta\alpha + u_3 \delta\beta = Q_\theta \delta\theta + Q_\alpha \delta\alpha + Q_\beta \delta\beta \end{aligned}$$

Using $\mathcal{L} = T - V$ and the generalized forces evident in the above virtual work expression, and making use of Eqs.(4.32)-(4.35), leads to the following nonlinear equations of motion.

Nonlinear Equations of Motion

θ equation

$$\begin{aligned} I_h \ddot{\theta} &+ \int_{l_{0_1}}^{l_1} \rho_1 \frac{\partial}{\partial t}(v_1^1 A_1^1 + v_1^2 B_1^1) dx_1 + \int_{l_{0_2}}^{l_2} \rho_2 \frac{\partial}{\partial t}(v_2^1 A_2^2 + v_2^2 B_2^2) dx_2 \\ &+ \int_{l_{0_3}}^{l_3} \rho_3 \frac{d}{dt}(v_3^1 A_3^2 + v_3^2 B_3^2) dx_3 + \int_{l_{0_4}}^{l_4} \rho_4 \frac{d}{dt}(v_4^1 A_4^2 + v_4^2 B_4^2) dx_4 \\ &+ m_2 \frac{d}{dt}(v_5^1 A_5^1 + v_5^2 B_5^1) + m_3 (v_6^1 A_6^2 + v_6^2 B_6^2) = u_1 + u_2 + u_3 \end{aligned}$$

w_1 equation

$$\rho_1 (-v_1^1 A_1^2 + \frac{d}{dt}(v_1^2 B_1^2)) + E_1 I_1 \frac{\partial^4 w_1}{\partial x_1^4} = 0$$

$w_1(l_1)$ equation

$$\int_{l_{0_2}}^{l_2} \rho_2[\frac{d}{dt}(v_2^2 B_2^1) - v_1^1 A_2^1]dx_2 + \int_{l_{0_3}}^{l_3} \rho_3[\frac{d}{dt}(v_3^2 B_3^1 - v_3^1 A_3^1]dx_3$$
$$+ \int_{l_{0_4}}^{l_4} \rho_4[\frac{d}{dt}(v_4^2 B_4^1) - v_4^1 A_4^1]dx_4 + m_2[\frac{d}{dt}(v_5^2 B_5^2) - v_5^1 A_5^2]$$
$$+ m_3[\frac{d}{dt}(v_6^2 B_6^1) - v_6^1 A_6^1] = E_1 I_1 \frac{\partial^3 w_1}{\partial {x_1}^3}\bigg|_{l_1} + f_1$$

$\alpha \equiv \frac{\partial w_1}{\partial x_1}\Big|_{l_1}$ equation

$$\int_{l_{0_2}}^{l_2} \rho_2[\frac{d}{dt}(v_2^1 A_2^3 + v_2^2 B_2^3) - v_2^1 A_2^4 - v_2^2 B_2^4]dx_2$$
$$+ \int_{l_{0_3}}^{l_3} \rho_3[\frac{d}{dt}(v_3^1 A_3^3 + v_3^2 B_3^3) - v_3^1 A_3^4 - v_3^2 B_3^4]dx_3$$
$$+ \int_{l_{0_4}}^{l_4} \rho_4[\frac{d}{dt}(v_4^1 A_4^3 + v_4^2 B_4^3) - v_4^1 A_4^4 - v_4^2 B_4^4]dx_4$$
$$+ m_3[\frac{d}{dt}(v_6^1 A_6^3 + v_6^2 B_6^3) - v_6^1 A_6^4 - v_6^2 B_6^4] = -E_1 I_1 \frac{\partial^2 w_1}{\partial {x_1}^2}\bigg|_{l_1} + u_2 + u_3$$

w_2 equation

$$\rho_2[\frac{d}{dt}(v_2^1 A_2^5 + v_2^2 B_2^5) - v_2^1 A_2^6 - v_2^2 B_2^6] + E_2 I_2 \frac{\partial^4 w_2}{\partial x_2^4} = 0$$

$w_2(l_2)$ equation

$$\int_{l_{0_3}}^{l_3} \rho_3[\frac{d}{dt}(v_3^1 A_3^5 + v_3^2 B_3^5) - v_3^1 A_3^6 - v_3^2 B_3^6]dx_3$$
$$+ \int_{l_{0_4}}^{l_4} \rho_4[\frac{d}{dt}(v_4^1 A_4^5 + v_4^2 B_4^5) - v_4^1 A_4^6 - v_4^2 B_4^6]dx_4$$
$$m_3[\frac{d}{dt}(v_6^1 A_6^5 + v_6^2 B_6^5) - v_6^1 A_6^6 - v_6^2 B_6^6] = E_2 I_2 \frac{\partial^3 w_2}{\partial x_2^3}\bigg|_{l_2} + f_2$$

$$\beta \equiv \left.\frac{\partial w_2}{\partial x_2}\right|_{l_2} \text{ equation}$$

$$\int_{l_{0_3}}^{l_3} \rho_3[\frac{d}{dt}(v_3^1A_3^7 + v_3^2B_3^7) - v_3^1A_3^8 - v_3^2B_3^8]dx_3$$

$$+\int_{l_{0_4}}^{l_4} \rho_4[\frac{d}{dt}(v_4^1A_4^7 + v_4^2B_4^7) - v_4^1A_4^8 - v_4^2B_4^8]dx_4 = -E_2I_2\left.\frac{\partial^2 w_2}{\partial x_2^2}\right|_{l_2} + u_3$$

w_3 equation

$$\rho_3[\frac{d}{dt}(v_3^1A_3^9 + v_3^2B_3^9) - v_3^1A_3^{10} - v_3^2B_3^{10}] + E_3I_3\frac{\partial^4 w_3}{\partial x_3^4} = 0$$

w_4 equation

$$\rho_4[\frac{d}{dt}(v_4^1A_4^9 + v_4^2B_4^9) - v_4^1A_4^{10} - v_4^2B_4^{10}] + E_4I_4\frac{\partial^4 w_4}{\partial x_4^4} = 0$$

Observe that the A_i^j and B_i^j are functions (Table 4.1) of the system state variables, geometric parameters, and structural parameters, and thus the above equations are nonlinear.

■

In Ref. 30, we use the above formulation to derive a rigorously stabilizing control law; the stability guarantee is based on the nonlinear hybrid ODE/PDE/integral description of the system dynamics. In Ref. 30, we also introduce a finite element discretization for the purpose of simulations and to accommodate tuning of the control gains over the stable region. Some laboratory demonstrations are reported in Ref. 29 for a special case of this configuration.

Finally, some qualitative remarks are offered on the above developments. Some significant generalizations of Lagrange's equations for distributed parameter systems are presented and illustrated by examples. Specifically, a topology of point-connected bodies is considered. The collection of bodies can consist of particles, rigid bodies or flexible bodies. Each flexible body can be described by any constitutive model function of inertial position and velocity, and an internal deformation state consisting of a deformation field, and a finite number of the space-time derivatives thereof. This collection of bodies may undergo large overall motions and deformations, under the influence of general distributed and point boundary loading. Consistent with the Lagrangian philosophy, any set of independent

Table 4.1 Definition of Functions Appearing in the Equations of Motion for the Multibody Flexible Spacecraft Example 4.7

$A_1^1 = -w_1 \quad A_1^2 = -\dot{\theta} \quad B_1^1 = x_1 \quad B_1^2 = 1$

$A_2^1 = -\dot{\theta} \quad A_2^2 = -w_1(l_1) - x_2 \cos\alpha + w_2 \sin\alpha \quad A_2^3 = -x_2\cos\alpha + w_2 \sin\alpha$

$A_2^4 = x_2\dot{\alpha}\sin\alpha + \dot{w}_2\sin\alpha + w_2\dot{\alpha}\cos\alpha + x_2\dot{\theta}\sin\alpha + w_2\dot{\theta}\cos\alpha$

$A_2^5 = -\cos\alpha \quad A_2^6 = (\dot{\alpha} + \dot{\theta})\sin\alpha$

$B_2^1 = 1 \quad B_2^2 = l_1 - x_2\sin\alpha - w_2\cos\alpha \quad B_2^3 = -x_2\sin\alpha - w_2\cos\alpha$

$B_2^4 = -x_2\dot{\alpha}\cos\alpha - \dot{w}_2\cos\alpha + w_2\dot{\alpha}\sin\alpha - x_2\dot{\theta}\cos\alpha + w_2\dot{\theta}\sin\alpha$

$B_2^5 = -\sin\alpha \quad B_2^6 = -(\dot{\alpha} + \dot{\theta})\cos\alpha$

$A_3^1 = -\dot{\theta} \quad A_3^2 = -w_1(l_1) - l_2\cos\alpha + w_2(l_2)\sin\alpha + x_3\cos(\alpha+\beta) + w_3\sin(\alpha+\beta)$

$A_3^3 = -l_2\cos\alpha + w_2(l_2)\sin\alpha + x_3\cos(\alpha+\beta) + w_3\sin(\alpha+\beta)$

$A_3^4 = l_2\dot{\alpha}\sin\alpha + \dot{w}_2(l_2)\sin\alpha + w_2(l_2)\dot{\alpha}\cos\alpha + l_2\dot{\theta}\sin\alpha + w_2(l_2)\dot{\theta}\cos\alpha$
$\quad -x_3(\dot{\alpha}+\dot{\beta})\sin(\alpha+\beta) + \dot{w}_3\sin(\alpha+\beta) + w_3(\dot{\alpha}+\dot{\beta})\cos(\alpha+\beta)$
$\quad -x_3\dot{\theta}\sin(\alpha+\beta) + w_3\dot{\theta}\cos(\alpha+\beta)$

$A_3^5 = -\cos\alpha \quad A_3^6 = (\dot{\alpha}+\dot{\theta})\sin\alpha \quad A_3^7 = x_3\cos(\alpha+\beta) + w_3\sin(\alpha+\beta)$

$A_3^8 = x_3(\dot{\alpha}+\dot{\beta})\sin(\alpha+\beta) + \dot{w}_3\sin(\alpha+\beta) + w_3(\dot{\alpha}+\dot{\beta})\cos(\alpha+\beta)$
$\quad +x_3\dot{\theta}\sin(\alpha+\beta) + w_3\dot{\theta}\cos(\alpha+\beta)$

$A_3^9 = -\cos(\alpha+\beta) \quad A_3^{10} = (\dot{\alpha}+\dot{\beta}+\dot{\theta})\sin(\alpha+\beta)$

$B_3^1 = 1 \quad B_3^2 = l_1 - l_2\sin\alpha - w_2(l_2)\cos\alpha + x_3\sin(\alpha+\beta) - w_3\cos(\alpha+\beta)$

$B_3^3 = -l_2\sin\alpha - w_2(l_2)\cos\alpha + x_3\sin(\alpha+\beta) - w_3\cos(\alpha+\beta)$

$B_3^4 = -l_2\dot{\alpha}\cos\alpha - \dot{w}_2(l_2)\cos\alpha + w_2(l_2)\dot{\alpha}\sin\alpha - l_2\dot{\theta}\cos\alpha + w_2(l_2)\dot{\theta}\sin\alpha$
$\quad +x_3(\dot{\alpha}+\dot{\beta})\cos(\alpha+\beta) - \dot{w}_3\cos(\alpha+\beta) + w_3(\dot{\alpha}+\dot{\beta})\sin(\alpha+\beta)$
$\quad +x_3\dot{\theta}\cos(\alpha+\beta) + w_3\dot{\theta}\sin(\alpha+\beta)$

$B_3^5 = -\sin\alpha \quad B_3^6 = -(\dot{\alpha}+\dot{\theta})\cos\alpha \quad B_3^7 = +x_3\sin(\alpha+\beta) - w_3\cos(\alpha+\beta)$

$B_3^8 = +x_3(\dot{\alpha}+\dot{\beta})\cos(\alpha+\beta) - \dot{w}_3\cos(\alpha+\beta) + w_3(\dot{\alpha}+\dot{\beta})\sin(\alpha+\beta)$
$\quad +x_3\dot{\theta}\cos(\alpha+\beta) + w_3\dot{\theta}\sin(\alpha+\beta)$

$B_3^9 = -\sin(\alpha+\beta) \quad B_3^{10} = -(\dot{\alpha}+\dot{\beta}+\dot{\theta})\cos(\alpha+\beta)$

Table 4.1 Definition of Functions Appearing in the Equations of Motion for the Multibody Flexible Spacecraft Example 4.7, (Continued)

$A_4^1 = -\dot{\theta} \quad A_4^2 = -w_1(l_1) - l_2\cos\alpha + w_2(l_2)\sin\alpha - x_4\cos(\alpha+\beta) + w_4\sin(\alpha+\beta)$

$A_4^3 = -l_2\cos\alpha + w_2(l_2)\sin\alpha - x_4\cos(\alpha+\beta) + w_4\sin(\alpha+\beta)$

$A_4^4 = l_2\dot{\alpha}\sin\alpha + \dot{w}_2(l_2)\sin\alpha + w_2(l_2)\dot{\alpha}\cos\alpha + l_2\dot{\theta}\sin\alpha + w_2(l_2)\dot{\theta}\cos\alpha$
$\quad +x_4(\dot{\alpha}+\dot{\beta})\sin(\alpha+\beta) + \dot{w}_4\sin(\alpha+\beta) + w_4(\dot{\alpha}+\dot{\beta})\cos(\alpha+\beta)$
$\quad +x_4\dot{\theta}\sin(\alpha+\beta) + w_4\dot{\theta}\cos(\alpha+\beta)$

$A_4^5 = -\cos\alpha \qquad A_4^6 = (\dot{\alpha}+\dot{\theta})\sin\alpha \quad A_4^7 = -x_4\cos(\alpha+\beta) + w_4\sin(\alpha+\beta)$

$A_4^8 = x_4(\dot{\alpha}+\dot{\beta})\sin(\alpha+\beta) + \dot{w}_4\sin(\alpha+\beta) + w_4(\dot{\alpha}+\dot{\beta})\cos(\alpha+\beta)$
$\quad +x_4\dot{\theta}\sin(\alpha+\beta) + w_4\dot{\theta}\cos(\alpha+\beta)$

$A_4^9 = -\cos(\alpha+\beta) \qquad A_4^{10} = (\dot{\alpha}+\dot{\beta}+\dot{\theta})\sin(\alpha+\beta)$

$B_4^1 = 1 \quad B_4^2 = l_1 - l_2\sin\alpha - w_2(l_2)\cos\alpha - x_4\sin(\alpha+\beta) - w_4\cos(\alpha+\beta)$

$B_4^3 = -l_2\sin\alpha - w_2(l_2)\cos\alpha - x_4\sin(\alpha+\beta) - w_4\cos(\alpha+\beta)$

$B_4^4 = -l_2\dot{\alpha}\cos\alpha - \dot{w}_2(l_2)\cos\alpha + w_2(l_2)\dot{\alpha}\sin\alpha - l_2\dot{\theta}\cos\alpha + w_2(l_2)\dot{\theta}\sin\alpha$
$\quad -x_4(\dot{\alpha}+\dot{\beta})\cos(\alpha+\beta) - \dot{w}_4\cos(\alpha+\beta) + w_4(\dot{\alpha}+\dot{\beta})\sin(\alpha+\beta)$
$\quad -x_4\dot{\theta}\cos(\alpha+\beta) + w_4\dot{\theta}\sin(\alpha+\beta)$

$B_4^5 = -\sin\alpha \qquad B_4^6 = -(\dot{\alpha}+\dot{\theta})\cos\alpha \quad B_4^7 = -x_4\sin(\alpha+\beta) - w_4\cos(\alpha+\beta)$

$B_4^8 = -x_4(\dot{\alpha}+\dot{\beta})\cos(\alpha+\beta) - \dot{w}_4\cos(\alpha+\beta) + w_4(\dot{\alpha}+\dot{\beta})\sin(\alpha+\beta)$
$\quad -x_4\dot{\theta}\cos(\alpha+\beta) + w_4\dot{\theta}\sin(\alpha+\beta)$

$B_4^9 = -\sin(\alpha+\beta) \qquad B_4^{10} = -(\dot{\alpha}+\dot{\beta}+\dot{\theta})\cos(\alpha+\beta)$

$A_5^1 = -w_1(l_1) \quad A_5^2 = -\dot{\theta} \qquad B_5^1 = l_1 \quad B_5^2 = 1$

$A_6^1 = -\dot{\theta} \qquad A_6^2 = -w_1(l_1) - l_2\cos\alpha + w_2(l_2)\sin\alpha \quad A_6^3 = -l_2\cos\alpha + w_2(l_2)\sin\alpha$

$A_6^4 = l_2\dot{\alpha}\sin\alpha + \dot{w}_2(l_2)\sin\alpha + w_2(l_2)\dot{\alpha}\cos\alpha + l_2\dot{\theta}\sin\alpha + w_2(l_2)\dot{\theta}\cos\alpha$

$A_6^5 = -\cos\alpha \qquad A_6^6 = (\dot{\alpha}+\dot{\theta})\sin\alpha \qquad B_6^1 = 1 \qquad B_6^2 = l_1 - l_2\sin\alpha - w_2(l_2)\cos\alpha$

$B_6^3 = -l_2\sin\alpha - w_2(l_2)\cos\alpha$

$B_6^4 = -l_2\dot{\alpha}\cos\alpha - \dot{w}_2(l_2)\cos\alpha + w_2(l_2)\dot{\alpha}\sin\alpha - l_2\dot{\theta}\cos\alpha + w_2(l_2)\dot{\theta}\sin\alpha$

$B_6^5 = -\sin\alpha \qquad B_6^6 = -(\dot{\theta}+\dot{\alpha})\cos\alpha$

generalized coordinates may be chosen to describe the hybrid system's configuration. For the main results of Eqs. (4.32)-(4.35), the deformable bodies consist of beamlike and rodlike substructures having one independent spatial variable each, but a vector of dependent spatial variables. These equations can be specialized to generate the governing equations and associated boundary conditions for a variety of systems and can be extended to admit more general configurations. These formulations are suitable for computerized symbol manipulation (via Mathematica or MACSYMA, for example) to make possible the determination of rigorous hybrid differential equation models for an extremely large family of dynamical systems. Since the associated hybrid coordinate methods for deriving (for example, Lyapunov stable) control formulations and solution methods are rapidly advancing, these developments promise to spawn many applications and extensions.

4.2.3 The Differential Eigenvalue Problem

Basic Considerations

The response of a finite-dimensional model is usually obtained by modal analysis, which often makes use of the solution of an *algebraic eigenvalue problem* associated with linearized equations of motion in the vicinity of a nominal state or trajectory. Development of a finite-dimensional model from a nonlinear hybrid coordinate system of ordinary and partial differential equations is a non-trivial process that necessarily involves approximation. Evaluating the accuracy of the resulting approximate models cannot be addressed in general; however, for the case of linear distributed parameter systems, it is possible to obtain a more rigorous analytical closure. Also, the solution of linear distributed parameter systems provides key insights necessary to deal with the nonlinear case, and so we consider the linear case in detail.

To determine the exact form of the response of a linear distributed parameter system, we must solve a *differential eigenvalue problem.* In contrast to the algebraic eigenvalue problem encountered in the linear finite-dimensional approximate solutions, the differential eigenvalue problem will be seen to yield a solution involving an *infinite* set of eigenvalues and corresponding eigenfunctions. Thus distributed parameter systems are correctly considered as infinite-dimensional.

As a most fundamental example, let us consider the free transverse vibration of a beam, p=0; then Eq. (4.12) reduces to

$$\frac{\partial^2}{\partial x^2}\left(EI\frac{\partial^2 y}{\partial x^2}\right) + \rho\frac{\partial^2 y}{\partial t^2} = 0 \tag{4.36}$$

Since Eq. (4.36) is a linear differential equation, we seek a separation of variables form for the solution as

$$y(x,t) = Y(x)F(t) \tag{4.37}$$

where $Y(x)$ represents the spatially varying beam deformation shape function and $F(t)$ represents the time-varying amplitude of the motion.

Substituting Eq. (4.37) into Eq. (4.36), and dividing through by $\rho(x)Y(x)F(t)$, we obtain the desired separation of variables

$$\frac{1}{\rho(x)Y(x)}\frac{d^2}{dx^2}\left[EI\frac{d^2Y(x)}{dx^2}\right] = -\frac{1}{F(t)}\frac{d^2F(t)}{dt^2} \tag{4.38}$$

Note that the left side of this equation is independent of t, whereas the right side is independent of x. Since this equation must hold for all t and x, it therefore follows that each side must equal the same constant value. Letting the (to-be-determined) constant be a positive value, ω^2, on the basis of physical arguments (if we choose a negative constant, the solution diverges with time and is inconsistent with the stable free oscillation of an obviously conservative system), we thus obtain the following two ordinary differential equations:

$$\frac{d^2}{dx^2}\left[EI\frac{d^2Y(x)}{dx^2}\right] - \omega^2\rho(x)Y(x) = 0 \tag{4.39}$$

$$\frac{d^2F(t)}{dt^2} + \omega^2F(t) = 0 \tag{4.40}$$

The general solution of Eq. (4.40) is the simple harmonic motion

$$F(t) = C\cos(\omega t - \phi) \tag{4.41}$$

where C is an arbitrary constant, ω is the frequency of the harmonic motion, and ϕ is a phase angle.

Now the problem of determining the value (or values) of the parameter ω^2, for which Eq. (4.39) has a nontrivial solution satisfying boundary conditions is known as the *characteristic problem* or the *differential eigenvalue problem* (Refs. 2-6). Equation (4.39) can also be solved for $Y(x,\rho,EI,\omega^2)$ (if EI and ρ are constant) in terms of complex exponential (or real sin, cos, sinh, cosh) functions, as shown below in the examples. In exceptionally simple cases, these transcendental *characteristic equations* have exact analytical solutions for the roots; however, they most often must be determined by iterative numerical methods and, in many cases, the higher roots are difficult to determine because these equations often oscillate through zero with divergent amplitude. As shown below, graphical methods are useful in establishing approximate root locations to essentially guarantee convergence of the numerical root-solving algorithms to the usually finite set of characteristic roots of interest.

Notice that ω^2 is a parameter in the solution for Y. The set of values of the parameter ω^2 that satisfy Eqs. (4.39) and (4.40) and the boundary conditions are called *characteristic values* or *eigenvalues*, and the associated nontrivial solutions for the shape functions Y(x) are called the *characteristic functions* or *eigenfunctions.* Note that the solution of the characteristic equation generally consists of an infinite set of eigenvalues, and the corresponding infinite set of eigenfunctions is unique except for their amplitudes. The details of this discussion cannot be carried to completion without specification of boundary conditions, and this is best illustrated by examples.

Example 4.8

Consider the free transverse vibration of the uniform (EI and ρ are constant) fixed-free beam, as shown in Fig. 4.2. For a uniform beam, Eq. (4.39) becomes

$$\frac{d^4Y(x)}{dx^4} - \beta^4 Y(x) = 0, \qquad \beta^4 \equiv \frac{\omega^2 \rho}{EI} \tag{4.42}$$

The boundary conditions at the clamped end, x = 0, are

$$Y(0) = 0, \qquad \left.\frac{dY}{dx}\right|_{x=0} = 0 \tag{4.43}$$

On the other hand, at the free end, x = L, the boundary conditions are

$$\left.\frac{d^2Y}{dx^2}\right|_{x=L} = 0, \qquad \left.\frac{d^3Y}{dx^3}\right|_{x=L} = 0 \tag{4.44}$$

Note that the two sets of boundary conditions (4.43) and (4.44) are the geometric (*essential*) boundary conditions and the physical (*natural*) boundary conditions, respectively.

The general solution of Eq. (4.42) is

$$Y(x) = C_1 \sin(\beta x) + C_2 \cos(\beta x) + C_3 \sinh(\beta x) + C_4 \cosh(\beta x) \tag{4.45}$$

Evaluating the boundary conditions at x = 0 and x = L, we obtain the following homogeneous system of algebraic equations:

$$\begin{bmatrix} 0 & 1 & 0 & 1 \\ \beta & 0 & \beta & 0 \\ -\beta^2 \sin(\beta L) & -\beta^2 \cos(\beta L) & \beta^2 \sinh(\beta L) & \beta^2 \cosh(\beta L) \\ -\beta^3 \cos(\beta L) & \beta^3 \sin(\beta L) & \beta^3 \cosh(\beta L) & \beta^3 \sinh(\beta L) \end{bmatrix} \begin{Bmatrix} C_1 \\ C_2 \\ C_3 \\ C_4 \end{Bmatrix} = \begin{Bmatrix} 0 \\ 0 \\ 0 \\ 0 \end{Bmatrix} \tag{4.46}$$

For this set of homogeneous equations to have a nontrivial solution, the determinant of the coefficients must vanish. It follows that the determinant vanishes if and only if β is such that the following condition holds:

$$\cos \beta L \cosh \beta L = -1 \tag{4.47}$$

which is recognized as the *characteristic equation*. No simple, explicit expression for the roots of this characteristic equation is available; therefore, these roots must be determined using some numerical method, generally yielding an infinite set of eigenvalues β_r ($r = 1, 2, \ldots$). The first few eigenvalues are given approximately as follows:

$$\beta_1 L = 1.875, \qquad \beta_2 L = 4.694, \qquad \beta_3 L = 7.855$$

From Eq. (4.42),

$$\omega_i = (\beta_i L)^2 \sqrt{\frac{EI}{\rho L^4}}$$

Therefore, the approximate values of the first three natural frequencies are obtained as follows:

$$\omega_1 = 1.875^2 \sqrt{\frac{EI}{\rho L^4}}, \qquad \omega_2 = 4.694^2 \sqrt{\frac{EI}{\rho L^4}}, \qquad \omega_3 = 7.855^2 \sqrt{\frac{EI}{\rho L^4}}$$

The corresponding eigenfunctions have the form

$$Y_r(x) = C\left[\cosh(\beta_r x) - \cos(\beta_r x) - k_r\left\{\sinh(\beta_r x) - \sin(\beta_r x)\right\}\right] \tag{4.48}$$

where the constant k_r is defined as

$$k_r = \frac{\cosh(\beta_r L) + \cos(\beta_r L)}{\sinh(\beta_r L) + \sin(\beta_r L)}$$

and C can be chosen arbitrarily according to some normalization scheme.

The solution of the characteristic equation (4.47) can be seen graphically, as shown in Figure 4.11. The first three eigenvalues ($\beta_i L$, i=1,2,3) are visible from the first two figures (Figs. 4.11a,b). On account of the presence of the hyperbolic functions, the characteristic equation behaves in a rapidly divergent, oscillatory fashion; the zero crossings for the higher eigenvalues are nearly perpendicular as shown in Figures 4.11c-e. Since the usual form of the characteristic equation includes the divergent hyperbolic functions, it is very hard to obtain the higher eigenvalues

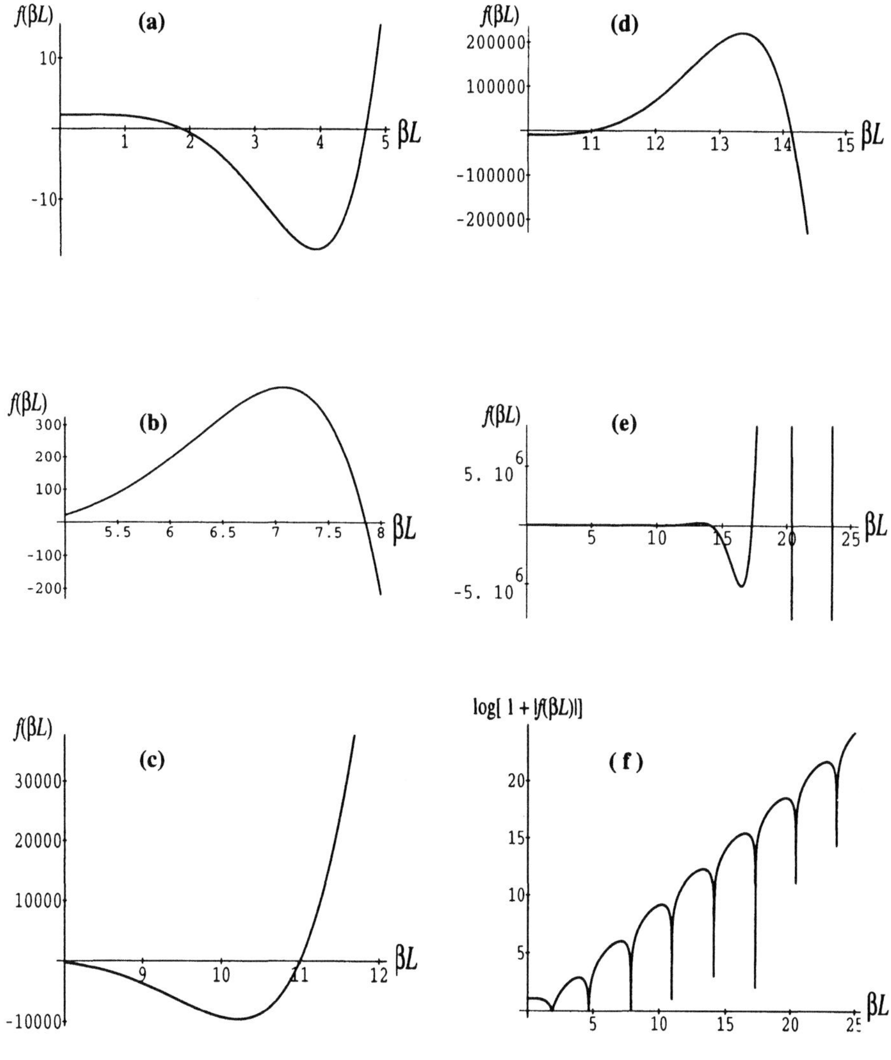

Figure 4.11 Eigenvalues of the Clamped-Free Beam: Roots of $f(\beta \mathrm{L}) = \cos(\beta \mathrm{L}) \cosh(\beta \mathrm{L}) + 1 = 0$.

numerically (for one thing, the values of the characteristic equation very near a higher characteristic root are sufficiently large to challenge even double precision floating point arithmetic). Note from the scale of $f(\beta L)$ in Figure 4.11e that even though we determine the value $\beta_r L$ that is nearest (accurate to 16 digits) to the zero crossing of the characteristic function (i.e., $f(\beta_r L - \epsilon)\ f(\beta_r L + \epsilon) < 0$ with $\epsilon = 1.0 \times 10^{-16}$), $f(\beta_r L)$ has a nonzero value that cannot usually be considered negligible and, therefore, an unsophisticated root-solving algorithm may have difficulty finding $\beta_r L$ as a sufficiently good approximation of an eigenvalue unless we are using very high precision.

As a consequence of this divergent oscillatory behavior, it is often difficult to obtain the higher eigenvalues without using graphical methods and, as might be expected, these higher eigenvalues are often highly sensitive to system parameter and boundary condition variations. The final graph (Figure 4.11f) illustrates a useful device for graphically displaying the location of zeros for divergent characteristic equations; we have found that these graphically located root approximations are almost always sufficiently accurate to start Newton's method or some other interval subdivision method for locating the eigenvalues to adequate precision.

■

Note that Eq. (4.39) is a special case of the general eigenvalue problem of the differential operator form

$$\mathcal{L}(Y) = \lambda \mathcal{M}(Y) \tag{4.49}$$

where λ denotes the eigenvalue, $\mathcal{L}$ and $\mathcal{M}$ denote the linear homogeneous differential operators of (spatial differentiation) orders $2p$ and $2q$ (with $p > q$), respectively, and the boundary conditions are

$$\mathcal{B}_i(Y) = 0, \qquad i = 1, 2, \ldots, p \tag{4.50}$$

where $\mathcal{B}_i$ are linear homogeneous differential operators of maximum order $2p - 1$.

As an illustration, let us consider the eigenvalue problem in Example 4.8. The differential operators for the general eigenvalue problem, Eq. (4.49), and the boundary conditions, Eqs. (4.50), for this special problem are represented for Eq. (4.39) [instead of Eq. (4.42)], Eqs. (4.43) and (4.44) as follows:

$$\mathcal{L} = \frac{d^2}{dx^2}\left[EI\frac{d^2}{dx^2}\right], \qquad \mathcal{M} = \rho(x), \qquad \lambda = \omega^2$$

with

$$\mathcal{B}_1 = 1, \qquad \mathcal{B}_2 = \frac{d}{dx}, \qquad \text{at } x=0$$

$$\mathcal{B}_1 = \frac{d^2}{dx^2}, \qquad \mathcal{B}_2 = \frac{d^3}{dx^3}, \qquad \text{at } x=L$$

The eigenvalue problem as defined by Eqs. (4.49) and (4.50) is said to be *self-adjoint* if the following conditions are satisfied for any two arbitrary eigenfunctions $Y_r(x)$ and $Y_s(x)$ (Refs. 2-4):

$$\int_0^L Y_r(x)\mathcal{L}\Big(Y_s(x)\Big)dx = \int_0^L Y_s(x)\mathcal{L}\Big(Y_r(x)\Big)dx \tag{4.51}$$

$$\int_0^L Y_r(x)\mathcal{M}\Big(Y_s(x)\Big)dx = \int_0^L Y_s(x)\mathcal{M}\Big(Y_r(x)\Big)dx \tag{4.52}$$

The self-adjoint property can be established by integrating Eqs. (4.51) and (4.52) by parts.

Orthogonality of Natural Modes

The most useful group property of the eigenfunctions is the property of orthogonality. Consider the general eigenvalue problem of the self-adjoint system, Eq. (4.49), for the two distinct eigenvalues λ_r, λ_s and the corresponding eigenfunctions $Y_r(x)$ and $Y_s(x)$, i.e.,

$$\mathcal{L}\Big(Y_r(x)\Big) = \lambda_r\mathcal{M}\Big(Y_r(x)\Big) \tag{4.53}$$

$$\mathcal{L}\Big(Y_s(x)\Big) = \lambda_s\mathcal{M}\Big(Y_s(x)\Big) \tag{4.54}$$

Multiplying Eq. (4.53) by $Y_s(x)$ and Eq. (4.54) by $Y_r(x)$, subtracting one result from the other, and integrating both sides of the equation over the domain $0 \le x \le L$ yield

$$\begin{aligned}\int_0^L \Big\{Y_s(x)\mathcal{L}\Big(Y_r(x)\Big) - Y_r(x)\mathcal{L}\Big(Y_s(x)\Big)\Big\}dx \\ = \int_0^L \Big\{\lambda_r Y_s(x)\mathcal{M}\Big(Y_r(x)\Big) - \lambda_s Y_r(x)\mathcal{M}\Big(Y_s(x)\Big)\Big\}dx\end{aligned} \tag{4.55}$$

Since eigenfunctions $Y_r(x)$ and $Y_s(x)$ are solutions of the eigenvalue problem of the self-adjoint system, Eq. (4.55) can be reduced using Eqs. (4.51) and (4.52) as

$$(\lambda_r - \lambda_s)\int_0^L Y_r(x)\mathcal{M}\Big(Y_s(x)\Big)dx = 0 \tag{4.56}$$

In general, for the distinct pairs of eigenvalues, we obtain

$$\int_0^L Y_r(x)\mathcal{M}\Big(Y_s(x)\Big)dx = 0, \qquad \text{for} \quad \lambda_r \ne \lambda_s \tag{4.57}$$

which is known as the *generalized orthogonality condition*, and the eigenfunctions $Y_r(x)$ and $Y_s(x)$ are said to be orthogonal. Therefore, we have an important property for a self-adjoint system (Refs. 2-4): *if a dynamical system is self-adjoint and the eigenvalues are distinct, then the corresponding eigenfunctions are orthogonal.* Note that, for the repeated eigenvalue case, the corresponding eigenfunctions are not generally orthogonal to each other, although they are orthogonal to the remaining eigenfunctions.

Equation (4.57) can be rewritten as follows:

$$\int_0^L Y_r(x)\mathcal{M}\Big(Y_s(x)\Big)dx = c_r\delta_{rs}, \qquad r,s = 1,2,\ldots \tag{4.58}$$

where δ_{rs} denotes the Kronecker delta function.

If the eigenfunctions are normalized so that they satisfy

$$\int_0^L Y_r(x)\mathcal{M}\Big(Y_r(x)\Big)dx = 1, \qquad r = 1,2,\ldots \tag{4.59}$$

then, they become orthonormal, and Eq. (4.58) specializes to the *orthonormality conditions* represented as

$$\int_0^L Y_r(x)\mathcal{M}\Big(Y_s(x)\Big)dx = \delta_{rs}, \qquad r,s = 1,2,\ldots \tag{4.60}$$

Moreover, since eigenfunctions $Y_r(x)$ and $Y_s(x)$ satisfy the general eigenvalue problem, Eq. (4.49), we also have

$$\int_0^L Y_r(x)\mathcal{L}\Big(Y_s(x)\Big)dx = \lambda_r\delta_{rs}, \qquad r,s = 1,2,\ldots \tag{4.61}$$

Normalized eigenfunctions are called *normal modes.* Note that the differential eigenvalue problem for distributed parameter systems, Eq. (4.49), and the eigenfunction orthonormality properties, Eqs. (4.60) and (4.61), are the counterparts of the algebraic eigenvalue problem and eigenvector orthonormality conditions, Eqs. (2.44)-(2.46), for discrete coordinate dynamical systems [for self-adjoint discrete coordinate systems, see also Eq. (2.51)].

Example 4.9

Consider the free transverse vibration of the uniform fixed-free beam as in Example 4.8. The eigenvalue problem is

$$\frac{d^2}{dx^2}\left[EI\frac{d^2Y_r(x)}{dx^2}\right] = \omega_r^2\rho(x)Y_r(x) \tag{4.62}$$

Multiply Eq. (4.62) by $Y_s(x)$ $\big(\neq Y_r(x)\big)$ and integrate from 0 to L:

$$\int_0^L \frac{d^2}{dx^2}\left[EI\frac{d^2Y_r(x)}{dx^2}\right]Y_s(x)dx = \omega_r^2 \int_0^L \rho(x)Y_r(x)Y_s(x)dx \tag{4.63}$$

Integrating by parts and noting that the same boundary conditions, Eqs. (4.65) and (4.66), apply to modes r and s, we can reduce Eq. (4.63) to

$$\int_0^L EI\frac{d^2Y_r(x)}{dx^2}\frac{d^2Y_s(x)}{dx^2}dx = \omega_r^2 \int_0^L \rho(x)Y_r(x)Y_s(x)dx \tag{4.64}$$

Starting with Eq. (4.62) written for mode s, multiplying by eigenfunction $Y_r(x)$, and integrating from 0 to L, we get

$$\int_0^L EI\frac{d^2Y_r(x)}{dx^2}\frac{d^2Y_s(x)}{dx^2}dx = \omega_s^2 \int_0^L \rho(x)Y_r(x)Y_s(x)dx \tag{4.65}$$

Subtracting Eq. (4.64) from Eq. (4.65) yields

$$(\omega_r^2 - \omega_s^2)\int_0^L \rho(x)Y_r(x)Y_s(x)dx = 0 \tag{4.66}$$

For pairs of modes having distinct eigenvalues, the eigenfunctions satisfy the orthogonality property

$$\int_0^L \rho(x)Y_r(x)Y_s(x)dx = c_r\delta_{rs} \tag{4.67}$$

These eigenfunctions are said to be *orthogonal with respect to the mass distribution*; i.e., the mass density distribution $\rho(x)$ is the weight function of the inner product. By substituting Eq. (4.67) into Eq. (4.64), we see that the eigenfunctions $Y_r(x)$ and $Y_s(x)$ are also *orthogonal with respect to the stiffness distribution* EI(x) as given by

$$\int_0^L EI\frac{d^2Y_r(x)}{dx^2}\frac{d^2Y_s(x)}{dx^2}dx = c_r\omega_r^2\delta_{rs} \tag{4.68}$$

■

The family of normal modes $Y_r(x)$ constitutes a complete set of orthonormal modes. An *expansion theorem* for distributed parameter systems can be stated as follows (Refs. 2-4): *Any function* y(x,t), *that satisfies the homogeneous boundary* *conditions of the problem and that is such that $\mathcal{L}\big(y(x,t)\big)$ is a continuous function can be represented by an absolutely and uniformly convergent series of the eigenfunctions.* In particular, the deflection y(x,t) can be written as a linear combination of the system eigenfunctions as

$$y(x,t) = \sum_{r=1}^{\infty} F_r(t) Y_r(x) \tag{4.69}$$

where the coefficients $F_n(t)$ are efficiently determined by using orthogonality condition, i.e., multipling Eq. (4.69) by $\mathcal{M}\big(Y_n(x)\big)$ and integrating from 0 to L, and using the orthogonality condition of Eq. (4.57), we get

$$F_n(t) = \frac{\int_0^L y(x,t)\mathcal{M}\big(Y_n(x)\big)dx}{\int_0^L Y_n(x)\mathcal{M}\big(Y_n(x)\big)dx} \tag{4.70}$$

By using Eq. (4.59) when we use normal modes as the basis functions in the expansion theorem, Eq. (4.70) is reduced to

$$F_n(t) = \int_0^L y(x,t)\mathcal{M}\big(Y_n(x)\big)dx \tag{4.71}$$

As an illustration, let us consider again the transverse vibration of the uniform fixed-free beam in Example 4.9. For this special problem, the coefficients $F_n(t)$ in the expansion theorem can be written as

$$F_n(t) = \frac{\int_0^L \rho(x)y(x,t)Y_n(x)dx}{\int_0^L \rho(x)Y_n(x)^2 dx} \tag{4.72}$$

and we note that the eigenfunctions in Eq. (4.72) have not been normalized with respect to the mass distribution.

The expansion theorem, Eqs. (4.69)-(4.71), plays an important role in the vibration of distributed systems such that it is used to obtain the system response by modal analysis (Refs. 2-4).

4.3 Approximate Methods for Finite-Dimensional Models

Since large flexible space structures are most rigorously modeled as distributed parameter systems, their motion is described by a coupled system of ordinary and partial differential equations. Since coupled partial differential equation models are difficult to deal with both analytically and computationally, approximate finite-dimensional equations of motion are usually used (Refs. 2-6). In the most common approximation technique, a spatial discretization is used, whereby the continuous elastic systems are replaced by spatially discrete systems. Note that while spatial discretization is introduced, the resulting ordinary differential equations are still continuous with respect to time. Time discretization can also be introduced, of course, and the resulting equations would be of the difference type. In this section, the two most common spatial discretization methods, the *assumed modes method* (a Ritz approximation) and the *finite element method*, will be discussed. Both of these approaches lead to linear, finite-dimensional, continuous time equations of motion for approximating the dynamics of flexible structures. The assumed modes approach is shown to be useful for problems whose geometry is sufficiently simple to permit the insight necessary to select a "good" set of global displacement functions. The finite element approach, on the other hand, is more broadly applicable to arbitrary geometry, but results in high-dimensioned systems of differential equations.

4.3.1 Assumed Modes Method

In the *assumed modes method*, the deflection of continuous elastic structures is modeled by a finite series of space-dependent functions that are multiplied by specified time-dependent amplitude functions (Refs. 2-6). The amplitudes become a set (or subset) of generalized configuration coordinates, in the usual Lagrangian interpretation of generalized coordinates. The space-dependent functions are typically chosen as a complete set of linearly independent functions, selected to satisfy at least the geometric boundary conditions, and differentiable at least half as many times as the order of the system. We refer to all such functions as *admissible functions.* Admissible functions are the most widely encountered choice for the basis functions in the assumed modes method. However, if the problem formulation and a priori insight permit, the basis functions may be selected to satisfy, in addition, some or all of the physical (natural) boundary conditions, and are differentiable as many times as the order of the system. We refer to this more restricted subset of admissible functions (satisfying both geometric and physical boundary conditions) as *comparison functions.* When we use comparison functions, more efficient convergence can be obtained as a general rule, but cannot be guaranteed for every forced response application. That is, for a given number of functions, we expect better accuracy in approximating the true system dynamics when using comparison functions instead of admissible functions. While better accuracy is expected, a particular forced response may, in fact, be more accurately approximated by some judicious admissible functions (this is because of the obvious truth that the forced response depends not only on the nature of the free response eigenfunctions but also on the time and space "shape" of the forcing functions). While the best space-dependent functions are usually the system's eigenfunctions, which satisfy not only all boundary conditions but also the rigorous spatial differential equations (resulting from the separation of variable technique applied to the governing partial differential equations), these are not always available. This is because the rigorous partial differential equations must usually be formulated and solved exactly to obtain the eigenfunctions and, obviously, it is impossible to carry these developments to completion for complicated structures. It is difficult to converge exactly on all of the eigenvalues and the corresponding eigenfunctions, even when an analytically derived characteristic equation can be obtained, because the number of eigenfunctions which can be retained is finite, and the arithmetic errors made in solving the characteristic equation to obtain the eigenvalues cannot always be ignored.

To generate an N degree-of-freedom approximate differential equation model for a continuous system, the displacement of the continuous system is expanded as a linear combination of N prescribed shape functions. Thus the deformation y(x,t) is approximated by

$$y(x,t) = \sum_{i=1}^{N} \phi_i(x) q_i(t) \tag{4.73}$$

where $\phi_i(x)$ denotes ith assumed mode shape, $q_i(t)$ denotes the ith generalized coordinate, and N denotes the number of terms retained in the approximation. The widely adopted term "assumed mode" is sometimes confusing, because *the actual system normal modes of vibration are obviously approximated by a linear combination of the* $\phi_i(x)$ *"assumed modes."* The approximate expansion Eq. (4.73) will be used in conjunction with Lagrange's equations to obtain a finite-dimensional approximate system of ordinary differential equations of motion that govern the time evolution of the amplitudes $q_i(t)$.

The kinetic and potential energy of a continuous system contain spatial integral expressions depending on space and time partial derivatives of y(x,t). Using the approximation (4.73), and performing the corresponding integration with respect to x, we can write the kinetic energy and potential energy in the quadratic forms

$$T(t) = \frac{1}{2}\sum_{i=1}^{N}\sum_{j=1}^{N}[M]_{ij}\dot{q}_i(t)\dot{q}_j(t) = \frac{1}{2}\dot{\mathbf{q}}^T M\dot{\mathbf{q}} \tag{4.74}$$

$$V(t) = \frac{1}{2}\sum_{i=1}^{N}\sum_{j=1}^{N}[K]_{ij}q_i(t)q_j(t) = \frac{1}{2}\mathbf{q}^T K\mathbf{q} \tag{4.75}$$

where $[M]_{ij}$ denotes the (i,j)th element of the symmetric mass matrix M, which depends on the mass distribution of the system and the assumed mode shapes, and $[K]_{ij}$ denotes the (i,j)th element of the symmetric stiffness matrix K, which depends on the stiffness distribution and the assumed mode shapes.

The equations of motion follow on introducing T and V into Lagrange's equations and carrying out the indicated derivatives:

$$\frac{d}{dt}\left(\frac{\partial T}{\partial \dot{q}_r}\right) - \frac{\partial T}{\partial q_r} + \frac{\partial V}{\partial q_r} = Q_r, \qquad r = 1, 2, \ldots, N \tag{4.76}$$

where Q_r denotes the generalized nonconservative forces.

Since T does not depend on the coordinates $q_i(t)$ and V does not depend on the velocities $\dot{q}_i(t)$, we can efficiently substitute Eqs. (4.74) and (4.75) into Eq. (4.76) and obtain the following equations of motion:

$$\sum_{j=1}^{N}[M]_{rj}\ddot{q}_j(t) + \sum_{j=1}^{N}[K]_{rj}q_j(t) = Q_r, \qquad r = 1, \ldots, N \tag{4.77}$$

which can be written compactly in the matrix form

$$M\ddot{\mathbf{q}}(t) + K\mathbf{q}(t) = \mathbf{Q}(t) \tag{4.78}$$

Equation (4.78) provides the desired equations of motion in which the time-varying amplitudes are the generalized coordinates. Clearly, given the instantaneous vector of generalized coordinates $\mathbf{q}(t)$, the instantaneous deformation of the structure is aproximated by the assumed modes expansion of Eq. (4.73). Notice, for example, that an initial value problem is obtained if $\mathbf{q}(0)$, $\dot{\mathbf{q}}(0)$ are known, and that the initial conditions can be approximated from any given $y(x, 0)$, $\dot{y}(x, 0)$ by inverting Eq. (4.73) and its time derivative at initial time.

Example 4.10

Consider a free transverse vibration of Euler-Bernoulli beam with one end clamped and the other free, as shown in Fig. 4.2. The kinetic energy of the system is

$$T(t) = \frac{1}{2}\int_0^L \rho(x)[\dot{y}(x,t)]^2 dx$$

Introducing Eq. (4.73), we find that the kinetic energy of the system is approximated as

$$\begin{aligned} T(t) &= \frac{1}{2}\int_0^L \rho(x)\left[\sum_{i=1}^{N}\phi_i(x)\dot{q}_i(t)\right]\left[\sum_{j=1}^{N}\phi_j(x)\dot{q}_j(t)\right]dx \\ &= \frac{1}{2}\sum_{i=1}^{N}\sum_{j=1}^{N}\dot{q}_i(t)\dot{q}_j(t)\left[\int_0^L \rho(x)\phi_i(x)\phi_j(x)dx\right] \end{aligned}$$

from which we conclude, comparing the above with Eq. (4.74), that the mass coefficients are obtained via the spatial integrals,

$$[M]_{ij} = \int_0^L \rho(x)\phi_i(x)\phi_j(x)dx \tag{4.79}$$

On the other hand, the potential energy can be written as

$$V(t) = \frac{1}{2}\int_0^L EI(x)[y''(x,t)]^2 dx$$

and approximated as

$$\begin{aligned} V(t) &= \frac{1}{2}\int_0^L EI(x)\Big[\sum_{i=1}^{N}\phi_i''(x)q_i(t)\Big]\Big[\sum_{j=1}^{N}\phi_j''(x)q_j(t)\Big]dx \\ &= \frac{1}{2}\sum_{i=1}^{N}\sum_{j=1}^{N} q_i(t)q_j(t)\Big[\int_0^L EI(x)\phi_i''(x)\phi_j''(x)dx\Big] \end{aligned}$$

so that the stiffness coefficients are

$$[K]_{ij} = \int_0^L EI(x)\phi_i''(x)\phi_j''(x)dx \tag{4.80}$$

Since there are no external forces, generalized forces Q_i are all zero. Therefore, the matrix form of the equations of motion is as follows:

$$M\ddot{\mathbf{q}} + K\mathbf{q} = 0$$

To carry out the integrands of Eqs. (4.79) and (4.80), it is necessary to adopt specific functions. In particular, if the comparison function $\phi_j(x)$ in Eqs. (4.79) and (4.80) are chosen as in Refs. 8 and 9, then,

$$\phi_j(x) = 1 - \cos\left(\frac{j\pi x}{L}\right) + \frac{1}{2}(-1)^{j+1}\left(\frac{j\pi x}{L}\right)^2$$

which satisfy the geometric and physical boundary conditions of a clamped-free beam.

For the above particular choices of comparison functions, the (i, j)th elements of the mass and stiffness matrices are then determined from integration [Eqs. (4.79) and (4.80)] as follows:

$$[M]_{ij} = \rho L\Big[1 + (-1)^{i+j}\big[(i/j)^2 + (j/i)^2\big] + \frac{1}{2}\delta_{ij}$$

$$+ (-1)^{i+1}\left(\frac{i^2\pi^2}{6}\right) + (-1)^{j+1}\left(\frac{j^2\pi^2}{6}\right) + (-1)^{i+j}\left(\frac{i^2j^2\pi^4}{20}\right)\Big]$$

$$[K]_{ij} = EI\Big[(-1)^{i+j} + \frac{1}{2}\delta_{ij}\Big]\left(\frac{i^2j^2\pi^4}{L^3}\right)$$

where δ_{ij} is the Kronecker delta.

Note that we have implemented the computation of the above M and K matrices as the MATLAB operation **beamasm**, with the syntax:

> **[M, K] = beamasm(N, rho, L, E, I)**

where N is the number of assumed mode shapes retained in Eq. (4.73).

■

Assumed Mode Application for a Rotating Spacecraft

As an illuminating example for the assumed modes (and other) approximation methods for flexible space structures, we consider the planar rotational/vibrational dynamics of a flexible structure consisting of a rigid hub with four identical cantilevered flexible appendages and tip masses as shown in Figure 4.12. Table 4.2 summarizes the configuration parameters of this flexible structure (Ref. 29). The kinematics assumptions are restricted such that the opposing appendages are constrained, for simplicity, to deflect antisymmetrically such that the deformed system's mass center always coincides with the hub center. For the flexible appendage, the Euler-Bernoulli beam model is used and, therefore, sectional rotary inertia and shear deformation of the beam are neglected. The control actuators are modeled as massless, concentrated, torque generating devices. The control system is assumed to generate a torque u_1 acting on the hub, a torque u_2 acting on the end of appendages 1 and 3, and a torque u_3 acting on the end of the appendages 2 and 4.

In order to derive the equations of motion, consider one (the ith) of the elastic appendages with tip mass m_t as in Figure 4.13. Let $\{\underline{n}_1, \underline{n}_2, \underline{n}_3\}$ denote right-handed inertial reference axes and $\{\underline{b}_1, \underline{b}_2, \underline{b}_3\}$ right-handed body-fixed axes. Note that r is the hub radius, x_i is the coordinate of a typical mass element measured along

Table 4.2 Configuration Parameters

Parameter	Symbol	Value
Hub radius	r	1 ft
Rotary inertia of hub	J_h	8 slug-ft^2
Mass density of beams	ρ	0.0271875 slug/ft
Elastic modulus of arms	E	0.1584×10^{10} lb/ft^2
Arm length	L	4.0 ft
Arm thickness	t	0.125 in.
Arm height	h	6.0 in.
Tip mass	m_t	0.156941 slug
Rotary inertia of tip masses	J_t	0.0018 slug·ft^2

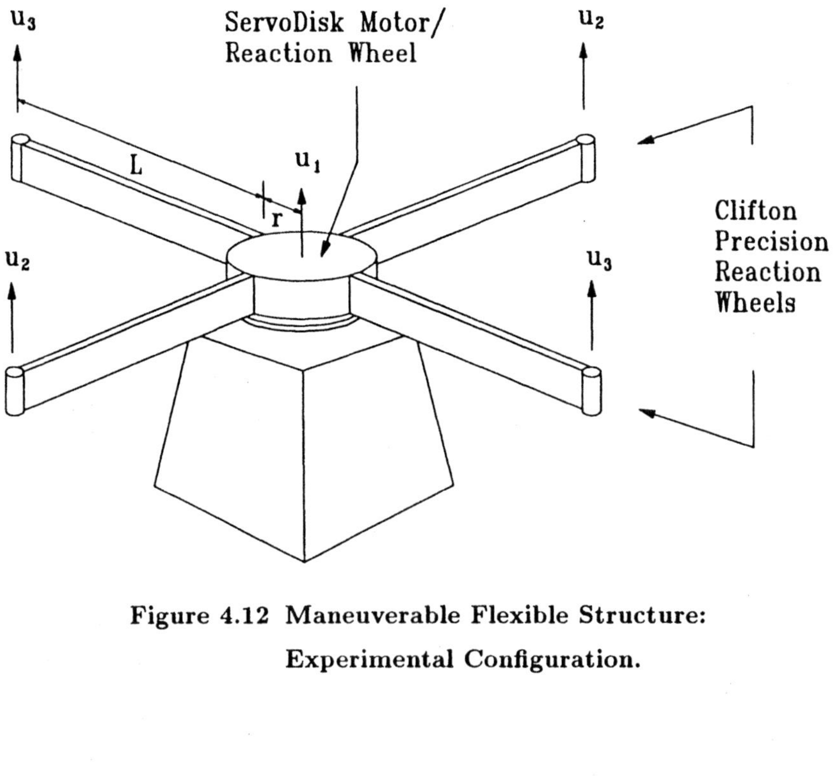

Figure 4.12 Maneuverable Flexible Structure: Experimental Configuration.

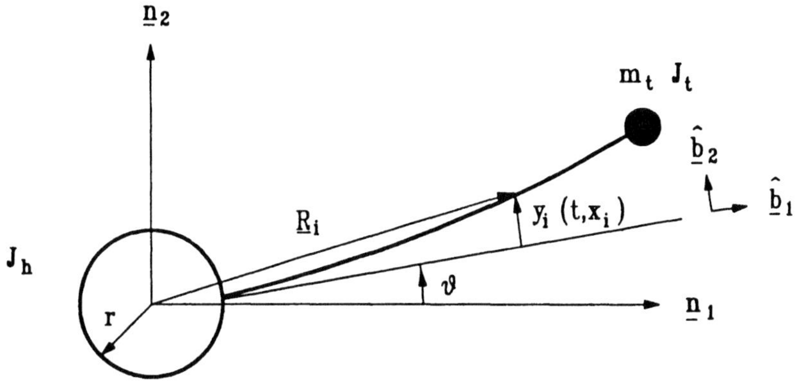

Figure 4.13 Reference Axes for Draper/RPL Configuration

the undeformed ith beam (measured from the outer radius of the hub along the $\underline{b}_1$ axis), and $y_i(t, x_i)$ is the local deformation measured perpendicular to the $\underline{b}_1$ axis.

The inertial position vector $\underline{R}_i$, to a typical deformed point in the ith appendage is given by

$$\underline{R}_i = (r + x_i)\hat{\underline{b}}_1 + y_i\hat{\underline{b}}_2 \tag{4.81}$$

where $(\hat{\ })$ denotes a unit vector.

The velocity of the typical deformed point is then

$$\dot{\underline{R}}_i = \frac{d}{dt}\left(\underline{R}_i\right)_N = \frac{d}{dt}\left(\underline{R}_i\right)_B + \underline{\omega} \times \underline{R}_i \tag{4.82}$$

where $\frac{d}{dt}(\cdot)_N$, $\frac{d}{dt}(\cdot)_B$ denote the velocity in the inertial reference frame and the velocity in the hub-fixed reference frame, respectively, and $\underline{\omega} = \dot{\theta}\hat{\underline{n}}_3$ is the angular velocity of the hub-fixed frame relative to the inertial frame.

Using Eq. (4.81) and $\underline{\omega} = \dot{\theta}\underline{\hat{n}}_3$ in Eq. (4.82), neglecting the small radial motion required to enforce zero elongation give

$$\underline{\dot{R}}_i(t, x_i) = -\dot{\theta} y_i \underline{\hat{b}}_1 + [\dot{y}_i + \dot{\theta}(r + x_i)]\underline{\hat{b}}_2 \tag{4.83}$$

Since we are considering a special class of antisymmetric deformations, we impose the kinematic constraint that appendages 1 and 3 (set 1) have the same deflection profiles, $y_3(t, x) = y_1(t, x)$, and appendages 2 and 4 (set 2) have the same deflection profiles, $y_4(t, x) = y_2(t, x)$, the kinetic energy for the Draper/RPL model is

$$T = T_{hub} + T_{appendages} + T_{tip\ masses} \tag{4.84}$$

with

$$T_{hub} = \frac{1}{2} J_h \dot{\theta}^2$$

$$T_{appendages} = \sum_{i=1}^{2} \int_0^{L_i} \rho \underline{\dot{R}}_i \cdot \underline{\dot{R}}_i dx_i$$

$$T_{tip\ masses} = \sum_{i=1}^{2} \left[m_{t_i} \underline{\dot{R}}_i(L_i) \cdot \underline{\dot{R}}_i(L_i) + J_{t_i} \{ \dot{\theta} + \dot{y}_i'(L_i) \}^2 \right]$$

where J_h is the rotary inertia of hub, J_{t_i} is the rotary inertia of the ith tip mass, ρ is the mass density of the appendages, m_{t_i} is mass of the ith tip mass, L_i is the length of the ith appendage, and $(\cdot)'$ denotes the partial derivative of $(\cdot)$ with respect to x_i.

Note that some higher-order terms in $T_{tip\ masses}$ have been neglected because of the assumed small deformation of the appendages.

Using Eq. (4.83) in Eq. (4.84) leads to a quadratic expression for the kinetic energy (as a function of spatial integrals involving y(x, t) and derivatives thereof) in the form

$$\begin{aligned} T = \frac{1}{2}\hat{J}\dot{\theta}^2 + \sum_{i=1}^{2} \Big[& \int_0^{L_i} \rho \dot{y}_i{}^2 dx_i + m_{t_i} \{\dot{y}_i(L_i)\}^2 + J_{t_i} \{\dot{y}_i'(L_i)\}^2 \\ & + 2\dot{\theta} \Big\{ \int_0^{L_i} \rho(r + x_i)\dot{y}_i dx_i + m_{t_i}(r + L_i)\dot{y}_i(L_i) + J_{t_i}\dot{y}_i'(L_i) \Big\} \Big] \end{aligned} \tag{4.85}$$

where the total moment of inertia $\hat{J}$ is

$$\hat{J} = J_h + 2 \sum_{i=1}^{2} \Big\{ \int_0^{L_i} \rho(r + x_i)^2 dx_i + m_{t_i}(r + L_i)^2 + J_{t_i} \Big\}$$

The total potential energy, under the Euler-Bernoulli assumption, is

$$V = \sum_{i=1}^{2} \int_0^{L_i} EI_i (y_i'')^2 dx_i \tag{4.86}$$

where EI_i is the appendage flexural rigidity of the ith beam and $(\cdot)''$ denotes the second partial derivative of $(\cdot)$ with respect to x_i.

Note that we have treated each appendage as a simple (Euler-Bernoulli) beam ignoring shear deformation and other higher-order effects.

Central to applying the Lagrangian approach, we introduce generalized coordinates that locate uniquely all mass in the system. Using the assumed modes method, these coordinates are the amplitudes of the chosen basis functions; in this case, the elastic displacements are modeled as the following series:

$$y_i(t, x_i) = \sum_{j=1}^{N} q_{ji}(t)\phi_j(x_i), \qquad i = 1, 2, \qquad 0 \leq x_i \leq L \tag{4.87}$$

where $q_{ji}(t)$ denotes the (j,i)th generalized coordinates, $\phi_j(x_i)$ denotes the set of comparison functions whose linear combination approximates the actual appendage mode shapes (we assume that all appendage mode shapes can be expressed as a linear combination of the same comparison function set), and N denotes the number of terms retained in the approximation.

As a particular example, consider the admissible functions [$\phi_j(x)$ in Eq. (4.87)]:

$$\phi_j(x) = 1 - \cos\left(\frac{j\pi x}{L}\right) + \frac{1}{2}(-1)^{j+1}\left(\frac{j\pi x}{L}\right)^2 \tag{4.88}$$

These functions satisfy the geometric and physical boundary conditions of a clamped-free beam (clamped to the hub), and have been found (Refs. 9 and 15) to be excellent admissible functions for clamped-free appendages.

Upon substituting Eqs. (4.87) and (4.88) into Eqs. (4.85) and (4.86), the finite-dimensional system of ordinary differential equations of motion for $\theta(t), andq_ij(t)$ can be derived by using the Lagrangian approach:

$$\frac{d}{dt}\left(\frac{\partial T}{\partial \dot{x}_i}\right) - \frac{\partial T}{\partial x_i} + \frac{\partial V}{\partial x_i} = F_i, \qquad i = 1, \ldots, n \tag{4.89}$$

where F_i is the ith generalized force, and x_i is the ith element of the *configuration vector* ($\mathbf{x}$):

$$\mathbf{x} = \{\theta \;\; q_{11} \;\; q_{21} \cdots q_{N1} \;\; q_{22} \cdots q_{N2}\}^T = \mathbf{col}\{\theta, \underline{q}_1, \underline{q}_2\} \tag{4.90}$$

The generalized force F_i, can be determined by finding the virtual work, δW, associated with the control torques, u_1, u_2, and u_3. For a general motion of the system, the total work W, done by control torques is

$$W = \int_{\mathbf{x}_1}^{\mathbf{x}_2} \left[u_1\, d\theta + 2[u_2\{d\theta + dy_1'(L)\} + u_3\{d\theta + dy_2'(L)\}]\right] \tag{4.91}$$

where $\mathbf{x}_1$ and $\mathbf{x}_2$ denote arbitrary initial and final configurations of the system. The virtual work δW of the control torques (with the system configuration undergoing arbitrary virtual displacements away from an arbitrary configuration $\mathbf{x}$) has the same form as the integrand of the work integral with the actual differential displacements d( ) replaced by virtual variations δ( ). Recollecting the integrand of Eq. (4.91) on the differential virtual variations δ( ) and, using Eq. (4.87), the virtual work can be written as

$$\delta W = [u_1 + 2u_2 + 2u_3]\delta\theta + 2\sum_{j=1}^{N}[u_2\phi_j'(L)\delta q_{j1} + u_3\phi_j'(L)\delta q_{j2}] \tag{4.92}$$

or

$$\delta W = F_\theta \delta\theta + \sum_{j=1}^{N}\left[F_{j1}\delta q_{j1} + F_{j2}\delta q_{j2}\right] \tag{4.93}$$

where F_θ and F_{ji} are the elements of the generalized force vector $\underline{F}$. Utilizing Eq. (4.92), we determine the generalized force vector $\underline{F}$ in matrix form as

$$\underline{F} = \begin{Bmatrix} F_1 \\ F_2 \\ F_3 \\ \vdots \\ \vdots \\ \vdots \\ \vdots \\ F_n \end{Bmatrix} = \begin{Bmatrix} F_\theta \\ F_{11} \\ F_{21} \\ \vdots \\ F_{N1} \\ F_{12} \\ \vdots \\ F_{N2} \end{Bmatrix} = \begin{bmatrix} 1 & 2 & 2 \\ 0 & 2\underline{\phi}'(L) & 0 \\ 0 & 0 & 2\underline{\phi}'(L) \end{bmatrix} \begin{pmatrix} u_1 \\ u_2 \\ u_3 \end{pmatrix} \tag{4.94}$$

where $\underline{\phi}'(L)$ is the vector formed by using Eq. (4.88) for $j = 1, 2, \ldots, N$.

Using Eqs. (4.85)–(4.87), (4.90), and (4.94) in Eq. (4.89) leads to the following matrix form for the system equation of motion:

$$\begin{bmatrix} \hat{J} & M_{\theta q_1}^T & M_{\theta q_2}^T \\ M_{\theta q_1} & M_{q_1 q_1} & 0 \\ M_{\theta q_2} & 0 & M_{q_2 q_2} \end{bmatrix} \ddot{\mathbf{x}} + \begin{bmatrix} 0 & 0 & 0 \\ 0 & K_{q_1 q_1} & 0 \\ 0 & 0 & K_{q_2 q_2} \end{bmatrix} \mathbf{x} = \begin{bmatrix} 1 & 2 & 2 \\ 0 & 2\underline{\phi}'(L) & 0 \\ 0 & 0 & 2\underline{\phi}'(L) \end{bmatrix} \mathbf{u} \tag{4.95}$$

where elements of the mass and stiffness matrices are

$$
\begin{aligned}
\hat{J} =& J_h + 4\int_0^L \rho(r+x)^2 dx + 4\left[m_t(r+L)^2 + J_t\right] \\
\left[M_{\theta q_1}\right]_i = \left[M_{\theta q_2}\right]_i =& 2\left[\int_0^L \rho(r+x)\phi_i(x)dx + m_t(r+L)\phi_i(L) + J_t\phi_i'(L)\right] \\
\left[M_{q_1 q_1}\right]_{ij} = \left[M_{q_2 q_2}\right]_{ij} =& 2\left[\int_0^L \rho\phi_i(x)\phi_j(x)dx + m_t\phi_i(L)\phi_j(L) + J_t\phi_i'(L)\phi_j'(L)\right] \\
\left[K_{q_1 q_1}\right]_{ij} = \left[K_{q_2 q_2}\right]_{ij} =& 2\int_0^L EI\phi_i''\phi_j'' dx
\end{aligned}
\tag{4.96}
$$

where $[\cdot]_i$ denotes the ith element of the vector $[\cdot]$, and $[\cdot]_{ij}$ denotes the (i,j) element of the matrix $[\cdot]$. The above Eqs. (4.96) are specialized for the case in which $m_{t_i} = m_t$, $J_{t_i} = J_t$, $L_i = L$, and $E_i = E$, when the admissible functions (4.88) are used; the results are listed in Table 4.3.

For obvious reasons, the matrix

$$
D = \begin{bmatrix} 1 & 2 & 2 \\ 0 & 2\underline{\phi}'(L) & 0 \\ 0 & 0 & 2\underline{\phi}'(L) \end{bmatrix}
$$

is known as the *control influence matrix*. In this case, it is evident that the three torques all cause a generalized force in the rotational θ degree of freedom, however the toques acting on the tips of the beams cause generalized forces only on the respective beams' subset of generalized coordinates q_{ij}. Note further [see Eq. (4.95)], however, that due to the coupling in the mass matrix, that any of the three torques being nonzero can cause a fairly general acceleration in all elements of $\mathbf{x}$.

Note that we provide a MATLAB operator for generating the (M, K, D) matrices for the above model. The syntax is

$> \; [\mathbf{M}, \mathbf{K}, \mathbf{D}] = \mathbf{rplasm}(\mathbf{n}, \mathbf{rho}, \mathbf{EI}, \mathbf{L}, \mathbf{r}, \mathbf{Jh}, \mathbf{mt}, \mathbf{Jt})$

where **n** is the number of assumed mode shapes.

Numerical experience with the above formulation [for the admissible functions of Eq.(4.88)] indicates that this model typically converges on the low frequency dynamics quite well (i.e., for modest dimensioned approximations). However for N>20, we have observed that the mass matrix becomes increasingly poorly conditioned. These issues are illustrated below in the numerical studies.

Table 4.3 Analytic Formulation of Mass and Stiffness Submatrices (Using the Assumed Modes Method)

$$\hat{J} = J_h + 4\rho L\left[r^2 + rL + \frac{L^2}{3}\right] + 4\left[m_t(r+L)^2 + J_t\right]$$

$$\begin{aligned}
\left[M_{\theta q_1}\right]_i &= \left[M_{\theta q_2}\right]_i \\
&= 2\rho L^2\left[\frac{1}{2} + [1 + (-1)^{i+1}]\left(\frac{1}{i^2\pi^2}\right) + (-1)^{i+1}\left(\frac{i^2\pi^2}{8}\right)\right] \\
&\quad + 2\rho r L\left[1 + (-1)^{i+1}\left(\frac{i^2\pi^2}{6}\right)\right] \\
&\quad + 2m_t(r+L)\left[1 + (-1)^{i+1} + (-1)^{i+1}\left(\frac{i^2\pi^2}{2}\right)\right] + 2(-1)^{i+1}J_t\left(\frac{i^2\pi^2}{L}\right)
\end{aligned}$$

$$\begin{aligned}
\left[M_{q_1 q_1}\right]_{ij} &= \left[M_{q_2 q_2}\right]_{ij} \\
&= 2\rho L\Bigg[1 + (-1)^{i+j}\left[\left(\frac{i}{j}\right)^2 + \left(\frac{j}{i}\right)^2\right] + \frac{1}{2}\delta_{ij} \\
&\qquad + (-1)^{i+1}\left(\frac{i^2\pi^2}{6}\right) + (-1)^{j+1}\left(\frac{j^2\pi^2}{6}\right) + (-1)^{i+j}\left(\frac{i^2 j^2\pi^4}{20}\right)\Bigg] \\
&\quad + 2m_t\left[1 + (-1)^{i+1} + (-1)^{i+1}\left(\frac{i^2\pi^2}{2}\right)\right]\left[1 + (-1)^{j+1} + (-1)^{j+1}\left(\frac{j^2\pi^2}{2}\right)\right] \\
&\quad + 2(-1)^{i+j}J_t\left(\frac{i^2 j^2\pi^4}{L^2}\right)
\end{aligned}$$

$$\begin{aligned}
\left[K_{q_1 q_1}\right]_{ij} &= \left[K_{q_2 q_2}\right]_{ij} \\
&= 2EI\left[(-1)^{i+j} + \frac{1}{2}\delta_{ij}\right]\left(\frac{i^2 j^2\pi^4}{L^3}\right)
\end{aligned}$$

with

$$\phi'(L) = (-1)^{i+1}\left(\frac{i^2\pi^2}{L}\right) \qquad \text{for } \underline{F} \text{ of Eq. (4.94)}$$

where δ_{ij} is a Kronecker delta.

4.3.2 Finite Element Method

In the previous section the assumed modes method for generating a finite dimensional model of a continuous system was introduced. This method is especially attractive for the case in which approximate mode shapes are available. For a structure having complex geometry, however, it is extremely difficult to establish the insight necessary to choose a judicious set of admissible functions that lead to efficient and accurate convergence (Refs. 4-6). Also, there is little carry over from one problem to the next; that is, even though we succeed in choosing the admissible functions for the given structure, we may have to begin the search for a new set of admissible functions for a modified structure (even for apparently slight modifications of a given system, such as adding tip masses to appendages). Therefore, for a structure having complex, irregular geometric shapes, a more general modeling technique is needed. The finite element method(FEM) is a broadly used family of techniques in engineering and applied mathematics (Refs. 10,11), only certain of the most fundamental aspects of applications to structural dynamics will be treated here.

In the finite element approach, the vehicle structure (however complex it may be) is regarded as an assembly of many (prescribed) geometrically simple discrete elements where, in each element, simple local interpolation functions (of motion on the element boundaries) are used to model deformations interior to the finite element. The family of piecewise continuous elements constitutes a finite element model, and the boundary coordinates become the system's generalized coordinates. In the finite element method, the admissible functions are *local* interpolation functions defined over small subdomains, and they are usually extremely simple functions such as low-degree polynomials. Note that, the admissible functions of the assumed modes method are *global* functions defined over the entire domain (or, more generally, over the domain of a significant substructure). The differences between the assumed modes method and the finite element method are listed in Table 4.4.

Using a set of admissible functions, the Ritz-Galerkin type of approximations lead to governing equations for the typical elements in the mesh. The most popular interpolation functions correspond to static deformations of the element under a linearly independent set of boundary forces and moments (most typically, unit forces and/or moments are applied at each translational/rotational degree of freedom while the other boundary forces are zero). Alternatively (and equivalently), a linearly independent set of unit boundary displacements and rotations can be used to generate a judicious set of local interpolation functions. This process is easily justified for static deformation, but may be more difficult to grasp for the case of time varying deformations. Qualititively, if we adopt as coordinates discrete points in the structure, then as the number of discrete points become large, there should

Table 4.4 Qualitative Differences Between the Assumed Modes Method and the Finite Element Method

	Assumed Modes Method	Finite Element Method
Admissible functions	• Global functions defined over the entire domain • Complicated functions • Hard to choose for a structure having complex geometry	• Local functions defined over small subdomain • Simple and low-degree polynomials
System order	• Low order (typically)	• High order
Coordinates	• Usually highly coupled M, K • Usually not sparse, although often diagonally dominant	• Sparsely populated M, K • Physical coordinates
Application	• Useful for problems where geometry is simple	• Broadly applicable to arbitrary geometries
For a different geometry	• A new set of functions must be selected	• Same functions are used (typically)

be many acceptable ways to interpolate the deformation between points. The finite element method based upon static local interpolation is the most common approach, with the added motivation that it is often observed that constitutive forces dominate inertial forces locally, and the most common static finite element formulation is contained as a special case of such dynamic formulations.

Having selected a set of admissible functions, the equations of motion for each element under general boundary loading can be approximated. Finally, the locally valid equations for all elements of the collection are connected by imposing the inter-element constraints which guarantee continuity of the *primary* (deflections, forces, and moments) variables. After the boundary conditions are imposed, the globally valid finite element model is obtained.

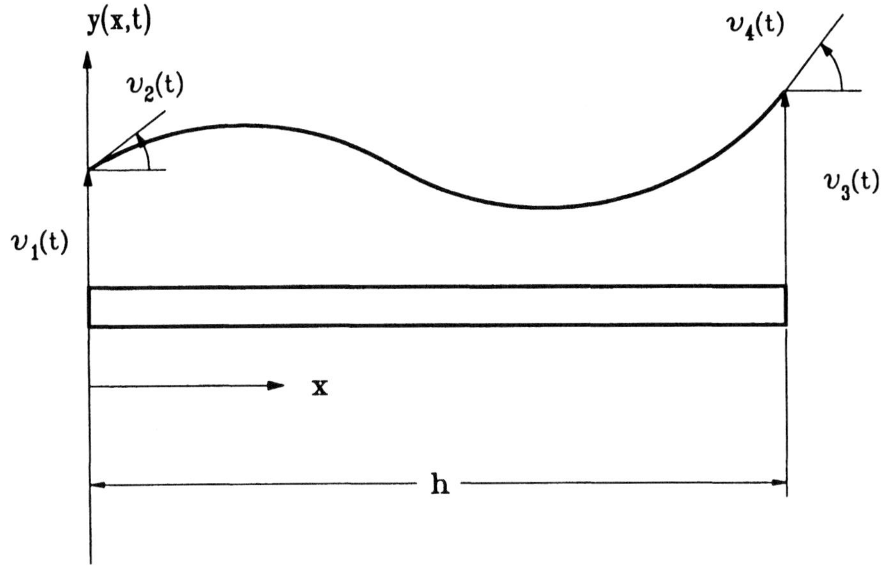

Figure 4.14 A Uniform Element Undergoing Transverse Deflection.

To illustrate the finite element approach, we derive the equations of motion for the transverse vibration of the beam element shown in Figure 4.14. The problem is simplified in the sense that it is one-dimensional and the joints consists simply of the two endpoints. The element is of length h, and assumed as a typical element of the given system shown in Figure 4.2. The transverse displacement is assumed to be given by

$$y(x,t) = \sum_{j=1}^{4} \phi_j(x)\nu_j(t) \tag{4.97}$$

where $\phi_j(x)$ are the so-called shape functions which must be selected to satisfy prescribed boundary conditions. From Figure 4.14, it follows that y(x,t) must be selected such that

$$\begin{aligned} y(0,t) &= \nu_1(t), \qquad y'(0,t) = \nu_2(t) \\ y(h,t) &= \nu_3(t), \qquad y'(h,t) = \nu_4(t) \end{aligned} \tag{4.98}$$

Subject to Eq. (4.98), considering the simplest possible case of a single element model, it follows from Eq. (4.97) that $\phi_j(x)$ must satisfy the boundary conditions

$$\begin{aligned} &\phi_1(0) = 1, \qquad \phi_1'(0) = \phi_1(h) = \phi_1'(h) = 0 \\ &\phi_2'(0) = 1, \qquad \phi_2(0) = \phi_2(h) = \phi_2'(h) = 0 \\ &\phi_3(h) = 1, \qquad \phi_3(0) = \phi_3'(0) = \phi_3'(h) = 0 \\ &\phi_4'(h) = 1, \qquad \phi_4(0) = \phi_4'(0) = \phi_4(h) = 0 \end{aligned} \tag{4.99}$$

Since there are four boundary conditions per interpolation function, the simplest interpolation functions we can select are linear polynomials of the form

$$\phi_i(x) = C_{1i} + C_{2i}x + C_{3i}x^2 + C_{4i}x^3, \qquad i = 1, \dots, 4 \tag{4.100}$$

on evaluating the constants in Eq. (4.100), subject to the respective boundary conditions of Eqs. (4.99), we find that the following Hermite-cubic polynomial functions are admissible shape functions for the ith finite element (Refs. 6,10,11):

$$\begin{aligned} &\phi_1(x) = 1 - 3\left(\frac{x}{h}\right)^2 + 2\left(\frac{x}{h}\right)^3, \qquad \phi_3(x) = 3\left(\frac{x}{h}\right)^2 - 2\left(\frac{x}{h}\right)^3 \\ &\phi_2(x) = x - 2h\left(\frac{x}{h}\right)^2 + h\left(\frac{x}{h}\right)^3, \qquad \phi_4(x) = -h\left(\frac{x}{h}\right)^2 + h\left(\frac{x}{h}\right)^3 \end{aligned} \tag{4.101}$$

Finite element modeling of a structure may be considered to be a many fold local application of the assumed modes method wherein linear combinations of the locally valid $\phi_i(x)$s are used to represent the deflection shape over a portion (finite element) of the structure, and with the element equations and boundary constraints being *assembled* to form a global model of the structural system. Equations (4.79) and (4.80) are the general expressions for $[M]_{ij}$ and $[K]_{ij}$. For the Euler-Bernoulli beam elements, we obtain

$$[M^i]_{jk} = \int_0^h \rho(x)\phi_j(x)\phi_k(x)dx, \quad \text{and} \quad [K^i]_{jk} = \int_0^h EI\phi_j''(x)\phi_k''(x)dx \tag{4.102}$$

Also, note that the generalized forces are given by

$$\{f^i\}_j = \int_0^h p(x,t)\phi_j dx \tag{4.103}$$

Observe that the integration domain above is the small subdomain $(0 \le x \le h)$ of the typical (generic) finite element.

Substituting Eqs. (4.101) into Eqs. (4.102) and (4.103), we obtain the differential equations governing the local dynamics of the ith element:

$$\mathrm{M}^{\mathrm{i}}\,\ddot{\underline{\nu}} + \mathrm{K}^{\mathrm{i}}\,\underline{\nu} = \mathbf{f}^{\mathrm{i}} \tag{4.104}$$

where the local (*element*) mass and stiffness matrices of the ith element are defined as follows:

$$\mathrm{M}^{\mathrm{i}} = \frac{\rho \mathrm{h}}{420}\begin{bmatrix} 156 & 22\mathrm{h} & 54 & -13\mathrm{h} \\ 22\mathrm{h} & 4\mathrm{h}^2 & 13\mathrm{h} & -3\mathrm{h}^2 \\ 54 & 13\mathrm{h} & 156 & -22\mathrm{h} \\ -13\mathrm{h} & -3\mathrm{h}^2 & -22\mathrm{h} & 4\mathrm{h}^2 \end{bmatrix} \equiv \begin{bmatrix} \mathrm{m}^{\mathrm{i}}_{11} & \mathrm{m}^{\mathrm{i}}_{12} \\ \mathrm{m}^{\mathrm{i}}_{21} & \mathrm{m}^{\mathrm{i}}_{22} \end{bmatrix} \tag{4.105}$$

$$\mathrm{K}^{\mathrm{i}} = \frac{\mathrm{EI}}{\mathrm{h}^3}\begin{bmatrix} 12 & 6\mathrm{h} & -12 & 6\mathrm{h} \\ 6\mathrm{h} & 4\mathrm{h}^2 & -6\mathrm{h} & 2\mathrm{h}^2 \\ -12 & -6\mathrm{h} & 12 & -6\mathrm{h} \\ 6\mathrm{h} & 2\mathrm{h}^2 & -6\mathrm{h} & 4\mathrm{h}^2 \end{bmatrix} \equiv \begin{bmatrix} \mathrm{k}^{\mathrm{i}}_{11} & \mathrm{k}^{\mathrm{i}}_{12} \\ \mathrm{k}^{\mathrm{i}}_{21} & \mathrm{k}^{\mathrm{i}}_{22} \end{bmatrix} \tag{4.106}$$

The equation for the generalized force components should be evaluated using Eq. (4.103) with Eqs. (4.101) and the particular distributed load of the given problem. For a uniform (in space, but variable in time) transverse load f(t), the generalized forces are represented by

$$\mathbf{f}^{\mathrm{i}} = \frac{\mathrm{f\,h}}{12}\begin{Bmatrix} 6 \\ \mathrm{h} \\ 6 \\ -\mathrm{h} \end{Bmatrix} \tag{4.107}$$

Example 4.11

Let us consider the motion of the rod-shaped element shown in Figure 4.15 (which is a typical element of the rod in axial vibration shown in Figure 4.3). In Example 4.2, we obtained the kinetic, potential energy and virtual work done by external forces. For the typical element considered here these functions are

$$\mathrm{T} = \frac{1}{2}\int_0^{\mathrm{h}} \rho\Big(\frac{\partial \mathrm{u}(\mathrm{x},\mathrm{t})}{\partial \mathrm{t}}\Big)^2 \mathrm{dx}, \quad \mathrm{V} = \frac{1}{2}\int_0^{\mathrm{h}} \mathrm{EA}\Big(\frac{\partial \mathrm{u}(\mathrm{x},\mathrm{t})}{\partial \mathrm{x}}\Big)^2 \mathrm{dx}$$
$$\text{and}\quad \delta \mathrm{W}_{\mathrm{nc}} = \int_0^{\mathrm{h}} \mathrm{p}(\mathrm{x},\mathrm{t})\delta \mathrm{u}(\mathrm{x},\mathrm{t})\,\mathrm{dx} \tag{4.108}$$

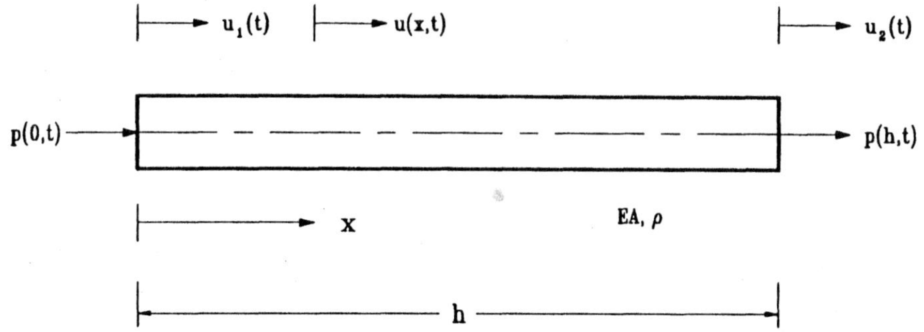

Figure 4.15 Finite Element Free-Body Diagram for the Rod.

The axial displacement is assumed to be given by

$$u(x,t) = \sum_{j=1}^{2} \phi_j(x) u_j(t) \tag{4.109}$$

Since $u(0,t) = u_1(t)$ and $u(h,t) = u_2(t)$, the shape functions $\phi_i(x)$ must satisfy the boundary conditions

$$\phi_1(0) = 1, \qquad \phi_1(h) = 0; \qquad \phi_2(0) = 0, \qquad \phi_2(h) = 1 \tag{4.110}$$

The simplest functions satisfying the above four boundary conditions are linear polynomials of the form

$$\phi_i(x) = C_{1i} + C_{2i}x, \qquad i = 1, \ldots, 2 \tag{4.111}$$

Applying the boundary conditions fo Eqs. (4.110) to the deflection approximation given by Eqs. (4.109), the following shape functions are obtained for the ith finite element:

$$\phi_1 = 1 - \left(\frac{x}{h}\right), \qquad \phi_2 = \left(\frac{x}{h}\right) \tag{4.112}$$

Expressions for the mass coefficients $[M]_{ij}$, the stiffness coefficients $[K]_{ij}$, and the generalized forces $\mathbf{f}^i$ for axial motion are given by dealing with Eqs. (4.108) using approximate method (in a manner similar manner to the developments in Example 4.10 for the transverse vibration of beam). Thus,

$$
\begin{aligned}
[M^i]_{jk} &= \int_0^h \rho(x)\phi_j(x)\phi_k(x)dx \\
[K^i]_{jk} &= \int_0^h EA\phi_j'(x)\phi_k'(x)dx \\
\{\mathbf{f}^i\}_j &= \int_0^h p(x,t)\phi_j dx
\end{aligned}
\tag{4.113}
$$

When Eqs. (4.112) are substituted into Eqs. (4.113), the local mass and stiffness matrices of the ith element for a uniform rod are represented as follows (Refs. 6,10,11):

$$
M^{(i)} = \frac{\rho h}{6}\begin{bmatrix} 1 & 1 \\ 1 & 2 \end{bmatrix}, \qquad K^{(i)} = \frac{EA}{h}\begin{bmatrix} 1 & -1 \\ -1 & 1 \end{bmatrix} \tag{4.114}
$$

■

So far, we have considered processes at the element level – generating the finite element mass, stiffness, and force matrices. There remains the *assembly* of system matrices (putting elements together through the use of displacement compatibility conditions at the joints) and enforcing system boundary conditions. To show the assembly procedure, we consider free transverse vibration of a uniform beam, and select a four-element model as shown in Figure 4.16. When several elements are combined to form a structure, the common set of local coordinates (primary variables) are defined as the system displacement coordinates. By considering the continuity of the primary variables, we can assemble the system mass and stiffness matrices as follows:

$$
M' = \begin{bmatrix}
m_{11}^1 & m_{12}^1 & & & \\
m_{21}^1 & m_{22}^1 + m_{11}^2 & m_{12}^2 & & \\
 & m_{21}^2 & m_{22}^2 + m_{11}^3 & m_{12}^3 & \\
 & & m_{21}^3 & m_{22}^3 + m_{11}^4 & m_{12}^4 \\
 & & & m_{21}^4 & m_{22}^4
\end{bmatrix} \tag{4.115a}
$$

$$
K' = \begin{bmatrix}
k_{11}^1 & k_{12}^1 & & & \\
k_{21}^1 & k_{22}^1 + k_{11}^2 & k_{12}^2 & & \\
 & k_{21}^2 & k_{22}^2 + k_{11}^3 & k_{12}^3 & \\
 & & k_{21}^3 & k_{22}^3 + k_{11}^4 & k_{12}^4 \\
 & & & k_{21}^4 & k_{22}^4
\end{bmatrix} \tag{4.115b}
$$

where m_{jk}^i and k_{jk}^i are 2×2 matrices defined as in Eqs. (4.105) and (4.106).

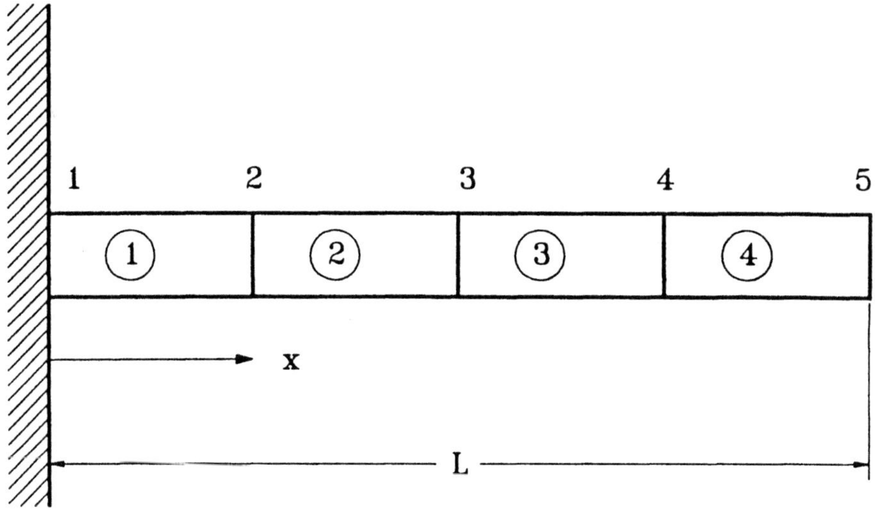

Figure 4.16 Discretization of Fixed-Free Beam (Four Elements).

Note that, in most problems of practical interest, elements are encountered in many arbitrary orientations; however, we do not address these straightforward generalizations in order not to obscure basic issues. However, in the more general circumstances, we should consider transverse vibration and axial vibration of the beam simultaneously, and transformation of all element matrices to a common (global) coordinate system is necessary before assembly is carried out.

It is necessary to incorporate boundary conditions of two types: prescribed displacements (essential boundary conditions) and prescribed forces (natural boundary conditions). Using the finite element method, the force boundary conditions are taken into account in the evaluation of the externally applied nodal point force vector. Displacement boundary conditions should be imposed after the assembling process. Assume that the equilibrium equations of a finite element system without the imposition of the displacement boundary conditions are

$$\begin{bmatrix} M_{aa} & M_{ab} \\ M_{ba} & M_{bb} \end{bmatrix} \begin{Bmatrix} \ddot{U}_a \\ \ddot{U}_b \end{Bmatrix} + \begin{bmatrix} K_{aa} & K_{ab} \\ K_{ba} & K_{bb} \end{bmatrix} \begin{Bmatrix} U_a \\ U_b \end{Bmatrix} = \begin{Bmatrix} \mathbf{f}_a \\ \mathbf{f}_b \end{Bmatrix} \tag{4.116}$$

where the U_a are the unknown displacements and the U_b are the prescribed displacements. Solving for U_a, we obtain

$$M_{aa}\ddot{U}_a + K_{aa}U_a = \mathbf{f}_a - K_{ab}U_b - M_{ab}\ddot{U}_b \tag{4.117}$$

Therefore, in the solution for U_a, only the stiffness and mass matrices of the complete assemblage corresponding to the unknown degrees of freedom U_a need be assembled, but the load vector $\mathbf{f}_a$ must be modified to include the effect of imposed nonzero displacements. Once the displacements U_a have been evaluated from Eq. (4.117), the nodal point forces corresponding to U_b are obtained using Eq. (4.116); i.e., we have

$$\mathbf{f}_b = M_{ba}\ddot{U}_a + M_{bb}\ddot{U}_b + K_{ba}U_a + K_{bb}U_b \tag{4.118}$$

Returning to our simple problem, since the beam is clamped at global node 1 as shown in Figure 4.16, it follows that the deflection and the slope are zero at node 1. When Eqs. (4.116) and (4.117) are considered, the equations of motion for this structure becomes

$$M\ddot{\mathbf{x}} + K\mathbf{x} = 0 \tag{4.119}$$

where M and K are 8×8 matrices such that

$$M = \begin{bmatrix} m^1_{22} + m^2_{11} & m^2_{12} & & \\ m^2_{21} & m^2_{22} + m^3_{11} & m^3_{12} & \\ & m^3_{21} & m^3_{22} + m^4_{11} & m^4_{12} \\ & & m^4_{21} & m^4_{22} \end{bmatrix} \tag{4.120a}$$

$$K = \begin{bmatrix} k^1_{22} + k^2_{11} & k^2_{12} & & \\ k^2_{21} & k^2_{22} + k^3_{11} & k^3_{12} & \\ & k^3_{21} & k^3_{22} + k^4_{11} & k^4_{12} \\ & & k^4_{21} & k^4_{22} \end{bmatrix} \tag{4.120b}$$

and global coordinate vector $\mathbf{x}$ includes vertical deflections and slopes of each nodes (except node 1).

The MATLAB operator syntax for computing (M, K) for the above model is

$> \ [\mathbf{M}, \mathbf{K}] = \mathbf{beamfem}(\mathbf{N}, \mathbf{rho}, \mathbf{L}, \mathbf{E}, \mathbf{I})$

where N is the number of finite elements.

Note that to apply the FEM to geometrically complicated space structures, two- and three-dimensional discrete elements should be used. For example, for two-dimensional problems such as heat transfer, the transverse deflection of membranes, the torsional deflection of cylindrical members, and the finite element mesh (discretization) consist of simple two-dimensional elements, such as triangles and rectangles, that are connected to each other at nodal points on the boundaries of the elements.

Finite Element Application: Spacecraft Undergoing Large Rotations

Using the FEM, the partial differential equations of motion are transformed into a finite-dimensional set of second-order differential equations in terms of the displacements, velocities, and accelerations of the finite element coordinates, and the external forcing functions. Several finite element models for flexible arms are presented in Refs. 12 and 13. In this section, we will present a finite element model for a hub with four appendages and tip masses by using the extended Hamilton's principle, which provides the variational method on which the finite element weak form is built. In most FEM codes and formulations, the elastic deflections are introduced to measure displacement of each mass element from its inertially fixed equilibrium point. It is significant that we carefully introduce the finite element in such a way that large hub rotations are admitted; the FEM is used to model small elastic displacements *measured with respect to moving hub-fixed axes.* Note specifically that, if we insist on small motions with respect to the inertial position of each mass element, then, to obtain linear equations, we would be restricted to small rotational motions. However, by defining elastic deformations in a moving frame, the resulting equations are greatly simplified; large rigid body motions coupled with small elastic motions can often be approximated accurately by linear differential equations. More generally, it is easy to extend these ideas to any case wherein small elastic motions occur in a body undergoing general large "rigid body" displacements, but certain nonlinear geometric and constitutive effects may arise as consequences of gyroscopic coupling, nonlinear kinematics, and rotational stiffening.

Let us first consider only the hub with one appendage and tip mass for convenience, and assemble the four appendages later. The coordinates we use are shown in Figure 4.17. The total transverse velocity (of a mass element on the beam) is

$$\mathrm{v}(\mathrm{x},\mathrm{t}) = \dot{\mathrm{w}}(\mathrm{x},\mathrm{t}) + (\mathrm{x}+\mathrm{r})\dot{\theta}(\mathrm{t}) \tag{4.121}$$

where x is the auxiliary variable measured from the outer radius of the hub along the undeformed appendage axis, $\theta(\mathrm{t})$ is the hub angle, $\mathrm{w}(\mathrm{x},\mathrm{t})$ is the deflection measured from the x axis, and overdots denote derivatives with respect to time t. It is obvious that the first term of Eq. (4.121) is the elastic vibratory velocity seen in a hub-fixed frame, whereas the second term is the rigid body velocity of the mass element in the absence of deformation. In Eq. (4.121), it is evident that we have neglected radial deformations and the nonlinear radial velocity correction required to rigorously enforce zero elongation of the deformed beam.

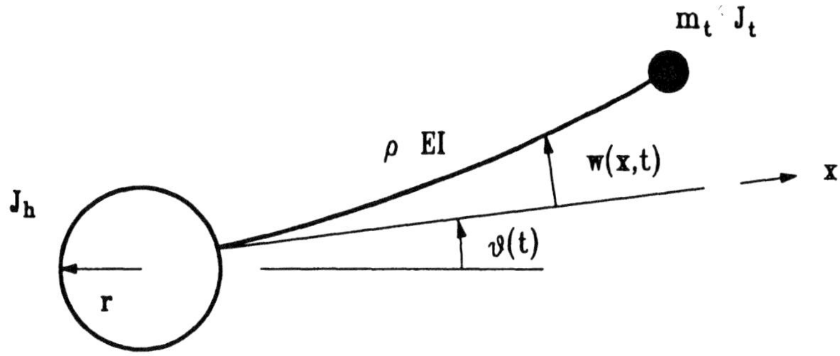

Figure 4.17 Deformation in a Flexible Appendage with Rigid Body Motion.

Assuming Euler-Bernoulli beam theory and small deformation, we can express the total kinetic and potential energy as follows:

$$T = \frac{1}{2}J_h\dot{\theta}^2 + \frac{1}{2}\int_0^L \rho v^2 dx + \frac{1}{2}m_t v^2(L) + \frac{1}{2}J_t\left(\frac{\partial v}{\partial x}\bigg|_L\right)^2 \tag{4.122}$$

and

$$V = \frac{1}{2}\int_0^L EI\left(\frac{\partial^2 w}{\partial x^2}\right)^2 dx \tag{4.123}$$

The virtual work done by the external torques u_1 and u_2 are given by

$$\delta W_{nc} = u_1\delta\theta(t) + u_2\left[\delta\theta + \delta\left(\frac{\partial w}{\partial x}\bigg|_L\right)\right] \tag{4.124}$$

Substitution of Eqs. (4.121)–(4.124) into Eq. (4.1) (extended Hamilton's principle) gives

$$\int_{t_1}^{t_2}\Bigg[\int_0^L \rho\big(\dot{w} + (x+r)\dot{\theta}\big)\big(\delta\dot{w} + (x+r)\delta\dot{\theta}\big)dx - \int_0^L EI\Big(\frac{\partial^2 w}{\partial x^2}\Big)\delta\Big(\frac{\partial^2 w}{\partial x^2}\Big)dx$$
$$+ m_t\big(\dot{w}(L) + (L+r)\dot{\theta}\big)\big(\delta\dot{w}(L) + (L+r)\delta\dot{\theta}\big) + J_t\Big(\frac{\partial \dot{w}}{\partial x}\Big|_L + \dot{\theta}\Big)\Big[\delta\Big(\frac{\partial \dot{w}}{\partial x}\Big|_L\Big) + \delta\dot{\theta}\Big]$$
$$+ J_h\dot{\theta}\delta\dot{\theta} + u_1\delta\theta + u_2\Big[\delta\theta + \delta\Big(\frac{\partial w}{\partial x}\Big|_L\Big)\Big]\Bigg]dt = 0 \tag{4.125}$$

Integrating Eq. (4.125) by parts and some manipulations to collect on δw and $\delta\theta$ gives

$$\int_{t_1}^{t_2}\Bigg[\int_0^L \Big[\rho\big(\ddot{w} + (x+r)\ddot{\theta}\big)\delta w + EI\Big(\frac{\partial^2 w}{\partial x^2}\Big)\delta\Big(\frac{\partial^2 w}{\partial x^2}\Big)\Big]dx$$
$$+ \Big\{\int_0^L \rho(x+r)\big(\ddot{w} + (x+r)\ddot{\theta}\big)dx + J_h\ddot{\theta} + m_t(L+r)\{\ddot{w}(L) + (L+r)\ddot{\theta}\}$$
$$+ J_t\Big(\frac{\partial \ddot{w}}{\partial x}\Big|_L + \ddot{\theta}\Big) - (u_1 + u_2)\Big\}\delta\theta + m_t\{\ddot{w}(L) + (L+r)\ddot{\theta}\}\delta w(L)$$
$$+ \Big[J_t\Big(\frac{\partial \ddot{w}}{\partial x}\Big|_L + \ddot{\theta}\Big) - u_2\Big]\delta\Big(\frac{\partial w}{\partial x}\Big|_L\Big)\Bigg]dt = 0 \tag{4.126}$$

Note that $\ddot{w}(L)$ and $\frac{\partial \ddot{w}}{\partial x}\Big|_L$ in the above equation are the accelerations of the transverse deflection and the rotation at the right end of the beam. We have not included the foreshortening effect or nonlinear kinematics in the present discussion, because we are interested in linear differential equations with $\dot{\theta}$ retained as a variable generalized coordinate. If $\dot{\theta}$ is prescribed (which implicitly requires an external constraint torque), it is possible to include, for example, the foreshortening and rotational stiffness effects and still retain linear equations.

The displacement $w(x,t)$ can be discretized using a finite element expansion

$$w(x,t) = \sum_{i=1}^{4} \phi_i^{(e)}(x)\nu_i^{(e)}(t) \tag{4.127}$$

where $\nu_1^{(e)}$, $\nu_2^{(e)}$ ($\nu_3^{(e)}$, $\nu_4^{(e)}$) are transverse deflection and rotation at the left (right) end of the finite element, and $\phi_i^{(e)}$ are the Hermite-cubic polynomial shape functions that satisfy the conditions for admissibility and that are defined over the finite element.

The acceleration and curvature are expressed as follows:

$$
\begin{aligned}
\ddot{w} &= \sum_{i=1}^{4} \phi_i^{(e)}(x)\ddot{\nu}_i^{(e)}(t) \\
\frac{\partial^2 w}{\partial x^2} &= \sum_{i=1}^{4} \frac{\partial^2}{\partial x^2}\Big(\phi_i^{(e)}(x)\Big)\nu_i^{(e)}(t)
\end{aligned}
\tag{4.128}
$$

The following cubic functions are adopted as the shape functions for the ith finite element (note that global coordinates are used in this case):

$$
\begin{aligned}
\phi_1 &= 1 - 3\left(\frac{x - x_i}{h}\right)^2 + 2\left(\frac{x - x_i}{h}\right)^3 \\
\phi_2 &= (x - x_i) - 2h\left(\frac{x - x_i}{h}\right)^2 + h\left(\frac{x - x_i}{h}\right)^3 \\
\phi_3 &= 3\left(\frac{x - x_i}{h}\right)^2 - 2\left(\frac{x - x_i}{h}\right)^3 \\
\phi_4 &= -h\left(\frac{x - x_i}{h}\right)^2 + h\left(\frac{x - x_i}{h}\right)^3
\end{aligned}
\tag{4.129}
$$

where x_i is the distance from the root of the appendage to the left end of the ith finite element, and h is the length of the finite element.

Substitution of Eqs. (4.127)–(4.129) into Eq. (4.126) and integration over the spatial domains leads to the global mass, stiffness, and forcing matrices. The assembled set of matrix differential equations is as follows:

$$
\begin{bmatrix} J_h + M_{\theta\theta} & M_{\theta\nu} \\ M_{\nu\theta} & M_{\nu\nu} \end{bmatrix}
\begin{Bmatrix} \ddot{\theta} \\ \ddot{\underline{\nu}} \end{Bmatrix}
+
\begin{bmatrix} 0 & 0 \\ 0 & K_{\nu\nu} \end{bmatrix}
\begin{Bmatrix} \theta \\ \underline{\nu} \end{Bmatrix}
=
\begin{bmatrix} 1 & 1 \\ 0 & \mathbf{f} \end{bmatrix}
\begin{Bmatrix} u_1 \\ u_2 \end{Bmatrix}
\tag{4.130}
$$

where $\underline{\nu}$ is the coordinate vector which consists of the transverse deflection and rotation at each node of the appendage, as shown in Figure 4.18, and the submatrices in Eq. (4.130) are defined as follows:

$$
M_{\theta\theta} = \sum_{i=1}^{N} M_{11}^{i} + M_{11}^{t}
$$

$$M_{\theta\nu} = [\,M_{13}^1 + M_{12}^2 \quad M_{13}^2 + M_{12}^3 \quad M_{13}^3 + M_{12}^4 \quad \dots \quad M_{13}^{N-1} + M_{12}^{N} \quad M_{13}^{N} + M_{12}^{t}\,]$$

$$M_{\nu\nu} = \begin{bmatrix} M_{33}^1 + M_{22}^2 & M_{23}^2 & & & & \\ M_{32}^2 & M_{33}^2 + M_{22}^3 & M_{23}^3 & & & \\ & M_{32}^3 & M_{33}^3 + M_{22}^4 & M_{23}^4 & & \\ & & & \ddots & & \\ & & & M_{32}^{N-1} & M_{33}^{N-1} + M_{22}^{N} & M_{23}^{N} \\ & & & & M_{32}^{N} & M_{33}^{N} + M_{22}^{t} \end{bmatrix}$$

$$K_{\nu\nu} = \begin{bmatrix} K_{33}^1 + K_{22}^2 & K_{23}^2 & & & & \\ K_{32}^2 & K_{33}^2 + K_{22}^3 & K_{23}^3 & & & \\ & K_{32}^3 & K_{33}^3 + K_{22}^4 & K_{23}^4 & & \\ & & & \ddots & & \\ & & & K_{32}^{N-1} & K_{33}^{N-1} + K_{22}^{N} & K_{23}^{N} \\ & & & & K_{32}^{N} & K_{33}^{N} \end{bmatrix}$$

$$\mathbf{f} = [0 \quad 0 \quad 0 \quad \dots \quad 0 \quad 0 \quad 1]^T \tag{4.131}$$

Note that boundary conditions are already imposed in the above matrices, and the matrix elements of the corresponding local mass and stiffness matrices are listed in Table 4.5.

By using Eq. (4.130), the equations of motion of the hub with four appendages and tip masses can be assembled. Since we are considering antisymmetric deformations such that appendage 1 and 2 (set 1) have the same deflection profile, $w_1(x_1,t) = w_2(x_1,t)$, and appendages 3 and 4 (set 2) have the same deflection profile, the assembled set of differential equations of motion can be expressed as follows (Ref. 14):

$$M\ddot{\mathbf{x}} + K\mathbf{x} = d\mathbf{u} \tag{4.132}$$

where

$$M = \begin{bmatrix} J_h + 4M_{\theta\theta} & 2M_{\theta\nu} & 2M_{\theta\nu} \\ 2M_{\nu\theta} & 2M_{\nu\nu} & 0 \\ 2M_{\nu\theta} & 0 & 2M_{\nu\nu} \end{bmatrix}, \quad K = \begin{bmatrix} 0 & 0 & 0 \\ 0 & 2K_{\nu\nu} & 0 \\ 0 & 0 & 2K_{\nu\nu} \end{bmatrix}, \quad D = \begin{bmatrix} 1 & 2 & 2 \\ 0 & 2\mathbf{f} & 0 \\ 0 & 0 & 2\mathbf{f} \end{bmatrix}$$

with configuration and control vectors $\mathbf{x} = [\theta \;\; \underline{\nu}_1^T \;\; \underline{\nu}_3^T]^T$ and $\mathbf{u} = [u_1 \; u_2 \; u_3]^T$, respectively.

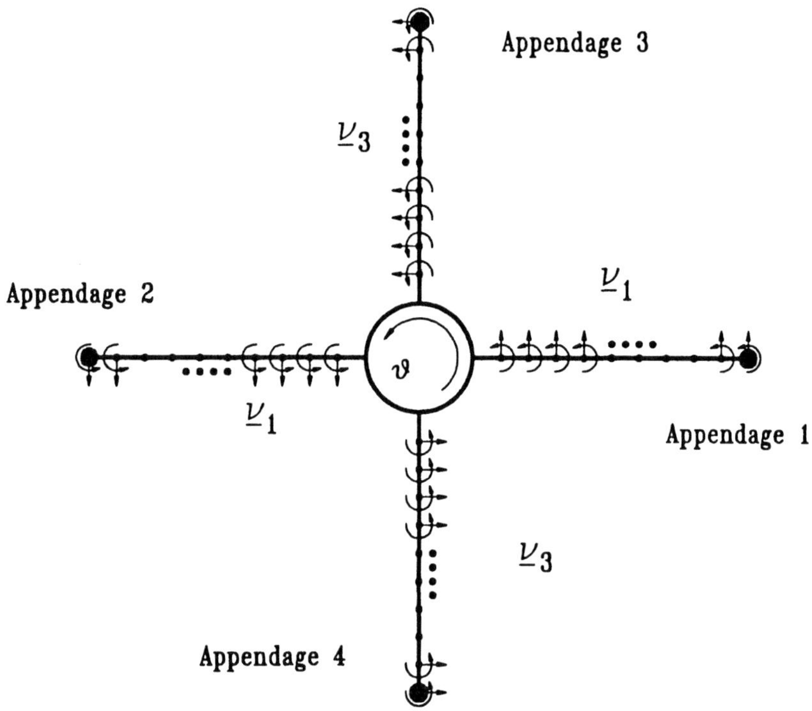

Figure 4.18 Finite Element Degree of Freedom (Draper/RPL Model)

The matrices M and K are symmetric, so obviously $M_{\nu\theta} = M^T_{\theta\nu}$, and the dimension of the configuration vector is 4N+1, when we discretize the appendage using N elements. Note that the syntax for the MATLAB operator we have developed for the above configuration is as follows:

> $[\mathbf{M}, \mathbf{K}, \mathbf{D}] = \mathbf{rplfem}(\mathbf{n}, \mathbf{rho}, \mathbf{EI}, \mathbf{L}, \mathbf{r}, \mathbf{Jh}, \mathbf{mt}, \mathbf{Jt})$

where **n** is the number of finite elements per flexible appendage.

For a more general and broadly applicable treatment of the finite element method, readers may refer to one of many excellent finite element textbooks (e.g., Refs. 10,11).

Table 4.5 Submatrix Elements (Finite Element Method)

The local mass, stiffness matrices (5 × 5) of the ith element of the appendage, and mass matrix (3 × 3) due to the tip mass are defined as follows:

$$M_e^{(i)} \equiv \begin{bmatrix} M_{11}^i & M_{12}^i & M_{13}^i \\ M_{21}^i & M_{22}^i & M_{23}^i \\ M_{31}^i & M_{32}^i & M_{33}^i \end{bmatrix}, \quad K_e^{(i)} \equiv \begin{bmatrix} 0 & 0 & 0 \\ 0 & K_{22}^i & K_{23}^i \\ 0 & K_{32}^i & K_{33}^i \end{bmatrix}, \quad M_t \equiv \begin{bmatrix} M_{11}^t & M_{12}^t \\ M_{21}^t & M_{22}^t \end{bmatrix}$$

where

$$M_{11}^i = \frac{\rho h}{3}\{(x_i + r)^2 + (x_i + r + h)(x_i + r) + (x_i + r + h)^2\}$$

$$M_{12}^i = [M_{21}^i]^T = \rho h\,[\,\tfrac{3}{20}h + \tfrac{1}{2}(x_i + r) \qquad \tfrac{1}{30}h^2 + \tfrac{1}{12}h(x_i + r)\,]$$

$$M_{13}^i = [M_{31}^i]^T = \rho h\,[\,\tfrac{7}{20}h + \tfrac{1}{2}(x_i + r) \qquad -\tfrac{1}{20}h^2 - \tfrac{1}{12}h(x_i + r)\,]$$

$$M_{22}^i = \frac{\rho h}{420}\begin{bmatrix} 156 & 22h \\ 22h & 4h^2 \end{bmatrix}$$

$$M_{23}^i = [M_{32}^i]^T = \frac{\rho h}{420}\begin{bmatrix} 54 & -13h \\ 13h & -3h^2 \end{bmatrix}$$

$$M_{33}^i = \frac{\rho h}{420}\begin{bmatrix} 156 & -22h \\ -22h & 4h^2 \end{bmatrix}$$

$$K_{22}^i = \frac{EI}{h^3}\begin{bmatrix} 12 & 6h \\ 6h & 4h^2 \end{bmatrix}$$

$$K_{23}^i = [K_{32}^i]^T = \frac{EI}{h^3}\begin{bmatrix} -12 & 6h \\ -6h & 2h^2 \end{bmatrix}$$

$$K_{33}^i = \frac{EI}{h^3}\begin{bmatrix} 12 & -6h \\ -6h & 4h^2 \end{bmatrix}$$

$$M_{11}^t = J_t + m_t(r + L)^2$$

$$M_{12}^t = [M_{21}^t]^T = [\,m_t(r + L) \quad J_t\,]$$

$$M_{22}^t = \begin{bmatrix} m_t & 0 \\ 0 & J_t \end{bmatrix}$$

where x_i is the distance from the root of the appendage to the left end of the ith finite element, r is the radius of the hub, and h is the length of the finite element.

Table 4.6 Natural Frequencies of Fixed-Free Beam

Exact Solution	Assumed Modes Method			Finite Element Method		
	N=8	N=16	N=80	N=4	N=8	N=40
ω	n=8	n=16	n=80	n=8	n=16	n=80
3.5160	3.5162	3.5160	3.5160	3.5161	3.5160	3.5160
22.0345	22.0456	22.0360	22.0345	22.0602	22.0363	22.0345
61.6972	61.7860	61.7089	61.6973	62.1749	61.7347	61.6973
120.9019	121.2115	120.9461	120.9023	122.6576	121.1728	120.9024
199.8595	200.9301	199.9839	199.8606	228.1374	201.0159	199.8617
298.5555	300.4822	299.8224	298.5579	366.3896	302.1051	298.5627
416.9908	423.5355	417.5504	416.9954	580.8491	425.2757	417.0102
555.1652	563.8152	556.0902	555.1733	953.0510	564.9725	555.2110

4.3.3 Comparison Between Two Spatial Discretization Models

It is easy to understand that the accuracy of a dynamical model using the assumed modes method depends not only on a suitable selection of admissible functions but also on the number of terms retained in the approximation. Likewise, the accuracy of the finite element model obviously depends on the number of the elements and the interpolation functions used. Based on well-known results (Ref. 6), we expect the FEM model to converge such that the computed frequencies, for example, approach the true natural frequencies from above, as we increase the number of elements.

Table 4.6 shows the first eight natural frequencies of the fixed-free beam. In the first column, the natural frequencies of the exact solution are listed, and the second through fourth columns are the natural frequencies computed using the assumed modes models (for the case in which N = eight, sixteen, and eighty assumed modes are retained in the approximation), and the fifth through seventh columns are using the finite element models (for the case that N = four, eight and forty elements are used per beam). For both cases, we list the number (n) of degrees of freedom (which is the dimension of the mass and stiffness matrices). We used the assumed modes model discussed in Section 4.3.1 (MATLAB code **beamasm**), and the finite element model discussed in Section 4.3.2 (MATLAB code **beamfem**) for simulation. For convenience in the convergence study, all material properties are taken as one, that is, EI=1, ρ=1, and L=1. As shown in Table 4.6, both models converge to the exact frequencies from above as we increase the number N (or n), and the assumed modes method converges with many fewer degrees of freedom (n) than the finite element representation, which has been mentioned in Ref. 5. The reason why the assumed modes model has a faster convergence rate than the finite element model is that the judicious assumed functions in this special example are comparison functions that satisfy both natural and essential boundary conditions of the given structure. This conjecture is based upon experience, rather than a formal proof. It is possible to find a linearly independent set functions satisfying all geometric and natural boundary conditions for this problem since the given structure has very simple boundary conditions. Even though we have the same geometry, if a system has some complication introduced into the boundary conditions (for example, lumped mass is attached at the end of the beam), it becomes much harder to determine comparison functions, and we would usually use admissible functions that satisfy only the essential (geometric) boundary conditions.

For a second example and to illustrate some practical implications of the above discussion, the Draper/RPL configuration discussed in the previous sections is considered again. This structure has finite tip masses, and we adopt the same basis functions of Eq. (4.88); now these are only admissible (not comparison) functions for this case because they do not satisfy the natural boundary conditions associated with the finite tip masses. For simulations, we use our MATLAB code **rplasm** for the assumed modes model, and **rplfem** for the finite element model. Table 4.7 shows the computed natural frequencies of the rigid body mode, the first 6 flexural modes, and the 7th, and 12th flexural modes for two sets of structural models:
(1) The models constructed by the assumed modes method for the case that 20, 40, and 100 assumed modes are retained in the approximation respectively, and
(2) The finite element models that 10, 20, and 50 elements are used per appendage.

It is important to note that the very accurate frequencies and eigenfunctions (essentially exact) can be computed using the approach of Ref. 15, that obtains the rigorous eigenfunctions and characteristic equation which can be determined by solving the governing partial differential equations and boundary conditions, corresponding to finite tip mass. However, the numerical solution of the exact characteristic equation is complicated and, obviously, it is impossible to obtain the higher eigenvalues exactly. The characteristic equation is sharply oscillatory, making accurate location of the zero crossings to determine the higher frequencies a difficult undertaking. However, we adopt this method and have succesfully computed the first twelve natural frequencies for Draper/RPL model, and we consider these values as "exact" numbers (converged to better than twelve significant digits), in order to compare above two approximation models (obtained using the assumed modes method and the finite element method). In this example, we use boundary conditions corresponding to one end clamped, the other end having a lumped-mass for obtaining the eigenfunctions, and seven of these "exact" eigenfunctions are retained in the approximation. The computed eigenvalues are also listed in Table 4.7. It is evident by inspection of Table 4.7 that the 50-element FEM solution is accurately converged, whereas the assumed mode solutions are not accurate for modes eleven and twelve.

In order to study the convergence rate further, the following relative errors for the eigenvalues (corresponding natural frequencies) are introduced:

$$r_i = \frac{\omega_i - \omega_{Ti}}{\omega_{Ti}} \tag{4.133}$$

where ω_i denotes the ith natural frequencies of the approximation model, and ω_{Ti} denotes the ith "exact" or "true" natural frequency. The true natural frequencies were determined accurate to better than twelve digits, as mentioned earlier.

To evaluate the convergence rate of the first s natural frequencies, the following root-mean-square (RMS) natural frequency error measure is introduced

$$r_{RMS} = \sqrt{\frac{1}{s}\sum_{i=1}^{s} r_i^2} \tag{4.134}$$

Note that relative errors are much larger for higher modes than those of lower modes. Therefore, we take the same number s of lower modes for different degree-of-freedom models to compare the convergence rate according to the total number of degrees of freedom N, even though more computed frequencies are available. The convergence of the first twelve modes (i.e., s = 12) is considered for this case. Figure 4.19 shows the relative errors of the 12th natural frequency of both models as we increase the number of degrees of freedom, and Fig. 4.20 shows the root-mean-square error of both models versus the number (n) of degrees of freedom. From Table 4.7 and Figures 4.19 and 4.20, we see that, as expected, both models approach the exact model from above as we increase n. We can also see that the convergence rate of the assumed modes method is much slower than the FEM, and the model using 100 assumed modes (which has 201 degrees of freedom) has almost the same accuracy with the model using only 10 finite elements (which has 41 degrees of freedom).

The "apparent" convergence of the assumed mode solution (the eigenvalues start changing slowly with increasing n) is a potential source of false confidence and an erroneous conclusion that the eigenvalues are accurately approximated. This is not a surprising result, since the relatively simple admissible functions – which we used in the assumed modes method – satisfy only the essential boundary conditions of the given problem and are not accurate approximations of the eigenfunctions.

Table 4.7 Natural Frequency[a] (rad/s)

	Assumed Modes Method[b]	Assumed Modes Method[c]			Finite Element Method		
Mode No.[e]	$N_1=7$ (15)[d]	$N_1=20$ (41)[d]	$N_1=40$ (81)[d]	$N_1=100$ (201)[d]	$N_2=10$ (41)[d]	$N_2=20$ (81)[d]	$N_2=50$ (201)[d]
0	0	0	0	0	0	0	0
1	4.3722	4.3722	4.3722	4.3722	4.3722	4.3722	4.3722
2	7.9066	7.9067	7.9066	7.9066	7.9066	7.9066	7.9066
3	51.3980	51.4129	51.4031	51.3995	51.3987	51.3980	51.3980
4	52.7505	52.7663	52.7560	52.7522	52.7513	52.7506	52.7505
5	155.7180	156.4543	156.0507	155.8438	155.7386	155.7193	155.7181
6	156.5071	157.2539	156.8451	156.6350	156.5278	156.5084	156.5071
11	697.6433	750.9556	724.3782	708.3044	699.4434	697.7589	697.6449
12	698.1619	751.4823	724.8989	708.8233	699.9633	698.2775	698.1635

[a] This table shows the natural frequencies of the rigid body mode, the first six flexible modes, the eleventh and twelveth flexible modes. N_1 denotes the number of functions retained in the approximation, and N_2 denotes the number of finite elements per appendage.

[b] Uses the exact eigenfunctions as assumed modes functions.

[c] Uses admissible functions of Eq. (4.66) as assumed modes functions.

[d] These numbers denote the total number of degrees of freedom.

[e] The zeroth mode is the rigid body mode.

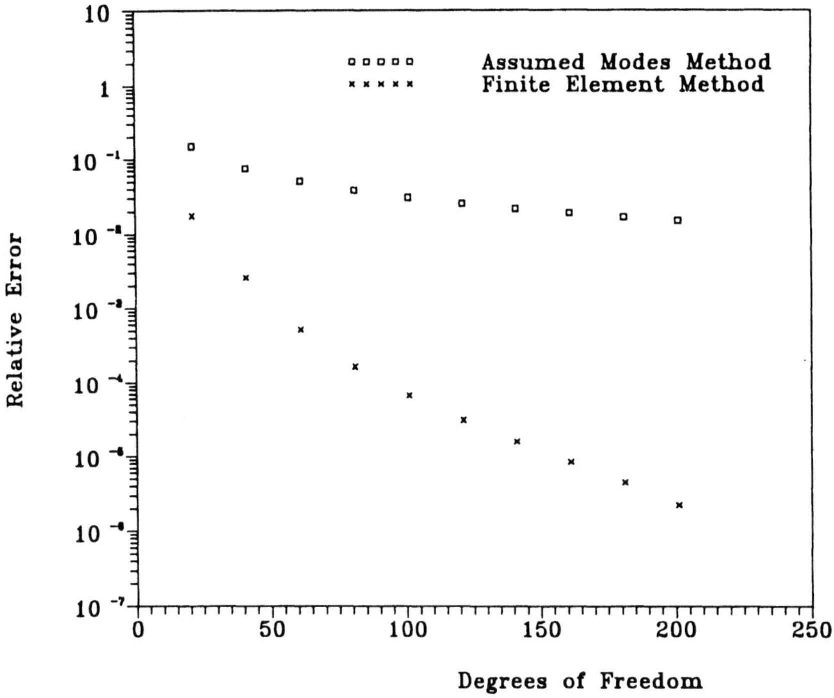

Figure 4.19 Relative Errors of the Twelfth Natural Frequency.

Therefore, these admissible functions do not inherently "know" the that the model considered has a finite tip mass although the finite tip mass is correctly included in the kinematics and the kinetic energy of the system. For a system that has complicated boundary conditions, in order to obtain the functions that satisfy not only the essential boundary conditions but also the natural boundary conditions, approximate eigenfunctions may be used for assumed mode functions (Ref. 15).

Based on this example and similar studies, we have found that the finite element model usually converges to the true model with fewer degrees of freedom than the model of the assumed modes method, in the most common circumstances in which we use configuration admissible functions that satisfy only essential boundary conditions, but do not satisfy the natural boundary conditions.

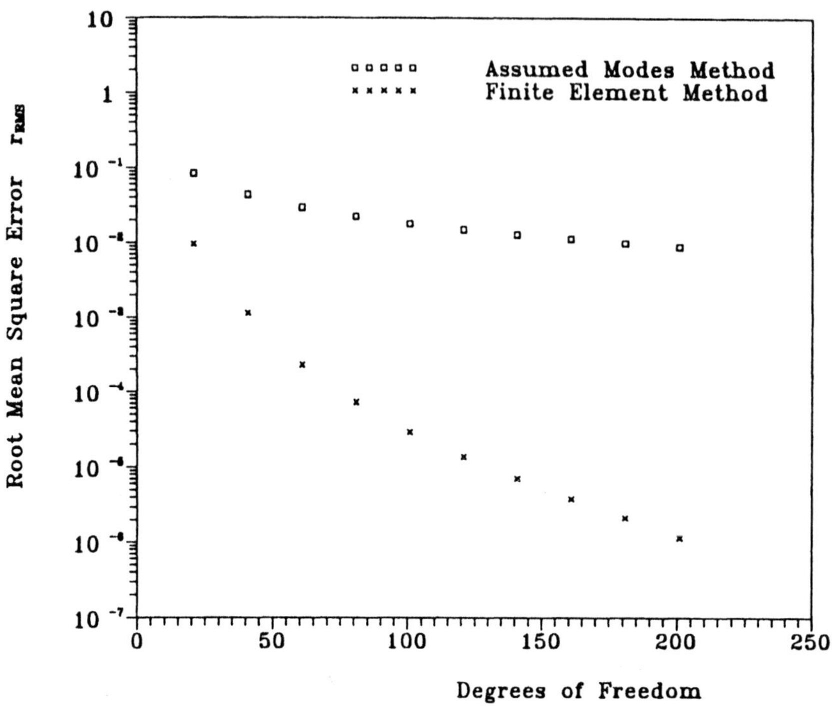

Figure 4.20 RMS Errors of Natural Frequencies.

4.4 Case Study: Consequences of Neglecting Coupling Between Rigid Motion and Elastic Motion

In order to describe the linear motion of large flexible structures that consists of rigid body and elastic motion, the coupling between rigid body motion and elastic motion is very important, since one of the primary objectives is to control the rigid body attitude. In several recent papers and monographs, in which it is assumed that the rigid motion is too large to be in the linear range, the flexural contribution of the system dynamics is obtained using the assumed modes method or the FEM, and the rigid body mode portion is obtained by linearizing the nonlinear rigid motion (Refs. 16, 17). In other words, in this approach, the flexible equations and rigid body equations are written independently, and then are simply combined to construct the uncoupled model. In other papers, properly coupled equations are

Table 4.8 Change of Natural Frequency[a] (Percent)

	Assumed Mode Method (N=20)			Finite Element Method (N=10)		
Mode No.	Coupled	Uncoupled[b]	Change (%)	Coupled	Uncoupled[b]	Change (%)
0	0	0	0	0	0	0
1	4.3722	4.3722	0.0000	4.3722	4.3722	-0.0000
2	7.9067	4.3722	44.7024	7.9066	4.3722	44.7023
3	51.4129	51.4129	0.0000	51.3987	51.3987	-0.0000
4	52.7663	51.4129	2.5649	52.7513	51.3987	2.5641
5	156.4543	156.4543	0.0000	155.7386	155.7386	-0.0000
6	157.2539	156.4543	0.5085	156.5278	155.7386	0.5042

[a] This table shows the natural frequencies of the rigid body mode (mode 0) and the first six flexible modes.
[b] $M_{\theta q}=0$ in the assumed modes method and $M_{\theta \nu}=0$ in the FEM.

initially formulated, rigid body and elastic coupling terms are then neglected as an approximate simplification (Refs. 18, 19). With this approach (Refs. 16-19), the rigid motion of this model is often (incorrectly) uncoupled from the flexible motion in the dynamic equations. This uncoupled approximation is potentially dangerous (Ref. 14), especially when we use this model without careful examination of the consequences for the given structure. We investigate the consequences of these approximations, in the context of the hub-appendage configuration, in this and subsequent sections.

The "apparent" convergence of the assumed modes solution (the eigenvalues start changing slowly with increasing n) is a potential source of false confidence and an erroneous conclusion that the eigenvalues are accurately approximated. This is not a surprising result, since the relatively simple admissible functions – which we used in the assumed modes method – satisfy only the essential boundary conditions of the given problem and are not accurate approximations of the eigenfunctions.

Table 4.8 shows the open-loop natural frequencies of the uncoupled models and the coupled models that were obtained by both the assumed modes method and the FEM. We include the natural frequencies of the rigid body mode and the first six flexible modes, since the trends are almost the same and the errors (% change) become smaller for the higher frequencies. The natural modes of the coupled model consist of two sets of modes: (1) the symmetric modes and (2) the antisymmetric modes, and their frequencies occur in neighboring pairs and become very close at high frequency (where both approach the corresponding "clamped-free" frequencies of cantilevered beam).

As can be seen in Table 4.8, when we neglect the coupling terms (in the mass matrix) between the rigid motion and the elastic motions, the natural modes occur in exact pairs. Convergence at high frequency has disappeared in the uncoupled model, the natural modes do not occur in neighboring pairs, and there is zero hub motion in all modes except the rigid body mode. Thus the uncoupled model does not capture the most important physical property of the dynamics of the given structure which arise as a consequence of rotational/vibrational coupling. Note that the second mode's natural frequency is changed most significantly by neglecting the mass matrix partition $M_{\theta\nu}$ that couples appendage vibration and hub rotation. The same trends are shown for both the assumed modes and the finite element models. In the subsequent section, a feedback controller will be designed using the uncoupled model (i.e., neglecting the coupling submatrix $M_{\theta\nu}$ in the mass matrix), and we will examine consequences of neglecting these terms in the characteristics of the closed-loop system.

Controller Design

In order to further investigate the coupled and uncoupled models, we adopt several different configurations of varying flexibility obtained by varying the length of the appendages (which changes the dominance ratio of the rigid motion and the elastic motion); the shorter appendage configurations behave more like a rigid body. As a preview of the results in subsequent chapters, we design control laws for each of the different configurations using conventional linear quadratic regulator (LQR) feedback design approaches, which will be discussed in Chapter 5. Modal coordinates are used as the state vector for design of the the controller, and the control gain matrices can be obtained in physical coordinates via an inverse coordinate transformation.

The equation of motion [Eqs. (4.95) and (4.132)] can be written in linear second order matrix form,

$$\mathbf{M\ddot{x}} + \mathbf{Kx} = \mathbf{Du} \tag{4.135}$$

Since we use an approximate "uncoupled" model to design the controller, consider the following uncoupled model:

$$\mathbf{M_U\ddot{x}} + \mathbf{Kx} = \mathbf{Du} \tag{4.136}$$

where M_U denotes the mass matrix in which the coupling elements (between hub rigid body coordinate θ and flexible coordinates) are set to zero.

Note that the stiffness matrix of the uncoupled model is the same as the stiffness matrix of the coupled model, since there are rigorously zero values for the rotational/vibrational coupling elements in stiffness matrix.

To perform the modal coordinate transformation, the following open-loop eigenvalue problem should be solved (see section 2.2.1 and Ref. 20):

$$K\underline{\phi}_i = \lambda_i M_U \underline{\phi}_i, \qquad i = 1, 2, \ldots, n \tag{4.137}$$

with the normalization

$$\underline{\phi}_i^T M_U \underline{\phi}_i = 1, \qquad i = 1, 2, \ldots, n \tag{4.138}$$

We introduce the modal matrix of open-loop eigenvectors,

$$\Phi = [\underline{\phi}_1, \underline{\phi}_2, \ldots, \underline{\phi}_n] \tag{4.139}$$

The general modal coordinate transformation is then

$$\mathbf{x}(t) = \Phi \underline{\eta}(t) \tag{4.140}$$

where $\underline{\eta}(t)$ is the $n \times 1$ vector of modal coordinates.

The transformed equation of motion becomes

$$\ddot{\underline{\eta}} + \tilde{K}\underline{\eta} = \tilde{D}\mathbf{u} \tag{4.141}$$

where $\Phi^T M_U \Phi = I$, $\tilde{K} = \Phi^T K \Phi = \mathrm{diag}(0, \omega_1^2, \omega_2^2, \ldots, \omega_{n-1}^2)$, $\tilde{D} = \Phi^T D$. Note that the diagonal zero in $\tilde{K}$ corresponds to the rigid body mode.

For control applications, the system dynamics are usually modeled in first order state space differential equations. Let us introduce the $2n \times 1$ dimensional modal state vector

$$\mathbf{z} = \begin{Bmatrix} \underline{\eta} \\ \dot{\underline{\eta}} \end{Bmatrix} \tag{4.142}$$

Equation (4.141) can be written as the first-order system

$$\dot{\mathbf{z}} = A\mathbf{z} + B\mathbf{u} \tag{4.143}$$

where

$$A = \begin{bmatrix} 0 & I \\ -\tilde{K} & 0 \end{bmatrix}, \qquad B = \begin{bmatrix} 0 \\ \tilde{D} \end{bmatrix}$$

We adopted the following performance index for LQR:

$$J = \int_0^\infty (\mathbf{z}^T Q \mathbf{z} + \mathbf{u}^T R \mathbf{u}) dt \tag{4.144}$$

with

$$Q = \begin{bmatrix} \Omega & 0 \\ 0 & I_n \end{bmatrix}, \qquad R = I_m$$

where $\Omega = \text{diag}[1, \omega_1^2, \omega_2^2, \ldots, \omega_{n-1}^2]$

The above performance index is an energy type, since the first term and the second term in the performance index correspond to the state energy and the control energy, respectively. By solving the Riccati equation (see section 5.1), we obtain the optimal feedback control,

$$\mathbf{u} = -\mathbf{g}\mathbf{z} = -\begin{bmatrix} g_1 & g_2 \end{bmatrix} \begin{Bmatrix} \underline{\eta} \\ \underline{\dot{\eta}} \end{Bmatrix} \tag{4.145}$$

Optimal feedback control may be represented in physical coordinates instead of modal coordinates by using Eq. (4.140) with the property ($\Phi^T M_U \Phi = I$) as follows:

$$\begin{aligned} \mathbf{u} &= -g_1 \Phi^T M_U \mathbf{x} - g_2 \Phi^T M_U \dot{\mathbf{x}} \\ &= -G_1 \mathbf{x} - G_2 \dot{\mathbf{x}} \end{aligned} \tag{4.146}$$

where

$$G_i \equiv g_i \Phi^T M_U$$

Therefore, the closed-loop system of the uncoupled model using the optimal control [Eq. (4.145)] can be written

$$M_U \ddot{\mathbf{x}} + DG_2 \dot{\mathbf{x}} + (K + DG_1)\mathbf{x} = 0 \tag{4.147}$$

We might anticipate that the controller designed using the uncoupled model will stabilize the uncoupled model; however, the more significant issue is whether or not it will stabilize the correctly coupled model. The closed-loop system of the original (coupled, physically correct) model using the same control can be represented as follows:

$$M\ddot{\mathbf{x}} + DG_2 \dot{\mathbf{x}} + (K + DG_1)\mathbf{x} = 0 \tag{4.148}$$

Since we designed a controller using an uncoupled model, the closed-loop system is, not surprisingly, stable when this control is applied to the uncoupled model. When we implement the designed controller on the coupled model, however, we find that the second flexible mode becomes unstable. This is a consequence of the fact that the controller was designed using the (physically incorrect) uncoupled model in which the second mode was especially erroneous. We mentioned in

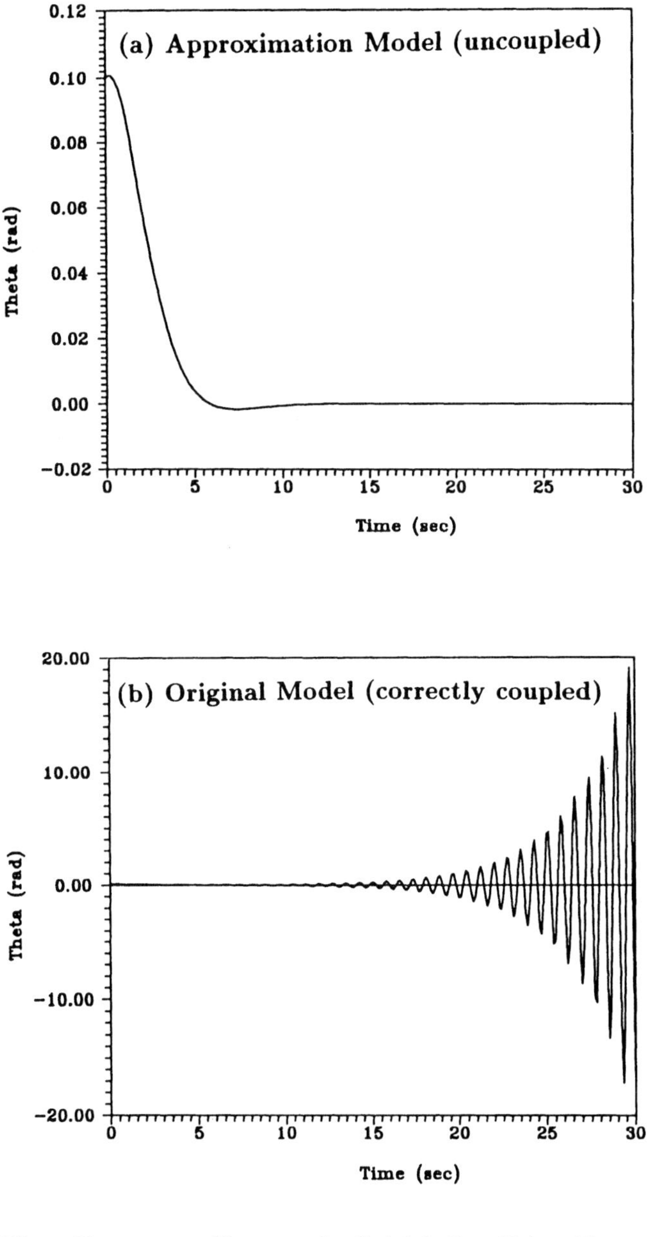

Figure 4.21 Time Responses Due to the Initial Condition Error $\theta(0) = 0.1$ rad (Both Cases Use Control Gains Based on Uncoupled Model).

the previous section that the changes in this mode (i.e., the perturbations of the frequency and mode shape) are most severe when we neglect the coupling terms, and this is consistent with the above result. Initial condition responses ($\theta(0) = 0.1$ rad, $\dot{\theta}(0) = 0.01$ rad/s) for both models are shown in Figure 4.21. A direct consequence of the model errors is a control law (which stabilizes the physically incorrect model) that destabilizes the correct model. It is very interesting to study the characteristics of the approximate closed-loop system [Eq. (4.147)] and correct system [Eq. (4.148)] for the different configurations with a family of structures having different flexibility (L = 4, 3.5, 3, 2.5, and 2 ft). Only the results from the finite element model (with ten elements per appendage) are presented here, since almost identical results for both models (using the assumed modes method and FEM) have been obtained. In other words, the key issue here is not how one does the spatial discretization, but the consequences of neglecting rotational/vibrational coupling in the "rigid body" coordinates.

Table 4.9 shows the open-loop and closed-loop eigenvalues (of the second mode) for the family of configurations having different flexibility. Note that the closed-loop eigenvalues (approximation model and original model) of all the other modes remain in the stable region. As shown in Table 4.9, the second mode remains unstable for model configurations with appendage length 3.5 and 3 ft. However, for the shorter appendage length, 2.5 and 2 ft, the closed-loop system of the correctly coupled model is stable, even though we use the controller designed using the incorrect (uncoupled) model. These two cases behave more nearly as a rigid body, and this indicates that robustness to modeling assumptions increases as flexibility becomes less dominant. These results are consistent with the intuitive notion that, for a "sufficiently rigid" structure, one can make the uncoupling approximation without catastrophic consequences.

It is possible that, for the first three configurations (L = 4, 3.5, and 3 ft), a more robust controller can be designed using the uncoupled model that stabilizes the coupled model, by judicious choice of control gain design methods such as using an H_∞ controller design approach (or some other robust control approach). In fact, neglected coupling terms can be considered as unmodeled dynamics for the controller design. It is expected that these control design methods, however, will take significant analysis and computational time, thus sacrificing the original benefit of using the uncoupled model; and, of course, careful validation using a correctly coupled model is required. On the other hand, when the controller is designed using the correctly coupled model, these particularly important model errors do not exist, and the use of a robust controller design should provide us with larger robustness bounds for other generally unknown model errors and/or other neglected parameter variations.

As discussed earlier, by changing the length of the appendages, the qualitative

Table 4.9 Open-Loop and Closed-Loop Eigenvalues of the Second Mode[a]

L[ft]	Open-loop	Closed-Loop[b]	Closed-Loop[c]
4.0	± 7.9066i	- 0.8507 ± 4.3715i	0.2927 ± 8.1440i
3.5	± 9.0568i	- 0.9846 ± 5.3877i	0.1660 ± 9.2955i
3.0	± 10.6713i	- 1.1634 ± 6.8489i	0.0149 ± 10.9081i
2.5	± 13.0933i	- 1.4141 ± 9.0815i	- 0.1802 ± 13.3253i
2.0	± 17.0721i	- 1.7905 ± 12.7977i	- 0.4597 ± 17.2965i

[a]This table shows the closed-loop eigenvalues of the approximation model (uncoupled) and the original model (correctly coupled) when the designed controllers (using uncoupled model) are implemented.

[b]Closed-loop eigenvalues using approximate model which neglects rotational/vibrational coupling.

[c]Closed-loop eigenvalues using original model which includes rotational/vibrational coupling.

and quantitative properties (especially, the importance of flexibility effects) of the structure are changed. Furthermore, we have shown that several model configurations with moderate flexibility are not stabilized by the control law designed using the uncoupled model. The uncoupled approximation can be used without causing significant errors for systems where rigid motion is largely dominant; however, one cannot trust controllers designed using the uncoupled model without careful closed-loop evaluation of the correctly coupled model. A serious potential problem we see is that the coupling terms are often neglected in the literature without comment, and an unsophisticated user may not be aware of potentially catastrophic model error.

Of course, the above observations hold, with varying degrees, for any modeling approximation. For example, in all of the applications considered in this chapter we neglected shear deformation, rotatory inertia, rotational stiffening, and beam foreshortening. It is therefore extremely important to validate final control law designs obtained using model simplifications, using the most rigorous dynamical model possible consistent with the judgment and resources available.

Problems

Problem 4-1

Solve the eigenvalue problem associated with a free axial vibration of uniform bar (A, E, and ρ are constant in Example 4.2), and plot the first three eigenfunctions.

Problem 4-2 (Ref. 15)

Consider a beam with one end fixed and a rigid mass attached to the free end as in Figure P4.1, and let the mass center of the tip mass be coincident with the end of the beam. Assume that the deformation of the beam is small and perpendicular to the major axis of the beam, and neglect any shortening of the distance between points of the beam due to deformation. The deflection y of a point on the beam may be described by a mathematical function of position x and time t, i.e.,

$$y = y(x, t)$$

The partial differential equation describing the motion of such a system can be shown to be

$$EI \frac{\partial^4 y(x,t)}{\partial x^4} + \rho A \frac{\partial^2 y(x,t)}{\partial t^2} = 0$$

with boundary conditions

$$y(x,t)\Big|_{x=0} = 0, \qquad \frac{\partial y(x,t)}{\partial x}\Big|_{x=0} = 0$$

$$-EI\frac{\partial^2 y(x,t)}{\partial x^2}\Big|_{x=L} = I_t \frac{d^2}{dt^2}\left(\frac{\partial y(x,t)}{\partial x}\right)\Big|_{x=L}, \qquad EI\frac{\partial^3 y(x,t)}{\partial x^3}\Big|_{x=L} = m_t \frac{d^2 y(x,t)}{dt^2}\Big|_{x=L}$$

where I_t is the mass moment of inertia about a transverse axis through the mass center of the tip mass, m_t is the mass of the tip mass, and L is the length of the beam.

Using the method of separation of variables, the deflection function y(x,t) can be represented as

$$y(x,t) = \sum_{i=1}^{\infty} F_i(t) Y_i(x)$$

For free vibration, we look for harmonic motion with frequency ω_n, and find the following equation

$$\frac{d^4 Y_n(x)}{dx^4} - \gamma_n^4 Y_n(x) = 0, \qquad \gamma_n^4 \equiv \frac{\rho A \omega_n^2}{EI}$$

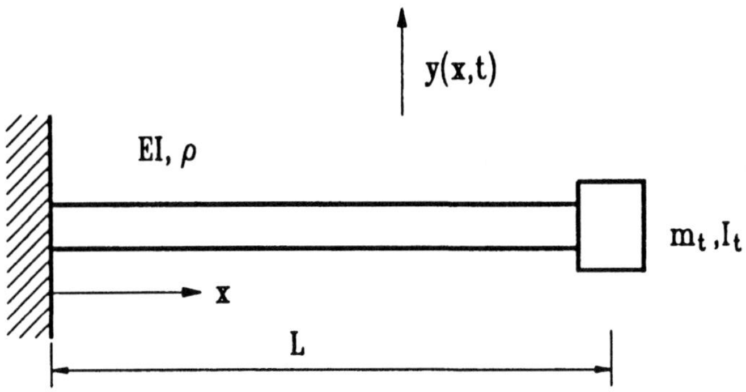

Figure P4.1 Clamped Beam with a Rigid Mass Attached to the Free End.

with boundary conditions

$$Y_n(0) = 0 \qquad\qquad Y'_n(0) = 0$$

$$Y''_n(L) = \left(\frac{\gamma_n^4 I_t}{\rho A}\right) Y'_n(L), \qquad\qquad Y'''_n(L) = -\left(\frac{\gamma_n^4 m_t}{\rho A}\right) Y_n(L)$$

where $(\;)' \equiv \frac{d}{dx}(\;)$.

Your tasks are to find the eigenfunctions $Y_n(x)$ by solving the above ordinary differential equation and applying boundary conditions, and also to find the characteristic (or transcendental) equation from which γ_n can be determined by root-solving methods.

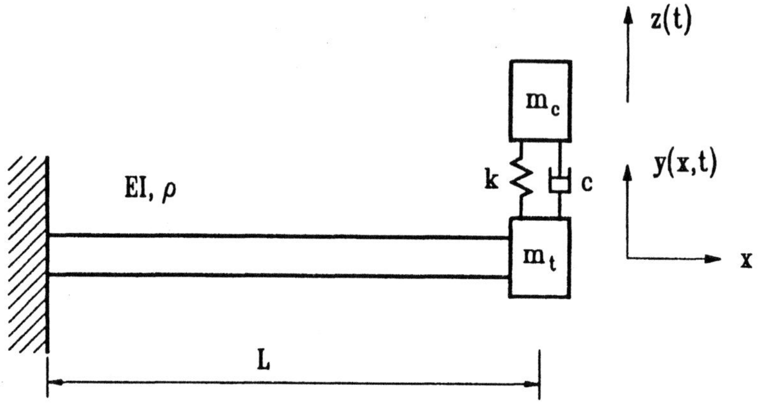

Figure P4.2 Beam with a Tip Vibration Absorber.

Problem 4-3 (Ref. 21)

Consider a system as shown in Figure P4.2 consisting of a flexible beam (one end clamped with a vibration absorber attached at the other end with a tip mass m_t). L is the length of the beam, ρ is the mass density of the beam per unit length, and EI denotes the bending rigidity of the beam. Note that y(x,t) is the deflection of the beam and z(t) is the deflection of m_c relative to m_t.

Your tasks are as follows:

(a) Using the Lagrangian approach, show that the equations of motion for the given system are

$$EIy''''(x,t) + \rho\ddot{y}(x,t) = 0$$

$$m_c\left[\ddot{z}(t) + \ddot{y}(L,t)\right] + c\dot{z}(t) + kz(t) = 0$$

where $(\ )' \equiv \frac{\partial}{\partial x}(\ )$ and $(\dot{\ }) \equiv \frac{\partial}{\partial t}(\ )$.

What are the boundary conditions?

(b) Discretize the partial differential equation obtained in part (a) by using the following assumed modes approximation:

$$y(x,t) = \sum_{i=1}^{3} \phi_i(x)\eta_i(t)$$

where the ϕ_i's satisfy the geometric boundary conditions.

Find the configuration state space representation in the following form

$$M\ddot{\mathbf{x}} + C\dot{\mathbf{x}} + K\mathbf{x} = 0$$

with $\mathbf{x}^T = [\eta_1 \quad \eta_2 \quad \eta_3 \quad z]$; i.e., determine the 4×4 M, C, K matrices.

Problem 4-4

Consider the mechanical second order system

$$M\ddot{\mathbf{x}} + C\dot{\mathbf{x}} + K\mathbf{x} = \mathbf{f}(t)$$

where $\mathbf{x}(t)$ is the $n \times 1$ state vector, M is the mass matrix, K is the stiffness matrix, C is the structural damping (i.e., can be diagonalized via modal coordinate transformation), and $\mathbf{f}(t)$ is the forcing function vector.

Using the modal coordinate transformation $\mathbf{x}(t) = \Phi\underline{\eta}(t)$ (where Φ is the modal matrix obtained solving the eigenvalue problem of mass and stiffness matrices), the system equations can be represented by

$$\begin{aligned} &\ddot{\eta}_i(t) = N_i(t), && i = 1, 2, \dots, r \\ &\ddot{\eta}_i(t) + 2\zeta_i\omega_i\dot{\eta}_i(t) + \omega_i^2\eta_i(t) = N_i(t), && i = r+1, \dots, n \end{aligned}$$

where r is the number of the rigid body modes, $\eta_i(t)$ is the ith modal coordinates, and $N_i(t)$ is the ith element of the column vector of $\Phi^T\mathbf{f}(t)$.

Your task is to show that vibration response can be obtained via modal superposition

$$\mathbf{x}(t) = \Phi\underline{\eta}(t)$$

where $\underline{\eta}(t)$ is the vector consisting of the following solutions:

$$\begin{aligned} \eta_i(t) &= \int_0^t \left[\int_0^\tau N_i(\sigma)d\sigma\right]d\tau + \eta_i(0) + t\dot{\eta}_i(0), && i = 1, \dots, r \\ \eta_i(t) &= \frac{1}{\omega_{di}}\int_0^t N_i(\tau)e^{-\zeta_i\omega_i(t-\tau)}\sin\big(\omega_{di}(t-\tau)\big)d\tau \\ &\quad + e^{-\zeta_i\omega_i t}\left[\cos(\omega_{di}t) + \frac{\zeta_i}{\sqrt{1-\zeta_i^2}}\sin(\omega_{di}t)\right]\eta_i(0) \\ &\quad + \left[\frac{1}{\omega_{di}}e^{-\zeta_i\omega_i t}\sin(\omega_{di}t)\right]\dot{\eta}_i(0), && i = r+1, \dots, n \end{aligned}$$

with $\omega_{di} = \omega_i\sqrt{1-\zeta_i^2}$.

Problem 4-5

Consider the mechanical second-order closed-loop system

$$M\ddot{\mathbf{x}} + C\dot{\mathbf{x}} + K\mathbf{x} = 0 \tag{$*$}$$

where $\mathbf{x} \in R^n$ is a configuration vector.

The corresponding eigenvalue problem is as follows

$$(\lambda_i^2 M + \lambda_i C + K)\underline{\alpha}_i = 0$$

where λ_i and $\underline{\alpha}_i$ are the ith eigenvalue and eigenvector, respectively.

The equivalent eigenvalue problem can also be obtained by considering a first-order state space form of the system differential equations that is equivalent to the second-order system of Eq. ($*$),

$$B\dot{\mathbf{z}} = A\mathbf{z} \tag{$\dagger$}$$

where

$$\mathbf{z} = \begin{pmatrix} \mathbf{x} \\ \dot{\mathbf{x}} \end{pmatrix}, \qquad A = \begin{bmatrix} 0 & M \\ -K & -C \end{bmatrix}, \qquad B = \begin{bmatrix} M & 0 \\ 0 & M \end{bmatrix}$$

Assume that the $2n \times 1$ right eigenvector $\underline{\phi}_i$ and $2n \times 1$ left eigenvector $\underline{\psi}_i$ (with biorthonormalization) of Eq. ($\dagger$) has the following form:

$$\underline{\phi}_i \equiv \begin{pmatrix} \underline{\phi}_i^{(1)} \\ \underline{\phi}_i^{(2)} \end{pmatrix}, \qquad \underline{\psi}_i \equiv \begin{pmatrix} \underline{\psi}_i^{(1)} \\ \underline{\psi}_i^{(2)} \end{pmatrix}$$

where $\underline{\phi}_i^{(j)}$ and $\underline{\psi}_i^{(j)}$, j=1,2, are $n \times 1$ vectors.

Your tasks are as follows:

(a) Verify the following relations:

(1) $\underline{\phi}_i^{(1)} = \underline{\alpha}_i$

(2) $\underline{\phi}_i^{(2)} = \lambda_i \underline{\alpha}_i$

(3) $\lambda_i M^T \underline{\psi}_i^{(1)} = -K^T \underline{\psi}_i^{(2)}$

(4) $(1 + \lambda_i^2)\underline{\alpha}_i^T M \underline{\alpha}_i = 1, \qquad i = 1, \ldots, 2n$

(5) $\left\{\underline{\psi}_j^{(2)}\right\}^T [\lambda_i \lambda_j M - K]\underline{\alpha}_i = \lambda_i \delta_{ij}, \qquad i,j = 1, \ldots, 2n$

(6) $\left\{\underline{\psi}_j^{(2)}\right\}^T \left[(\lambda_i + \lambda_j)K + \lambda_i \lambda_j C\right]\underline{\alpha}_i = -\lambda_i^2 \delta_{ij}, \qquad i,j = 1, \ldots, 2n$

(b) Show that the eigenvalue/eigenvector sensitivities for the second-order system are computed as follows:

$$\frac{\partial \lambda_i}{\partial p_k} = -\{\underline{\psi}_i^{(2)}\}^T \left(\lambda_i^2 \frac{\partial M}{\partial p_k} + \lambda_i \frac{\partial C}{\partial p_k} + \frac{\partial K}{\partial p_k}\right)\underline{\alpha}_i$$

$$\frac{\partial \underline{\phi}_i}{\partial p_k} = \sum_{j=1}^{2n} a_{ji}^k \underline{\phi}_j, \qquad \frac{\partial \underline{\psi}_i}{\partial p_k} = \sum_{j=1}^{2n} b_{ji}^k \underline{\psi}_j$$

where the "components" of the eigenvector derivatives (projections onto the eigenvectors) are

$$\begin{aligned} a_{ji}^k &= \frac{1}{\lambda_j - \lambda_i}\{\underline{\psi}_j^{(2)}\}^T \left(\lambda_i^2 \frac{\partial M}{\partial p_k} + \lambda_i \frac{\partial C}{\partial p_k} + \frac{\partial K}{\partial p_k}\right)\underline{\alpha}_i, && j \neq i \\ &= -\frac{1}{2}\left[(1+\lambda_i^2)\underline{\alpha}_i^T \frac{\partial M}{\partial p_k}\underline{\alpha}_i + \sum_{\substack{l=1 \\ l \neq i}}^{2n} a_{li}^k (1+\lambda_i \lambda_l)\underline{\alpha}_l^T (M + M^T)\underline{\alpha}_i\right], && j = i \\ b_{ji}^k &= -\{\underline{\psi}_i^{(1)}\}^T \frac{\partial M}{\partial p_k}\underline{\alpha}_j + \frac{1}{\lambda_j - \lambda_i}\{\underline{\psi}_i^{(2)}\}^T \left(\lambda_i \lambda_j \frac{\partial M}{\partial p_k} + \lambda_j \frac{\partial C}{\partial p_k} + \frac{\partial K}{\partial p_k}\right)\underline{\alpha}_j, && j \neq i \\ &= -\{\underline{\psi}_i^{(1)} + \lambda_i \underline{\psi}_i^{(2)}\}^T \frac{\partial M}{\partial p_k}\underline{\alpha}_i - a_{ii}^k, && j = i \end{aligned}$$

References

1. Greenwood, D.T., *Classical Dynamics,* Prentice-Hall, Inc., Englewood Cliffs, NJ, 1977.
2. Meirovitch, L., *Analytical Methods in Vibrations,* Macmillan Company, Toronto, Canada, 1967.
3. Meirovitch, L., *Elements of Vibration Analysis,* 2nd Ed. McGraw-Hill Book Company, New York, NY, 1986.
4. Meirovitch, L., *Dynamics and Control of Structures,* John Wiley & Sons, New York, NY, 1990.
5. Junkins, J.L. and Turner, J.D., *Optimal Spacecraft Rotational Maneuvers,* Studies in Astronautics, Vol. 3, Elsevier Scientific Publishing Company, New York, NY, 1985.
6. Craig, R.R. Jr., *Structural Dynamics - an Introduction to Computer Methods,* John Wiley & Sons, New York, NY, 1981.
7. Pars, L.A., *A Treatise on Analytical Dynamics,* Heinemann, Ltd., London, 1965.
8. Bodden, D.S. and Junkins, J.L., "Eigenvalue Optimization Algorithms for Structural/Controller Design Iterations," *Journal of Guidance, Control, and Dynamics,* Vol. 8, No. 6., Nov.-Dec. 1985.
9. Turner, J.D. and Chun, H.M., "Optimal Feedback Control of a Flexible Spacecraft During a Large Angle Rotational Maneuver," *Journal of Guidance, Control, and Dynamics,* Vol. 7, No. 3, 1984, pp. 257-264.
10. Reddy, J.N., *An Introduction to the Finite Element Method,* McGraw-Hill Book Company, New York, NY, 1984.
11. Bathe, K.J., *Finite Element Procedure in Engineering Analysis,* Prentice-Hall, Inc., Englewood Cliffs, NJ, 1982.
12. Bayo E., "A Finite-Element Approach to Control the End-Point Motion of a Single-Link Flexible Robot," *Journal of Robotic Systems,* Vol. 4, No. 1, 1987, pp. 63-75.
13. Naganathan G. and Soni A.H., "Coupling Effects of Kinematics and Flexibility in Manipulators," *The International Journal of Robotics Research,* Vol. 6, No. 1, 1987, pp. 75-84.
14. Kim, Y., Junkins, J.L., and Kurdila, A.J., "On the Consequences of Certain Modeling Approximations in Dynamics and Control of Flexible Space Structures," Preprint, *Proceedings of the 33th SDM Conference,* AIAA, Dallas, TX, April 1992, pp. 1173-1184.

15. Thomson, R.C., *A Perturbation Approach to Control of Rotational Translational Maneuvers of Flexible Space Vehicles,* M.S. Thesis, Virginia Polytechnic Institute and State University, Blacksburg, VA, April 1985.

16. Joshi, S.M., *Control of Large Flexible Space Structures*, Lecture Notes in Control and Information Sciences, Vol. 131, Springer-Verlag, New York, NY, 1989.

17. Halbani, H.B., "Modal Identities for Multibody Elastic Spacecraft – An Aid to Selecting Modes for Simulations," AIAA Paper 89–0544, *27th Aerospace Science Meeting,* Reno, NV, Jan. 1989.

18. Turcic, D.A. and Midha, A., "Dynamic Analysis of Elastic Mechanism Systems. Part I: Applications," *Journal of Dynamic Systems, Measurement, and Control,* Transactions of the ASME, Vol. 106, Dec. 1984, pp. 249-254.

19. Gevarter, W.B., "Basic Relations for Control of Flexible Vehicles," *AIAA Journal,* Vol. 8, No. 4, April 1970, pp. 666-672.

20. Meirovitch, L., *Computational Methods in Structural Dynamics*, Sijthoff & Noordhoff, Leyden, The Netherlands, 1980.

21. Juang, J.N., "Optimal Design of a Passive Vibration Absorber for a Truss Beam," AIAA Paper 83-2291, 1983.

22. Lee, S. and Junkins, J.L., "Explicit Generalizations of Lagrange's Equations for Hybrid Coordinate Dynamical Systems," to appear, *AIAA Journal of Guidance, Control and Dynamics*, 1992.

23. Lee, S. and Junkins, J.L., "Explicit Generalizations of Lagrange's Equations for Hybrid Coordinate Dynamical Systems," Department of Aerospace Engineering, *Technical Report AERO 91-0301*, Texas A&M University, College Station, TX, March 1991.

24. Meirovitch, L., "Hybrid State Equations of Motion for Flexible Bodies in Terms of Quasi-Coordinates," *AIAA Journal of Guidance, Control, and Dynamics*, Vol. 14, No. 5, 1991, pp. 1008-1013.

25. Berbyuk, V.E., and Demidyuk, M.V., "Controlled Motion of an Elastic Manipulator with Distributed Parameters," *Mechanics of Solids*, Vol. 19, No. 2, 1984, pp. 57-65.

26. Low, K.H., and Vidyasagar, M., "A Lagrangian Formulation of the Dynamic Model for Flexible Manipulator Systems," *ASME Journal of Dynamic Systems, Measurement, and Control*, Vol. 110, June 1988, pp. 175-181.

27. Pars, L.A., *A Treatise on Analytical Dynamics*, Cambridge University Press, Cambridge, U. K., 1965.

28. Junkins, J.L., Rahman, Z., and Bang, H., "Near-Minimum-Time Maneuvers of Flexible Vehicles: A Liapunov Control Law Design Method," *Mechanics and Control of Large Flexible Structures*, edited by J. L. Junkins,*Progress in Astronautics and Aeronautics*, Vol.129, AIAA, Washington, DC, 1990, pp.565-593.
29. Das, Alok, "Large-Angle Maneuver Experiments in Ground-Based Laboratories," *Mechanics and Control of Large Flexible Structures*, edited by J. L. Junkins,*Progress in Astronautics and Aeronautics*, Vol.129, AIAA, Washington, DC, 1990, pp.491-506.
30. Junkins, J.L., and Bang, H., "Maneuver and Vibration Control of Nonlinear Hybrid Coordinate System using Liapunov Stability Theory," AIAA Paper 92-4458, Guidance, Navigation, and Control Conference, Hilton Head, SC, Aug., 1992.
31. Hailey, J. A., Sortun, C. A., and Agrawal, B. N., "Experimental Verification of Attitude Control Techniques for Slew Maneuvers of Flexible Spacecraft," AIAA Paper 92-4456, Guidance, Navigation, and Control Conference, Hilton Head, SC, Aug., 1992.

CHAPTER 5

Design of Linear State Feedback Control Systems

Among the most fundamental results in modern control theory are the two sets of analytical developments that underlie the linear-quadratic regulator (LQR) and eigenstructure assignment regulator (EAR). Design and implementation of practical control systems for flexible structures have been accomplished using both design techniques. Many other promising, but less computationally mature, techniques have emerged during the past five years (mostly in the quest to better accommodate model error-induced degradation of stability margins and performance); these can be collectively viewed as sophistications and enhancements of the methodology that supplement, and in many cases, provide particular gain tuning strategies for the LQR and EAR design approaches. The central aspects of these two sets of approaches are presented in this chapter, and the traditional developments of these topics are integrated with selected recent innovations from the research literature. We also consider several examples and a case study that provide some practical illustrations and pragmatic insight into implementation issues.

In the LQR approach, the central feature is the minimization of a quadratic performance index, subject to satisfaction of a linearized system differential equation model. This represents the most fundamental application of Pontryagin's minimum principle to linear dynamical systems, and can be viewed as a cornerstone of optimal control theory for dynamical systems generally. The integral quadratic performance measure appeals to the intuitive energy-based notion that we should seek a weighted compromise between the closed-loop output error energy and a measure of the control effort required to regulate the system. As is evident in our presentations herein, the fundamental practical difficulties in applying the LQR approach arise because it is difficult to capture many practical performance, robustness, and admissible design inequality constraint issues (which must actually be considered in real-world control law designs) in a single scalar performance measure, simply through a judicious a priori choice of weight matrices. These weight matrices should be viewed as parameterizing an infinite family of stabilizing control laws, and the analysts must, in practice, tune these weight matrices to impose the feasibility and robustness constraints for their particular control law design problems.

Since the complete response of a given linear system can be fully determined by the closed-loop eigensolution, another attractive approach is to find a control law that satisfies prescribed constraints on the closed-loop eigenvalues and eigenvectors. It is also possible to determine many of the candidate measures of performance and stability robustness (insensitivity with respect to system model errors and external disturbances) as simple functions of the closed-loop eigenvalues and eigenvectors.

The class of methods that directly impose prescribed constraints on the eigenvalues and eigenvectors are known as *eigenstructure assignment regulator* design methods. While the EAR approaches allow some immediate insight that narrows the search for practical designs, some additional difficulties arise. For controllable systems, we will see that the closed-loop eigenvalues can be placed exactly at an infinity of locations while, generally, the closed-loop eigenvectors can be placed only in certain feasible subspaces. The implications of these truths are explored through several alternative approaches, and are studied further through the use of case studies.

The quadratic regulator and eigenstructure assignment approaches are among the most important representatives of modern control theory, which is essentially the set of approaches based on the time-domain state space description of linear dynamical systems. As will be evident in the discussion below, the EAR formulations are actually a diverse family of recently developed concepts and the associated gain design algorithms that subsume and greatly expand upon the methodology historically known as *pole placement*.

In section 5.1, we review the fundamental aspects of the LQR approach, and present some alternative LQR formulations based on Lyapunov stability theory. Several variations of the central ideas, formulations, and algorithms of EAR are discussed in section 5.2, including several novel formulations and associated algorithms. In section 5.3, computational results (a case study using EAR gain design algorithms) are presented for an illustrative sixth-order discrete system, and a more representative twenty-fourth-order (reduced-order) model of a flexible spacecraft.

5.1 Linear Optimal Control

5.1.1 Necessary Conditions for Optimality

If a dynamical system is *completely controllable*, these exists at least one control that will transfer the system from any/all given initial states to any/all desired states. In the same fashion as in local/global stability definitions, we can introduce analogous restrictive interpretations of controllability. If the system is not controllable, at least in some local or restricted sense, it is obviously not meaningful to search for the optimal control. However, we could seek system modifications, such as a different number and/or locations for the actuators, to render the system more or less controllable.

Consider a completely controllable and observable dynamical system whose behavior is modeled by a nonlinear system of differential equations in the first-order state variable form

$$\dot{\mathbf{x}}(t) = \mathbf{F}\big(\mathbf{x}(t), \mathbf{u}(t), t\big) \tag{5.1}$$

where $\mathbf{x}(t) \in R^n$ is the state vector, $\mathbf{u}(t) \in R^m$ is the control vector, and $\mathbf{F}$ is a given function of the state vector, the control vector, and ttime.

The objective of optimal control theory is to determine the control law that satisfies the differential equation constraints – i.e. all trajectories are consistent with our model of the system dynamics [Eq. (5.1)] – and at the same time minimizes (or maximizes) some performance criterion. The most general performance index functional to be minimized (in the current discusion) is defined as

$$J = h\big(\mathbf{x}(t_f), t_f\big) + \int_{t_0}^{t_f} g\big(\mathbf{x}(t), \mathbf{u}(t), t\big)\, dt \tag{5.2}$$

where h and g are given functions, t_0 is the initial time, and t_f is the final time. For the present discussion, we consider the case in which the initial state $\mathbf{x}(t_0)$ is specified, and the initial time t_0 is specified. Thus the objective is to find the optimal control $\mathbf{u}^*(t)$ that drive the solution of the differential equations of Eq. (5.1) from the given initial conditons to time t_f such that the performance index of Eq. (5.2) is minimized.

The derivation of the necessary conditions for optimal control is relatively lengthy and straightforward and is not presented here; the reader may refer to Refs. 1-3 for details of the derivations. For the present discussion, suffice it to say that a candidate trajectory and the associated control history satisfy necessary conditions for optimality if the constrained *path variation* δJ performance differential vanishes [δJ is obtained by subtracting the performance index of Eq. (5.2) evaluated along a candidate trajectory from the value obtained along an arbitrary neighboring trajectory satisfying Eq. (5.1); for the *first variation* necessary conditions, only first-order terms in all δ quantities are retained]. To summarize the most important results, let us define the *Hamiltonian*:

$$\mathcal{H} \equiv \mathbf{g}\big(\mathbf{x}(t), \mathbf{u}(t), t\big) + \underline{\lambda}^T \Big[\mathbf{f}\big(\mathbf{x}(t), \mathbf{u}(t), t\big)\Big] \tag{5.3}$$

where $\underline{\lambda}(t) \in R^n$ is a vector of Lagrange multipliers (which is also widely known as the *costate vector*).

The necessary conditions for $\mathbf{u}^*(t)$ to be an optimal control are

$$\dot{\mathbf{x}}^*(t) = \frac{\partial \mathcal{H}}{\partial \underline{\lambda}}\big(\mathbf{x}^*(t), \mathbf{u}^*(t), \underline{\lambda}^*(t), t\big) \tag{5.4}$$

$$\dot{\underline{\lambda}}^*(t) = -\frac{\partial \mathcal{H}}{\partial \mathbf{x}}\big(\mathbf{x}^*(t), \mathbf{u}^*(t), \underline{\lambda}^*(t), t\big) \tag{5.5}$$

$$\frac{\partial \mathcal{H}}{\partial \mathbf{u}} = 0, \qquad \text{if all smooth unbounded } \mathbf{u} \text{ are admitted} \tag{5.6}$$

and the boundary conditions must be consistent with the *transversality condition:*

$$\left[\frac{\partial h}{\partial \mathbf{x}}(\mathbf{x}^*(t_f), t_f) - \underline{\lambda}^*(t_f)\right]^T \delta \mathbf{x}_f + \left[\mathcal{H}(\mathbf{x}^*(t_f), \mathbf{u}^*(t_f), \underline{\lambda}^*(t_f), t_f) + \frac{\partial h}{\partial t}(\mathbf{x}^*, t_f)\right]\delta t_f = 0 \tag{5.7}$$

If all of the necessary conditions of Eqs. (5.4)-(5.7) are satisfied, $\mathbf{u}^*(t)$ is called an optimal control and $\mathbf{x}^*(t)$ is called an optimal trajectory.

Equation (5.6) is the optimality condition that ensures the stationarity of the Hamiltonian; due to implicit assumptions, this condition is often not valid when the admissible controls are discontinuous and subject to inequality constraints such as saturation bounds. These cases and the necessary modifications of the above conditions are discussed below. The transversality condition [Eq. (5.7)] must be considered if the final state $\mathbf{x}(t_f)$ and/or the final time t_f are free; note that this equation is trivially satisfied if both $\mathbf{x}(t_f)$ and t_f are specified, because only zero variations are admitted. If subsets of $\mathbf{x}(t_f)$ or t_f are free, (i.e., not specified or otherwise constrained), the corresponding bracketed coefficient of Eq. (5.7) must vanish as a necessary boundary condition. Note that these conditions are only necessary, but not (in general) sufficient, for optimality because they arise from consideration of first variations of the performance index of Eq. (5.2).

In many control applicationions, the above formulation suffers a serious shortcoming because of the implicit assumption that the admissible controls $\mathbf{u}(t)$ are smooth, and otherwise unconstrained functions. Generalization of the optimal control formulation, to accommodate restrictions on the set of admissible controls, leads to *Pontryagin's minimum principle* (Refs. 1-4), which enables us to solve discontinuous control problems with inequality constraints on the admissible control variables such that

$$\mathbf{u}(t) \in \Omega \tag{5.8}$$

where Ω is a given set that includes all *admissible* controls.

Pontryagin's principle requires, in lieu of Eq. (5.6), that the Hamiltonian of Eq. (5.3) be *minimized with respect to* $\mathbf{u}(t)$ *over the set of all admissible controls* $\mathbf{u}(t) \in \Omega$. We state Pontryagin's necessary condition without proof; this principle replaces the optimality condition of Eq. (5.6) for the unbounded control case: The Hamiltonian must be minimized with respect to $\mathbf{u}$ at every instant t along the optimal trajectory, i.e,

$$\mathcal{H}(\mathbf{x}^*(t), \mathbf{u}^*(t), \underline{\lambda}^*(t), t) \leq \mathcal{H}(\mathbf{x}^*(t), \mathbf{u}(t), \underline{\lambda}^*(t), t), \qquad \mathbf{u} \in \Omega \tag{5.9}$$

Pontryagin's minimum principle, although derived to accommodate the more general case of bounded controls, can also be applied to problems in which the admissible controls are continuous and not bounded. For example, Pontryagin's principle requires, for an interior minimum of $\mathcal{H}$, that

$$\frac{\partial \mathcal{H}}{\partial \mathbf{u}}\left(\mathbf{x}^*(t), \mathbf{u}^*(t), \underline{\lambda}^*(t), t\right) = 0 \tag{5.10}$$

and also that the matrix of second partials must be positive definite, that is,

$$\frac{\partial^2 \mathcal{H}}{\partial \mathbf{u}^2}\left(\mathbf{x}^*(t), \mathbf{u}^*(t), \underline{\lambda}^*(t), t\right) > 0 \tag{5.11}$$

Thus, Pontryagin's principle is fully consistent with the classical variational calculus necessary conditions of optimality, since it specializes to yield Eq. (5.6). The additional second-order information that $\partial^2\mathcal{H}/\partial \mathbf{u}^2$ must be positive definite is also obtained, which is sufficient to guarantee that $\mathbf{u}^*(t)$ causes $\mathcal{H}$ to be a local minimum. To obtain this insight from the classical calculus of variations, one must investigate the *second variation.*

We have not considered above the extensions and the additional details associated with imposing inequality constraints on the state variables. The interested reader may pursue further details in Refs. 1-5.

5.1.2 Linear Regulator Problem

The regulator problem is defined as the problem of designing a control input (feedback form) to stabilize the plant about the zero state equilibrium point. The linear regulator problem is the most obvious special case of a linear system with the feedback control law being parameterized as a linear function of the instantaneous state. Consider class of systems described by the linear first-order state space differential equation

$$\dot{\mathbf{x}}(t) = A(t)\mathbf{x}(t) + B(t)\mathbf{u}(t) \tag{5.12}$$

where state vector $\mathbf{x}(t) \in R^n$, control vector $\mathbf{u}(t) \in R^m$, and system matrix A and control influence matrix B may contain time-varying elements.

The object is to determine an optimal feedback control gain G(t) for a linear law of the form

$$\mathbf{u}(t) = -G(t)\mathbf{x}(t) \tag{5.13}$$

which minimizes the quadratic performance index

$$J = \frac{1}{2}\mathbf{x}^T(t_f)S\mathbf{x}(t_f) + \frac{1}{2}\int_{t_0}^{t_f}\left[\mathbf{x}^T(t)Q(t)\mathbf{x}(t) + \mathbf{u}^T(t)R(t)\mathbf{u}(t)\right]dt \tag{5.14}$$

where S and Q are real symmetric positive semidefinite matrices, and R is a real symmetric positive definite matrix. It is assumed that the states and controls are not bounded. Moreover, we assume that all elements of the final state vector $\mathbf{x}(t_f)$ are free and the final time t_f is fixed.

Note that the physical and geometric motivation for this performance measure is as follows: *It is desired to maintain the time-varying state close to the origin without incurring an excessive expenditure of control effort.* The balance between small errors and small control effort is achieved through judicious assignment of the weight matrices S, Q, and R. The flexibility/responsibility for selecting S, Q, and R is simultaneously the major advantage and disadvantage of the LQR approach. Obviously, by exercising our freedom to choose the positive symmetric matrices S, Q, and R, we can generate an infinite family of controllers. Note that there are often other issues such as actuator saturation constraints and robustness considerations that are not explicitly addressed in the LQR formulation, and the particular numbers adopted in the S, Q, R matrices affect the suitability of the controller vis-à-vis these considerations.

The necessary conditions for the linear regulator problem are obtained by introducing the Hamiltonian [see Eqs. (5.1)-(5.3)], specialized for Eqs. (5.12) and (5.14),

$$\mathcal{H} = \frac{1}{2}\left\{\mathbf{x}^T(t)Q(t)\mathbf{x}(t) + \mathbf{u}^T(t)R(t)\mathbf{u}(t)\right\} + \underline{\lambda}^T(t)\left\{A(t)\mathbf{x}(t) + B(t)\mathbf{u}(t)\right\} \tag{5.15}$$

The necessary conditions [Eqs. (5.4)-(5.6)] for optimality specialize for the above Hamiltonian to

$$\dot{\mathbf{x}}^*(t) = \frac{\partial \mathcal{H}}{\partial \underline{\lambda}} = A(t)\mathbf{x}^*(t) + B(t)\mathbf{u}^*(t) \tag{5.16}$$

$$\dot{\underline{\lambda}}^*(t) = -\frac{\partial \mathcal{H}}{\partial \mathbf{x}} = -Q(t)\mathbf{x}^*(t) - A^T(t)\underline{\lambda}^*(t) \tag{5.17}$$

$$0 = \frac{\partial \mathcal{H}}{\partial \mathbf{u}} = R(t)\mathbf{u}^*(t) + B^T(t)\underline{\lambda}^*(t) \tag{5.18}$$

Moreover, note that, for fixed final time, the corresponding final variations are zero ($\delta t_f = 0$), and thus the first term of the transversality condition [Eq. (5.7)] yields

$$\underline{\lambda}^*(t_f) = \frac{\partial h}{\partial \mathbf{x}}\left(\mathbf{x}^*(t_f), t_f\right) = S\mathbf{x}^*(t_f) \tag{5.19}$$

From Eq. (5.18), we can solve for the optimal control in terms of the costate to obtain

$$\mathbf{u}^*(t) = -R^{-1}(t)B^T(t)\underline{\lambda}^*(t) \tag{5.20}$$

By observing the final condition (5.19), a full state feedback structure for the control is determined by assuming that the costate solution has the form

$$\underline{\lambda}^*(t) = P(t)\mathbf{x}^*(t) \tag{5.21}$$

where P(t) is a time-varying matrix function that is to be determined.

Differentiating Eq. (5.21) with respect to time, we have

$$\dot{\underline{\lambda}}^*(t) = \dot{P}(t)\mathbf{x}^*(t) + P(t)\dot{\mathbf{x}}^*(t) \tag{5.22}$$

Substituting Eqs. (5.16), (5.17), (5.20), and (5.21) into Eq. (5.22) leads to the *matrix differential Riccati equation*:

$$\dot{P}(t) = -A^T(t)P(t) - P(t)A(t) + P(t)B(t)R^{-1}(t)B^T(t)P(t) - Q(t) \tag{5.23}$$

The solution P(t) (an $n \times n$ positive semidefinite symmetric matrix), called the *Riccati matrix*, is subject to the following final boundary condition from Eqs. (5.19) and (5.21):

$$P(t_f) = S \tag{5.24}$$

Since P(t) is an $n \times n$ matrix, carrying out the algebra to obtain the scalar elements of Eq. (5.23) generates a coupled system of n^2 first-order differential equations. Notice that if P satisfies Eq. (5.23), P^T also satisfies it, because the transpose of Eq. (5.23) yields exactly Eq. (5.23). Since P(t) is symmetric, there are not n^2 but rather $n(n+1)/2$ first-order differential equations that must be solved. The Riccati equation, Eq. (5.23), is a nonlinear matrix differential equation that is often a source of computational difficulties for particular system and weight matrix specifications. This equation is famous for being difficult to integrate because of stiffness and divergent behavior, especially near the final time (t_f). We will discuss several algorithms for solving the Riccati equation in the next section.

Substituting Eq. (5.21) into Eq. (5.20), we obtain the optimal feedback control

$$\begin{aligned}\mathbf{u}^*(t) &= -R^{-1}(t)B^T(t)P(t)\mathbf{x}(t)\\ &\equiv -G(t)\mathbf{x}(t)\end{aligned} \tag{5.25}$$

Note that, even though the plant matrices are constant, the optimal feedback gain matrix G is time-varying for the finite time control case. Also, note that measurements (or estimates) of all of the state variables must be available to implement the optimal control law. For obvious reasons, such a law is known as a *full state feedback* control law.

The most important special case of the linear regulator problem is the one for which:

(1) The plant is completely controllable

(2) The plant dynamics are time invariant, i.e., A, B, Q, and R are constant matrices

(3) The final state error weight matrix is zero, S=0 in Eq. (5.14)

(4) The final time t_f is infinite.

In other words, the problem is to find the optimal control minimizing the performance index

$$J = \frac{1}{2}\int_{t_0}^{\infty}\left[\mathbf{x}^T Q\mathbf{x} + \mathbf{u}^T R\mathbf{u}\right]dt \tag{5.26}$$

subject to

$$\dot{\mathbf{x}}(t) = A\mathbf{x}(t) + B\mathbf{u}(t), \qquad \mathbf{x}(t_0) = \mathbf{x}_0 \tag{5.27}$$

In this case, it can be established (see Ref. 1, and the developments below) that the optimal control law for minimizing the quadratic index over an infinite time duration is stationary, and the feedback gain matrix is a constant,

$$\begin{aligned} \mathbf{u}^*(t) &= -G\mathbf{x}(t) \\ G &= R^{-1}B^T P \end{aligned} \tag{5.28}$$

determined by the constant solution P of the matrix Riccati equation, Eq. (5.23), reduced to a set of quadratically nonlinear algebraic equations,

$$A^T P + PA - PBR^{-1}B^T P + Q = 0 \tag{5.29}$$

which is called the *algebraic Riccati equation* (ARE). The fact that P approaches a constant can be established from the general time-varying P(t) solution using, for example, the solution form of Eq. (5.61) established below, and letting $t_f \to \infty$. Numerical algorithms for solving the ARE will be discussed in the next section.

We substitute the Q matrix from the ARE, Eq. (5.29), into the performance index of Eq. (5.26) to get the minimal performance,

$$\begin{aligned} J^* &= \frac{1}{2}\int_{t_0}^{\infty}\left[\mathbf{x}^T Q\mathbf{x} + \mathbf{u}^T R\mathbf{u}\right]dt \\ &= \frac{1}{2}\int_{t_0}^{\infty}\left[\mathbf{x}^T(-PA - A^T P + PBR^{-1}B^T P)\mathbf{x} + \mathbf{u}^T R\mathbf{u}\right]dt \end{aligned} \tag{5.30}$$

The above equation can be rewritten by using Eqs. (5.27) and (5.28) as follows:

$$
\begin{aligned}
J^* &= -\frac{1}{2}\int_{t_0}^{\infty}\Big\{[\mathbf{A}\mathbf{x}+\mathbf{B}\mathbf{u}]^T\mathbf{P}\mathbf{x}+\mathbf{x}^T\mathbf{P}[\mathbf{A}\mathbf{x}+\mathbf{B}\mathbf{u}]\Big\}dt \\
&= -\frac{1}{2}\int_{t_0}^{\infty}\Big\{\dot{\mathbf{x}}^T\mathbf{P}\mathbf{x}+\mathbf{x}^T\mathbf{P}\dot{\mathbf{x}}\Big\}dt = -\frac{1}{2}\int_{t_0}^{\infty}\frac{d}{dt}\Big[\mathbf{x}^T\mathbf{P}\mathbf{x}\Big]dt \qquad (5.31) \\
&= \frac{1}{2}\Big\{\mathbf{x}^T\mathbf{P}\mathbf{x}\Big|_{t=t_0} - \mathbf{x}^T\mathbf{P}\mathbf{x}\Big|_{t=\infty}\Big\}
\end{aligned}
$$

Since we consider only *stable* admissible controllers for which

$$\mathbf{x}(t) \rightarrow 0 \qquad \text{as} \qquad t \rightarrow \infty \tag{5.32}$$

and using the given initial condition $\mathbf{x}(t_0) = \mathbf{x_0}$ and the final condition, Eq. (5.32), the minimal performance index of Eqs. (5.26), (5.31) becomes

$$J^* = \frac{1}{2}\mathbf{x_0}^T\mathbf{P}\mathbf{x_0} \tag{5.33}$$

which depends on the initial condition and the algebraic Riccati solution P. The developments in this section are widely known as the linear quadratic regulator (LQR). Obviously "linear" refers to the linear system dynamics [Eq. (5.12)] and "quadratic" refers to the quadratic performance index [Eq. (5.26)]. Note that the MATLAB (Control System Toolbox) operator for the steady-state LQR problem has the following syntax:

$$> \ [\mathbf{G},\mathbf{P}] = \mathbf{lqr}(\mathbf{A},\mathbf{B},\mathbf{Q},\mathbf{R})$$

which returns the constant optimal gain G and the ARE solution P.

Example 5.1

To appreciate some elementary features of the LQR, we consider the simplest case of a scalar dynamical system

$$\dot{x}(t) = ax(t) + bu(t), \qquad x(0) = x_0$$

We seek the feedback control u which minimizes

$$J = \frac{1}{2}\int_{t_0}^{\infty} [qx^2 + ru^2]\ dt$$

where q and r are positive.

From the ARE of Eq. (5.29), we obtain the scalar algebraic quadratic equation for P,

$$b^2P^2 - 2arP - qr = 0$$

The analytical solution for the two roots of the above equation are simply

$$P = \frac{ar \pm \sqrt{(ar)^2 + qrb^2}}{b^2}$$

Since, for stability, P should be positive definite, in this case a positive scalar, we therefore choose the positive sign to obtain the solution for P,

$$P = \frac{ar + \sqrt{(ar)^2 + qrb^2}}{b^2}$$

and, from Eq. (5.28), the optimal gain is

$$G = \frac{b}{r}P = \frac{1}{b}\left(a + \sqrt{a^2 + \frac{q}{r}b^2}\right)$$

The value of the minimal performance index for this problem is obtained from Eq. (5.33) as

$$J^* = Px_0^2 = \frac{ar + \sqrt{(ar)^2 + qrb^2}}{b^2}x_0^2$$

Notice that G depends on the *ratio* (q/r) of the state-to-control performance weights. Thus we could arbitrarily set q or r to unity without loss of generality. More generally, for the multidimensional generalization, the Q or R matrices can be scaled by a free parameter without changing the optimal gain.

■

Example 5.2 Linear Tracking Problem

In this example, we generalize the above results obtained for the linear regulator problem to the closely related *linear tracking problem*, that is defined as the problem of designing a feedback control law which will cause the plant state to follow a given reference trajectory $\mathbf{r}(t)$. The state equations are

$$\dot{\mathbf{x}}(t) = A(t)\mathbf{x}(t) + B(t)\mathbf{u}(t) \tag{5.34}$$

In this case, we wish the plant to follow the reference trajectory $\mathbf{r}(t)$, and so the performance measure to be minimized is modified to give a positive measure of the time history and final value of the tracking error as

$$\begin{aligned} J = {} & \frac{1}{2}\left[\mathbf{x}(t_f) - \mathbf{r}(t_f)\right]^T S\left[\mathbf{x}(t_f) - \mathbf{r}(t_f)\right] \\ & + \frac{1}{2}\int_{t_0}^{t_f}\left\{\left[\mathbf{x}(t) - \mathbf{r}(t)\right]^T Q(t)\left[\mathbf{x}(t) - \mathbf{r}(t)\right] + \mathbf{u}^T(t)R(t)\mathbf{u}(t)\right\}dt \end{aligned} \tag{5.35}$$

where S and Q are real symmetric positive semidefinite matrices, and R is a real symmetric positive definite matrix; the final time t_f is fixed, and the final state $\mathbf{x}(t_f)$ is free.

The Hamiltonian is given by

$$\begin{aligned}\mathcal{H} =& \frac{1}{2}[\mathbf{x}(t) - \mathbf{r}(t)]^T Q(t)[\mathbf{x}(t) - \mathbf{r}(t)] + \frac{1}{2}\mathbf{u}^T(t)R(t)\mathbf{u}(t) \\ &+ \underline{\lambda}^T(t)\Big\{A(t)\mathbf{x}(t) + B(t)\mathbf{u}(t)\Big\}\end{aligned} \tag{5.36}$$

Using the necessary conditions (5.4)-(5.6) for optimality, we obtain

$$\dot{\mathbf{x}}^*(t) = \frac{\partial \mathcal{H}}{\partial \underline{\lambda}} = A(t)\mathbf{x}^*(t) + B(t)\mathbf{u}^*(t) \tag{5.37}$$

$$\dot{\underline{\lambda}}^*(t) = -\frac{\partial \mathcal{H}}{\partial \mathbf{x}} = -Q(t)\mathbf{x}^*(t) - A^T(t)\underline{\lambda}^*(t) + Q(t)\mathbf{r}(t) \tag{5.38}$$

$$0 = \frac{\partial \mathcal{H}}{\partial \mathbf{u}} = R(t)\mathbf{u}^*(t) + B^T(t)\underline{\lambda}^*(t) \tag{5.39}$$

Moreover, note that $\delta t_f = 0$; the first term of the transversality condition [Eq. (5.7)] yields

$$\underline{\lambda}^*(t_f) = S\mathbf{x}^*(t_f) - S\mathbf{r}(t_f) \tag{5.40}$$

From Eq. (5.39), we can solve for the optimal control in terms of the costate to obtain

$$\mathbf{u}^*(t) = -R^{-1}(t)B^T(t)\underline{\lambda}^*(t) \tag{5.41}$$

Substituting Eq. (5.41) into Eq. (5.37) yields the coupled state and costate equations

$$\begin{Bmatrix}\dot{\mathbf{x}}^*(t) \\ \dot{\underline{\lambda}}^*(t)\end{Bmatrix} = \begin{bmatrix} A(t) & -B(t)R^{-1}(t)B^T(t) \\ -Q(t) & -A^T(t)\end{bmatrix}\begin{Bmatrix}\mathbf{x}^*(t) \\ \underline{\lambda}^*(t)\end{Bmatrix} + \begin{Bmatrix} 0 \\ Q(t)\mathbf{r}(t)\end{Bmatrix} \tag{5.42}$$

The general solution of Eq. (5.42) is given by

$$\begin{Bmatrix}\mathbf{x}^*(t_f) \\ \underline{\lambda}^*(t_f)\end{Bmatrix} = \Phi(t_f, t)\begin{Bmatrix}\mathbf{x}^*(t) \\ \underline{\lambda}^*(t)\end{Bmatrix} + \int_t^{t_f} \Phi(t_f, \tau)\begin{Bmatrix} 0 \\ Q(\tau)\mathbf{r}(\tau)\end{Bmatrix} d\tau \tag{5.43}$$

where Φ is the transition matrix of the state/costate system (5.42), which can be partitioned into $n \times n$ submatrices Φ_{ij} as

$$\Phi(t_f, t) = \begin{bmatrix}\Phi_{11}(t_f, t) & \Phi_{12}(t_f, t) \\ \Phi_{21}(t_f, t) & \Phi_{22}(t_f, t)\end{bmatrix} \tag{5.44}$$

where Φ_{ij} are $n \times n$ matrices. Then Eq. (5.43) can be rewritten as the following pair of equations:

$$\mathbf{x}^*(t_f) = \Phi_{11}(t_f, t)\mathbf{x}^*(t) + \Phi_{12}(t_f, t)\underline{\lambda}^*(t) + \mathbf{f}_1(t) \tag{5.45}$$

$$\underline{\lambda}^*(t_f) = \Phi_{21}(t_f, t)\mathbf{x}^*(t) + \Phi_{22}(t_f, t)\underline{\lambda}^*(t) + \mathbf{f}_2(t) \tag{5.46}$$

where

$$\mathbf{f}_1(t) = \int_t^{t_f} \Phi_{12}(t_f, \tau)Q(\tau)\mathbf{r}(\tau)d\tau, \qquad \mathbf{f}_2(t) = \int_t^{t_f} \Phi_{22}(t_f, \tau)Q(\tau)\mathbf{r}(\tau)d\tau \tag{5.47}$$

Substituting Eqs. (5.45) and (5.46) into Eq. (5.40), we obtain

$$\begin{aligned} S\big[\Phi_{11}(t_f, t)\mathbf{x}^*(t) + \Phi_{12}(t_f, t)\underline{\lambda}^*(t) + \mathbf{f}_1(t)\big] - S\mathbf{r}(t_f) \\ = \Phi_{21}(t_f, t)\mathbf{x}^*(t) + \Phi_{22}(t_f, t)\underline{\lambda}^*(t) + \mathbf{f}_2(t) \end{aligned} \tag{5.48}$$

Solving for $\mathbf{x}^*(t)$, we obtain the following linear relationship between the instantaneous optimal costate vector $\underline{\lambda}^*(t)$ and the instantaneous optimal state vector $\mathbf{x}^*(t)$:

$$\begin{aligned} \underline{\lambda}^* &= \big[\Phi_{22}(t_f, t) - S\Phi_{12}(t_f, t)\big]^{-1}\big[S\Phi_{11}(t_f, t) - \Phi_{21}(t_f, t)\big]\mathbf{x}^*(t) \\ &\quad + \big[\Phi_{22}(t_f, t) - S\Phi_{12}(t_f, t)\big]^{-1}\Big\{S[\mathbf{f}_1(t) - \mathbf{r}(t_f)] - \mathbf{f}_2(t)\Big\} \\ &\equiv P(t)\mathbf{x}^*(t) + \mathbf{s}(t) \end{aligned} \tag{5.49}$$

where

$$\begin{aligned} P(t) &= \big[\Phi_{22}(t_f, t) - S\Phi_{12}(t_f, t)\big]^{-1}\big[S\Phi_{11}(t_f, t) - \Phi_{21}(t_f, t)\big] \\ \mathbf{s}(t) &= \big[\Phi_{22}(t_f, t) - S\Phi_{12}(t_f, t)\big]^{-1}\Big\{S[\mathbf{f}_1(t) - \mathbf{r}(t_f)] - \mathbf{f}_2(t)\Big\} \end{aligned} \tag{5.50}$$

It can be proven that $\big[\Phi_{22}(t_f, t) - S\Phi_{12}(t_f, t)\big]$ is generally nonsingular.

Substituting Eq. (5.49) into Eq. (5.41), we are led to the optimal (LQR) tracking control law

$$\begin{aligned} \mathbf{u}^*(t) &= -R^{-1}(t)B^T(t)P(t)\mathbf{x}^*(t) - R^{-1}(t)B^T(t)\mathbf{s}(t) \\ &\equiv -G(t)\mathbf{x}^*(t) - \mathbf{v}(t) \end{aligned} \tag{5.51}$$

Note that $G(t) = R^{-1}(t)B^T(t)P(t)$ is the optimal feedback gain matrix and that $\mathbf{v}(t) = R^{-1}(t)B^T(t)\mathbf{s}(t)$ is the command signal that depends on the system parameters and the reference signal $\mathbf{r}(t)$. Obviously, if we set $\mathbf{r}(t) = \mathbf{s}(t) = \mathbf{v}(t) = 0$, then the tracking law contains the classical LQR control law as the most obvious special case.

To establish an alternative solution process that yields P(t) and s(t) without computing the state transition matrix, let us differentiate Eq. (5.49) with respect to time t.

$$\dot{\underline{\lambda}}^*(t) = \dot{P}(t)\mathbf{x}^*(t) + P(t)\dot{\mathbf{x}}^*(t) + \dot{\mathbf{s}}(t) \tag{5.52}$$

Substituting Eq. (5.42) into Eq. (5.52), and using Eq. (5.49) to eliminate $\underline{\lambda}^*(t)$, we obtain

$$\begin{aligned}\Big[\dot{P}(t)+Q(t) + P(t)A(t) + A^T(t)P(t) - P(t)B(t)R^{-1}(t)B^T(t)P(t)\Big]\mathbf{x}^*(t)\\ + \Big[\dot{\mathbf{s}}(t) + A^T(t)\mathbf{s}(t) - P(t)B(t)R^{-1}(t)B^T(t)\mathbf{s}(t) - Q(t)\mathbf{r}(t)\Big] = 0\end{aligned} \tag{5.53}$$

Because this equation must be satisfied for all $\mathbf{x}^*(t)$, we conclude that P(t) and s(t) must satisfy the differential equations,

$$\dot{P}(t) = -P(t)A(t) - A^T(t)P(t) - Q(t) + P(t)B(t)R^{-1}(t)B^T(t)P(t) \tag{5.54}$$

$$\dot{\mathbf{s}}(t) = -\left[A^T(t) - P(t)B(t)R^{-1}(t)B^T(t)\right]\mathbf{s}(t) + Q(t)\mathbf{r}(t) \tag{5.55}$$

From observing Eqs. (5.40) and (5.49), we obtain the boundary conditions,

$$P(t_f) = S \tag{5.56}$$

$$\mathbf{s}(t_f) = -S\mathbf{r}(t_f) \tag{5.57}$$

Therefore, to obtain the optimal control, Eqs. (5.54) and (5.55) should be integrated *backward in time* starting with the terminal boundary conditions of Eqs. (5.56) and (5.57), and the results [P(t), $\mathbf{s}(t)$] stored to determine G(t) and $\mathbf{v}(t)$ in Eq. (5.51). Note that Eq. (5.54) is the same Riccati equation for the linear regulator problem, and the form obtained in Eq. (5.49) justifies the introduction of P as in Eq. (5.21) of our previous developments (i.e., for optimal feedback control, the costate vector is linearly related to the instantaneous state vector).

■

5.1.3 Numerical Algorithms for Solving the Riccati Equation

The differential Riccati equation of Eq. (5.23) is a nonlinear matrix differential equation. By backward integration of Eq. (5.23) with the final condition of Eq. (5.24), the solution can be obtained. However, because of nonlinearity and stiffness, a particular numerical method may encounter computational difficulties.

Numerous efficient numerical algorithms for solving this equation have been developed, and two of them will be discussed in this section. While satisfactorily accurate solutions of the Riccati differential equation can most often be obtained, there is no known method that universally guarantees stable and accurate computation. The situation is not unlike numerical difficulties encountered in matrix inversion, the conditioning of the solution process is dependent upon the particular system matrices and becomes increasingly difficult as the order of the system is increased.

Method I

In order to avoid solving the nonlinear Eq. (5.54), this algorithm introduces a judicious nonlinear matrix transformation,

$$P(t) = F(t)E^{-1}(t) \tag{5.58}$$

It can be shown [e.g., Ref. 6] that time-varying matrices E and F satisfy the following system of $2n^2$ linear differential equations:

$$\begin{Bmatrix} \dot{E}(t) \\ \dot{F}(t) \end{Bmatrix} = \begin{bmatrix} A(t) & -B(t)R^{-1}(t)B^T(t) \\ -Q(t) & -A^T(t) \end{bmatrix} \begin{Bmatrix} E(t) \\ F(t) \end{Bmatrix} \tag{5.59}$$

with the boundary conditions

$$E(t_f) = I, \qquad F(t_f) = S \tag{5.60}$$

Equation (5.59) can be integrated backward in time using boundary conditions (5.60) and, therefore, the Riccati solution can be obtained from the solution of the linear differential equation system (5.59) using the transformation of Eq. (5.58). Thus the nonlinear transformation of Eq. (5.58) maps the system of nonlinear equations into a system of linear equations, but doubles the order of the system of differential equations.

Method II

This algorithm introduces another judicious nonlinear variable change so as to avoid numerical integration (Refs. 3, 7, 8)

$$P(t) = P_{ss} + Z^{-1}(t), \qquad P(t_f) = S \tag{5.61}$$

where P_{ss} is the solution to the algebraic Riccati equation (5.29), which is computed (for example) using Potter's method (this will be discussed subsequently).

This solution is especially attractive when the time variation of P is a first-order perturbation of the steady-state solution, because first-order approximations or numerical errors made in solving for the perturbation $Z^{-1}(t)$ will obviously introduce only second-order perturbations of the solution for P.

For the case of constant A and B matrices, simple matrix manipulations provide the following linear constant coefficient matrix differential equation and associated final boundary condition for Z(t):

$$\dot{Z}(t) = \bar{A}Z(t) + Z\bar{A}^T - E, \qquad Z(t_f) = (S - P_{ss})^{-1} \tag{5.62}$$

where

$$\begin{aligned} \bar{A} &= A - EP_{ss} \\ E &= BR^{-1}B^T \end{aligned} \tag{5.63}$$

Equation (5.62) is known as the *differential Lyapunov equation*, and many algorithms exist for solving this equation (see Refs. 3, 9). As a typical illustration of this class of methods, we present an interesting algorithm (for solving the differential Riccati equation), utilizing the results of Lacoss and Shakel (Ref. 9), which was developed for solving the differential Lyapunov equation

Algorithm 5.1: Solution for Differential Riccati Equation

(1) Compute the solution for the algebraic matrix Riccati equation for P_{ss}.

(2) Solve the eigenvalue problem for the closed-loop system $\bar{A}$ $(= A - BR^{-1}B^TP_{ss})$

$$\bar{A}R = R\Lambda, \qquad \bar{A}^TL = L\Lambda$$

subject to the normalization equation $L^TR = I$.

(3) Compute

$$\rho_{ij} = \Lambda_i + \Lambda_j, \qquad \Sigma = L^T(BR^{-1}B^T)L, \qquad M = L^T(S - P_{ss})^{-1}L$$

(4) For each time t, compute T(t). (Note: Take advantage of symmetry in T(t).)

$$[T(t)]_{ij} = [F_1]_{ij}\, e^{-\rho_{ij}(t_f - t)} + [F_2]_{ij}$$

where $[F_1]_{ij} = [M]_{ij} - \dfrac{\Sigma_{ij}}{\rho_{ij}}$ and $[F_2]_{ij} = \dfrac{\Sigma_{ij}}{\rho_{ij}}$.

(5) Compute P(t):

$$P(t) = P_{ss} + [RT(t)R^T]^{-1}$$

while taking advantage of the structure of $RT(t)R^T$ and observing that the product is real, thus yielding a real symmetric matrix to be inverted.

■

Note that this algorithm does not execute time integration; instead, the algebraic Riccati equation (Step 1) and eigenvalue problem (Step 2) must be solved. The interested reader may refer to Ref. 3 for details of the derivation of the above algorithm.

Solution for ARE (Potter's Method)

Clearly, the algebraic Riccati equation, Eq. (5.29), is an algebraic equation of second degree; a number of methods exist for solving it. Among the most attractive and most widely used is a method developed by Potter (Ref. 10). We summarize this method as follows;

Algorithm 5.2 Potter's Method

(1) Construct the $2n \times 2n$ Hamiltonian matrix,

$$\mathcal{H} = \begin{bmatrix} A & -BR^{-1}B^T \\ -Q & -A^T \end{bmatrix} \tag{5.64}$$

(2) Solve the eigenvalue problem associated with matrix $\mathcal{H}$,

$$\mathcal{H}\{\underline{\phi}_j\} = \lambda_j \underline{\phi}_j, \qquad j = 1, 2, \ldots, 2n \tag{5.65}$$

where $\underline{\phi}_j$ is $2n \times 1$ eigenvectors. It is not necessary to normalize the eigenvectors.

(3) Rearrange and partition the computed eigenvectors such that

$$\Phi = \begin{bmatrix} \Phi_{11} & \Phi_{12} \\ \Phi_{21} & \Phi_{22} \end{bmatrix} \tag{5.66}$$

where the $n \times n$ matrices Φ_{12}, Φ_{22} are partitions of the eigenvector matrix corresponding to the n eigenvalues with negative real (stable) parts.

(4) The steady-state solution (P_{ss}) is obtained by solving the following equation:

$$P_{ss} = \Phi_{22}\Phi_{12}^{-1} \tag{5.67}$$

■

The above algorithm is relatively well known, and P_{ss} have been calculated for many systems with $n > 100$. Numerical difficulties are often encountered in Step 2, however, depending on the particular matrices A, B, Q, and R. A more numerically stable version of this algorithm is obtained by modifying the above discussion using Schur decomposition (Ref. 11).

Solution for ARE (Matrix Sign Function)

Recently, fast and stable algorithms have been developed for the computation of the matrix sign function (Refs. 12-14), and a sign function iterative ARE solution has been demonstrated to yield numerical robustness, accurate results, rapid (superlinear) convergence, algorithmic simplicity, and modest storage requirements (Refs. 15-18). We discuss the essential features of the ARE solution using the matrix sign function here.

The matrix sign function has been introduced as a generalization of the scalar sign function,

$$\mathrm{sgn(s)} = \begin{cases} +1, & \text{if } \mathrm{Re(s)}>0 \\ -1, & \text{if } \mathrm{Re(s)}<0 \end{cases} \tag{5.68}$$

with sgn(s) undefined for s along the imaginary axis of the complex plane.

The formal definition of the matrix sign function of a square matrix A with $\mathrm{Re}\{\lambda(\mathrm{A})\} \neq 0$ is as follows (Ref. 12):

$$\mathrm{sign(A)} = 2\ \mathrm{sign}^+(\mathrm{A}) - \mathrm{I_n} \tag{5.69}$$

where $\mathrm{I_n}$ is an $n \times n$ identity matrix and

$$\mathrm{sign}^+(\mathrm{A}) = \frac{1}{2\pi \mathrm{i}} \oint_{\mathrm{c}+} (\lambda \mathrm{I_n} - \mathrm{A})^{-1} d\lambda \tag{5.70}$$

c^+ is a simple closed contour in right half-plane of λ and encloses all the right-half-plane eigenvalues of A.

The matrix sign function of A can be represented by a matrix continued fraction (Refs. 13, 14) such as

$$\mathrm{sign(A)} = \mathrm{A}[\mathrm{I_n} + (\mathrm{A}^2 - \mathrm{I_n})[2\mathrm{I_n} + (\mathrm{A}^2 - \mathrm{I_n})[2\mathrm{I_n} + (\mathrm{A}^2 - \mathrm{I_n})[\cdots]^{-1}]^{-1}]^{-1}]^{-1} \tag{5.71}$$

The matrix sign function can be defined in a different way by using the definition of the principal square root (which can also be represented by a continued fraction) and Eq. (5.71) as follows (Ref. 13):

$$\mathrm{sign(A)} = \mathrm{A}(\sqrt{\mathrm{A}^2})^{-1} = \mathrm{A}^{-1}(\sqrt{\mathrm{A}^2}) \tag{5.72}$$

where the matrix $\sqrt{A^2}$ denotes the principal square root of A^2. A fast and stable algorithm for computing the principal square root of a matrix has been developed (Ref. 15), and it can be applied to compute the matrix sign function, Eq. (5.72).

The matrix sign function has the following useful properties (Ref. 16):

(1) $[\text{sign}(A)]^2 = I_n$

(2) The eigenvalues of sign(A) are ± 1.

(3) $\text{sign}(cA)=\text{sgn}(c)\ \text{sign}(A)$

(4) $\text{sign}(TAT^{-1}) = T\ \text{sign}(A)\ T^{-1}$

where c is nonzero scalar and T is any nonsingular square matrix.

By using the above properties, a solution of the algebraic Riccati equation that uses the matrix sign function has been derived (Refs. 16-18). Since matrix Φ of Eq. (5.66) is the modal matrix of the Hamiltonian $\mathcal{H}$ of Eq. (5.64), $\mathcal{H}$ can be decomposed as follows (spectral decomposition; section 2.2.1):

$$\mathcal{H} = \Phi\Lambda\Phi^{-1} = \Phi \begin{bmatrix} \Lambda_n & 0 \\ 0 & -\Lambda_n \end{bmatrix} \Phi^{-1} \tag{5.73}$$

where Λ_n $(-\Lambda_n)$ has all eigenvalues in the right- (left-) half-plane.

The matrix sign function of the Hamiltonian matrix, $\text{sign}(\mathcal{H})$, can be represented using the matrix sign function property (4) as

$$\text{sign}(\mathcal{H}) = \Phi\ \text{sign} \begin{bmatrix} \Lambda_n & 0 \\ 0 & -\Lambda_n \end{bmatrix} \Phi^{-1} \tag{5.74}$$

Since Λ_n and $-\Lambda_n$ are diagonal matrices, and because all elements of Λ_n lie in the left-half-plane, note

$$\begin{aligned} \text{sign}(\Lambda_n) &= I_n \\ \text{sign}(-\Lambda_n) &= -I_n \end{aligned} \tag{5.75}$$

Substituting Eq. (5.75) into Eq. (5.74) yields

$$\text{sign}(\mathcal{H}) = \Phi \begin{bmatrix} I_n & 0 \\ 0 & -I_n \end{bmatrix} \Phi^{-1} \tag{5.76}$$

Define a new matrix $\mathcal{W}$ as

$$\mathcal{W} \equiv \frac{1}{2}\Big[\text{sign}(\mathcal{H}) - I_{2n}\Big] \tag{5.77}$$

Substituting Eq. (5.76) into Eq. (5.77), along with some simple matrix manipulations, gives the elegant result

$$\mathcal{W} = -\Phi \begin{bmatrix} 0 & 0 \\ 0 & I_n \end{bmatrix} \Phi^{-1} \tag{5.78}$$

The submatrices of the inverse of the modal matrix Φ are defined as

$$\Phi^{-1} = \begin{bmatrix} \overline{\Phi}_{11} & \overline{\Phi}_{12} \\ \overline{\Phi}_{21} & \overline{\Phi}_{22} \end{bmatrix} \tag{5.79}$$

$\mathcal{W}$ can be represented in terms of the submatrices Φ_{ij} and $\overline{\Phi}_{ij}$ by using the definitions of Φ and Φ^{-1} [Eqs. (5.66) and (5.79)] in Eq. (5.78):

$$\mathcal{W} = -\begin{bmatrix} \Phi_{11}\overline{\Phi}_{21} & \Phi_{12}\overline{\Phi}_{22} \\ \Phi_{21}\overline{\Phi}_{21} & \Phi_{22}\overline{\Phi}_{22} \end{bmatrix} \equiv \begin{bmatrix} W_{11} & W_{12} \\ W_{21} & W_{22} \end{bmatrix} \tag{5.80}$$

Thus, the algebraic Riccati equation solution P_{ss} in Eq. (5.67) can be indirectly determined using the two submatrices W_{22} and W_{12} as a simple function of Φ_{22} and Φ_{12}, as follows

$$P_{ss} = W_{22}W_{12}^{-1} = \left(\Phi_{22}\overline{\Phi}_{22}\right)\left(\Phi_{12}\overline{\Phi}_{22}\right)^{-1} = \Phi_{22}\Phi_{12}^{-1} \tag{5.81}$$

As a result, Eqs. (5.64), (5.77), and (5.81) establish a new algorithm for solving algebraic Riccati equations. Observe, for poorly conditioned systems, and especially for high-dimensioned systems, the inverse indicated in Eq. (5.81) should be computed using singular value decomposition (the Moore-Penrose generalized inverse, see Chapter 2).

Based on the above developments and the matrix sign function, we have developed a MATLAB operator for solving the algebraic Riccati equation and the optimal gains for the steady-state LQR problem. The syntax is as follows:

$$> \ [\mathbf{G}, \mathbf{P}] = \mathbf{lqrmsf}(\mathbf{A}, \mathbf{B}, \mathbf{Q}, \mathbf{R})$$

which returns the constant optimal gain G and algebraic Riccati solution P.

5.1.4 Generalized Linear Quadratic Regulator Formulations

In this section, three recently developed generalized linear quadratic regulator formulations are discussed that introduce modifications of the usual quadratic performance index (including cross-product coupling weights between state and control).

Generalized LQR I

Consider the quadratic criterion in the form (Ref. 4)

$$J = \frac{1}{2}\mathbf{x}^T(t_f)S\mathbf{x}(t_f) + \frac{1}{2}\int_{t_0}^{t_f}\begin{Bmatrix}\mathbf{x}(t)\\ \mathbf{u}(t)\end{Bmatrix}^T\begin{bmatrix}Q(t) & N(t)\\ N^T(t) & R(t)\end{bmatrix}\begin{Bmatrix}\mathbf{x}(t)\\ \mathbf{u}(t)\end{Bmatrix} \tag{5.82}$$

By defining the Hamiltonian [Eq. (5.3)], and applying the necessary conditions for optimality [Eqs. (5.4)-(5.6)], we find that the optimal control $\mathbf{u}^*(t)$ has the form

$$\mathbf{u}^*(t) = -R^{-1}(t)\Big\{N^T(t) + B^T(t)P(t)\Big\}\mathbf{x}(t) \tag{5.83}$$

where matrix P satisfies the following matrix differential equation:

$$\dot{P} = -(A - BR^{-1}N^T)^T P - P(A - BR^{-1}N^T) + PBR^{-1}B^T P - (Q - NR^{-1}N^T) \tag{5.84}$$

with the final condition $P(t_f) = S$.

Note that the above results are equivalent to the "prefeedback" optimal regulator problem, wherein we seek an optimal feedback control $\mathbf{u}(t)$ for the system

$$\dot{\mathbf{x}}(t) = \Big\{A(t) - B(t)R^{-1}(t)N^T(t)\Big\}\mathbf{x}(t) + B(t)\mathbf{u}(t) \tag{5.85}$$

such that, for specified system and weight matrices (A, B, N, R, Q, S), the following performance index is minimized

$$J = \frac{1}{2}\mathbf{x}^T(t_f)S\mathbf{x}(t_f) + \frac{1}{2}\int_{t_0}^{t_f}\begin{Bmatrix}\mathbf{x}(t)\\ \mathbf{u}(t)\end{Bmatrix}^T\begin{bmatrix}Q(t) - N(t)R^{-1}(t)N^T(t) & 0\\ 0 & R(t)\end{bmatrix}\begin{Bmatrix}\mathbf{x}(t)\\ \mathbf{u}(t)\end{Bmatrix} \tag{5.86}$$

It is evident that the off-diagonal weight matrix N can be thought of as some arbitrary feedback gain matrix associated with a control input $N\mathbf{x}(t)$ that was applied a priori; then the LQR approach was used to design a superimposed "optimal" feedback for this system, which already has a feedback loop closed. The LQR design with the cross-coupling weight matrix N offers more choices of the feedback gains than the standard LQR design. However, the N matrix introduces redundancy in the sense that the number of weight parameters are greatly increased compared to the number of elements in the gain matrix.

At this point, it remains to be seen whether or not introduction of this off-diagonal weight matrix has any advantages. As is evident below, it does lead to some useful insights. Now, let us consider the steady-state linear quadratic regulator.

Generalized LQR II

Consider the modified performance index that is an extension of Eq. (5.82).

$$J = \frac{1}{2}\int_{t_0}^{\infty} \begin{Bmatrix} \mathbf{x}(t) \\ \mathbf{u}(t) \end{Bmatrix}^T \begin{bmatrix} Q + N^T R N & N^T R \\ RN & R \end{bmatrix} \begin{Bmatrix} \mathbf{x}(t) \\ \mathbf{u}(t) \end{Bmatrix} \tag{5.87}$$

Equations. (5.83) and (5.84) lead to the generalized Riccati equation

$$P(A - BN) + (A - BN)^T P - PBR^{-1}B^T P + Q = 0 \tag{5.88}$$

and the optimal gain matrix

$$G = R^{-1}B^T P + N \tag{5.89}$$

For this case, it has been shown (Ref. 19) that an *arbitrary* $m \times n$ cross-coupling weight matrix N can be chosen without destabilizing the closed-loop system. Note that N is equivalent to a prefeedback gain, and so it is evident that redundancy must exist.

Generalized LQR III

LQR-based control synthesis algorithms often suffer from the computational burden associated with solving the algebraic Riccati equation. This is especially troublesome for high-dimensioned systems for which numerous design iterations may be required to satisfy saturation constraints (for example) and to accommodate other issues not explicitly captured in the optimal control considerations leading to the LQR. In order to reduce the cost and numerical difficulties of these design iterations, a modified LQR algorithm based on Lyapunov stability theory has been developed (Ref. 19). The solution of the nonlinear Riccati equation can be eliminated in favor of an LQR formulation in which the optimal gains are parameterized in terms of a much simpler linear Lyapunov equation.

Motivated by the performance index of Eq. (5.87), we introduce the generalized quadratic criterion

$$J = \frac{1}{2}\int_{t_0}^{\infty} \begin{Bmatrix} \mathbf{x}(t) \\ \mathbf{u}(t) \end{Bmatrix}^T \begin{bmatrix} Q + PBR^{-1}B^T P + N^T R N & N^T R \\ RN & R \end{bmatrix} \begin{Bmatrix} \mathbf{x}(t) \\ \mathbf{u}(t) \end{Bmatrix} \tag{5.90}$$

The motivation for adding the final term $(PBR^{-1}B^TP)$ in the (1,1) partition of weighting matrix is that it results in an important simplification of the necessary conditions. Specifically, this modification results in the vanishing of the quadratic term of the Riccati matrix, and we find that the P matrix satisfies the (much simpler) linear algebraic Lyapunov equation

$$P(A - BN) + (A - BN)^T P + Q = 0 \tag{5.91}$$

instead of the nonlinear algebraic Riccati equation (5.88). The optimal gain then follows from Eq. (5.89), as before.

It should be noted that the above solution process encounters one conceptually unusual feature: We must solve Eq. (5.91) for the P matrix before we can fully construct the associated performance index of Eq. (5.90), because P appears in the weight matrix. The above weight matrix parameterization appears rather contrived, however, given the degree of arbitrariness implicit in specifying the weight matrix, and given the consequences of this parameterization (solve a Lyapunov equation instead of a Riccati equation); this device appears worth mentioning. The optimal gain solution of Eq. (5.89) holds so long as a positive definite solution of Eq. (5.91) exists. However, when an arbitrary N matrix is chosen, a positive definite solution is not guaranteed. One way to guarantee the solution to the Lyapunov equation of Eq. (5.91) is to construct any N matrix such that all the eigenvalues of the matrix (A-BN) lie in the left half-plane of the complex plane. Thus any stabilizing choice for the prefeedback gain matrix N will guarantee the existence of P.

While this gain parameterization involves redundancy, it is obvious that solving Eq. (5.91) is much more attractive computationally than solving the usual Riccati equation. In many cases, simply choosing N to be some small (low-authority) stabilizing feedback gain provides an attractive basis for subsequent design iterations, because of the elimination of the necessity to solve the matrix Riccati equation.

5.2 Robust Eigenstructure Assignment

Typically, for a controllable system, a given set of eigenvalues can be assigned by an infinity of gain matrices. This fact can be exploited to exercise the freedom "beyond" placement of the eigenvalues to achieve robustness or to extremize some given performance measure. During the past decade, a number of eigenstructure assignment regulator (EAR or "pole placement methods") formulations and algorithms have been developed and have been shown (Refs. 20-26) to be useful tools for state and output feedback system designs. The EAR design approach allows the analyst/designer to directly choose eigenvalues and to explore the arbitrariness of admissible eigenvectors.

These ideas were first introduced by Brogan in Ref. 20. They utilize formulations based on either Sylvester's equation (Refs. 21, 22) or projections to a subspace of admissible eigenvectors (Refs. 23, 24). These formulations are conceptually equivalent in the sense that they share the same parameterization scheme for eigenvectors. Differences arise among the several algorithms because of choices of criteria that generate the closed-loop eigenvectors, as well as the implementation details. The success of these approaches depends mainly on the selection of criteria or parameter sets that explicitly or implicitly determine the closed-loop eigenvectors.

It is well known that near orthogonality of the closed-loop eigenvectors is very desirable to minimize sensitivity of eigenvalue placement to model errors. Arbitrary selection of feasible eigenvectors may cause the eigenvector modal matrix to be illconditioned. If this is the case, the associated gain matrix may not be accurately calculated (if computed via the usual EAR gain design algorithms which would require solving poorly conditioned linear systems) and, further, the stability and response of the closed-loop system may be highly sensitive to parameter variations. Therefore, generation of well-conditioned eigenvectors is a key issue for this family of pole placement algorithms with regard to achieving both insensitive feedback system design and numerical stability.

In this section, we describe a pole placement algorithm based on defining subspaces of admissible eigenvectors. A specialization for a mechanical second-order system is also provided.

5.2.1 Sylvester's Equation

Consider the linear dynamical system in the state space form

$$\dot{\mathbf{x}}(t) = A\mathbf{x}(t) + B\mathbf{u}(t), \qquad \mathbf{x}(0) = \mathbf{x}_0 \tag{5.92}$$

with the linear feedback control

$$\mathbf{u}(t) = -G\mathbf{x}(t) \tag{5.93}$$

where state vector $\mathbf{x}(t) \in R^n$, and control vector $\mathbf{u}(t) \in R^m$. We assume, for simplicity, that the full state is measurable and the pair (A,B) is completely controllable. Also, we assume that rank(B)=m.

From Eqs. (5.92) and (5.93), we form the closed-loop system

$$\dot{\mathbf{x}}(t) = (A - BG)\mathbf{x}(t) \tag{5.94}$$

and corresponding eigenvalue problems

$$\begin{aligned} (A - BG)\underline{\phi}_i &= \lambda_i \underline{\phi}_i \\ (A - BG)^T \underline{\psi}_i &= \lambda_i \underline{\psi}_i \end{aligned} \tag{5.95}$$

where $\underline{\phi}_i$ and $\underline{\psi}_i$ are the right and left eigenvectors, respectively, corresponding to the eigenvalue λ_i. We adopt the normalization for the eigenvectors by scaling them such that

$$\underline{\phi}_i^H \underline{\phi}_i = 1, \qquad \underline{\psi}_i^T \underline{\phi}_j = \delta_{ij} \tag{5.96}$$

Then, the central constraint imposed in the eigenvalue assignment problem is to determine a gain matrix G that results in a prescribed set of eigenvalues. Note that G is an $m \times n$ dimensional matrix; it is evident that the solution for G is *underdetermined* so that, for controllable systems, an infinity of gain matrices exist for given eigenvalue locations. We could choose $n \times (m-1)$ gain parameters arbitrarily for n prescribed eigenvalues. However, we exploit this redundancy in other ways, as is evident in the following discussions.

The pole placement algorithm of Ref. 22 introduces the parameter vector $\mathbf{h}_i \in C^m$ defined by

$$\mathbf{h}_i = G\underline{\phi}_i \tag{5.97}$$

Then Eq. (5.95) is put in the form

$$(A - \lambda_i I)\underline{\phi}_i = B\mathbf{h}_i \tag{5.98}$$

or, in matrix form, Eq. (5.98) is a generalized Lyapunov equation known as Sylvester's equation,

$$A\Phi - \Phi\Lambda = BH \tag{5.99}$$

where $\Phi = [\phi_1, \phi_2, \ldots, \phi_n]$, $\Lambda = \text{diag}[\lambda_1, \lambda_2, \ldots, \lambda_n]$, and $H = [\mathbf{h}_1, \mathbf{h}_2, \ldots, \mathbf{h}_n]$.

The pole placement scheme based on Sylvester's equation [Eq. (5.98) or (5.99)] can be summarized as follows: For given set of A, B matrices, and for a prescribed Λ matrix, we can choose a parameter matrix H and solve for Φ from Eq. (5.99). Then, we can solve for G from the linear system [which is simply the matrix version of Eqs. (5.97)]:

$$G\Phi = H \tag{5.100}$$

Unfortunately, an arbitrary choice for the complex H matrix does not usually generate an attractive set of closed-loop eigenvectors; occasionally the resulting eigenvectors are so poorly conditioned that computing an accurate gain matrix G from Eq. (5.100) is not possible. In essence, the advantage of "guessing Λ and H" instead of "guessing G" is that the exact prescribed eigenvalue positions are guaranteed if we specify Λ and choose H. The H matrix generates [through the solution of (5.99) for Λ specified] all infinity of achievable eigenvector matrices.

Note, from inversion of Eq. (5.98), that the closed-loop eigenvectors corresponding to given λ_i and $\mathbf{h}_i$ are simply

$$\underline{\phi}_i = (A - \lambda_i I)^{-1} B\mathbf{h}_i \tag{5.101}$$

So, if the λ_is are distinct from their open-loop positions, the columns of H directly generate all possible corresponding closed-loop eigenvectors. Note that our MATLAB operator **assign**(for eigenstructure assignment) based on the above is as follows:

$$> \mathbf{G} = \mathbf{assign}(\mathbf{A}, \mathbf{B}, \mathbf{lambdad}, \mathbf{Phi})$$

where **lambdad** is a vector containing the desired closed-loop eigenvalues, **Phi** denotes the matrix of *target eigenvectors*, and the operator **assign** returns gain matrix **G**. In the operator *assign*, we determine the H matrix to obtain the closed-loop eigenvectors in least square agreement with the corresponding assigned target eigenvectors (the columns of **Phi**). Given this H, the assign algorithm then solves Eq. (5.99) for the achieved eigenvectors, then solves Eq. (5.100) to determine the gain matrix **G**.

5.2.2 Projection Method for Eigenstructure Assignment

Since an arbitrary selection of the $m \times n$ complex matrix H is not appropriate, we are motivated to consider those choices that have a high probability of generating attractive gain matrices. In our review (Chapter 2), we showed that the condition number of the closed-loop modal matrix is a measure of robustness (sensitivity) of the placement of the closed-loop eigenvalues and, for fixed eigenvalue positions, minimizing the condition number of the closed-loop modal matrix is equivalent to maximizing (one measure of) the system robustness with respect to variation of the system parameters. This motivates our search for an H matrix to give as small as possible a condition number for the closed-loop modal matrix. An attractive algorithm results if we seek the H matrix which makes the closed-loop modal matrix lie as close as possible to a prescribed, well-conditioned matrix. Notice that, if we select some target set of well-conditioned closed-loop eigenvectors

$$\widehat{\Phi} = [\widehat{\phi}_1, \widehat{\phi}_2, \ldots, \widehat{\phi}_n] \tag{5.102}$$

then, we can use Eq. (5.98), or equivalently Eq. (5.99), to solve for the $\widehat{\mathrm{H}}$ that most nearly (e.g., in the least square sense) produces the desired eigenvectors $\widehat{\Phi}$. Upon substituting this solution for the H matrix and re-solving Eq. (5.99) for the actually achieved eigenvector matrix Φ, we will find $\Phi \neq \widehat{\Phi}$, exactly, with the degree of approximation being problem-dependent. However, the Φ matrix lies as near $\widehat{\Phi}$ as possible (in the least square sense) and is typically well conditioned. Of significance, however, *the arbitrary complex scaling of the columns of Φ and $\widehat{\Phi}$ make "closeness" of Φ and $\widehat{\Phi}$ impossible to measure, unless a consistent normalization is imposed.* The gain matrix G calculated from solution of Eq. (5.100) with Φ and $\widehat{\mathrm{H}}$ will, however, place the eigenvalues exactly, to within arithmetic errors. We now address the crucial issue of how to select good target eigenvectors $\widehat{\Phi}$.

From Eq. (5.101), we observe that the admissible eigenvectors must lie in the m-dimensional column space of $(A - \lambda_i I)^{-1}B$. This space can be spanned by an $n \times m$ unitary matrix U_i. Through appropriate choice of $\mathbf{h}_i$, we can generate all admissible eigenvectors from the projection

$$\underline{\phi}_i = U_i \mathbf{h}_i \tag{5.103}$$

Denoting the target eigenvectors $\widehat{\underline{\phi}}_i$, we formulate the least squares problem

$$\widehat{\underline{\phi}}_i = U_i \mathbf{h}_i + \mathbf{e}_i \tag{5.104}$$

The $\mathbf{h}$ vector that minimizes the residual errors $|\mathbf{e}_i|$ is given by orthogonal projection as

$$\widehat{\mathbf{h}} = U_i^H \widehat{\underline{\phi}}_i \tag{5.105}$$

Then the feasible eigenvectors $\underline{\phi}_i$ nearest target eigenvectors $\widehat{\underline{\phi}}_i$ are obtained by substituting Eq. (5.105) into Eq. (5.103), which gives

$$\underline{\phi}_i = U_i U_i^H \widehat{\underline{\phi}}_i \tag{5.106}$$

and the corresponding minimized residual error $\mathbf{e}_i$,

$$\mathbf{e}_i = \widehat{\underline{\phi}}_i - \underline{\phi}_i = (I - U_i U_i^H)\widehat{\underline{\phi}}_i \tag{5.107}$$

There are several computationally stable algorithms (Refs. 25, 26) available for calculation of the unitary basis matrices U_i:

(1) Decompose $(A - \lambda_i I)^{-1}B$ using the singular value decomposition:

$$(A - \lambda_i I)^{-1}B = [\,U_{1i} \quad U_{2i}\,] \begin{bmatrix} \Sigma_i & 0 \\ 0 & 0 \end{bmatrix} \begin{bmatrix} V_{1i}^H \\ V_{2i}^H \end{bmatrix}$$

and define $U_i = U_{1i}$.

(2) Decompose $(A - \lambda_i I)^{-1}B$ using the QR decomposition:

$$(A - \lambda_i I)^{-1}B = [\,Q_{1i} \quad Q_{2i}\,] \begin{bmatrix} R \\ 0 \end{bmatrix}$$

and define $U_i = Q_{1i}$.

(3) Compute the orthogonal complement $B^{\perp}$ of the column space of B (i.e., $B^{\perp}B = 0$), and compute the complement of the column space of $B^{\perp}(A - \lambda_i I)$ using singular value decomposition:

$$B^{\perp}(A - \lambda_i I) = [U_{1i} \quad U_{2i}] \begin{bmatrix} \Sigma_i & 0 \\ 0 & 0 \end{bmatrix} \begin{bmatrix} V_{1i}^H \\ V_{2i}^H \end{bmatrix}$$

and define $U_i = V_{2i}$.

The third approach is a good choice, since it does not require inversion of $(A - \lambda_i I)$. It should be noted that when B is a full-rank square matrix, the orthogonal complement does not exist. If this is the case, Eq. (5.101) can be replaced by

$$B^{-1}(A - \lambda_i i)\underline{\phi}_i = \mathbf{h}_i \tag{5.108}$$

which means that, for any given vector $\underline{\phi}_i$, there exists an exact solution for a unique vector $\mathbf{h}_i$. For this case, the present algorithm reduces to a first order version of *independent modal space control* (Ref. 28), *if the $\underline{\phi}_i$s are selected as the open-loop eigenvectors.*

The above developments still do not tell us how to choose the $\widehat{\underline{\phi}}_i$ judiciously. With Eq. (5.107), in a fashion analogous to the developments in Ref. 26, we define an optimization problem as follows:

$$\text{Find} \quad \widehat{\underline{\phi}}_i \; (i = 1, \ldots, n), \qquad \text{to minimize } J = \sum_{j=1}^{n} \mathbf{e}_j^H \mathbf{e}_j \tag{5.109}$$

subject to

$$\widehat{\underline{\phi}}_i^H \, \widehat{\underline{\phi}}_i = \delta_{ij}, \qquad i, j = 1, 2, \ldots, n \tag{5.110}$$

Unfortunately, as a consequence of the nonlinear unitary parameterization constraints of Eq. (5.110), solutions of this problem requires iterative numerical optimization; it can, in principle, be solved by using nonlinear programming algorithms. A more direct, more computationally attractive, but less rigorous approach, is to simply choose some judicious unitary basis (i.e., a well-conditioned, physically motivated set of target eigenvectors $\widehat{\underline{\phi}}_i$), and solve the unconstrained optimization problem of Eq. (5.109). Thus, we present an algorithm (Ref. 27) to generate a target unitary basis vectors from the subspace matrices U_i ($i = 1, 2, \ldots, n$) for admissible eigenvectors.

We define the global matrix S by arranging the n sets of unitary basis matrices as

$$S = \left[U_1, U_2, \ldots, U_n\right] \tag{5.111}$$

and take the singular value decomposition of S to obtain the SVD factorization

$$S = \widehat{U}\Sigma V^H \tag{5.112}$$

where $\widehat{U}$, V are the left and right matrices of singular vectors and Σ is the diagonal matrix of singular values. Note that U_i is a unitary matrix that spans the space of achievable eigenvectors corresponding to the ith eigenvalue. We hypothesize that since $\widehat{U}$ is a unitary matrix spanning the intersection space of all admissible eigenvectors' achievable spaces, it will prove an attractive set of target eigenvectors. Now, it remains to determine which of the unitary basis vectors (columns of $\widehat{U}$) should be assigned as the target eigenvector corresponding to each particular assigned eigenvalue, such that the performance index J of Eq. (5.109) is minimized. Based on this approach, we summarize the main steps of the algorithm (Ref. 27) as follows.

Algorithm 5.3 Robust Eigenstructure Assignment

(1) For $i = 1, 2, \ldots, n$, compute the unitary basis matrix U_i of the column space of the $(A - \lambda_i I)^{-1}B$ matrix

(2) Find the left singular vectors $\widehat{U} = [\widehat{\mathbf{u}}_1, \widehat{\mathbf{u}}_2, \ldots, \widehat{\mathbf{u}}_n]$ of the matrix S defined by Eq. (5.111), and define these as the set of target vectors; $\widehat{\underline{\phi}}_i = \widehat{\mathbf{u}}_i$, for $i = 1, 2, \ldots, n$.

(3) For $i = 1, 2, \ldots, n$, determine the index k for the ith target eigenvector $\widehat{\underline{\phi}}_k$ that minimizes the eigenvector assignment residuals $\|e_i\| = \|(I - U_i U_i^H)\widehat{\underline{\phi}}_k\|$. Store this index in an array, index(i)=k, and remove $\widehat{\underline{\phi}}_k$ from the set of desired basis vectors.

(4) Calculate admissible eigenvectors as the least squares projection:

$$\underline{\phi}_i^a = U_i U_i^H \widehat{\underline{\phi}}_k, \qquad k = \text{index}(i)$$

(5) Determine the H matrix by solving Eq. (5.99):

$$H = B^{\dagger}(A\Phi^a - \Phi^a\Lambda)$$

where $B^{\dagger}$ is the Moore-Penrose pseudoinverse of the B matrix and, using H matrix, re-solve the following Lyapunov equation (i.e., Sylvester's equation [Eq. (5.99)]) to get the target eigenvector Φ:

$$A\Phi - \Phi\Lambda = BH$$

Now, compute the gain matrix G by solving Eq. (5.100):

$$G\Phi = H$$

■

It should be noted that the computed gain matrix G should contain only real values. If the gain calculation yields complex control gains, then assignment of the closed-loop eigenvalues to the desired locations has failed. Failure indicates that either (1) the system is uncontrollable, or (2) the eigenvalues and eigenvectors have been inconsistently paired. Note especially that, for complex conjugate closed-loop eigenvalues, the corresponding mode's target eigenvectors must also be complex conjugates. Failure to correctly assign the eigenvectors as complex conjugate pairs corresponding to the eigenvalues will reliably result in an erroneous, complex gain matrix. By considering this property, if the algorithm is modified such that only real numbers are treated for solving Sylvester's equation (by decomposing all complex modes' matrices and vectors into real parts and imaginary parts), we can avoid the possibility of this inconsistency and the ensuing undesired results (complex gains). However, there is still a possibility that we may get large gains, which may come from the fact that we specified some hard-to-achieve target eigenvectors (note that since we obtain the target eigenvectors from singular value decomposition, they are not necessarily near the corresponding physical open-loop eigenvectors, even though they may have desirable mathematical properties).

Specifically, the correspondence between target eigenvalues and target eigenvectors is extremely important and cannot be chosen at random. One approach is to specify the correspondence between the target eigenvectors and (say) the ith eigenvalue, which results in the target eigenvector being the smallest perturbation of the ith open-loop eigenvector. See steps 2-4 in the above algorithm. It is known that even if the eigenvalues change significantly upon closing a control loop, the corresponding eigenvectors often do not change very much and most of the mode shapes (eigenvectors) resemble themselves before and after closing the control loop. This is especially true for the high-frequency eigenvectors. For a flexible vibrating structure, this is easy to understand, because the normal vibration mode shapes are a function of the overall mass and stiffness distribution; the control effectively introduces local stiffness perturbations of these inherent structural properties. "Large" changes in the eigenvectors are usually associated with "large" control forces. Therefore, from a physical point of view, open-loop eigenvectors are one good heuristic choice of a family of admissible target vectors. We have found that this choice usually leads to robust, small gain control laws and provides a good nominal choice to initiate optimization. An exception to this rule of thumb is when we seek to design active joint control for multiple flexible bodies having inner-body joints with very low stiffness. See the case study in section 5.3. This exception occurs because the open-loop modal matrix is usually very poorly conditioned for such systems, and attempting to preserve the near-zero joint stiffness for the closed-loop control is obviously not an appropriate solution path when active stiffening of the joint is probably required. When the open-loop eigenvectors are considered as the desired closed-loop eigenvectors, Step 2 should be replaced by computation of the right eigenvectors of the open-loop system matrix.

Example 5.3

Consider the mechanical system

$$\mathrm{M}\ddot{\mathbf{x}}(t) + \mathrm{K}\dot{\mathbf{x}}(t) = \mathrm{D}\mathbf{u}(t)$$

where the mass, stiffness, and control-influence matrices are

$$\mathrm{M} = \begin{bmatrix} 1 & 0 & 0 \\ 0 & 1 & 0 \\ 0 & 0 & 1 \end{bmatrix}, \qquad \mathrm{K} = \begin{bmatrix} 20 & -10 & 0 \\ -10 & 30 & -20 \\ 0 & -20 & 20 \end{bmatrix}, \qquad \mathrm{D} = \begin{bmatrix} 1 & 0 & 0 \\ 0 & 1 & 0 \\ 0 & 0 & 1 \end{bmatrix}$$

For the first order-state space form, the coefficient matrices are

$$\mathrm{A} = \begin{bmatrix} 0 & \mathrm{I} \\ -\mathrm{M}^{-1}\mathrm{K} & 0 \end{bmatrix}, \qquad \mathrm{B} = \begin{bmatrix} 0 \\ \mathrm{M}^{-1}\mathrm{D} \end{bmatrix}$$

The open-loop eigenvalues of the above system are

$$\lambda_{1,2} = \pm 1.4447i, \qquad \lambda_{3,4} = \pm 4.4721i, \qquad \lambda_{5,6} = \pm 6.9219i$$

and the corresponding right eigenvectors are

$$\phi_{1,2} = \begin{Bmatrix} \mp 0.1986i \\ \mp 0.3558i \\ \mp 0.3973i \\ 0.2870 \\ 0.5141 \\ 0.5740 \end{Bmatrix}, \qquad \phi_{3,4} = \begin{Bmatrix} \pm 0.1952i \\ \pm 0.0000i \\ \mp 0.0976i \\ -0.8729 \\ -0.0000 \\ 0.4364 \end{Bmatrix}, \qquad \phi_{5,6} = \begin{Bmatrix} 0.0400 \\ -0.1116 \\ 0.0800 \\ \pm 0.2767i \\ \mp 0.7724i \\ \pm 0.5535i \end{Bmatrix}$$

where the eigenvectors have been normalized so that Eq. (5.96) is satisfied.

The desired locations of the closed-loop eigenvalues are specified here as

$$\lambda^c_{1,2} = -0.5 \pm 1.4447i, \qquad \lambda^c_{3,4} = -1.0 \pm 4.4721i, \qquad \lambda^c_{5,6} = -1.5 \pm 6.9219i$$

and we take the open-loop eigenvector matrix as the target eigenvector matrix T. To solve the Sylvester equation (5.99) for computing an approximate H matrix, we need

$$B^{\dagger} = \begin{bmatrix} 0 & 0 & 0 & 1 & 0 & 0 \\ 0 & 0 & 0 & 0 & 1 & 0 \\ 0 & 0 & 0 & 0 & 0 & 1 \end{bmatrix} = (B^T B)^{-1} B^T$$

and the H matrix that is in least square agreement with the target eigenvalues (Λ_c) and eigenvectors (T) is then

$$\begin{aligned} H =& B^{\dagger}(T\Lambda_c - AT) \\ =& \begin{bmatrix} -0.1435 & -0.1435 & 0.8729 & 0.8729 & -0.0000 & -0.0000 \\ -0.2570 & -0.2570 & 0.0000 & 0.0000 & 0.0000 & 0.0000 \\ -0.2870 & -0.2870 & -0.4364 & -0.4364 & -0.0000 & -0.0000 \end{bmatrix} \\ &+ \begin{bmatrix} -0.0000 & 0.0000 & -0.0000 & 0.0000 & -0.4151 & 0.4151 \\ 0.0000 & -0.0000 & 0.0000 & -0.0000 & 1.1587 & -1.1587 \\ -0.0000 & 0.0000 & -0.0000 & 0.0000 & -0.8302 & 0.8302 \end{bmatrix} i \end{aligned}$$

With this H matrix, we solve the Sylvester equation Eq. (5.99) to get the achieved eigenvector Φ. That is, we solve the following Sylvester equation for Φ:

$$A\Phi - \Phi\Lambda = BH$$

The solution for Φ gives the complex modal matrix

$$\begin{aligned} \Phi =& \begin{bmatrix} -0.0167 & -0.0167 & 0.0108 & 0.0108 & 0.0198 & 0.0198 \\ -0.0299 & -0.0299 & 0.0000 & 0.0000 & -0.0551 & -0.0551 \\ -0.0344 & -0.0344 & -0.0054 & -0.0054 & 0.0395 & 0.0395 \\ 0.1477 & 0.1477 & -0.4418 & -0.4418 & -0.0148 & -0.0148 \\ 0.2645 & 0.2645 & -0.0000 & -0.0000 & 0.0414 & 0.0414 \\ 0.2953 & 0.2953 & 0.2209 & 0.2209 & -0.0296 & -0.0296 \end{bmatrix} \\ &+ \begin{bmatrix} -0.0964 & 0.0964 & 0.0964 & -0.0964 & -0.0021 & 0.0021 \\ -0.1727 & 0.1727 & 0.0000 & -0.0000 & 0.0060 & -0.0060 \\ -0.1929 & 0.1929 & -0.0482 & 0.0482 & -0.0043 & 0.0043 \\ 0.0241 & -0.0241 & -0.0482 & 0.0482 & 0.1400 & -0.1400 \\ 0.0432 & -0.0432 & -0.0000 & 0.0000 & -0.3907 & 0.3907 \\ 0.0482 & -0.0482 & 0.0241 & -0.0241 & 0.2799 & -0.2799 \end{bmatrix} i \end{aligned}$$

The gain matrix can be obtained by solving Eq. (5.100):

$$\begin{aligned} G =& H\Phi^{-1} \\ =& \begin{bmatrix} 1.0064 & -0.4364 & 0.0127 & 1.9564 & -0.4364 & -0.0873 \\ -0.4364 & 1.4682 & -0.8729 & -0.4364 & 2.2182 & -0.8729 \\ 0.0127 & -0.8729 & 1.0254 & -0.0873 & -0.8729 & 1.8254 \end{bmatrix} \end{aligned}$$

Note that the Frobenius norm of the gain matrix G is 4.4861, the condition number of the open-loop eigenvector matrix (T) is 6.9219, and the condition number of the closed-loop eigenvector matrix (Φ) is only slightly increased to 7.2536.

■

5.2.3 Specialization for Mechanical Second-Order Systems

When we deal with the frequently encountered case of n second-order differential equations (for mechanical vibrating systems), it is not necessary to determine the $2n \times 2n$ eigenvector matrix for the corresponding $2n$ first-order state space equations. In fact, we can reduce dimensionality by half and consider the $n \times n$ modal matrix associated with the self-adjoint second order system. Suppose that a mechanical system is defined by the n second-order equations of motion

$$\mathrm{M}\ddot{\mathbf{x}}(t) + \mathrm{C}\dot{\mathbf{x}}(t) + \mathrm{K}\mathbf{x}(t) = \mathrm{D}\mathbf{u}(t) \tag{5.113}$$

where $\mathbf{x}(t)$ and $\mathbf{u}(t)$ are the configuration and control vectors, respectively. M is an $n \times n$ positive definite mass matrix, K and C are $n \times n$ positive semi definite stiffness and damping matrices, respectively, and D is an $n \times m$ control influence matrix (assumed to be of maximum rank m).

For this system, we introduce the full state feedback control in the form

$$\mathbf{u}(t) = -\left\{\mathrm{G}_1\mathbf{x}(t) + \mathrm{G}_2\dot{\mathbf{x}}(t)\right\} \tag{5.114}$$

Substituting Eq. (5.114) into Eq. (5.113), we obtain the closed-loop system

$$\mathrm{M}\ddot{\mathbf{x}}(t) + \mathrm{C}\dot{\mathbf{x}}(t) + \mathrm{K}\mathbf{x}(t) = -\mathrm{D}\left[\mathrm{G}_1 \quad \mathrm{G}_2\right]\left\{\begin{matrix}\mathbf{x}(t)\\ \dot{\mathbf{x}}(t)\end{matrix}\right\} \tag{5.115}$$

and the corresponding eigenvalue problem

$$\left(\lambda_i^2\mathrm{M} + \lambda_i\mathrm{C} + \mathrm{K}\right)\underline{\phi}_i = -\mathrm{D}\left[\mathrm{G}_1 \quad \mathrm{G}_2\right]\left\{\begin{matrix}\underline{\phi}_i\\ \lambda_i\underline{\phi}_i\end{matrix}\right\} \tag{5.116}$$

In a fashion similar to Eq. (5.97), we define the parameter vector $\mathbf{h}_i$ as

$$\mathbf{h}_i = -\left[\mathrm{G}_1 \quad \mathrm{G}_2\right]\left\{\begin{matrix}\underline{\phi}_i\\ \lambda_i\underline{\phi}_i\end{matrix}\right\} \tag{5.117}$$

Assuming that the inverse of the coefficient matrix of the left-hand side of Eq. (5.116) exists, we can write the modal subvector $\underline{\phi}_i$ with Eq. (5.117) as

$$\underline{\phi}_i = \left(\lambda_i^2\mathrm{M} + \lambda_i\mathrm{C} + \mathrm{K}\right)^{-1}\mathrm{D}\mathbf{h}_i \tag{5.118}$$

Obviously, as before, existence of the inverse in Eq. (5.118) depends on assigning the eigenvalues λ_i to positions other than the open-loop eigenvalue position. It should be noted, because the solution is of the form $\mathbf{x} = \sum_i \underline{\phi}_i e^{\lambda_i t}$, $\dot{\mathbf{x}} = \sum_i \lambda_i \underline{\phi} e^{\lambda_i t}$ that the modal vector for the velocity vector $\dot{\mathbf{x}}$ is simply $\lambda_i \underline{\phi}_i$. Also, note that if the subvectors $\underline{\phi}_i$ $(i = 1, 2, \ldots, n)$, are a unitary set, then the $2n \times 1$ eigenvectors for the corresponding closed-loop system in the state space have the structure

$$\left\{ \begin{matrix} \underline{\phi}_i \\ \lambda_i \underline{\phi}_i \end{matrix} \right\}, \quad (i = 1, 2, \ldots, n)$$

and are guaranteed to be linearly independent. Therefore, we conclude that the same procedure developed in the previous section can be used for the system of Eq. (5.113) with the formulation of Eq. (5.118). That is, after completing the calculation of admissible modal vectors $\underline{\phi}_i$ $(i = 1, 2, \ldots, n)$, we form the corresponding $2n \times 2n$ eigenvector matrix such that

$$\Phi_{2n} = \begin{bmatrix} \Phi & \overline{\Phi} \\ \Phi\Lambda & \overline{\Phi\Lambda} \end{bmatrix} \tag{5.119}$$

where Λ is the diagonal eigenvalue matrix, Φ is the $n \times n$ modal matrix for the displacement vector $\mathbf{x}$, and $(\bar{\ })$ denotes the complex conjugate.

With this, we compute the gain matrix $[\,G_1 \quad G_2\,]$ by solving the linear system

$$-[\,G_1 \quad G_2\,]\,\Phi_{2n} = H \tag{5.120}$$

where H is an $m \times 2n$ matrix defined by

$$H = [\,\mathbf{h}_1 \quad \mathbf{h}_2 \quad \ldots \quad \mathbf{h}_n \quad \overline{\mathbf{h}}_1 \quad \overline{\mathbf{h}}_2 \quad \ldots \quad \overline{\mathbf{h}}_n\,] \tag{5.121}$$

Equation (5.118) provides useful insight into the independent modal space control (IMSC) algorithm of Ref. 28 for the case in which $D \in R^n$ and $C=0$. For the special case in which the number of actuators are assumed equal to the number of controlled modes $(m = n)$, and for the special case in which the target eigenvectors are taken to be the open-loop undamped eigenvectors, the analytical and numerical equivalence of the above to IMSC has been demonstrated (Ref. 27). In the next section, we review a pole placement technique based on the IMSC method.

5.2.4 Eigenstructure Assignment by IMSC Method

One conceptually attractive choice for the target eigenvectors (for the case of a mechanical system) is the open-loop eigenvectors, and this leads directly to the

IMSC algorithm. A key feature of IMSC is the determination of a feedback control that preserves the open-loop modal matrix, so that the closed-loop system becomes totally decoupled. For the system of Eq. (5.113) with C=0 and square full-rank control influence matrix D, we introduce the modal coordinate vector $\underline{\eta}(t)$ defined by

$$\mathbf{x}(t) = \Phi\underline{\eta}(t) \tag{5.122}$$

where Φ is the modal matrix that satisfies

$$K\Phi = M\Phi\Omega \tag{5.123}$$

with the diagonal natural frequency matrix Ω given by

$$\Omega = \text{diag}\,[\omega_1^2, \omega_2^2, \ldots, \omega_n^2] \tag{5.124}$$

Note that the modal vectors $\underline{\phi}_i$ are normalized by

$$\underline{\phi}_i^T M \underline{\phi}_j = \delta_{ij}, \qquad \underline{\phi}_i^T K \underline{\phi}_i = \omega_i^2, \qquad \text{i,j} = 1, 2, \ldots, n \tag{5.125}$$

The transformation brings Eq. (5.113) to the modal space form as

$$\ddot{\underline{\eta}}(t) + \Omega\underline{\eta}(t) = \mathbf{w}(t) \tag{5.126}$$

where the modal force vector $\mathbf{w}$ is related to the actual force $\mathbf{u}$ by $\mathbf{w}(t) = \Phi^T D\mathbf{u}(t)$.

If a feedback control with diagonal gain matrices is introduced in Eq. (5.126), then the uncoupled structure of independent modal space equations is maintained. Therefore, with a proper choice of diagonal gain matrices, we can place the closed-loop eigenvalues arbitrarily. For this case, we can write the eigenvectors for open- and closed-loop systems in the state space form as

$$\underline{\phi}_i^o = \left\{ \begin{matrix} \underline{\phi}_i \\ j\omega\underline{\phi}_i \end{matrix} \right\}, \qquad \underline{\phi}_i^c = \left\{ \begin{matrix} \underline{\phi}_i \\ \lambda_i\underline{\phi}_i \end{matrix} \right\}, \qquad j^2 = -1 \tag{5.127}$$

where the superscripts o and c represent open-loop and closed-loop, respectively.

For the case with D=I, we rewrite Eq. (5.118) as

$$\underline{\phi}_i = \left(\lambda_i^2 M + \lambda_i C + K\right)^{-1} \mathbf{h}_i \tag{5.128}$$

Observing that the columns of the coefficient matrix in the right-hand side span a complete n-dimensional space, we conclude that the closed-loop eigenvectors for the displacement **x** can be arbitrarily assigned. In other words, Eq. (5.128) has an exact solution for any $\underline{\phi}_i$ vector. Therefore, the closed-loop modal vector obtained by projecting an open-loop one will be exactly the same vector as obtained by the IMSC method for each desired eigenvalue. If M is the identity matrix, from Eq. (5.125), we notice that the modal matrix whose columns contain $\underline{\phi}_i$ is a unitary matrix. For this special case, the projection methods based on unitary basis vectors and open-loop modal vectors and the IMSC method will generate equally well-conditioned closed-loop eigenvectors. However, the projection algorithm based on unitary basis may not necessarily produce the same gain matrix since the unitary basis vectors selected for this algorithm do not generally coincide with the orthogonal open-loop modal vectors. It should be noted that it is usually impossible to have perfectly conditioned closed-loop eigenvectors for pairs of complex conjugate eigenvalues. Note, that even if Φ of Eq. (5.119) is unitary, Φ_{2n} is not generally unitary.

5.2.5 Sequential Robust Eigenstructure Assignment

It is very important to select a good set of target eigenvectors, since this decision is directly related to the robustness of the closed-loop system. Based upon historical studies, and as discussed above, a family of good candidates has been established for target eigenvectors: (1) open-loop eigenvectors that are usually good for the case tin which the closed-loop damped frequencies are close to the their open-loop values, (2) unitary vectors nearest to the open-loop eigenvectors, and (3) unitary vectors chosen to maximize their projections on the good admissible eigenvectors (section 5.2.2). While these target eigenvector sets usually lead to good nominal designs, the resulting gains from either of these choices may require optimization to accommodate the requirements of a particular applicationion.

Continuing the discussion of eigenstructure methods, we now discuss a sequential EAR gain design algorithm (Ref. 29), which is an extension of the ideas (Refs. 26 and Refs. 27) discussed in sections 5.2.2 and 5.2.3. The key new ingredient is the focus on the intersection of the assignable (or achievable) closed-loop eigenvector space and the space of desired eigenvectors. This approach consists of two main steps. First, an assignable (achievable) eigenvector subspace is determined for each closed-loop eigenvector, using the singular value decomposition (or alternatively, the QR decomposition). Second, a sequential algorithm is used to choose the closed-loop eigenvectors using a subspace intersection concept.

Closed-Loop Eigenvalue Problem

We begin with the closed-loop eigenvalue problem, Eq. (5.116), for the second order mechanical system, Eq. (5.113), and with the full state feedback control law, Eq. (5.114). Equation (5.116) can be written in a compact, homogeneous form as follows:

$$\left[(\lambda_i^2 \, M + \lambda_i \, C + K) \;\; \vdots \;\; B \right] \left\{ \begin{array}{c} \underline{\phi}_i \\ (\lambda_i \, G_2 + G_1)\underline{\phi}_i \end{array} \right\} = 0 \tag{5.129}$$

Note that M, C, K are $n \times n$ matrices, B is an $n \times m$ matrix, and G_1, G_2 are $m \times n$ gain matrices. It is evident by inspection that the vector in braces lies in the null space of the matrix $\left[(\lambda_i^2 \, M + \lambda_i \, C + K) \vdots B\right]$. This observation leads to an attractive algorithm as is discussed below.

Let us define an $n \times (n+m)$ matrix Γ_i and an $(n+m) \times 1$ vector $\mathbf{h}_i$ for the ith eigenvalue as

$$\Gamma_i = \left[(\lambda_i^2 \, M + \lambda_i \, C + K) \;\; \vdots \;\; B \right] \tag{5.130}$$

and

$$\mathbf{h}_i = \left\{ \begin{array}{c} \underline{\phi}_i \\ (\lambda_i \, G_2 + G_1)\underline{\phi}_i \end{array} \right\} \tag{5.131}$$

Then, Eq. (5.129) becomes

$$\Gamma_i \mathbf{h}_i = 0 \tag{5.132}$$

It is easy to see from Eq. (5.132) that the vector $\mathbf{h}_i$ must lie in the null space of the matrix Γ_i and, therefore, we anticipate that gain matrices G_i can be obtained after computing the null space of the matrix Γ_i for each mode. The null space of the matrix Γ_i can be determined by use of the singular value decomposition (section 2.2.2):

$$\Gamma_i = U_i S_i V_i^H = U_i \left[\Sigma_i \quad 0 \right] \begin{bmatrix} V_{\Sigma_i}^H \\ V_{oi}^H \end{bmatrix} \tag{5.133}$$

Since the rank of Γ_i is n, Σ_i contains n nonzero singular values. From the unitary property of matrix V_i, Eq. (2.59), the following condition can be obtained

$$V_i^H V_i = \begin{bmatrix} V_{\Sigma_i}^H \\ V_{oi}^H \end{bmatrix} \left[V_{\Sigma_i} \quad V_{oi} \right] = I \qquad \Longrightarrow \qquad V_{\Sigma_i}^H V_{oi} = 0$$

That means that the $(n+m) \times m$ matrix V_{oi} contains a set of unitary vectors that span the null space of the matrix Γ_i, i.e.,

$$\Gamma_i V_{oi} \mathbf{c}_i = \left(U_i \Sigma_i V_{\Sigma_i}^H \right) V_{oi} \mathbf{c}_i = 0 \tag{5.134}$$

where $\mathbf{c}_i$ is an arbitrary vector with length m.

By comparing Eq. (5.132) and Eq. (5.134), the vector $\mathbf{h}_i$ can be partitioned as

$$\mathbf{h}_i = V_{oi}\mathbf{c}_i \equiv \begin{bmatrix} \overline{V}_{oi} \\ \hat{V}_{oi} \end{bmatrix} \mathbf{c}_i \equiv \begin{Bmatrix} \bar{\mathbf{h}}_i \\ \hat{\mathbf{h}}_i \end{Bmatrix} \tag{5.135}$$

where $\bar{\mathbf{h}}_i$ is an $n \times 1$ vector, $\hat{\mathbf{h}}_i$ is an $m \times 1$ vector and, similarly $\overline{V}_{oi}$ is an $n \times m$ matrix, and $\hat{V}_{oi}$ is an $m \times m$ matrix.

Let us assume that vectors $\mathbf{h}_i$, $i = 1, 2, \ldots, n$, in Eq. (5.135) are determined by selecting judicious $\mathbf{c}_i$ vectors for each mode (the process will be discussed in detail later). Substituting Eq. (5.135) into Eq. (5.131) gives

$$\left(\lambda_i G_2 + G_1\right)\bar{\mathbf{h}}_i = \hat{\mathbf{h}}_i, \qquad i = 1, \ldots, n \tag{5.136}$$

or, in matrix form,

$$G_2\bar{H}\Lambda + G_1\bar{H} = \hat{H} \tag{5.137}$$

with the $n \times n$ matrix $\bar{H} = [\bar{\mathbf{h}}_1 \;\; \bar{\mathbf{h}}_2 \;\; \ldots \;\; \bar{\mathbf{h}}_n]$, the $m \times n$ matrix $\hat{H} = [\hat{\mathbf{h}}_1 \;\; \hat{\mathbf{h}}_2 \;\; \ldots \;\; \hat{\mathbf{h}}_n]$, and the eigenvalue matrix $\Lambda = \text{diag}(\lambda_1, \lambda_2, \ldots, \lambda_n)$.

Since gain matrices G_i in Eq. (5.137) are real, Eq. (5.137) can be written by decomposing real and imaginary parts as follows:

$$G\mathcal{H} \equiv [G_1 \;\; G_2] \begin{bmatrix} \Re(\bar{H}) & \Im(\bar{H}) \\ \Re(\bar{H}\Lambda) & \Im(\bar{H}\Lambda) \end{bmatrix} = [\Re(\hat{H}) \;\; \Im(\hat{H})] \tag{5.138}$$

where $\Re(\;)$ and $\Im(\;)$ denote the real and imaginary parts of the matrix ( ), respectively.

Once we obtain matrices $\bar{H}$ and $\hat{H}$, then gain matrices G_1 and G_2 can be obtained by inverting Eq. (5.138). If the number of eigenvalues to be assigned is less than n, then Eq. (5.138) becomes an underdetermined problem, and we use the minimum norm solution for this case (section 2.1.1). Note from Eq. (5.135) that matrices $\bar{H}$ and $\hat{H}$ consist of matrices $\overline{V}_{oi}$, $\hat{V}_{oi}$ [which can be obtained from Eq. (5.133) by using the singular value decomposition of the matrix Γ_i] and the vectors $\mathbf{c}_i$ should be selected carefully. Note that the choice of the $\mathbf{c}_i$ in Eq. (5.135) directly affects the conditioning of $\mathcal{H}$, and also the closed-loop eigenvectors in Eq. (5.131). Thus we see that the orthogonality of the eigenvectors and the conditioning of $\mathcal{H}$ both depend on $\mathbf{c}_i$, and we expect that a judicious choice of $\mathbf{c}_i$ leads to both a robust algorithm to compute gain matrices and a robust eigenvalue placement. Since the matrix $\mathcal{H}$, defined in Eq. (5.131), consists of the desired eigenvalues and the closed-loop eigenvectors, we return to the important issue: how to select the target closed-loop eigenvector set to be used for eigenvector assignment; this is equivalent to selecting judicious vectors c_i [in Eq. (5.135)] for each mode. The following method is suggested, Ref. 29, for selecting vectors c_i considering the above discussion.

Sequential Eigenvector Assignment

In order to obtain a well-conditioned closed-loop matrix, we focus on the *intersection subspace between the space of assignable closed-loop eigenvectors and the space spanned by the target eigenvectors*, as introduced in Ref. 29. The algorithm consists of four steps, which we overview as follows. First, define a well-conditioned target eigenvector matrix. Second, obtain an orthogonal basis spanning the intersection space of the assignable closed-loop eigenvector space and the target eigenvectors; if the intersection space is not null, we can choose any vector from the intersection space as the first closed-loop eigenvector and, if the intersection space is null, we can choose the achievable eigenvector closest to the target eigenvector space as the first closed-loop eigenvector. Third, reduce the column space of the desired matrix by the first closed-loop eigenvector or its closest vector in the desired space. Finally, the second and third steps are repeated for the remaining eigenvectors. The above four steps are now discussed in greater detail.

Step 1

Form a target set of well-conditioned (desired) eigenvectors; as possibilities, the open-loop eigenvectors, any unitary matrix, or the closest unitary matrix to the open-loop eigenvectors may be used. It has been shown in Ref. 29 that the closest unitary matrix to the open-loop eigenvectors is often a very attractive choice. The open-loop eigenvector matrix Φ_o is computed from the following open-loop eigenvalue problem:

$$\left(\lambda_{oi}^2\, M + \lambda_{oi}\, C + K\right)\underline{\phi}_{oi} = 0 \tag{5.139}$$

or

$$\lambda_{oi}\begin{bmatrix} M & 0 \\ 0 & M \end{bmatrix}\begin{Bmatrix} \underline{\phi}_{oi} \\ \lambda_{oi}\underline{\phi}_{oi} \end{Bmatrix} = \begin{bmatrix} 0 & M \\ -K & -C \end{bmatrix}\begin{Bmatrix} \underline{\phi}_{oi} \\ \lambda_{oi}\underline{\phi}_{oi} \end{Bmatrix} \tag{5.140}$$

and the $n \times n$ open-loop eigenvector matrix Φ_o is defined as

$$\Phi_o = [\underline{\phi}_{o1} \quad \underline{\phi}_{o2} \quad \cdots \quad \underline{\phi}_{on}] \tag{5.141}$$

Φ_o is partitioned using singular value decomposition as

$$\Phi_o = U_o \Sigma_o V_o^H \tag{5.142}$$

The closest unitary matrix to the open-loop eigenvectors is obtained as follows:

$$\Psi_o = U_o V_o^H \tag{5.143}$$

Step 2

Observing Eq. (5.135), we find that the assignable closed-loop eigenvector space for the first closed-loop eigenvalue may be spanned by the matrix $\overline{V}_{o1}$. What we want is a vector that lies in the assignable closed-loop eigenvector space (which can be represented by $\overline{V}_{o1}$), and is as close as possible to the range space of the columns of the target space (Ψ_o in the present discussion). That vector can be obtained by computing *principal angles* and *principal vectors* of a subspace pair $\overline{V}_{o1}$ and Ψ_o (Ref. 30). QR decomposition (Theorem 2.1) of matrices $\overline{V}_{o1}$ and Ψ_o gives

$$\overline{V}_{o1} = Q_{\overline{V}_{o1}} R_{\overline{V}_{o1}} \tag{5.144}$$

$$\Psi_o = Q_{\Psi_o} R_{\Psi_o} \tag{5.145}$$

where $Q_{\overline{V}_{o1}}$, Q_{Ψ_o} are $n \times m$ and $n \times n$ real orthonormal (or complex unitary) matrices, and R_{Ψ_o}, $R_{\overline{V}_{o1}}$ are $m \times m$ and $n \times n$ upper triangular matrices, respectively.

The principal angles of the subspace pair are obtained (Refs. 30, 31) by singular value decomposition of the matrix

$$Z = \left(Q_{\overline{V}_{o1}}\right)^T Q_{\Psi_o} \tag{5.146}$$

Singular value decomposition of the Z matrix yields

$$Z = U_Z \Sigma_Z V_Z^H \tag{5.147}$$

Note that the singular values σ_i are all positive and less than or equal to 1, since matrix Z is formed by orthonormal (or unitary) matrices as shown in Eq. (5.146).

$$1 \geq \sigma_1 \geq \sigma_2 \geq \ldots \geq \sigma_m \tag{5.148}$$

The principal angles θ_i, $i = 1, \ldots, m$, are defined to parameterize the singular values of the matrix Z as (Refs. 29-31)

$$\sigma_i = \cos\theta_i \tag{5.149}$$

The ith column of $Q_{\overline{V}_{o1}} U_Z$ and the ith column of $Q_{\Psi_o} V_Z$ are a pair of vectors in $\overline{V}_{o1}$ and Ψ_o, respectively, with abstract angle θ_i between them.

Therefore, the closest vector to the column space of $\overline{V}_{o1}$ is the first column of $Q_{\overline{V}_{o1}} U_Z$, i.e.,

$$\bar{\mathbf{h}}_1 = Q_{\overline{V}_{o1}} \mathbf{u}_1 \tag{5.150}$$

where $\mathbf{u}_1$ is the first column vector of the matrix U_Z.

Since the $m \times m$ matrix $R_{\overline{V}_{o1}}$ is nonsingular, Eq. (5.144) can be rewritten as follows:

$$Q_{\overline{V}_{o1}} = \overline{V}_{o1} \left(R_{\overline{V}_{o1}}\right)^{-1} \tag{5.151}$$

Substituting Eq. (5.151) into Eq. (5.150) gives

$$\bar{\mathbf{h}}_1 = \overline{V}_{o1} \left(R_{\overline{V}_{o1}}\right)^{-1} \mathbf{u}_1 \tag{5.152}$$

Comparison of Eqs. (5.135) and (5.152) yields the coefficient vector $\mathbf{c}_1$ as follows:

$$\mathbf{c}_1 = \left(R_{\overline{V}_{o1}}\right)^{-1} \mathbf{u}_1 \tag{5.153}$$

Step 3

To make the closed-loop eigenvectors linearly independent, the column space of matrix Ψ_o should be reduced by the computed vector $\mathbf{h}_1$. If $\mathbf{h}_1$ is not in the column space of Ψ_o, then reduce the column space by the closest vector in the space to the vector $\mathbf{h}_1$. As mentioned in Step 2, the closest vector to the column space of $\overline{V}_{o1}$ is $\bar{\mathbf{h}}_1$ $(= Q_{\overline{V}_{o1}} \mathbf{u}_1)$, and the pair vector in the space Ψ_o is the first column vector of $Q_{\Psi_o} V_Z$. Consider the reduced space that is spanned by the columns of matrix Ψ_1, which is defined as

$$\Psi_1 = \Psi_o - \Psi_o^{\perp} \tag{5.154}$$

The columns of matrix Ψ_1 should be orthogonal to the first column vector of $Q_{\Psi_o} V_Z$, i.e.,

$$\left(Q_{\Psi_o} \mathbf{v}_1\right)^{*} \Psi_1 = 0 \tag{5.155}$$

Using Eq. (5.154) in Eq. (5.155), we have

$$\left(\mathbf{v}_1{}^{*} Q_{\Psi_o}^{*}\right) \Psi_o^{\perp} = \left(\mathbf{v}_1{}^{*} Q_{\Psi_o}^{*}\right) \Psi_o \tag{5.156}$$

where $\mathbf{v}_1$ is the first column vector of the matrix V_Z.

By observing Eq. (5.156) and noting the fact that matrix Q_{Ψ_o} and the vector $\mathbf{v}_1$ are unitary, we obtain the matrix $\Psi_o^{\perp}$ as

$$\Psi_o^{\perp} = Q_{\Psi_o} \mathbf{v}_1 \mathbf{v}_1{}^{*} Q_{\Psi_o}^{*} \Psi_0 \tag{5.157}$$

which satisfies Eq. (5.156) and therefore Eq. (5.155).

Substitution of Eq. (5.157) into Eq. (5.154) gives

$$\Psi_1 = \left[I - \left(Q_{\Psi_o} v_1 \right) \left(Q_{\Psi_o} v_1 \right)^* \right] \Psi_0 \tag{5.158}$$

and Ψ_1 is the matrix which spans the reduced space.

Step 4

Repeat Step 2 and Step 3 for the remaining $n - 1$ eigenvalues. Note that the recursion that replaces Eq. (5.158) for the ith eigenvalue (which reduces the column space of matrix Ψ_{i-1} to ensure the linear independence of the closed-loop eigenvectors) is as follows:

$$\Psi_i = \left[I - \left(Q_{\Psi_{i-1}} v_1 \right) \left(Q_{\Psi_{i-1}} v_1 \right)^* \right] \Psi_{i-1} \tag{5.159}$$

From Steps 2 through 4, the closed-loop eigenvectors are chosen as close as possible to the column space of the well-conditioned target matrix Ψ_o. The gain matrix for a robust closed-loop system can now be obtained using Eq. (5.138). Additional discussion of these ideas and numerical demonstrations can be found in Ref. 29.

5.2.6 Sequential Eigenvalue Assignment

This section introduces an efficient eigenvalue assignment algorithm (Ref. 32) that assigns specified eigenvalue(s) at each step. The method consists of three steps: (1) a Schur decomposition is used to triangularize the state matrix; (2) a series of coordinate rotations (Givens rotations) is used to place the assignable eigenvalues at the lower right corner of the diagonal of its Schur form; and (3) a feedback gain matrix (which operates on the transformed coordinates) is computed for exactly assigning the specific eigenvalue(s) to the desired places *without affecting the remaining eigenvalues.* By repeating the second and third steps, all eigenvalues to be assigned can be moved to the desired place. Moreover, the extra freedom can be used on each iteration to, as examples, minimize the norm of the gain matrix and/or optimize a robustness index of the closed-loop system. We now consider the details of this approach.

Schur Transformation

Consider the fully controllable linear dynamical system in the state space form

$$\dot{x}(t) = Ax(t) + Bu(t) \tag{5.160}$$

with the linear feedback control

$$\mathbf{u}(t) = G\mathbf{x}(t) \tag{5.161}$$

The following coordinate transformation is introduced:

$$\mathbf{x}(t) = U\underline{\eta}(t) \tag{5.162}$$

where U is an $n \times n$ unitary Schur factor of the state matrix A, and $\underline{\eta}(t)$ is the vector of transformed coordinates.

The transformed equation of motion becomes

$$\dot{\underline{\eta}}(t) = U^H A U \underline{\eta}(t) + U^H B \mathbf{u}(t) \tag{5.163}$$

Since U is a Schur factor of A, $U^H AU$ becomes an upper triangular matrix with the eigenvalues of the open-loop system A on the diagonal (see section 2.2.4). Using the full state feedback control law, Eq. (5.161), in Eq. (5.163), the closed-loop system is obtained in the transformed coordinates as follows:

$$\begin{aligned} \dot{\underline{\eta}}(t) &= U^H A U \underline{\eta}(t) + U^H B G U \underline{\eta}(t) \\ &\equiv \bar{A}\underline{\eta}(t) + \bar{B}\bar{G}\underline{\eta}(t) \end{aligned} \tag{5.164}$$

where $\bar{A} = U^H AU$, $\bar{B} = U^H B$, and $\bar{G} = GU$.

Assume that the gain matrix is constructed such that $U^H BGU$ is an upper triangular matrix; then the diagonal elements in the matrix $U^H BGU$ changes the eigenvalues of the system matrix. We can anticipate that, by choosing the gain matrix such that the differences between the open-loop eigenvalues and the desired eigenvalues are placed in the $U^H BGU$ matrix, the eigenvalue assignment can be successfully completed. However, it may not be easy, or in some cases it may be impossible, to find a G matrix that simultaneously assigns many eigenvalues in the desired positions utilizing the above idea. A recursive method has been formulated in Ref. 32, that places only one eigenvalue in the desired region on each step without affecting the remaining eigenvalues. To make the process simple and numerically robust, a sequence of unitary coordinate transformations (Givens rotations) is used; *each unitary transformation moves a particular eigenvalue (to be reassigned on this step) to the last position on the diagonal of the Schur form.*

Givens Rotation (Coordinate Transformation)

Consider the appropriate unitary coordinate transformation $G_{i,i+1}$ such that

$$\underline{\eta}(t) = G_{i,i+1}\, \underline{\eta}'(t) \tag{5.165}$$

which switches the position of λ_i and λ_{i+1} on the diagonal of the $U^H AU$ matrix in Eq. (5.164). In other words, the $G_{i,i+1}$ matrix will be determined to satisfy the unitary property

$$G_{i,i+1}^H \, G_{i,i+1} = I \tag{5.166}$$

and the switching operation

$$G_{i,i+1}^H \, \bar{A} \, G_{i,i+1} = \bar{A}' \tag{5.167}$$

where $\bar{A}$ is the original matrix in the Schur form, and $\bar{A}'$ is the transformed matrix; these matrices have the structure:

$$\bar{A} = \begin{bmatrix} \lambda_1 & X & \dots & & & \dots & X \\ 0 & \ddots & & & & & X \\ \vdots & & & \lambda_i & b & & \vdots \\ & & & & \lambda_{i+1} & & \\ & & & & & \ddots & \\ 0 & \dots & & & & & \lambda_n \end{bmatrix} \tag{5.168}$$

and

$$\bar{A}' = \begin{bmatrix} \lambda_1 & \dots & & & & \dots & X \\ 0 & \ddots & & & & & \vdots \\ \vdots & & & \lambda_{i+1} & & & \\ & & & & \lambda_i & & \\ & & & & & \ddots & \\ 0 & \dots & & & & & \lambda_n \end{bmatrix} \tag{5.169}$$

in which b is the (i, i+1) element of the matrix $\bar{A}$. Note that we do not concern ourselves with any of the other elements in the matrix $\bar{A}$.

A unitary transformation matrix that satisfies the conditions of Eqs. (5.166) and (5.167) is

$$G_{i,i+1} = \begin{bmatrix} 1 & 0 & 0 & \dots & & & \dots & 0 \\ 0 & 1 & 0 & \dots & & & & \vdots \\ \vdots & & \ddots & & & & & \\ & & & & c_i & s_i^* & & \\ & & & & -s_i & c_i & & \\ & & & & & & \ddots & \\ 0 & \dots & & & & & & 1 \end{bmatrix} \tag{5.170}$$

which is a unitary Givens rotation (Ref. 30). The parameter c_i is a real scalar and the parameter s_i is a complex scalar. From conditions (5.166) and (5.167) with matrices (5.168)-(5.170), c_i and s_i are obtained as follows:

$$c_i{}^2 = \frac{1}{1 + \left\{(\lambda_i - \lambda_{i+1})(\lambda_i - \lambda_{i+1})^* / (bb^*)\right\}} \tag{5.171}$$

$$s_i = \frac{(\lambda_i - \lambda_{i+1})c_i}{b} \tag{5.172}$$

Note that if b=0 (or, as a practical matter, is smaller in absolute value than a small tolerance), then c_i and s_i are set to 0 and 1, respectively. It can be readily verified by direct expansion that this simple unitary transformation accomplishes the desired diagonal element interchange operation.

We can shift the eigenvalue λ_i to the last position on the diagonal of its Schur form by simply repeating the above transformation $n - \mathrm{i}$ times, and the corresponding composite coordinate transformation can be written as

$$\underline{\eta}(t) = T\, \underline{\eta}_1(t) \tag{5.173}$$

where $\underline{\eta}_1(t)$ is the vector of new coordinates, and the unitary transformation matrix T can be obtained by the continued product of the elementary Givens rotations:

$$T = G_{i,i+1}\, G_{i+1,i+2}\, G_{i+2,i+3}\, \ldots\, G_{n-1,n} \tag{5.174}$$

Note that since T is unitary, the numerical computation is stable even for large-order systems. The dynamics of the system in the final transformed coordinates are written as follows:

$$\begin{aligned} \dot{\underline{\eta}}_1(t) &= T^H \bar{A} T\, \underline{\eta}_1(t) + T^H U^H B G U T\, \underline{\eta}_1(t) \\ &\equiv \bar{A}_1\, \underline{\eta}_1(t) + \bar{B}_1 \bar{G}_1\, \underline{\eta}_1(t) \end{aligned} \tag{5.175}$$

where $\bar{A}_1 = T^H \bar{A} T = (UT)^H A(UT)$, $\bar{B}_1 = T^H U^H B = (UT)^H B$, and $\bar{G}_1 = G(UT)$.

Eigenvalue Assignment Algorithm

As mentioned earlier, to assign the eigenvalue λ_i to the desired location without affecting the remaining eigenvalues, the gain matrix $\bar{G}_1$ in Eq. (5.175) should be selected such that $\bar{B}_1 \bar{G}_1$ is an upper triangular matrix with its diagonal terms zero except for its lower right element. The non zero lower right element should be precisely the desired perturbation of the eigenvalue. One approach is to seek a gain matrix in the form

$$\bar{G}_1 = \begin{bmatrix} O_{m\times(n-1)} & \mathbf{g_1} \end{bmatrix} \tag{5.176}$$

where $O_{m\times(n-1)}$ denotes the zero matrix, and $\mathbf{g_1}$ denotes an $m \times 1$ vector.

Because of the unitary properties of the transformation matrices T and U, the gain matrix in the physical coordinates is related to the $\bar{G}_1$ matrix as

$$G = \bar{G}_1(UT_1)^H \tag{5.177}$$

Note that since $\bar{G}_1$ has the special form of Eq. (5.176), only the last column of matrix UT_1 is needed for back-computing the gain matrix in Eq. (5.177). Assume that λ_{id} is the desired closed-loop eigenvalue for replacing λ_i; then $\mathbf{g_1}$ should be determined from Eqs. (5.175) and (5.176) such that

$$\mathbf{b}_1^n \mathbf{g_1} = \lambda_{id} - \lambda_i \tag{5.178}$$

where $\mathbf{b}_1^n$ denotes the last row of the matrix $\bar{B}_1$ in Eq. (5.175).

Any vector satisfying Eq. (5.178) will produce the gain matrix G from Eq. (5.177), which moves λ_i to the desired location λ_{id}. Note that, for the multi-input case, the problem becomes underdetermined, and we are free to impose some criterion (e.g., a minimum gain solution, or use some robustness enhancement measure) to select from the infinity of possibilities.

To make the iterative process complete, let us write the gain matrix G in Eq. (5.161) as the summation of the local gain matrices G_i

$$G = \sum_{i=1}^{r} G_i : \tag{5.179}$$

where r is the number of eigenvalues to be assigned, and G_i are the local gain matrices, which are obtained separately for assigning different eigenvalues at each step.

Now, Eq. (5.175) can be rewritten as

$$\begin{aligned} \dot{\underline{\eta}}_1(t) &= T_1^H \bar{A} T_1\ \underline{\eta}_1(t) + T_1^H U^H B\Big(\sum_{i=1}^{r} G_i\Big) UT_1\ \underline{\eta}_1(t) \\ &= \bar{A}_1\ \underline{\eta}_1(t) + \bar{B}_1 \bar{G}_1\ \underline{\eta}_1(t) + \bar{B}_1\Big(\sum_{i=2}^{r} G_i\Big) UT_1\ \underline{\eta}_1(t) \end{aligned} \tag{5.180}$$

The gain matrix $\bar{G}_1$ can be obtained for assigning eigenvalue λ_i to the desired place, and the local gain matrix G_1 is obtained from Eq. (5.177). The current global gain matrix is introduced as

$$G_g = G_1 \tag{5.181}$$

After the gain matrix G_1 is computed to assign the eigenvalue λ_i, Eq. (5.180) can be written as

$$\dot{\underline{\eta}}_1(t) = \bar{A}_1^c \, \underline{\eta}_1(t) + \bar{B}_1\Big(\sum_{i=2}^{r} G_i\Big)UT_1 \, \underline{\eta}_1(t) \tag{5.182}$$

where $\bar{A}_1^c = \bar{A}_1 + \bar{B}_1\bar{G}_1$.

To move the next assignable eigenvalue λ_j to the end of the diagonal, a sequence of Givens rotations is again used, and thus we use the following coordinate transformation:

$$\underline{\eta}_1(t) = T_2 \, \underline{\eta}_2(t) \tag{5.183}$$

Equation (5.182) may be written as

$$\begin{aligned}
\dot{\underline{\eta}}_2(t) &= T_2^H\bar{A}_1^cT_2 \, \underline{\eta}_2(t) + T_2^H\bar{B}_1\Big(\sum_{i=2}^{r} G_i\Big)UT_1T_2 \, \underline{\eta}_2(t) \\
&= T_2^H\bar{A}_1^cT_2 \, \underline{\eta}_2(t) + T_2^H\bar{B}_1G_2UT_1T_2 \, \underline{\eta}_2(t) + T_2^H\bar{B}_1\Big(\sum_{i=3}^{r} G_i\Big)UT_1T_2 \, \underline{\eta}_2(t) \\
&\equiv \bar{A}_2 \, \underline{\eta}_2(t) + \bar{B}_2\bar{G}_2 \, \underline{\eta}_2(t) + \bar{B}_2\Big(\sum_{i=3}^{r} G_i\Big)UT_1T_2 \, \underline{\eta}_2(t)
\end{aligned} \tag{5.184}$$

where $\bar{A}_2 = T_2^H\bar{A}_1^cT_2$, $\bar{B}_2 = T_2^H\bar{B}_1 = (UT_1T_2)^HB$, and $\bar{G}_2 = G_2UT_1T_2$.

The gain matrix $\bar{G}_2$ can be obtained for λ_j in a similar fashion as the process to determine $\bar{G}_1$ for placing λ_i, and the local gain matrix G_2 can be obtained as [see Eq. (5.177)]

$$G_2 = \bar{G}_2\Big(UT_1T_2\Big)^H \tag{5.185}$$

Note that, for computing the minimum gain $\bar{G}_2$ in this stage [Eq. (5.178)], we should seek to minimize the current *global* gain matrix, not merely the local perturbation introduced at this step of the recursive process. The current global gain matrix is written as

$$G_g = \sum_{i=1}^{2} G_i = G_1 + \bar{G}_2\Big(UT_1T_2\Big)^H \tag{5.186}$$

The above steps are repeated r times to move all the assignable eigenvalues to the desired locations.

During each local iteration, the solution for gain matrix G_i must be real. However, because of the optimization process with complex value constraints on the eigenvalues, the resulting gain matrix may generally be complex. This minor problem can be overcome by adding additional constraints to the above discussion (which insures that we assign conjugate values to the corresponding pairs of conjugate eigenvalues), or an alternative procedure may be considered such that a pair of complex conjugate eigenvalues (i.e., the conjugate eigenvalues corresponding to one mode) are assigned at each step. This latter device is recommended because of its simplicity. Also, robustness measures can be used as constraints (or as the performance measure), in order that the gains be designed to ensure a robust closed-loop system. Readers may refer to Ref. 32 for further discussions, derivation details, and implementations based on the above discussions.

5.3 Case Study: Robust Eigenstructure Assignment

The projection algorithms developed in section 5.2 have been tested for several models in two classes: (1) a sixth-order mass-spring system with three actuators, and (2) a twenty-fourth-order model of a flexible space structure with six actuators. Model variations for each class are introduced through various choices for the system mass and stiffness properties and through the use of order reduction.

Considering the second class of applicationions, six actuators on the flexible structure are assumed to impart three orthogonal forces and three orthogonal moments to the structure at a single point. Conceptually, this is an idealized model of an active joint between a large rigid body and a flexible substructure. The active joint has open-loop compliance modeled using small passive spring and damping constants in the absence of an active control command; thus the associated open-loop "rigid body" frequencies are very small, but nonzero because of the small stiffness and damping (Table 5.1). The travel limits of the active joint must also be considered in practical applicationions, but we do not address these in the present discussion. In all of the examples below, we consider the full state feedback case; we do not address sensing and estimation, but we have implicitly assumed that the six across-the-joint displacements and the first six modal amplitudes and their time derivatives were available from measurements or an estimation process.

The specification of models 1-3 with various mass matrices (the first class of models) is given in Table 5.2. For the second class, we generated four models (models 4-7) by varying the order of the reduced-order system (the order of the reduced-order system is twice the number of controlled modes) from 12 to 24. The open-loop and desired closed-loop frequencies and damping factors for each model in this class are given in Table 5.1.

Table 5.1 Test Examples (Flexible Space Structure)

	Open-Loop Eigenvalues		Desired Closed-Loop Eigenvalues	
Mode No.	Freq. (rad/s)	Damp. Factor	Freq. (rad/s)	Damp. Factor
1	4.90×10^{-4}	0.001	6.78	0.70
2	4.46×10^{-4}	0.001	8.17	0.70
3	4.27×10^{-4}	0.001	8.98	0.70
4	2.57×10^{-4}	0.001	8.79	0.70
5	1.95×10^{-4}	0.001	11.97	0.70
6	1.07×10^{-4}	0.001	15.26	0.70
7	22.36	0.001	16.46	0.70
8	20.49	0.001	18.40	0.70
9	20.86	0.001	20.10	0.70
10	42.28	0.001	21.86	0.70
11	42.28	0.001	23.87	0.70
12	61.95	0.001	67.23	0.70

For the seven structural systems, we tested the projection method based on unitary target basis vectors (algorithm I) and the projection method based on open-loop target eigenvectors (algorithm II). For the three models in the first class, we tested also an eigenvalue placement algorithm using the IMSC method (algorithm III). The classical IMSC method could not be tested for the higher-order models of the flexible structure without introducing additional actuators above the adopted number (m=6) that was held fixed in this study. In Tables 5.3 and 5.4, we report condition numbers of the closed-loop eigenvectors (generally, the closed-loop eigenvectors were not the open-loop modal matrix) and Frobenius norms of the gain matrices for each model.

As shown in Table 5.3, the same results were obtained by algorithms II and III for all models. For models with the same number of controllers as the number of controlled modes, algorithm II is obviously equivalent to the IMSC method, since the open-loop modal matrix is preserved. For the case with the unitary open-loop

Table 5.2 Test Examples (Mass-Spring Systems)

		Desired Eigenvalues	
Model[a] No.	Mass Matrix	Freq. (rad/s)	Damp. factor
1	M=diag[1, 1, 1]	$\omega_1 = 1.44$ $\omega_2 = 4.47$ $\omega_3 = 6.92$	$\zeta_1 = \zeta_2 = \zeta_3 = 0.5$
2	M=diag[10^2, 1, 1]	$\omega_1 = 0.89$ $\omega_2 = 2.33$ $\omega_3 = 6.76$	$\zeta_1 = \zeta_2 = \zeta_3 = 0.5$
3	M=diag[10^3, 10^2, 1]	$\omega_1 = 0.09$ $\omega_2 = 0.95$ $\omega_3 = 4.70$	$\zeta_1 = \zeta_2 = \zeta_3 = 0.5$

[a] All of the above models have the same stiffness matrix:

$$K = \begin{bmatrix} 20 & -10 & 0 \\ -10 & 30 & -20 \\ 0 & -20 & 20 \end{bmatrix}$$

modal matrix (model 1), the condition number obtained by all three algorithms is exactly the same. For this case, algorithm I, however, generates a different gain matrix, because the unitary basis employed in the projection step did not coincide with the open-loop modal vectors. For model 3, algorithm I performs better than algorithms II and III because the conditioning of open-loop modal matrices obviously depends on the mass and stiffness matrices. In this simple problem, it is obvious that large variations in either M or K serve to nearly pin or release one or two of the three masses and, of course, the physical significance of this simple example is of dubious value except for illustration purposes. However, it is not uncommon for the effective "modal mass" or "modal stiffness" associated with various degrees of freedom (in more complicated high-dimensioned examples) to vary by orders of magnitude. In fact, the eigenvalue spectrum (of the free vibration eigenvalue problem) often displays several orders of magnitude variation and therefore presents analogous conditioning issues in practical applicationions. It is important to note, for multibody flexible structures with one or more very compliant joints, that the

Table 5.3 Performance of Eigenstructure Assignment Algorithm

		Algorithm I[a]		Algorithm II[b]		Algorithm III[c]	
Model No.	$\mathcal{K}(\Phi^o)$[d]	$\mathcal{K}(\Phi^c)$[d]	$\|G\|_f$	$\mathcal{K}(\Phi^c)$[d]	$\|G\|_f$	$\mathcal{K}(\Phi^c)$[d]	$\|G\|_f$
1	6.92	9.26	50.91	9.26	19.83	9.26	19.83
2	7.10	9.04	68.39	10.39	45.08	10.39	45.08
3	20.79	10.08	121.79	22.31	128.96	22.31	128.96

[a] Projection method using unitary vectors as targets.
[b] Projection method using open-loop eigenvectors as targets.
[c] IMSC method.
[d] The superscript o and c denote open-loop and closed-loop, respectively.

Table 5.4 Performance of Eigenstructure Assignment Algorithm

		Algorithm I[a]		Algorithm II[b]	
Model No.	No. of States[c]	$\mathcal{K}(\Phi)$	$\|G\|_f$	$\mathcal{K}(\Phi)$	$\|G\|_f$
4	12	31	234,000	32	208,000
5	16	347	786,749	461	906,380
6	20	8,342	11,904,002	14,753	7,913,951
7	24	11,406	5,920,422	25,495	10,730,075

[a] Projection method using unitary vectors as targets.
[b] Projection method using open-loop eigenvectors as targets.
[c] All the models have the same number of controls ($m = 6$).

conditioning of the open-loop eigenvalue problem will always be poor, and that the lower frequency closed-loop eigenvectors will depart dramatically from the open-loop eigevectors, for any control law which has the effect of stiffening the compliant joint. From these observations and other numerical studies, we conclude that *the open-loop modal matrix is not generally the optimal choice of the closed-loop modal matrix, vis-à-vis conditioning of the closed-loop eigenvalue problem, especially for multibody flexible structural systems with compliant joints.* However, for active control of structures having a well-conditioned open-loop modal matrix, we have found that the open-loop eigenvectors (or the columns of the unitary matrix nearest the open-loop modal matrix) are an excellent nominal choice for the target closed-loop eigenvectors, usually leading to both small gains and good closed-loop eigenstructure conditioning.

Test results for models 4, 5, 6, and 7 are interesting since these are more consistent with the models encountered in practical applicationions and also reveal the usefulness of these results for the case ($m < n$) in which there are few actuators to control a larger number of modes. As an illustration, we seek to impose substantial damping ($\zeta = 0.7$) in all modes. In practice, for flexible structure control, we would almost certainly settle for much smaller stability margins on the seventh and higher modes (i.e., the controlled flexible modes). As shown in Table 5.4, algorithm I performs better than algorithm II, in these examples, with regard to conditioning (robustness) of the closed-loop eigenvectors.

Some Final Observations

All of the above developments on full state feedback gain design make the tacit assumption that the full state is available in order to implement the control law. Of course, the state vector must, in practice, be estimated, and these developments must be augmented by design of a state estimator that operates on available sensor output measurements to obtain the needed state estimates. Generally speaking, we can expect the full state feedback results to provide an optimistic assessment of the closed-loop dynamics, and we may expect the tuning of the feedback gains to be suboptimal, when we take into account the estimation issues.

It is possible to construct the estimator design optimization problem as the *dual* of the corresponding full state feedback gain design problems for both the LQG (the usual state estimator is the Kalman filter, Refs. 1–5) and the eigenstructure assignment algorithms. However, the dual eigenstructure assignment estimation algorithms are in a less mature state of development. In Refs. 3 and 35, eigenstructure assignment algorithms are presented for design of a general class of dynamic compensators and estimators that lead to overall parameter-insensitive feedback controllers. The *separation principle* (Refs. 1, 2, 4) indicates that the estimation and control gain optimization problems can be solved independently, however this principle's validity does not hold in practice. The separation principle is found to have limited utility, because this now-classical result does not take into account robustness in the presence of model errors, and a host of other issues which are not conveniently captured in the quadratic regulator and Kalman filter optimization formulations.

Achieving robust designs for estimator gains and control gains has been found to generally be a coupled nonlinear parameter optimization problem, although good nominal designs may often be determined using the separation principle. The basic idea underlying the *loop transfer recovery* approach (Refs. 34, 35) is to couple the compensator gain design to the feedback control gain design so that the estimator zeros are placed at the *invariant zeros* of the full state feedback transfer function. Exercising the additional design freedom to judiciously select the compensator eigenvectors can lead to very robust designs as is evident in Ref. 34.

Alternatively, we may choose to abandon the full state feedback approach altogether, and consider instead the class of less complicated controllers known as *direct output feedback controllers* (discussed in Chapter 7), these approaches are found to be very attractive, especially vis-à-vis robustness, for applicationions to high-dimensioned dynamical systems wherein establising state estimators compatible with on-line real time operation is a well-known difficulty.

Problems

Problem 5-1

Suppose we wish to minimize

$$J = \frac{1}{2}\int_0^\infty \Big\{\mathbf{x}^T Q\mathbf{x} + \mathbf{u}^T R\mathbf{u}\Big\} e^{2\alpha t} dt, \qquad \alpha > 0$$

subject to satisfying $\dot{\mathbf{x}} = A\mathbf{x} + B\mathbf{u}$.

Your tasks are as follows:

(a) Show that, by the change of variables: $\mathbf{x}_\alpha = \mathbf{x}e^{\alpha t}$, $\mathbf{u}_\alpha = \mathbf{u}e^{\alpha t}$, that the above problem can be converted to a standard LQ problem.

(b) For A=Q=R=B=1 (scalar system), determine closed-loop poles in terms of α, and show that the closed-loop poles have real parts less than $-\alpha$.

Problem 5-2

Consider the system

$$\dot{\mathbf{x}} = A\mathbf{x} + B\mathbf{u} + \mathbf{f}(t), \qquad \mathbf{x}(0) = \mathbf{x_0}$$

where $\mathbf{f}(t)$ is a known forcing function.

The performance measure to be minimized is

$$J = \frac{1}{2}\Big(\mathbf{x}^T Q_f \mathbf{x}\Big)_{t=t_f} + \frac{1}{2}\int_0^{t_f} \Big\{\mathbf{x}^T Q\mathbf{x} + \mathbf{u}^T R\mathbf{u}\Big\} dt$$

with the final time t_f fixed.

Show that the optimal control law is

$$\mathbf{u}(t) = -R^{-1}B^T\{P\mathbf{x}(t) + \mathbf{r}(t)\}$$

where P satisfies the Riccati equation with $P(t_f) = Q_f$, and $\mathbf{r}(t)$ satisfies the following equation:

$$\dot{\mathbf{r}}(t) = -(A^T - PBR^{-1}B^T)\mathbf{r}(t) - P\mathbf{f}(t), \qquad \mathbf{r}(t_f) = 0$$

Note that this is similar to the linear tracking problem (Example 5.2).

Problem 5-3

Consider a second-order system without damping

$$\mathbf{M\ddot{x} + Kx = u}$$

where $\mathbf{x}(t)$ is an $n \times 1$ state vector, and $\mathbf{u}$ is an $n \times 1$ control vector. Note that this system is the case in which the number of actuators is equal to the number of model degrees of freedom.

Your task is as follows:

Show that the following full state feedback control law:

$$\mathbf{u} = -\mathrm{G}(\mathbf{x} + \dot{\mathbf{x}})$$

with positive definite gain symmetric matrix G, minimizes the following performance index:

$$\mathrm{J} = \frac{1}{2}\int_0^\infty \left\{\mathbf{x}^\mathrm{T}\mathrm{G}\mathbf{x} + \dot{\mathbf{x}}^\mathrm{T}\mathrm{G}\dot{\mathbf{x}} + 2\mathbf{x}^\mathrm{T}\mathbf{u} + \mathbf{u}^\mathrm{T}\mathrm{G}^{-1}\mathbf{u}\right\}\mathrm{dt}$$

Problem 5-4 (Ref. 33)

Consider the undamped second-order systems (with no rigid body modes) for the case in which the number of actuators is equal to the number of model degrees of freedom (D is full rank),

$$\mathbf{M\ddot{x} + Kx = Du}$$

or, equivalently,

$$\dot{\mathbf{z}} = \mathbf{Az + Bu}$$

where

$$\mathbf{z} = \begin{Bmatrix} \mathbf{x} \\ \dot{\mathbf{x}} \end{Bmatrix}, \qquad \mathrm{A} = \begin{bmatrix} 0 & \mathrm{I} \\ -\mathrm{M}^{-1}\mathrm{K} & 0 \end{bmatrix}, \qquad \mathrm{B} = \begin{bmatrix} 0 \\ \mathrm{M}^{-1}\mathrm{D} \end{bmatrix}$$

For the regulator design, the following quadratic cost functional is chosen as the energy type such as

$$\mathrm{J} = \frac{1}{2}\int_0^\infty \left\{\mathbf{z}^\mathrm{T}\mathrm{Q}\mathbf{z} + \mathbf{u}^\mathrm{T}\mathrm{R}\mathbf{u}\right\}\mathrm{dt}$$

with

$$\mathrm{Q} = \begin{bmatrix} \alpha\mathrm{K} & 0 \\ 0 & \beta\mathrm{M} \end{bmatrix}, \qquad \mathrm{R} = \mathrm{D}^\mathrm{T}\mathrm{K}^{-1}\mathrm{D}$$

where α, β are arbitrary positive scalar constants.

Show that the following matrix satisfies the Riccati equation of the above problem:

$$P = \begin{bmatrix} \sqrt{1+\alpha}\zeta G & \eta M \\ \eta M & \zeta M K^{-1} G \end{bmatrix}$$

where $\eta = \sqrt{1+\alpha} - 1$, $\zeta = \sqrt{2\eta + \beta}$, and $G = M^{\frac{1}{2}}\left[M^{-\frac{1}{2}} K M^{-\frac{1}{2}}\right]^{\frac{1}{2}} M^{\frac{1}{2}}$ and the optimal solution (closed form) is represented by

$$D\mathbf{u} = -\eta K\mathbf{x} - \zeta G\dot{\mathbf{x}}$$

Problem 5-5

Consider the system

$$\dot{\mathbf{x}} = \begin{bmatrix} 0 & 1 \\ -1 & 0 \end{bmatrix}\mathbf{x} + \begin{bmatrix} 1 & 0 \\ 0 & 1 \end{bmatrix}\mathbf{u}$$

The performance index to be minimized is

$$J = \frac{1}{2}\int_0^{t_f}\left\{\mathbf{x}^T\begin{bmatrix} 1 & 0 \\ 0 & 0 \end{bmatrix}\mathbf{x} + \mathbf{u}^T\mathbf{u}\right\}dt$$

with the initial condition $\mathbf{x}(0) = [1 \quad 1]^T$, the final condition $\mathbf{x}(t_f)$ is free, and the final time is fixed $t_f = 5$ s.

Your tasks are:

(a) Determine the optimal control law and the regulation cost. Simulate the closed-loop system.

(b) Find the steady-state gain matrix, and simulate the system for $t_f = 15$ s. Find the regulation cost.

(c) Repeat part b by choosing

$$B = \begin{Bmatrix} 0 \\ 1 \end{Bmatrix}, \qquad R = \begin{bmatrix} 0 & 0 \\ 0 & 1 \end{bmatrix}$$

i.e., use one control only.

Find the steady-state gain matrix. Is the regulation cost more or less in comparison to that of part b?

Problem 5-6

Consider the Sylvester equation

$$A\Phi - \Phi\Lambda = BG\Phi$$

For the second-order system without a rigid body mode, it is known that the eigenvalues and eigenvectors occur in complex conjugate pairs which can be denoted

$$\Phi = [\underline{\phi}_1 \;\; \overline{\underline{\phi}}_1 \;\; \underline{\phi}_2 \;\; \overline{\underline{\phi}}_2 \;\; \cdots \;\; \underline{\phi}_n \;\; \overline{\underline{\phi}}_n]$$
$$\Lambda = \text{diag}\left\{\lambda_1,\ \overline{\lambda}_1,\ \lambda_2,\ \overline{\lambda}_2,\ \ldots,\ \lambda_n,\ \overline{\lambda}_n\right\}$$

where $(\bar{\ })$ denotes the conjugate of the ( ).

Let us define real eigenvector and eigenvalue matrices X and E as

$$X = [\Re(\underline{\phi}_1) \;\; \Im(\underline{\phi}_1) \;\; \Re(\underline{\phi}_2) \;\; \Im(\underline{\phi}_2) \;\; \ldots \;\; \Re(\underline{\phi}_n) \;\; \Im(\underline{\phi}_n)]$$
$$E = \text{diag}\left\{E_1,\ E_2,\ \ldots,\ E_n\right\}$$

where $\Re(\ )$ and $\Im(\ )$ denote the real part and imaginary part of the matrix ( ), respectively, and E_i denotes the 2×2 block matrices of the ith modes eigenvalues defined as follows:

$$E_i \equiv \begin{bmatrix} \Re(\lambda_i) & \Im(\lambda_i) \\ -\Im(\lambda_i) & \Re(\lambda_i) \end{bmatrix}$$

Your task is to show that Sylvester's equation can be rewritten as the real algebraic equation

$$AX - XE = BGX$$

Note that, by using the above equation in control design calculations such as eigenstructure assignment, the computed gain matrix is guaranteed to always be real valued, because only real arithmetic is needed.

Problem 5-7

Consider the mechanical second-order system with the control law

$$M\ddot{\mathbf{x}} + C\dot{\mathbf{x}} + K\mathbf{x} = D\mathbf{u}$$

$$\mathbf{u} = -(G_1\mathbf{x} + G_2\dot{\mathbf{x}})$$

where $\mathbf{x} \in R^n$ is a state vector, $\mathbf{u} \in R^m$ is a control vector, and C is a structural damping.

Suppose that the desired closed-loop system can be represented in the modal coordinates:

$$\ddot{\underline{\eta}} + \tilde{C}\dot{\underline{\eta}} + \tilde{K}\underline{\eta} = 0$$

where $\tilde{C}$ and $\tilde{K}$ include the desired closed-loop natural frequencies and damping factors.

Your tasks are as follows:

(a) Show that the gain matrices satisfy the following relations:

$$DG_1 = M\Phi\tilde{K}\Phi^T M - K$$

$$DG_2 = M\Phi\tilde{C}\Phi^T M - C$$

where Φ is the modal matrix obtained solving the eigenvalue problem of mass and stiffness matrix M and K.

(b) Discuss the existence and uniqueness of the solution (using the above control law) for the cases in which (1) $n < m$, (2) $n = m$, and (3) $n > m$.

Problem 5-8 (Ref. 6)

Consider the following undamped second-order system for which the number of actuators is equal to the number of coordinates:

$$\begin{bmatrix} 1 & 0 \\ 0 & 2 \end{bmatrix} \ddot{\mathbf{x}} + \begin{bmatrix} 5 & -4 \\ -4 & 4 \end{bmatrix} \mathbf{x} = \begin{bmatrix} 1 & 0 \\ 0 & 1 \end{bmatrix} \mathbf{u}$$

(a) Find the uncoupled equations of motion using modal coordinate transformation ($\mathbf{x} = \Phi\underline{\eta}$):

$$\tilde{M}\ddot{\underline{\eta}} + \tilde{K}\underline{\eta} = \Phi^T D\mathbf{u} \equiv \mathbf{w}$$

Compute the open-loop eigenvalues.

(b) Assume that the controller has the following form:

$$\mathbf{w} = -(G\underline{\eta} + H\dot{\underline{\eta}})$$

where G and H are diagonal matrices such as

$$G = \begin{bmatrix} g_1 & 0 \\ 0 & g_2 \end{bmatrix}, \qquad H = \begin{bmatrix} h_1 & 0 \\ 0 & h_2 \end{bmatrix}$$

Find the control gains such that the closed-loop poles are located at

$$\lambda_{1,2}^c = -1 \pm 0.6i, \qquad \lambda_{3,4}^c = -2 \pm 2.6i$$

(c) Compute the gain matrices G_1, G_2 for the physical coordinates such that

$$\mathbf{u} = -(G_1\mathbf{x} + G_2\dot{\mathbf{x}})$$

using the results of part b.

References

1. Athans, M. and Falb, P.L., *Optimal Control*, McGraw-Hill, New York, NY, 1966.
2. Kirk, D.E., *Optimal Control theory*, Prentice-Hall, Englewood Cliffs, NJ, 1970.
3. Junkins, J.L. and Turner, J.D., *Optimal Spacecraft Rotational Maneuvers*, Studies in Astronautics, Vol. 3, Elsevier Scientific Publishing Company, New York, NY, 1985.
4. Bryson, A.E. Jr. and Ho, Y.C., *Applied Optimal Control*, revised printing, John Wiley & Sons, New York, NY, 1975.
5. Sage, A.P., *Optimum Systems Control*, Prentice Hall, Englewood Cliffs, NJ, 1968.
6. Meirovitch, L., *Dynamics and Control of Structures*, Press, John Wiley & Sons, New York, NY, 1990.
7. Potter, J.E., "A Matrix Equation Arising in Statistical Filter Theory," NASA CR-270, 1965.
8. Serbin, S.M. and Serbin, C.A., "A Time-Stepping Procedure for $\dot{X} = A_1X + XA_2 + D$, $X(0)=C$," *IEEE Transactions on Automatic Control*, Vol. AC-25, No. 6, 1980, pp. 1138-1141.
9. Lacoss, R.T. and Shakel, A.E., "More $A_1E + EA_2 = -D$ and $\dot{X} = A_1X + XA_2 + D$, $X(0)=C$," *IEEE Transactions on Automatic Control,* Vol. AC-21, 1976, pp. 405-406.
10. Potter, J.E., "Matrix Quadratic Solutions," *Journal of SIAM*, Applied Mathematics, Vol. 14, No. 3, 1966, pp. 496-501.
11. Laub, A., "A Schur Method for Solving Algebraic Riccati Equations," *IEEE Transactions on Automatic Control,* Vol. AC-24, Jun. 1979, pp. 913-925.
12. Roberts, J.D., "Linear Model Reduction and Solution of the Algebraic Riccati Equation by Use of the Sign Function," *International Journal of Control,* Vol. 32, 1980, pp. 677-687.
13. Shieh, L.S., Tsay, Y.T., and Yates, R.E., "Some Properties of Matrix Sign Functions Derived from Continued Fractions," *IEEE Proceedings,* Vol. 130, Part D, No. 3, May 1983, pp. 111-118.
14. Shieh, L.S., Tsay, Y.T., Lin, S.W., and Coleman, N.P., "Block-Diagonalization and Block-Triangularization of a Matrix via the Matrix Sign Function," *International Journal of Systems Sciences,* Vol. 15, No. 11, 1984, pp. 1203-1220.
15. Shieh, L.S., Lian, S.R., and Mcinnis,B.C., "Fast and Stable Algorithms for Computing the Principal Square Root of a Complex Matrix," *IEEE Transactions on Automatic Control,* Vol. AC-32, No. 9, 1987, pp. 820-822.

16. Bierman, G.J., "Computational Aspects of the Matrix Sign Function to the ARE," *Proceedings of 23rd IEEE Conference on Decision and Control,* Las Vegas, NV, Dec. 1984, pp. 514-519.

17. Gardiner, J.D. and Laub, A.J., "A Generalization of the Matrix Sign Function Solution for the Algebraic Riccati Equation," *Proceedings of 24th Conference on Decision and Control,* Ft. Lauderdale, FL, Dec. 1985, pp. 1233-1235.

18. Sunkel, J.W. and Shieh, L.S., "Optimal Momentum Management Controller for the Space Station," *Journal of Guidance, Control, and Dynamics,* Vol. 13, No. 4, Aug. 1990, pp. 659-668.

19. Rew, D.W. and Junkins, J.L., "Multi-Criterion Approaches to Optimization of Linear Regulators," AIAA Paper 86-2198-CP, 1986.

20. Brogan, W.L., *Modern Control Theory,* Quantum Publishers, Inc., New York, NY, 1974.

21. Bhattacharyya, S.P. and deSouza, E., "Pole Assignment via Sylvester's Equation," *Systems and Control Letters,* Vol. 1, No. 4, Jan. 1982, pp. 261-263.

22. Cavin III, R.K. and Bhattacharyya, S.P., "Robust Well-Conditioned Eigenstructure Assignment via Sylvester's Equation," *Journal of Optimal Control Applications and Methods,* Vol. 4, 1983, pp. 205-212.

23. Porter, B. and D'Azzo, J.J., "Algorithm for Closed-loop Eigenstructure Assignment by State Feedback in Multivariable Linear Systems," *International Journal of Control,* Vol. 27, No. 6, 1978, pp. 943-947.

24. Moore, B.C., "On the Flexibility Offered by State Feedback in Multivariable systems Beyond Closed-Loop Eigenvalue Assignments," *IEEE Transactions on Control,* Vol. AC-21, 1976, pp. 689-692.

25. Kautsky, J., Nichols, N.K., and Van Dooren, P., "Robust Pole Assignment in Linear State Feedback," *International Journal of Control,* Vol. 41, No. 5, 1985, pp. 1129-1155.

26. Juang, J.N., Lim, K.B., and Junkins, J.L., "Robust Eigensystem Assignment for Flexible Structures," *Journal of Guidance, Control, and Dynamics,* Vol. 12, No. 3, May-June 1989, pp. 381-387.

27. Rew, D.W., Junkins, J.L., and Juang, J.N., "Robust Eigenstructure Assignment by a Projection Method: Application Using Multiple Optimization Criteria," *AIAA Journal of Guidance, Control, and Dynamics,* Vol. 12, No. 3, May-June 1989, pp. 396-403.

28. Oz., H. and Meirovitch, L., "Optimal Modal-Space Control of Flexible Gyroscopic Systems," *Journal of Guidance and Control,* Vol. 3, No. 3, 1980, pp. 218-226.

29. Juang, J.N. and Maghami, P.G., "Robust Eigensystem Assignment for Second-Order Dynamic System," *Mechanics and Control of Large Space Structures*, Progress in Aeronautics and Astronautics Series, Vol. 129, edited by J.L. Junkins, AIAA, New York, NY, 1990, Chapter 14.

30. Golub, G.H. and Van Loan, C.F., *Matrix Computations*, Johns Hopkins Univ. Press, Baltimore, MD, 1983.

31. Klema, V.C. and Laub, A.J., "The Singular Value Decomposition: Its Computation and Some Applications," *IEEE Transactions on Automatic Control*, Vol. AC-25, No. 2, April 1980, pp. 164-176.

32. Maghami, P.G. and Juang, J.N., "Efficient Eigenvalue Assignment for Large Space Structures," *Journal of Guidance and Control*, Vol. 13, No. 6, Nov.-Dec. 1990, pp. 1033-1039.

33. Belvin, W.K. and Park, K.C., "Structural Tailoring and Feedback Control Synthesis: An Interdisciplinary Approach," *Journal of Guidance and Control*, Vol. 13, No. 3, May-June 1990, pp. 424-429.

34. Fennell, R.E., Adams, W.M. Jr., and Christhilf, D.M., "Parameter Insensitive Control Utilizing Eigenspace Methods," *Journal of Guidance and Control*, Vol. 14, No. 6, Nov-Dec 1991, pp. 1208-1214.

35. Kazerooni, H. and Houpt, P.K., "On the Loop Transfer Recovery," *International Journal of Control*, Vol. 43, No. 3, 1986, pp. 981-996.

CHAPTER 6

Controllability and Observability of Finite-Dimensional Dynamical Systems

The purpose of this chapter is to introduce concepts of controllability, observability, and modal cost analysis. We develop these broadly applicable ideas in the context of the closely related issues of sensor and actuator placement optimization and consider, in particular, applications to vibration suppression problems for flexible structures. We see a wide class of civil and aerospace structures in which active vibration suppression will be required in the coming decades. In particular, the space structures envisioned for future applications may require a large number of sensors and actuators for attitude, shape, and vibration control.

The problem of choosing the appropriate number and locations of actuators and sensors to control large flexible structures is an important area of current research, since arbitrary decisions can be expected to degrade the system performance and directly limit the range of practical applications. Implicitly, by choosing the most effective positions for a given number of actuators (sensors), we can probably minimize the number of actuators (sensors) actually needed. As we recognize the fundamental role played by actuator (sensor) placement on system controllability (observability), it is natural that the problem of placing actuators (sensors) be considered simultaneously with the quest to define more meaningful measures of controllability (observability).

As a practical matter, actuators are typically more expensive and massive than sensors. Thus optimal placement of actuators, which implicitly permits minimization of the number of required actuators, is likely to have greater significance than sensor location optimization, especially for space applications. Also, as a matter of prudence, the desire to minimize the number of sensors and actuators must be balanced against an obvious need to have an adequate level of redundancy (in order to maintain system reliability to accommodate a realistic frequency of sensor/actuator failures). For the purpose of selecting actuator locations to control flexible structural systems, several concepts pertaining to the degree of controllability/observability (Refs. 1-15) and cost analysis (Refs. 16-26) have recently been developed.

After the controllability and observability concepts are defined in section 6.1, measures of controllability and the central results from modal cost analysis are reviewed in sections 6.2 and 6.3, respectively. Finally, in section 6.4, a new measure of controllability for linear time-invariant systems is introduced for guiding the placement of actuators to control vibrating structures. The new criterion combines a measure of modal controllability (Ref. 13) with the idea of modal cost analysis (Refs. 22, 26); this new criterion provides us with useful information on each mode's relative importance which takes into account disturbability, controllability, and participation of the modes in physically important system outputs.

6.1 Definitions of Controllability and Observability

The concepts of controllability and observability are very important in the study of dynamical system control and estimation problems. The concept of controllability is closely related to the existence of a feasible control law, since it is not always possible to find a control law of a given form that makes the closed-loop system stable with respect to the desired state or trajectory. Similarly, the concept of observability is related to the existence of a feasible algorithm for estimating the state variables from measurements. Controllability measures the particular actuator input configuration's ability to control all system states, whereas observability measures the particular sensor output configuration's ability to obtain the information needed to estimate all system states. These heuristic interpretations need to be made more rigorous. Following are the conventional definitions for linear dynamical systems.

Consider a linear system,

$$\begin{aligned} \dot{\mathbf{x}}(t) &= A\mathbf{x}(t) + B\mathbf{u}(t) \\ \mathbf{y}(t) &= C\mathbf{x}(t) \end{aligned} \tag{6.1}$$

where $\mathbf{x} \in R^n$, $\mathbf{u} \in R^m$, and $\mathbf{y} \in R^l$.

Definition 6.1: Controllability

A linear system is said to be *controllable* at time t_0 if there exists a control $\mathbf{u}(t)$ that will transfer the initial state $\mathbf{x}(t_0)$ to the origin at some finite time $t_1 > t_0$. If this is true for *all initial times* t_0 and *all initial states* $\mathbf{x}(t_0)$, the system is said to be *completely controllable.*

■

There are several necessary and sufficient conditions (or criteria) for controllability (Refs. 27 and 28).

Theorem 6.1: Criteria for Controllability

A system is completely controllable if and only if any one of the following equivalent conditions is satisfied:

(1) The $(n \times rn)$ controllability matrix $\mathcal{C}$ has rank n.

$$\mathcal{C} \equiv [B \quad AB \quad A^2B \quad \cdots \quad A^{n-1}B] \tag{6.2}$$

(2) The controllability grammian W_c is full-rank (in fact, positive definite).

$$W_c \equiv \int_0^\infty e^{A\tau} B B^T e^{A^T \tau} d\tau \tag{6.3}$$

(3) For every eigenvalue λ of A, the following $n \times (n+m)$ complex matrix has rank n [this is the *Popov, Belevitch, and Hautus (PBH) rank test*]:

$$\left[(\lambda I - A) \;\vdots\; B \right] \tag{6.4}$$

(4) There exists no left eigenvector $\mathbf{q}_i$ of A that is orthogonal to all the columns of B, i.e., if and only if [this is the *PBH eigenvector test*]:

$$\text{Given that } \mathbf{q}_i^T A = \lambda_i \mathbf{q}_i^T, \text{ then the condition : } \mathbf{q}_i^T B = 0 \text{ implies } \mathbf{q}_i \equiv 0 \tag{6.5}$$

■

The PBH eigenvector and rank tests are useful to efficiently evaluate the modal controllability of the system. Note that MATLAB operators for computing controllability matrix $\mathcal{C}$ and the controllability grammian W_c are as follows:

> **C = ctrb[A, B]**, > **Wc = gram[A, B]**

Definition 6.2: Observability

A linear system is said to be *observable* at t_0, if $\mathbf{x}(t_0)$ can be determined from the knowledge of $\mathbf{u}(\tau)$, $\mathbf{y}(\tau)$ for $t_0 \le \tau \le t_1$. If this is true for all t_0 and all $\mathbf{x}(t_0)$, the system is said to be *completely observable*.

■

Theorem 6.2: Criteria for Observability

A system is completely observable if and only if any one of the following equivalent conditions is satisfied:

(1) The $(nl \times n)$ observability matrix $\mathcal{O}$ has rank n.

$$\mathcal{O} \equiv \begin{bmatrix} C \\ CA \\ CA^2 \\ \vdots \\ CA^{n-1} \end{bmatrix} \tag{6.6}$$

(2) The observability grammian W_o is full-rank (in fact, positive definite).

$$W_o \equiv \int_0^\infty e^{A^T \tau} C^T C e^{A\tau} d\tau \tag{6.7}$$

(3) PBH rank test: For every eigenvalue λ of A, the following $(n+l) \times n$ complex matrix has rank n:

$$\begin{bmatrix} \lambda \mathrm{I} - \mathrm{A} \\ \mathrm{C} \end{bmatrix} \tag{6.8}$$

(4) PBH eigenvector test: There exists no eigenvector $\mathbf{p}_i$ of A that is orthogonal to all the rows of C, i.e., if and only if

$$\mathrm{A}\mathbf{p}_i = \lambda_i \mathbf{p}_i, \qquad \mathrm{C}\mathbf{p}_i = 0 \qquad \Longrightarrow \qquad \mathbf{p}_i \equiv 0 \tag{6.9}$$

■

The above criteria show the *duality* of the observability and controllability. It has been proven that the controllability and observability grammian matrices satisfy the following Lyapunov equations (Ref. 28):

$$\begin{aligned} \mathrm{AW_c} + \mathrm{W_c A^T} + \mathrm{BB^T} &= 0 \\ \mathrm{A^T W_o} + \mathrm{W_o A} + \mathrm{C^T C} &= 0 \end{aligned} \tag{6.10}$$

Note that MATLAB operators for computing the observability matrix $\mathcal{O}$ and the observability grammian $\mathrm{W_o}$ are as follows:

$$> \mathbf{O} = \mathbf{obsv}[\mathbf{A}, \mathbf{C}], \qquad > \mathbf{Wo} = \mathbf{gram}[\mathbf{A}', \mathbf{C}']$$

For discrete-time systems, an analogous operator is available: **dgram**(which utilizes the discrete matrix Lyapunov equation operator **dlyap**).

6.2 Measures of Controllability

The information from the criteria for controllability (Theorem 6.1) provides binary (yes/no) information, and it does not provide a graduated measure of "how controllable" a system is. A vibratory mode of a beam, for example, is not controllable by a force actuator placed exactly at one of the nodes, but it is theoretically controllable by a force actuator placed a small distance away from the node. One can expect that an actuator placed slightly farther from the node will result in greater controllability of this mode, and one might conjecture that this configuration would require needs less control energy than the former one. We have an obvious need for quantitative, graduated measures of controllability and we can anticipate that such measures will provide useful information for selecting actuator locations.

Motivated by these considerations, we first review and extend the recent research contributions of Longman, Moore, Hamdan, and Nayfeh, which are all focused on establishing quantitative measures of controllability. As discussed in the previous section, observability is the dual problem of controllability. Given the basic theoretical results for controllability, the corresponding results for observability can be obtained easily by analogy, as a consequence of the duality between these two concepts. Therefore, we will deal only with controllability measures in detail.

6.2.1 Degree of Controllability

We begin with a brief discussion of the recent research literature on this subject. Longman et al. (Refs. 1-10) have introduced several approaches to measure the degree of controllability of a linear dynamical system; they then develop numerical methods to generate approximate values of their measures for any linear time-invariant system. Longman introduces the concept of a "recovery region" to measure the degree of controllability, where the recovery region is defined as the set of initial states that can be returned to the origin in a given amount of time T using bounded controls (or bounded energy, or bounded fuel). Longman's degree of controllability is a scalar measure of the size of the recovery region, where the scalar is chosen as the shortest distance from the origin to an initial state that cannot be returned to the origin in time T. In Refs. 1-3, approximate computational procedures are developed for the time optimal degree of controllability (based on the minimum distance to the boundary of the parallelepiped that approximates the recovery region). Analogous energy optimal degree of controllability concepts are developed in Refs. 4 and 5, while fuel optimal controllability measures are studied in Refs. 5-7.

The approaches for computing the degree of controllability discussed above give approximations in the form of upper bounds. In Ref. 8, a lower bound is generated by discretizing the system equations; however, the upper bound approximation has been shown in Ref. 1 to be useful and easy to compute for modal systems with sufficiently large T. The use of degree of controllability for optimizing actuator locations is studied in Refs. 7, 9 and 10. One of the problems in these approaches is the greatly increased computational complexity encountered for the case of multiple actuators (Ref. 15). Since a highly desirable attribute of controllability measures is ease of computation, we are motivated to seek alternatives to these approaches.

Consider the linear system

$$\dot{\mathbf{x}}^*(t) = \mathcal{A}\mathbf{x}^*(t) + \mathcal{B}\mathbf{u}^*(t) \tag{6.11}$$

The coordinates of a state space will very rarely all have the same physical units, and hence, for comparing distances in the state space, it is advantageous to require that each coordinate be made nondimensional by a judicious normalization. The original state vector $\mathbf{x}^*$ must be normalized based on physical considerations so that unit disturbances from the origin in all directions are equally undesirable. This implies that the state vector should be normalized such that the need for recovery from an initial unit displacement away from the origin is considered to be equal for all displacement directions.

In terms of the normalized state vector $\mathbf{x}$, and the normalized control vector $\mathbf{u}$, the system equations are written in the following form;

$$\begin{aligned} \dot{\mathbf{x}}(t) &= \mathbf{A}\mathbf{x}(t) + \mathbf{B}\mathbf{u}(t) \\ |u_i(t)| &\leq 1, \qquad i = 1, \ldots, m \end{aligned} \tag{6.12}$$

When the normalized system (6.12) is used, the recovery region and degree of controllability are defined as follows.

Definition 6.3: Recovery Region

The *recovery region* associated with a specified time T for the normalized dynamical system of Eq. (6.12) is the set

$$\mathcal{R} = \left\{ \mathbf{x}(0) \middle| \; \exists \; \mathbf{u}(t), \; t \in [0, T], \; |u_i(t)| \leq 1, \; i = 1, 2, \ldots, m, \; \exists \; \mathbf{x}(T) = 0 \right\} \tag{6.13}$$

■

Definition 6.4: Degree of Controllability

The *degree of controllability* in time T of the $\mathbf{x} = 0$ solution of the normalized system is defined as

$$\rho = \inf \|\mathbf{x}(0)\|, \qquad \forall \; \mathbf{x}(0) \notin \mathcal{R} \tag{6.14}$$

where $\|\cdot\|$ represents the Euclidean norm.

■

Thus the recovery region identifies all of the initial conditions (or disturbed states) that can be returned to the origin in time T using the bounded controls. The degree of controllability is a scalar measure of the size of the recovery region, where the scalar ρ is chosen as the shortest distance from the origin to an initial state that cannot be returned to the origin in time T. To obtain a qualitative appreciation for the concepts of recovery region and degree of controllability, let us consider an example system with only two state variables. Suppose the recovery region is as shown by the area inside the ellipse in Figure 6.1; then the shortest distance ρ from the origin to the boundary of the recovery region (ellipse) is the degree of controllability.

In order to make the definition of the degree of controllability useful, it is necessary to develop a simple algorithm to generate at least an approximation of the distance ρ, and this necessitates approximating the recovery region $\mathcal{R}$. The recovery region $\mathcal{R}$ is approximated by the n-dimensional parallelepiped, which is constructed using column vectors of B and eigenvalues/eigenvectors of A. The

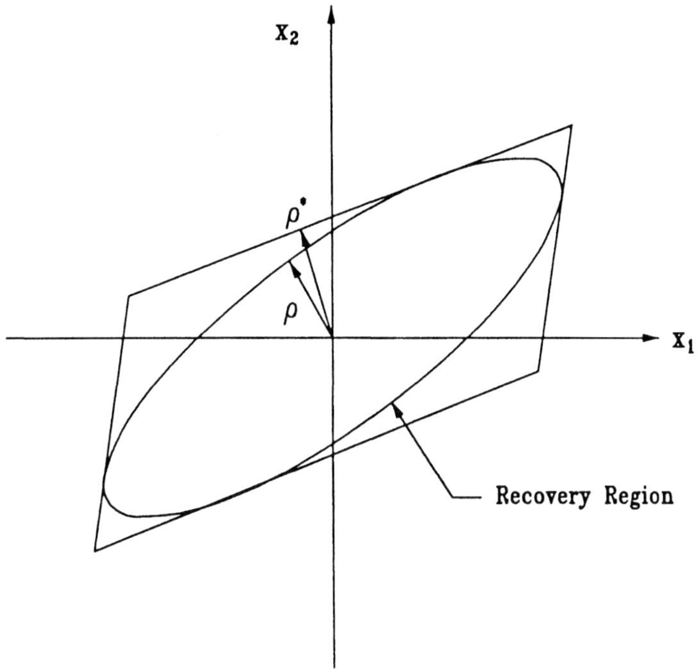

Figure 6.1 Recovery Region, Degree of Controllability, and Their Approximations.

minimum perpendicular distance to a side of the parallelepiped $\mathcal{R}^*$ can be used as an approximation ρ^* of the degree of controllability ρ. Only one method for approximating the recovery region and the degree of controllability will be treated in detail here.

It is well known that the solution of Eq. (6.12) is given by

$$\mathbf{x}(\mathrm{T}) = \phi(\mathrm{T},0)\mathbf{x}(0) + \phi(\mathrm{T},0)\int_0^{\mathrm{T}} \phi(0,\mathrm{t})\mathbf{B}\mathbf{u}(\mathrm{t})\mathrm{dt} \tag{6.15}$$

Since we are concerned with driving the system to the origin, then $\mathbf{x}(\mathrm{T}) = 0$.

Observe that Eq. (6.15) can be rewritten as

$$\delta \equiv -\mathbf{x}(0) = \int_0^{\mathrm{T}} \phi(0,\mathrm{t})\mathrm{B}\mathbf{u}(\mathrm{t})\mathrm{dt} \tag{6.16}$$

If $\mathbf{x}(0)$ can be returned to the origin using control vector $\mathbf{u}(t)$ in time T, then $\delta = -\mathbf{x}(0)$ lies also inside the recovery region since δ can be also returned to the origin using control vector $-\mathbf{u}(t)$. Therefore, the recovery region is the set of all such δ generated by admissible control functions,

$$|\mathrm{u_r(t)}| \leq 1; \quad \mathrm{t} \in [0,\mathrm{T}]; \quad \mathrm{r} = 1,\ldots,m \tag{6.17}$$

The state transition matrix can be determined using a spectral decomposition in terms of the eigenvalues/eigenvectors of matrix A (the Cayley-Hamilton theorem, see Refs. 27 and 28),

$$\phi(0,\mathrm{t}) = \mathrm{e}^{-\mathrm{At}} = \Phi \mathrm{e}^{-\Lambda \mathrm{t}} \Psi^{\mathrm{T}} = \sum_{\mathrm{i}=1}^{n} \underline{\phi}_{\mathrm{i}} \mathrm{e}^{-\lambda_{\mathrm{i}}\mathrm{t}} \underline{\psi}_{\mathrm{i}}^{\mathrm{T}} \tag{6.18}$$

Note that $\Lambda = \mathrm{diag}(\lambda_1,\ldots,\lambda_n)$, λ_{i} are eigenvalues of matrix A; further, Φ and Ψ denote the right and left modal matrices of As right and left eigenvectors, respectively.

Substitution of Eq. (6.18) into Eq. (6.16) yields

$$\delta = \sum_{\mathrm{i}=1}^{n} \underline{\phi}_{\mathrm{i}} \Big[\int_0^{\mathrm{T}} \mathrm{e}^{-\lambda_{\mathrm{i}}\mathrm{t}} \underline{\psi}_{\mathrm{i}}^{\mathrm{T}} \mathrm{B}\mathbf{u}(\mathrm{t})\mathrm{dt} \Big] \tag{6.19}$$

Let us define C as the set of j for which λ_{j} is real, and C* as the set of j for which λ_{j} is complex. If we use $\underline{\phi}_{\mathrm{j}} = \underline{\phi}_{\mathrm{j}}^{\mathrm{R}} + i\underline{\phi}_{\mathrm{j}}^{\mathrm{I}}$ for $\mathrm{j} \in \mathrm{C}^*$ ($\underline{\phi}_{\mathrm{j}}^{\mathrm{R}}$ and $\underline{\phi}_{\mathrm{j}}^{\mathrm{I}}$ denote the real and imaginary parts of $\underline{\phi}_{\mathrm{j}}$, respectively),, we rewrite δ as follows;

$$\begin{aligned}\delta = \sum_{\mathrm{j}\in \mathrm{C}} \Big[\sum_{\mathrm{r}=1}^{m} \int_0^{\mathrm{T}} (\mathrm{e}^{-\lambda_{\mathrm{j}}\mathrm{t}} \underline{\psi}_{\mathrm{j}}^{\mathrm{T}} \mathbf{b}_{\mathrm{r}}) \mathrm{u_r dt} \Big] \underline{\phi}_{\mathrm{j}} + \sum_{\mathrm{j}\in \mathrm{C}^*} 2 \Big[\sum_{\mathrm{r}=1}^{m} \int_0^{\mathrm{T}} \Re(\mathrm{e}^{-\lambda_{\mathrm{j}}\mathrm{t}} \underline{\psi}_{\mathrm{j}}^{\mathrm{T}} \mathbf{b}_{\mathrm{r}}) \mathrm{u_r dt} \Big] \underline{\phi}_{\mathrm{j}}^{\mathrm{R}} \\ - \sum_{\mathrm{j}\in \mathrm{C}^*} 2 \Big[\sum_{\mathrm{r}=1}^{m} \int_0^{\mathrm{T}} \Im(\mathrm{e}^{-\lambda_{\mathrm{j}}\mathrm{t}} \underline{\psi}_{\mathrm{j}}^{\mathrm{T}} \mathbf{b}_{\mathrm{r}}) \mathrm{u_r dt} \Big] \underline{\phi}_{\mathrm{j}}^{\mathrm{I}}\end{aligned} \tag{6.20}$$

where $\Re\{\ \}$ and $\Im\{\ \}$ denote the real and imaginary values of $\{\ \}$. Note that the vectors $\underline{\phi}_j$ for $j \in C$, $\underline{\phi}_j^R$ and $\underline{\phi}_j^I$ for $j \in C^*$ form a set of n linearly independent vectors and, therefore, the recovery region can be generated using those vectors.

In order to maximize the component of δ along $\underline{\phi}_j$ with bounded control effort, the control component u_r should be chosen as

$$\begin{aligned} u_r &= \text{sign}\left[e^{-\lambda_j t}\underline{\psi}_j^T \mathbf{b}_r\right], && \text{for direction } \underline{\phi}_j,\ j \in C \\ u_r &= \text{sign}\left[\Re\left(e^{-\lambda_j t}\underline{\psi}_j^T \mathbf{b}_r\right)\right], && \text{for direction } \underline{\phi}_j^R,\ j \in C^* \\ u_r &= -\text{sign}\left[\Im\left(e^{-\lambda_j t}\underline{\psi}_j^T \mathbf{b}_r\right)\right], && \text{for direction } \underline{\phi}_j^I,\ j \in C^* \end{aligned} \tag{6.21}$$

When each integral in Eq. (6.20) is maximized independently, then the set of points δ generated becomes the parallelepiped, and this n-dimensional parallelepiped can be considered as an approximation of the recovery region, as shown in Fig. 6.1.

Theorem 6.3: Approximation $\mathcal{R}^*$ of the Recovery Region $\mathcal{R}$

Let A have all distinct eigenvalues $\lambda_1, \ldots, \lambda_n$. The recovery region $\mathcal{R}$ can be approximated by the n-dimensional parallelepiped.

$$\mathcal{R}^* = \left\{\mathbf{x} \mid \mathbf{x} = \sum_{i=1}^{n} c_i \nu_i, \qquad \forall\ c_i \in |c_i| \le 1\right\} \tag{6.22}$$

where the vectors ν_i represent the following n vectors:

$$\begin{aligned} \nu_i &= \left\{\sum_{r=1}^{m}\int_0^T \left|e^{-\lambda_j t}\underline{\psi}_j^T \mathbf{b}_r\right| dt\right\} \underline{\phi}_j, && \text{for } j \in C \\ \nu_i &= 2\left\{\sum_{r=1}^{m}\int_0^T \left|\Re\left(e^{-\lambda_j t}\underline{\psi}_j^T \mathbf{b}_r\right)\right| dt\right\} \underline{\phi}_j^R, && \text{for } j \in C^* \\ \nu_i &= 2\left\{\sum_{r=1}^{m}\int_0^T \left|\Im\left(e^{-\lambda_j t}\underline{\psi}_j^T \mathbf{b}_r\right)\right| dt\right\} \underline{\phi}_j^I, && \text{for } j \in C^* \end{aligned} \tag{6.23}$$

■

Now, it is easy to understand that the minimum perpendicular distance to a side of the parallelepiped $\mathcal{R}^*$ can be used as an approximation ρ^* of the degree of controllability ρ. We define the matrix F, which includes all of the vectors ν_i:

$$F \equiv [\nu_1 \quad \nu_2 \quad \ldots \quad \nu_n] \tag{6.24}$$

Note that the ν_i are linearly independent and, for a controllable system all ν_i are nonzero. Since the columns of F are linearly independent, F can be inverted and we can write the inverse of F as follows:

$$F^{-1} \equiv [\,\mu_1 \quad \mu_2 \quad \ldots \quad \mu_n\,]^T \tag{6.25}$$

When $F^{-1}F = I$ is used, Eqs. (6.24) and (6.25) yield the following relations between vectors ν_j and μ_k:

$$\mu_j^T \nu_k = \delta_{jk} \tag{6.26}$$

where δ_{jk} denotes the Kronecker delta.

The geometrical meaning of Eq. (6.26) is that vector μ_j is perpendicular to all vectors ν_k except ν_j. Therefore, μ_j is normal to the surface of the parallelepiped parallel to all ν_k for $k \neq j$. We define d_j as the component of ν_j along this normal, which can be computed, using Eq. (6.26), by the inner product of vectors μ_j and ν_j:

$$d_j = \frac{\mu_j}{\|\mu_j\|} \cdot \nu_j = \frac{\mu_j^T \nu_j}{\|\mu_j\|} = \frac{1}{\|\mu_j\|} \tag{6.27}$$

Figure 6.2 shows the parallelopiped approximation of the recovery region for two degrees of freedom with a single input, and illustrates the geometry of the vectors ν_i, and μ_i, and their relation to the recovery region. Since the d_j are the normal distances from the origin to the parallelepiped that approximates the recovery region, the smallest distance d_j can be taken as an approximate degree of controllability. These ideas are captured in the following theorem.

Theorem 6.4: Approximation ρ^* of the Degree of Controllability ρ

From Theorem 6.3, let the columns of F be linearly independent and denote them as

$$F \equiv [\,\nu_1 \quad \nu_2 \quad \ldots \quad \nu_n\,] \tag{6.28}$$

and define the column vectors μ_j as the row vector partitions of F^{-1},

$$F^{-1} \equiv [\,\mu_1 \quad \mu_2 \quad \ldots \quad \mu_n\,]^T \tag{6.29}$$

Then the normal distances d_j, $j = 1, \ldots, n$, to the surface of the n-dimensional parallelepiped prescribed by F are given by

$$d_j = \frac{1}{\|\mu_j\|} \tag{6.30}$$

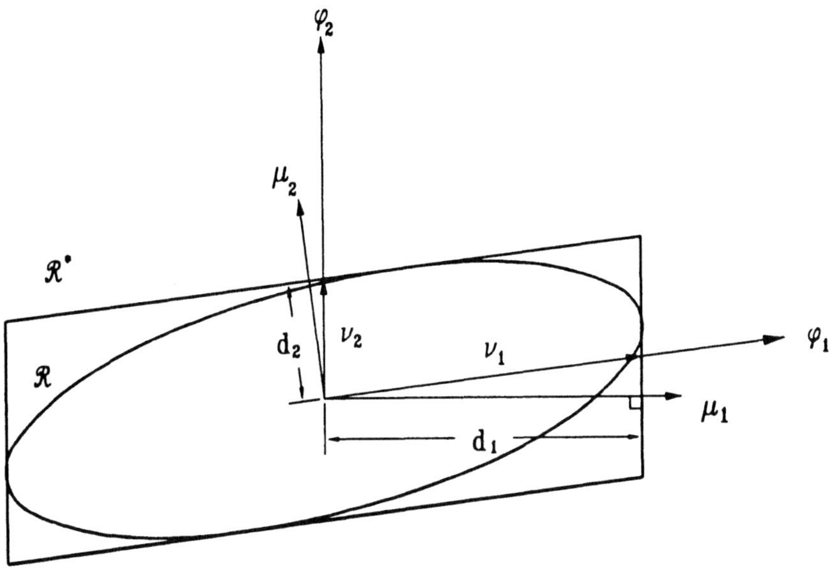

Figure 6.2 Approximation of the Recovery Region and the Degree of Controllability.

The approximation ρ^* of the degree of controllability ρ is then given as the smallest of the normal distances:

$$\rho^* = \min_j d_j \tag{6.31}$$

■

We now specialize the degree of controllability criteria for second-order systems (e.g., flexible structures), and we elect to describe the dynamics using modal coordinates. Consider a flexible spacecraft with dynamic equations expressed in terms of normal modes (for systems with repeated eigenvalues, and for the case of rigid body modes, see Ref. 1), the modal space differential equations are

$$\ddot{\eta}_i + 2\zeta_i\omega_i\dot{\eta}_i + \omega_i^2\eta_i = \Gamma_i^T \mathbf{u}^* \tag{6.32}$$

where Γ_i and $\mathbf{u}^*$ are defined so that each $|u_i{}^*|$ is bounded by the unnormalized actuator strength k_i.

When considering the relative importance of η_i and $\dot{\eta}_i$, it is useful to introduce normalized modal state variables as follows:

$$x_{2i-1} \equiv \frac{\eta_i}{N_i}, \qquad x_{2i} \equiv \frac{\dot{\eta}_i}{\overline{N}_i} \tag{6.33}$$

After some straightforward but tedious algebraic manipulations using Theorems 6.3 and 6.4 (see Ref. 1 for the details of this derivation), the distances from the origin to the boundary of the parallelepiped that corresponds to the ith mode's position and velocity coordinates are obtained as follows:

$$d_{2i-1} = \sigma_i, \qquad d_{2i} = \sigma_i \left(\frac{1 - \zeta_i^2}{\overline{N}_i^{\,2}/(N_i\omega_i)^2 + \zeta_i^2} \right)^{\frac{1}{2}} \tag{6.34}$$

where

$$\sigma_i \equiv \frac{2}{\pi} \frac{\|\Gamma_i\|_k}{N_i\omega_i} \left(\frac{e^{\zeta_i\omega_i T} - 1}{\zeta_i\omega_i\sqrt{1 - \zeta_i^2}} \right), \qquad \|\Gamma_i\|_k = \sum_{r=1}^{m} |\Gamma_{ir}| k_r \tag{6.35}$$

Therefore, the approximate degree of controllability ρ^* is given by

$$\rho^* = \min_i \frac{2}{\pi} \left\{ \frac{\|\Gamma_i\|_k}{N_i\omega_i} \left(\frac{e^{\zeta_i\omega_i T} - 1}{\zeta_i\omega_i} \right) \min\left[\left(1 - \zeta_i^2\right)^{-\frac{1}{2}}, \left(\overline{N}_i^{\,2}/(N_i\omega_i)^2 + \zeta_i^2\right)^{-\frac{1}{2}} \right] \right\} \tag{6.36}$$

Further simplifications can be obtained for the important special cases: (1) lightly damped systems, and (2) undamped systems. These results follow.

Special Case 1: Lightly Damped Systems

Over time intervals much smaller than the ith modal time constant (due to light natural damping), it is reasonable to approximate the modal displacement as $\eta_i \cong \sin(\omega_i t + \phi)$ and the modal velocity as $\dot{\eta}_i \cong \omega_i \cos(\omega_i t + \phi)$. Therefore, normalizations N_i and $\overline{N}_i$ should be approximately related by

$$\overline{N}_i = N_i\omega_i \tag{6.37}$$

Then, Eq. (6.36) becomes

$$\rho^* = \min_i \frac{2}{\pi} \left\{ \frac{\|\Gamma_i\|_k}{N_i\omega_i} \left[\frac{e^{\zeta_i\omega_i T} - 1}{\zeta_i\omega_i(1 + \zeta_i^2)^{\frac{1}{2}}} \right] \right\} \tag{6.38}$$

■

The MATLAB operator we provide for computing this degree of controllability (for lightly damped systems) is as follows:

$$> \mathbf{ind = doc(M, C, K, D, T, Ni, ki)}$$

where Ni, Ki are scale (normalization) factors for modal coordinates and actuators, respectively. Note that Ni, Ki, and final time T are parameters to be chosen carefully, it would be prudent to vary these over a range of reasonable values to study their effects upon the conclusions being drawn. In particular, the final T should be chosen significantly smaller than the smallest modal time constant $(\zeta_i\omega_i)$, since the approximation $T << \min\{\zeta_i\omega_i\}$ is implicit in the Eqs. (6.37) and (6.38).

Special Case 2: System Without Damping

An undamped approximation is a pessimistic (i.e, conservative) assumption often made for designing control laws for the lower frequency modes of large flexible spacecraft. Taking this limit $(\zeta_i \rightarrow 0)$, the approximate degree of controllability [Eq. (6.38)] reduces to the simple form

$$\rho^* = \frac{2T}{\pi} \min_i \left(\frac{\|\Gamma_i\|_k}{N_i\omega_i} \right) \tag{6.39}$$

■

If one is not particularly interested in controlling any specific mode, then a small value can be assigned to the corresponding N_i, and therefore that mode does not affect this approximate degree of controllability.

In this section, we have discussed the main features of one approach (due to Longman et al.) for measuring the degree of controllability. We see that relatively simple formulas can be obtained for lightly damped systems or undamped systems. However, several scale parameters (normalization of coordinates and final time T) must be chosen carefully based on physical considerations and heuristic reasoning. Some applications using these ideas to optimize actuator placement can be found in the recent literature (Refs. 2, 7, 9 and 10). There are several alternative ways to approach these issues, as is evident in the remainder of this chapter.

6.2.2 Balanced Realization

For the purpose of model reduction, Moore (Ref. 12) introduces the concept of *internal balancing*, making use of singular values to measure nearness of the controllability and observability grammians to rank deficiency. The smallest singular value of the controllability grammian (of Moore's balanced system) can be taken as the measure of controllability, which is similar to Friedland's suggestion (Ref. 11) that the condition number of the controllability matrix can be taken as the controllability index. Carrol and Hughes (Ref. 29) discuss a method of controller order reduction that accounts for the dynamics of actuators, and that also makes judicious use of Moore's balancing techniques (Ref. 12). Related concepts of "substate controllability" and "substate observability" are also introduced in Ref. 29.

The original objective of Moore's balanced realization development was model order reduction. Let us consider a system having some state variables that are both uncontrollable and unobservable. Those states can be deleted (in principle) without affecting the system characteristics. Such a *reduced-order system* (with all uncontrollable and unobservable states deleted) is known as a *minimal realization* (Ref. 27). Deleting the less controllable and observable states is one of the more elegant ways to obtain a reduced-order system. To accomplish this order reduction, we are motivated to convert a given system realization to another Moore chose controllability and observability grammians for the purpose of defining observability and controllability measures (Ref. 12). We now present the main features of these developments.

Consider a linear dynamical system

$$\begin{aligned}\dot{\mathbf{x}}(t) &= A\mathbf{x}(t) + B\mathbf{u}(t)\\ \mathbf{y}(t) &= C\mathbf{x}(t)\end{aligned} \tag{6.40}$$

For a systems having asymptotic stability, the controllability and observability grammians are defined as follows:

$$W_c \equiv \int_0^\infty e^{At}BB^Te^{A^Tt}dt, \qquad W_o \equiv \int_0^\infty e^{A^Tt}C^TCe^{At}dt \tag{6.41}$$

These grammians satisfy the following algebraic Lyapunov equations:

$$W_cA^T + AW_c + BB^T = 0, \qquad W_oA + A^TW_o + C^TC = 0 \tag{6.42}$$

Definition 6.5: Balanced System

The system (A_b, B_b, C_b) is said to be *internally balanced* if the controllability and observability grammians are equal and diagonal, i.e.,

$$(W_c)_b = (W_o)_b = \Sigma^2 = \text{diag}\{\sigma_1^2, \ldots, \sigma_n^2\} \tag{6.43}$$

■

Consider the linear coordinate transformation from the current state vector $\mathbf{x}$ to the new *balanced* state vector $\mathbf{x}_b$ generated by

$$\mathbf{x} = T\mathbf{x}_b \tag{6.44}$$

where T is a constant full rank transformation matrix.

It has been proven (Ref. 12) that one can always find T such that the transformed system (A_b, B_b, C_b) is internally balanced [i.e., the state variables (elements of $\mathbf{x}_b$) are equally controllable and observable].

For the case of a general nonsingular T, introducing the transformation of Eq. (6.45) into (6.40) gives the following transformed system:

$$\begin{aligned} \dot{\mathbf{x}}_b &= T^{-1}AT\mathbf{x}_b + T^{-1}B\mathbf{u} \equiv A_b\mathbf{x}_b + B_b\mathbf{u} \\ \mathbf{y} &= CT\mathbf{x}_b \equiv C_b\mathbf{x}_b \end{aligned} \tag{6.45}$$

Utilizing the singular value decomposition, Moore developed an algorithm (Ref. 12) to find a particular T matrix which results in a balanced realization. Laub developed an equivalent but numerically more efficient algorithm (Ref. 30), using Cholesky decomposition and that involves the solution of a symmetric eigenvalue problem. Laub's algorithm follows:

Algorithm 6.1: Balanced Realization Algorithm

Step 1. Solve for W_c and W_o from the following pair of Lyapunov equations:

$$W_cA^T + AW_c + BB^T = 0, \qquad W_oA + A^TW_o + C^TC = 0$$

Step 2. Compute the Cholesky factorization of W_c:

$$W_c = L_cL_c^T$$

Step 3. Form the symmetric positive definite matrix:

$$Y \equiv L_c^TW_oL_c$$

Step 4. Solve the symmetric eigenvalue problem:

$$YU = U\Lambda^2$$

Step 5. Form the transformation matrix $T \equiv L_cU\Lambda^{-\frac{1}{2}}$.

Step 6. Coordinate transformation. Use the linear transformation of Eq. (6.44) to generate the balanced dynamical system having the form of Eqs. (6.45).

■

Note that the above algorithm is applicable only if (1) the controllable modes of system (6.40) are asymptotically stable, (2) the observable modes of system (6.40) are asymptotically stable, and (3) the matrix A in Eq. (6.40) has no eigenvalues with the property

$$\lambda_i[A] + \lambda_j[A] = 0, \qquad \text{for any i and j} \tag{6.46}$$

A MATLAB operator is provided for computing the balanced realization with syntax as follows:

$$> \ [\mathbf{Ab}, \mathbf{Bb}, \mathbf{Cb}, \mathbf{gr}, \mathbf{T}] = \mathbf{balreal}(\mathbf{A}, \mathbf{B}, \mathbf{C})$$

This operator returns balanced versions of (A, B, C), a vector **gr** containing the diagonal of the grammian of the balanced realization, and matrix T, the similarity transformation used to map (A, B, C) into (Ab, Bb, Cb) as indicated in Eqs. (6.45).

To solve problems wherein one or more of the three assumptions for Algorithm 6.1 are not satisfied, Skelton suggests a modified balanced realization algorithm that uses the controllability matrix $\mathcal{C}$ (Eq. 6.2) and observability matrix $\mathcal{O}$ (Eq. 6.6), instead of controllability grammian W_c and observability grammian W_o in algorithm 6.1.

As mentioned earlier, we provide several examples and applications using balanced realization.

Application 6.1: Model Order Reduction

If the system is normalized properly, the diagonal of the joint grammian (of the balanced realization) can be used to reduce the order of the model. Since the joint grammian reflects the combined controllability and observability of individual balanced states, it is reasonable to remove from the dynamical model those states that have a small σ_i. Elimination of those states retains the most important input-output characteristics of the original system. In other words, by discarding state variables associated with the smallest of the σ_i, the most "observable and controllable" portion of the full-order system is retained in the reduced-order model of the dynamical system.

■

Example 6.1

Consider the system matrices

$$A = \begin{bmatrix} 0 & 0 & 0 & -150 \\ 1 & 0 & 0 & -245 \\ 0 & 1 & 0 & -113 \\ 0 & 0 & 1 & -19 \end{bmatrix}, \qquad B = \begin{bmatrix} 4 \\ 1 \\ 0 \\ 0 \end{bmatrix}, \qquad C = [0 \ \ 0 \ \ 0 \ \ 1]$$

The system parameters of the balanced realization are

$$A_b = \begin{bmatrix} -0.4378 & 1.1685 & 0.4143 & 0.0510 \\ -1.1685 & -3.1353 & -2.8352 & -0.3289 \\ 0.4143 & 2.8352 & -12.4753 & -3.2492 \\ -0.0510 & -0.3289 & 3.2492 & -2.9516 \end{bmatrix}, \qquad B_b = \begin{bmatrix} -0.1181 \\ -0.1307 \\ 0.05634 \\ -0.006875 \end{bmatrix}$$

$$C_b = [-0.1181 \quad 0.1307 \quad 0.05634 \quad 0.006875]$$

And the joint grammian matrix is

$$(W_c)_b = (W_o)_b = \Sigma^2 = \text{diag}\{\ 0.0159,\ \ 0.272 \times 10^{-2},\ \ 0.127 \times 10^{-3},\ \ 0.8 \times 10^{-5}\}$$

Figure 6.3 shows the impulse responses of the full-order system (n=4) and the reduced-order systems (n=3 and 2). The fourth balanced state is least controllable and, equivalently, least observable, as seen in joint grammian matrix. The impulse response of the reduced-order system with order 3 (by deleting the fourth balanced state) is almost identical to the impulse response of the full-order system.

■

Note that this example provides a basis for optimism that the balanced coordinates are indeed ranked vis-à-vis controllability and observability, so that attractive reduced-order models can be obtained by simple truncation. On the other hand, there are some missing ingredients in this approach: The key issue that needs to be addressed is some rank ordering of the coordinates that are most likely to be disturbed (in view of the location, direction, and spectral content of expected disturbances), so that, in addition to *observability* and *controllability*, we need to consider *disturbability* in seeking to establish attractive reduced-order models and the assoiciated issues which arise in control actuator location and controller design. These considerations motivate the subsequent developments in this chapter; these issues lie in an arena of active research, and definitive design tools are just beginning to emerge based on simultaneous consideration of controllability, observability, and disturbability.

Application 6.2: Measures of Controllability.

Motivated by some computational experience, such as the results of Application 6.1, perhaps the simplest measure of controllability/observability that we could adopt is the smallest singular value of the joint grammian matrix of the balanced realization. Note that this choice measures both the controllability and observability of the system. We will subsequently see that related measures can be introduced for disturbability.

■

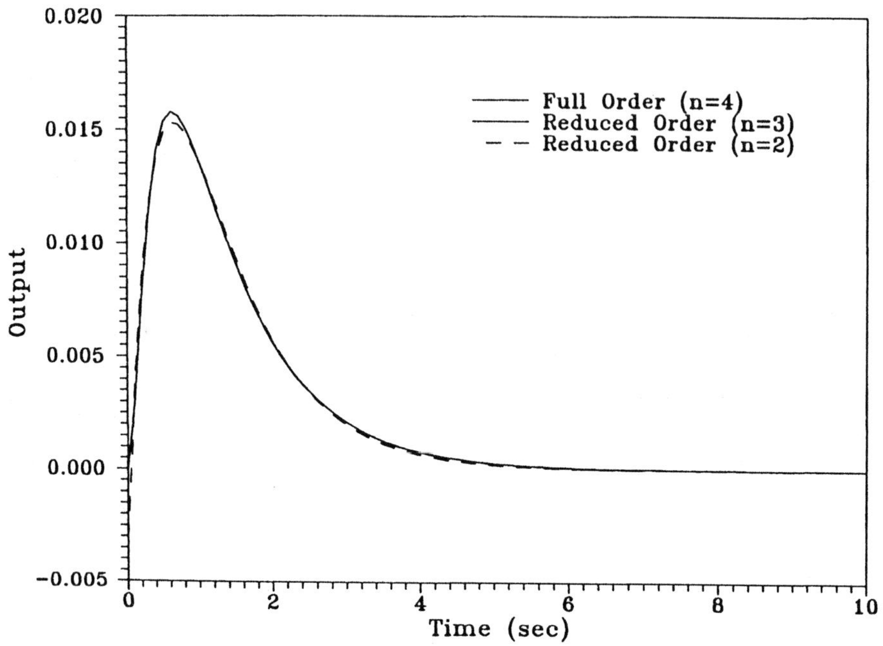

Figure 6.3 Impulse Responses

6.2.3 Measures of Modal Controllability

In this section we review a geometrically appealing measure of modal controllability introduced by Hamdam and Nayfeh (Ref. 13). This measure makes use of the generalized angles between the left eigenvectors of the system matrix A and the columns of the control input influence matrix B for the system described by the triple (A,B,C). It is shown in Ref. 14 that this measure of modal controllability has an interesting connection with Longman's degree of controllability, and it is can also be related to the singular values of Moore's balancing method. This measure of modal controllability provides useful information on each mode's controllability.

Consider the linear dynamical system

$$\dot{\mathbf{x}} = \mathrm{A}\mathbf{x} + \mathrm{B}\mathbf{u} \tag{6.47}$$

where $\mathbf{x} \in \mathrm{R}^n, \mathbf{u} \in \mathrm{R}^m$, and A, B are real matrices with appropriate dimensions.

It is easy to understand the mathematical and physical significance of Hamdan and Nayfeh's measure of modal controllability by considering the PBH eigenvector test. The PBH eigenvector test is presented here without proof.

Theorem 6.5: PBH Eigenvector Test

The ith mode is not controllable using the jth actuator input if and only if

$$\mathbf{q}_i^T[\lambda_i \mathrm{I} - \mathrm{A} \vdots \mathbf{b}_j] = 0^T \tag{6.48}$$

where $\lambda_i, \mathbf{q}_i$ are an eigenvalue/eigenvector pair of A^T, and $\mathbf{b}_j$ is the jth column of the control influence matrix B.

■

It is evident by inspection of the PBH eigenvector test, that any column vector $\mathbf{b}_j$ of the input matrix B cannot be orthogonal to the ith left eigenvector $\mathbf{q}_i$ of A, if the ith mode of the system is controllable. It is apparent that a practical problem exists with a yes/no application of the PBH eigenvector test, considering the case in which vector $\mathbf{b}_j$ is not orthogonal to the ith left eigenvector, but is very nearly so. It is also clear that this situation corresponds to the modal controllability of the ith mode from the jth input being nonzero, but nearly zero. In view of the need to replace the binary PBH test by a graduated measure, and considering the generalized angle θ_{ij} between the two vectors as a measure of orthogonality, Hamdan and Nayfeh proposed simply using $\cos\theta_{ij}$ as a measure of modal controllability. By introducing these ideas based on geometrical interpretations of the PBH eigenvector test, Hamdan and Nayfeh (Ref. 13) introduced the following theorems.

Theorem 6.6 (Hamdan and Nayfeh, Ref. 13)

Assume that the right eigenvectors $\mathbf{p}_i$ and a corresponding set of left eigenvectors $\mathbf{q}_i$ are normalized such that $\mathbf{q}_i^T \mathbf{p}_j = \delta_{ij}$; then,

(1) A measure of controllability of the ith mode from the jth actuator input of the given system is $\cos\theta_{ij}$, where θ_{ij} is the angle between $\mathbf{b}_j$ and $\mathbf{q}_i$.

$$\cos\theta_{ij} = \frac{|\mathbf{q}_i^T \mathbf{b}_j|}{\|\mathbf{q}_i\|\|\mathbf{b}_j\|} \tag{6.49}$$

(2) The Euclidean (root-sum-square) norm of the vector $\mathbf{f}_i$, where $\mathbf{f}_i^T = \mathbf{q}_i^T B/\|\mathbf{q}_i\|$, is a gross measure of controllability ρ_i of the ith mode from all inputs.

■

The scalar gross measure of controllability ρ_i of the ith mode from all actuators is related to the vector measure $\mathbf{f}_i$ of controllability as simply

$$\rho_i = \|\mathbf{f}_i\|, \qquad F = [\cos\Theta]\mathrm{diag}[\|\mathbf{b}_1\|, \|\mathbf{b}_2\|, \ldots, \|\mathbf{b}_m\|] = [\mathbf{f}_1 \;\; \mathbf{f}_2 \;\; \ldots \;\; \mathbf{f}_n]^T \tag{6.50}$$

where $\mathbf{f}_i^T$ is the ith row of F, and $[\cos\Theta]$ is the $m \times n$ matrix of modal controllability measures.

The second measure of the above theorem provides an attractive way to measure each mode's controllability, with full visibility of the controllability contributions of each actuator. When we compute this measure of controllability, however, we should be careful in dealing with coordinate transformations. Since these measures are invariant only under orthonormal coordinate transformations, the measures should be used consistently only after all transformations, including scaling, have been carried out. A MATLAB operator for Hamdan and Nayfeh's measure of modal controllability is provided with the following syntax:

$$> \;[\mathbf{MC}, \mathbf{gMC}] = \mathbf{moc}(\mathbf{A}, \mathbf{B})$$

which returns matrix MC containing the measure of modal controllability of the ith mode from the jth input of the given system, and a vector gMC that includes each mode's gross measure [Eq. (6.50)] of controllability. The results developed for computing Hamdan and Nayfeh's controllability measures are directly applicable to the second-order representations, as well, as reported in Ref. 13.

6.3 Cost Analysis

Skelton et al. (Refs. 16-26) also have made significant contributions to controllability, observability, and a related idea, "modal importance." In all of the subsequent discussion, a quadratic performance index is adopted as the cost function, and *input cost analysis, output cost analysis* (Refs. 16-19), and *component cost analysis* (Refs. 20-26). The input cost analysis (output cost analysis) concept is addressed to decomposition of the quadratic performance metric into contributions from each input (output) and, therefore, the actuator (sensor) that has larger input cost (output cost) contributes more to the system performance. This type of cost decomposition has found application in the selection of the best set of actuators and sensors in engineering control systems (Refs. 16-19). By deleting the actuator (sensor) that has the smallest input cost and repeating this procedure until acceptable performance can just be achieved, a minimal number of actuators and sensors can be obtained.

In Ref. 18, an algorithm to determine weighting matrices for the LQG problem is developed using input/output cost analysis and, in Refs. 19 and 20, the dynamics of sensors and actuators are also included. On the other hand, component cost analysis provides a means to determine the contribution of each element of the system state vector to the overall performance metric (Refs. 20-22), and provides an appealing framework for determining which coordinates to retain and which to discard. By choosing modal coordinates, component cost analysis can be applied to model reduction (Refs. 23-26). In Ref. 25, dynamics of typical actuators are included in the system model and, by truncation of the controller states that have the smallest contribution in the performance metric, reduced-order controllers are obtained. This is a modification/extension of the methods (Refs. 16-19) for selecting the optimal number of actuators. Cost analysis for second-order systems is treated in Refs. 21 and 26, and is applied to model reduction in Refs. 26.

In the following section, we review the basic concepts of component cost analysis and treat the modal cost analysis for second-order mechanical systems, which gives us some measure of each mode's relative importance.

6.3.1 Component Cost Analysis

Component cost analysis was introduced as an attempt to unify the model-reduction and control design problems. Cost is an integral quadratic performance measure that might represent the system error or a norm of the output errors over some interval of time (usually an infinite time is used). The component of the system is taken in an abstract sense, and can be (1) physical subsystems such as the sensors, actuators, or structural members of a large structure, or (2) mathematical subsystems such as one or more modal coordinates, balanced coordinates, or any convenient coordinates. The definition of the component cost is the fraction of the system performance due to the participation of the state variables or control inputs associated with modeling the dynamics of the entire system (with a particular underlying set of reference inputs and initial conditions). By taking the actuators as components of the system, for example, we can obtain the information which represents the relative importance of each actuator in the system performance. By retaining those most significant actuators (components) and removing the least significant actuators from the model, we can determine the number and/or location of actuators.

Component cost analysis follows from the intuitive decomposition of the performance (cost) V into the sum of contributions V_i (component cost) associated with each component and, therefore, component cost must satisfy the following cost decomposition property:

$$V = \sum_{i=1}^{n} V_i \tag{6.51}$$

where n is the number of components. Retaining "component" as an abstract term (i.e., ambiguous) allows us the freedom to consider a component as a physical coordinate, a modal coordinate, a physical substructure, or collections thereof.

In order to understand the concepts of cost decomposition, we consider a simple algebraic example. Let the performance of the system be defined as the quadratic function

$$V = \mathbf{x}^T Q \mathbf{x} = x_1 Q_{11} x_1 + x_1 Q_{12} x_2 + x_2 Q_{12} x_1 + x_2 Q_{22} x_2 \tag{6.52}$$

where x_i are components and the entries of the symmetric matrix Q represent the wights on the contribution of the i-th component.

The total contribution of x_1 involves terms 1 and 2 in Eq. (6.52), and the total contribution of x_2 involves terms 3 and 4. In order to satisfy the cost decomposition property (6.51), the costs of x_1 and x_2 must be defined by

$$\begin{aligned} V_1 &= x_1 Q_{11} x_1 + x_1 Q_{12} x_2 = \left[\mathbf{x}\mathbf{x}^T Q\right]_{11} \\ V_2 &= x_2 Q_{22} x_2 + x_2 Q_{12} x_1 = \left[\mathbf{x}\mathbf{x}^T Q\right]_{22} \end{aligned} \tag{6.53}$$

where the ith component cost V_i represents the contribution of state variable x_i to the total cost function V.

The following mathematical definition of component cost satisfies the cost decomposition property and yields the same results, Eq. (6.53), for the above example.

Definition 6.6: Component Cost

The definition of the component cost is the fraction of the system performance due to each component of the system. The component cost V_i associated with the component x_i can be related to the gradient of the total cost as follows:

$$V_i \equiv \frac{1}{2} \frac{\partial V}{\partial x_i} x_i \tag{6.54}$$

■

Consider the multi-input, multi-output system

$$\dot{\mathbf{x}}(t) = A\mathbf{x}(t) + B\mathbf{u}(t) \tag{6.55}$$

where $\mathbf{x} \in R^n$ and $\mathbf{u} \in R^m$.

The scalar function

$$\int_0^\infty \mathbf{y}^T(t) Q \mathbf{y}(t) dt \tag{6.56}$$

is widely used as an ("output signal energy") measure of system performance where the vector $\mathbf{y}(t)$ is composed of those output error variables that are of importance to the designer. The unit impulse response with zero initial conditions is also commonly used to discuss the transient response of the system. This impulse response can be computed assuming from one or more specified disturbance input locations. By considering the above two observations, we can take the cost function that represents the system performance as follows:

$$V = \sum_{i=1}^{m} \int_0^{\infty} \mathbf{y}_d^{i\,T}(t) Q_v \mathbf{y}_d^i(t) dt \tag{6.57}$$

$$\mathbf{y}_d^i(t) = C_d \mathbf{x}^i(t) \tag{6.58}$$

where $\mathbf{x}^i(t)$ is the response due to the unit impulse input $u_i(t) = \delta(t)$ (with $u_j(t) = 0$, $i \neq j$) applied at $t = 0$ with zero initial conditions, Q_v is a weighting matrix, and the sensor (output) influence matrix C_d is defined so that $\mathbf{y}_d(t)$ models an important output for the design objectives. Usually Q_v is simply taken as a diagonal matrix that effectively rank-orders the relative physical importance of the elements of $\mathbf{y}_d$. We may use the contribution of each modal state variable to the cost function as a measure of that mode's relative importance in the system output error performance. The following developments leading to the modal decomposition of this cost function were originally presented in Refs. 22 and 26.

Theorem 6.7: Cost Analysis

The cost function V and the contribution V_{x_i} of the state variable x_i in the cost function V can be computed as follows:

$$V = \text{trace}\{Q_v C_d X C_d^T\} \tag{6.59}$$

$$V_{x_i} = [X C_d^T Q_v C_d]_{ii} \tag{6.60}$$

where X is the controllability grammian that satisfies the following Lyapunov equation:

$$XA^T + AX + BB^T = 0 \tag{6.61}$$

■

Proof of the Theorem 6.7

Using Eq. (6.58) in Eq. (6.57), we obtain the following equation:

$$V = \sum_{i=1}^{m} \int_0^{\infty} \mathbf{x}^{i\,T}(t) C_d^T Q_v C_d \mathbf{x}^i(t) dt \tag{6.62}$$

The unique solution of Eq. (6.55) due to the unit impulse input (i-th input only) applied at t = 0 with zero initial conditions is given by

$$\mathbf{x}^i(t) = e^{At}\mathbf{x}(0) + \int_0^t e^{A(t-\tau)}\mathbf{B}\mathbf{u}(\tau)d\tau = e^{At}\mathbf{b}_i \tag{6.63}$$

Note that the ith integral in Eq. (6.62) can be manipulated as

$$\begin{aligned}\int_0^\infty \mathbf{x}^{i^T}(t)C_d^T Q_v C_d \mathbf{x}^i(t)dt &= \text{trace}\{\int_0^\infty Q_v C_d \mathbf{x}^i(t)\mathbf{x}^{i^T}(t)C_d^T dt\} \\ &= \text{trace}\{Q_v C_d[\int_0^\infty \mathbf{x}^i(t)\mathbf{x}^{i^T}(t)dt]C_d^T\}\end{aligned} \tag{6.64}$$

Substitution of Eqs. (6.63) and (6.64) into Eq. (6.62) gives

$$\begin{aligned}V &= \text{trace}\{Q_v C_d[\sum_{i=1}^m \int_0^\infty e^{At}\mathbf{b}_i\mathbf{b}_i^T e^{A^T t}dt]C_d^T\} \\ &= \text{trace}\{Q_v C_d[\int_0^\infty e^{At}BB^T e^{A^T t}dt]C_d^T\}\end{aligned} \tag{6.65}$$

We know that if the matrix

$$X \equiv \int_0^\infty e^{At}BB^T e^{A^T t}dt \tag{6.66}$$

exists, it satisfies the following Lyapunov equation:

$$XA^T + AX + BB^T = 0 \tag{6.67}$$

Making use of Eq. (6.66) in Eq. (6.65), the total quadratic cost function can be computed as

$$V = \text{trace}\{Q_v C_d X C_d^T\} \tag{6.68}$$

The decomposition of the total cost V into the contribution of state variable x_i is determined by (Definition 6.6)

$$V_{x_i} = \frac{1}{2}\frac{\partial V}{\partial x_i}x_i \tag{6.69}$$

and we know that V_{x_i} is finite if there are no unstable modes.

Making use of Eqs. (6.57), (6.58) in Eq. (6.69), V_{x_i} can be written as

$$V_{x_i} = \frac{1}{2}\sum_{k=1}^{m}\int_0^\infty \left[\frac{\partial}{\partial x_i^k}\{x^{k^T} C_d^T Q_v C_d x^k\} x_i^k\right] dt \tag{6.70}$$

Note the following matrix identity

$$\frac{\partial}{\partial x_i}\{x^T C_d^T Q_v C_d x\} x_i = 2[x x^T C_d^T Q_v C_d]_{ii} \tag{6.71}$$

where $[\cdot]_{ii}$ is the (i,i) element of matrix $[\cdot]$. Further, using Eq. (6.71) in Eq. (6.70),

$$V_{x_i} = \sum_{k=1}^{m}\int_0^\infty [x^k(t) x^{k^T}(t) C_d^T Q_v C_d]_{ii} dt \tag{6.72}$$

Since only $x^k(t)$ is time-varying in the integrand of Eq. (6.72), the following manipulations remove all constants from under the integration signs:

$$\begin{aligned}\sum_{k=1}^{m}\int_0^\infty [x^k(t) x^{k^T}(t) C_d^T Q_v C_d]_{ii} dt &= \sum_{k=1}^{m}\int_0^\infty \sum_{j=1}^{n}\left\{[x^k(t) x^{k^T}(t)]_{ij} [C_d^T Q_v C_d]_{ji}\right\} dt \\ &= \sum_{j=1}^{n}\left\{[\sum_{k=1}^{m}\int_0^\infty x^k(t) x^{k^T}(t) dt]_{ij} [C_d^T Q_v C_d]_{ji}\right\}\end{aligned} \tag{6.73}$$

We know from Eqs. (6.63) and (6.66) that

$$\sum_{k=1}^{m}\int_0^\infty x^k(t) x^{k^T}(t) dt = X \tag{6.74}$$

Substitution of Eq. (6.74) into Eq. (6.73) gives

$$V_{x_i} = \sum_{j=1}^{n}\{[X]_{ij}[C_d^T Q_v C_d]_{ji}\} \tag{6.75}$$

Finally, manipulation of Eq. (6.75), leads to a convenient formula for computing the component cost:

$$V_{x_i} = [X C_d^T Q_v C_d]_{ii} \tag{6.76}$$

Q.E.D.

The unit impulse response (with zero initial conditions) is used for the system performance in Eq. (6.57). A specific initial condition response (with zero control or force input) is also widely used to study the transient response of the system. Analogous cost function and component cost formulas for initial condition response (with zero input) may also be obtained. In order to eliminate the dependence of the cost function on the initial conditions, we can average the cost functions over a distribution of initial conditions. This modifies Eq. (6.61) such that the initial condition covariance matrix, [that is, the expectation of $\mathbf{x}(0)\mathbf{x}(0)^{\mathrm{T}}$], replaces the symmetric matrix BB^{T}.

We are seeking each mode's contribution to the cost function, which Skelton calls the "modal cost." When the modal coordinates are used as the elements of the state vector, it is obvious that each state's contribution becomes identically the modal cost. For a formulation of the system dynamics that uses "physical" (or "configuration") coordinates in the state vector, on the other hand, the modal cost can be obtained through the use of a modal coordinate transformation. In the following section, we discuss modal cost analysis for second order dynamical systems.

6.3.2 Modal Cost Analysis for Mechanical Second Order Systems

In the study of vibrating mechanical systems, we often encounter systems of n second-order equations having the form

$$\mathrm{M}\ddot{\mathbf{x}} + \mathrm{C}\dot{\mathbf{x}} + \mathrm{K}\mathbf{x} = \mathrm{D}\mathbf{u} \tag{6.77}$$

where $\mathbf{x} \in \mathrm{R}^n$ and $\mathbf{u} \in \mathrm{R}^m$ are configuration and control vectors, respectively, M is an $n \times n$ positive definite symmetric mass matrix, C is an $n \times n$ positive semidefinite symmetric structural damping matrix, K is an $n \times n$ positive semidefinite stiffness matrix, D is an $n \times m$ control influence matrix, and $(\dot{\ })$ represents differentiation with respect to time.

When modal coordinates are used, three well-known benefits are realized: (1) the computational processes become simplified because we have uncoupled (open-loop) equations, (2) each state's contribution to the cost function corresponds to the modal cost, and (3) the inherent "frequency ordering" of modal coordinates is often of invaluable importance in order reduction processes.

We introduce the modal coordinate transformation as follows:

$$\mathbf{x}(\mathrm{t}) = \Phi \underline{\eta}(\mathrm{t}) \tag{6.78}$$

where Φ is the open-loop modal matrix obtained by solving the eigenvalue problem associated with the nominal mass and stiffness matrices, and $\underline{\eta}(\mathrm{t})$ is the $n \times 1$ vector of modal coordinates.

Then the transformed equation of motion (closed-loop) becomes

$$\tilde{M}\ddot{\underline{\eta}} + \tilde{C}\dot{\underline{\eta}} + \tilde{K}\underline{\eta} = \tilde{D}\mathbf{u} \tag{6.79}$$

and where the transformed (*modal space*) mass, damping, stiffness, and control influence matrices are, respectively, given by:

$$\tilde{M} = \Phi^T M \Phi = I$$
$$\tilde{C} = \Phi^T C \Phi = \text{diag}(2\zeta_1\omega_1, 2\zeta_2\omega_2, \ldots, 2\zeta_n\omega_n)$$
$$\tilde{K} = \Phi^T K \Phi = \text{diag}(\omega_1^2, \omega_2^2 \ldots, \omega_n^2)$$
$$\tilde{D} = \Phi^T D.$$

If alternative normalizations of the eigenvectors are used, then the scalings of $\tilde{M}$, $\tilde{K}$, $\tilde{C}$, and $\tilde{D}$ are affected accordingly. The diagonal structure of $\tilde{C}$ requires that C be a linear combination of M and K; we make this traditional assumption (modal damping) for the sake of convenience. A modest generalization of the following developments are required to accommodate more general damping models. We note that in practice, the modal damping factors are almost always assigned values based upon experimental results rather than deriving a fully populated C matrix based on modeling assumptions.

For most modern control applications, the system dynamics are usually transformed to a system of first-order differential equations. For the case that the system is initially represented as a system of second-order equations in the form of Eq. (6.77), there are several routes to obtain a first-order state space model. One of the most useful is to define the state vector in terms of modal coordaintes. Let us introduce the $2n \times 1$ modal state vector

$$\mathbf{z} = \begin{Bmatrix} \underline{\eta} \\ \dot{\underline{\eta}} \end{Bmatrix} \tag{6.80}$$

Equation (6.79) can be written as the first-order system

$$\dot{\mathbf{z}} = A\mathbf{z} + B\mathbf{u} \tag{6.81}$$

where

$$A = \begin{bmatrix} 0 & I \\ -\tilde{K} & -\tilde{C} \end{bmatrix}, \qquad B = \begin{bmatrix} 0 \\ \tilde{D} \end{bmatrix} \tag{6.82}$$

In order to evaluate the modal cost for the system, we take the following system performance measure [Eq. (6.57)] as the cost function:

$$V = \sum_{i=1}^{m} \int_0^{\infty} \mathbf{y}_d^{i\,T}(t) Q_v \mathbf{y}_d^i(t) dt \tag{6.83}$$

where $y_d^i(t)$ is the output response due to the unit impulse input $u_i(t) = \delta(t)$ (with $u_j(t) = 0$, $i \neq j$) applied at $t = 0$ with zero initial conditions, and

$$\mathbf{y}_d(t) = \begin{bmatrix} C_{dx} & 0 \\ 0 & C_{d\dot{x}} \end{bmatrix} \begin{Bmatrix} \mathbf{x}(t) \\ \dot{\mathbf{x}}(t) \end{Bmatrix} \equiv C_d \begin{Bmatrix} \mathbf{x}(t) \\ \dot{\mathbf{x}}(t) \end{Bmatrix} \tag{6.84}$$

Note that choosing a suitable weight matrix Q_v, $\mathbf{y}_d(t)$ becomes a vector of physically important variables. For the output weighting matrix, we adopt the following structure:

$$Q_v = \begin{bmatrix} Q_x & 0 \\ 0 & Q_{\dot{x}} \end{bmatrix} \tag{6.85}$$

If the modal coordinate transformation is introduced, the output vector $\mathbf{y}_d(t)$ can be linearly related to the modal state vector as

$$\mathbf{y}_d(t) = C_d \begin{Bmatrix} \Phi \underline{\eta}(t) \\ \Phi \dot{\underline{\eta}}(t) \end{Bmatrix} = \begin{bmatrix} C_{dx}\Phi & 0 \\ 0 & C_{d\dot{x}}\Phi \end{bmatrix} \begin{Bmatrix} \underline{\eta}(t) \\ \dot{\underline{\eta}}(t) \end{Bmatrix} \equiv C_{d\eta} \begin{Bmatrix} \underline{\eta}(t) \\ \dot{\underline{\eta}}(t) \end{Bmatrix} \tag{6.86}$$

Therefore matrix C_d in Eqs. (6.59) and (6.60) must be replaced by $C_{d\eta}$ to evaluate the total cost and modal cost in the state space. Another advantage of using modal coordinates is that we can take the analytical solution of the $2n \times 2n$ Lyapunov equation (6.61), developed in Ref. 26 and also included in Appendix E:

$$X = \begin{bmatrix} X_{\eta\eta} & X_{\eta\dot{\eta}} \\ X_{\eta\dot{\eta}}^T & X_{\dot{\eta}\dot{\eta}} \end{bmatrix} \tag{6.87}$$

where the elements of $n \times n$ block matrices are

$$\begin{aligned}
[X_{\eta\eta}]_{ij} &= \frac{2(\zeta_i\omega_i + \zeta_j\omega_j)}{\Delta}[\tilde{D}\tilde{D}^T]_{ij}, & i,j = 1,2,\ldots,n \\
[X_{\eta\dot{\eta}}]_{ij} &= \frac{\omega_i^2 - \omega_j^2}{\Delta}[\tilde{D}\tilde{D}^T]_{ij}, & i,j = 1,2,\ldots,n \\
[X_{\dot{\eta}\dot{\eta}}]_{ij} &= \frac{2\omega_i\omega_j(\zeta_i\omega_j + \zeta_j\omega_i)}{\Delta}[\tilde{D}\tilde{D}^T]_{ij}, & i,j = 1,2,\ldots,n
\end{aligned} \tag{6.88}$$

with

$$\Delta = (\omega_i^2 - \omega_j^2)^2 + 4\omega_i\omega_j(\zeta_i\omega_i + \zeta_j\omega_j)(\zeta_i\omega_j + \zeta_j\omega_i) \tag{6.89}$$

Note that the ith mode's modal cost consists of two parts: the modal cost of the ith mode's displacement and the modal cost of the corresponding mode's velocity,

$$V_i = V_{\eta_i} + V_{\dot{\eta}_i} \tag{6.90}$$

where V_i is the ith mode's cost and $\underline{\eta}_i$ is the modal coordinate. V_{η_i} and $V_{\dot{\eta}_i}$ can be obtained by substituting Eqs. (6.85)-(6.89) into Eq. (6.60) as follows:

$$V_{\eta_i} = [X_{\eta\eta}\Phi^T C_{dx}^T Q_x C_{dx} \Phi]_{ii}, \qquad V_{\dot{\eta}_i} = [X_{\dot{\eta}\dot{\eta}}\Phi^T C_{d\dot{x}}^T Q_{\dot{x}} C_{d\dot{x}} \Phi]_{ii} \tag{6.91}$$

By judicious selection (problem-dependent, obviously) of the physically important variable vector $\mathbf{y}_d(t)$, the modal cost V_i represents the contribution of the ith mode to the total cost V. Also the underlying disturbance input location(s) can be altered (by varying the specification of the D matrix) as appropriate to reflect the system response to a variety of input or disturbance assumptions. Thus the normalized modal cost (V_i/V) provides a measure of each mode's relative contribution to the system performance, under particular disturbance assumptions.

Note that a MATLAB operator for computing modal cost for second order systems has been developed with the following syntax:

$$> \ [\mathbf{V}, \mathbf{Vi}] = \mathbf{modcst2}(\mathbf{M}, \mathbf{C}, \mathbf{K}, \mathbf{D}, \mathbf{Phi}, \mathbf{Cx}, \mathbf{Cxd}, \mathbf{Qx}, \mathbf{Qxd})$$

where Phi is the modal matrix in Eq. (6.78), Cx and Cxd are matrices, Eq. (6.84), for making the output vector $\mathbf{y}_d$ physically important, Qx and Qxd are weighting matrices, Eq. (6.85), and MATLAB operator **modcst2** returns the total cost V and the vector Vi, which contains all n of the modal costs.

From the recent work of Longman et al. (Refs. 1-10), Moore (Ref. 12), Hamdan and Nayfeh (Ref. 13), Skelton et al. (Refs. 16-26), Carrol and Hughes (Ref. 29), Kim and Junkins (Ref. 31), and Hurtado, Rahman, and Junkins (Ref. 36), it is evident that the concepts of controllability/observability are central issues and are closely related to the recent theoretical and computational research for model order reduction.

To investigate the modal cost analysis further, consider a system in the form

$$M\ddot{\mathbf{x}} + C\dot{\mathbf{x}} + K\mathbf{x} = D\mathbf{u} + D_w\mathbf{w} \tag{6.92}$$

where $\mathbf{w}(t)$ is a vector of disturbances.

If disturbances are likely to act on the structure in some known way, then *modal disturbability* can be defined analogously to modal controllability and, of course, the control influence matrix D should be replaced by the appropriate disturbance influence matrix D_w. More specifically, if disturbances enter only at actuator locations, then the disturbance influence matrix is the same as the control influence matrix, and the normalized modal cost (V_i/V) provides a measure of each mode's disturbability. Note that the disturbances induced by the actuators or other disturbance sources can be considered one at a time or simultaneously by zeroing appropriate entries in the influence matrix.

If the dynamics of the actuators and/or disturbances are accounted for, then the control vector $\mathbf{u}(t)$ and/or the disturbance vector $\mathbf{w}(t)$, can be contribute additional partitions to an augmented state vector, and the system dynamical model modified appropriately to reflect the sensor/actuator/disturbance dynamical models. For this case, for example, the commands to the actuators would represent the new inputs, and the actual forces and moments generated would would be 'state' variable functions of the actuator dynamical model. In this more realistic situation, making use of component cost analysis (where components are states which may now include control states or disturbance states), we can find optimal locations and a minimal number of sensors or actuators (Ref. 25). Special cases can obviously be considered for which subsets of the inputs, actuators, and sensors are prescribed, to reflect practical design considerations.

6.4 Case Study: Measures of Controllability for Actuator Placement

For a system having a given actuator placement configuration, we can determine each mode's gross measure of controllability and modal cost by using the approach presented in the previous sections. We now address the issue of how to combine the modal cost ideas with a gross measure of modal controllability, to define a new measure of controllability. We mainly address the issue of actuator placement, but these ideas have broader implications. The least controllable mode is often considered to be critical for the controllability of the system. Motivated by Ref. 11 and classical stability margin ideas regarding eigenvalue placement, we could conjecture the following rule: *The smallest value of the gross measures of controllability ρ_j is the controllability index for a given system.* We have found this intuitively attractive rule to be incorrect for most practical applications. The reason for this pitfall follows. Consider a system for which the *k-th* mode of the system is the least controllable. Then, using the conjectured rule, the *k-th* mode's gross measure of controllability ρ_k (i.e., the minimum ρ_k) would be taken as the system's controllability index. However, if the *k-th* mode does not participate significantly in the important physical outputs of the system, and if it is weakly disturbed initially, then using only the controllability of such a relatively unimportant and undisturbed mode to characterize the overall system controllability is clearly not suitable.

An index is needed that incorporates more information in order to properly measure the controllability of the system. In particular, some measure of the "relative importance of each mode" to the overall system performance is needed to weight modal controllability measures. It also seems important to consider "how disturbable" each mode is. One appealing approach is to combine the modal cost ideas of Skelton et al. (Refs. 22, 26) with the modal controllability ideas of Hamdan and Nayfeh (Ref. 13).

In view of these heuristic considerations, we introduce the following (α) index as a weighted measure of controllability (Ref. 31):

$$\alpha = \sum_{\mathrm{i}=1}^{n} \frac{\mathrm{V_i}}{\mathrm{V}} \rho_\mathrm{i}^2 \tag{6.93}$$

where

α denotes the weighted disturbability/controllability index

$\mathrm{V_i}$ is the ith mode's component cost due to disturbance input

ρ_i is the ith mode's gross measure of modal controllability [Eq. (6.50)]

Qualitatively, α is a measure of "output controllability" which reflects modal controllability, disturbability, and the modal participation of all modeled modes in the physically important cost function. "Controllability," in some sense, has been included twice, however, since each $\mathrm{V_i}$ is affected by B and is itself a measure of the ith mode's response. However, notice that B can be replaced by a disturbance influence matrix that reflects the locations where known disturbance sources act on the structure, so $\mathrm{V_i/V}$ is more generally viewed as the fractional participation of the specifically disturbed modes in important pyhsical outputs, i.e., an output disturbability measure. To study the utility of the α index [Eq. (6.93)], we consider a particular problem. To evaluate the possible coupling between the controllability index and the method used to design the feedback law, we evaluate two controllers: (i) a robust output feedback controller (Chapter 7), and (ii) a linear quadratic regulator. To provide some insight, numerical results are presented using these two controllers (for several actuator configurations), along with corresponding α controllability index values.

We adopt for an illustrative example, a 60 degree-of-freedom model of a three input, six output flexible structure (a cantilevered 5ft $\times$ 5ft aluminum grid). A finite element analysis of the grid was performed (Ref. 32). Finite element nodes were placed at each of the 20 joints on the grid, and each substructure of the grid was modeled using beam elements. Three degrees of freedom were considered at each node (one normal displacement, a transverse rotation, and a vertical rotation, appropriate for modeling out-of-plane vibration normal to the nominal undeformed vertical plane of the grid). Figure 6.4 shows the flexible grid experimental configuration. The material properties are listed in Table 6.1.

Feedback control torques are provided by three reaction wheel actuators. The actuator axes (about which positive or negative control torques are applied) lie in the plane of the grid as indicated by the arrows, for configuration No. 4, in Figure 6.4. The actuator mass is significant for such a lightweight structure, so in order to

Table 6.1 Material Properties of the Flexible Grid Model

EI	$3255.2\ \text{lb} \cdot \text{in.}^2$	$9.34982\ \text{N} \cdot \text{m}^2$
GJ	$4943.0\ \text{lb} \cdot \text{in.}^2$	$14.18550\ \text{N} \cdot \text{m}^2$
ρA	$0.0000648\ \text{lb} \cdot \text{s}^2/\text{in.}^2$	$0.44678\ \text{kg/m}$
ρJ	$0.0000217\ \text{lb} \cdot \text{s}^2$	$0.0000965\ \text{kg} \cdot \text{m}$

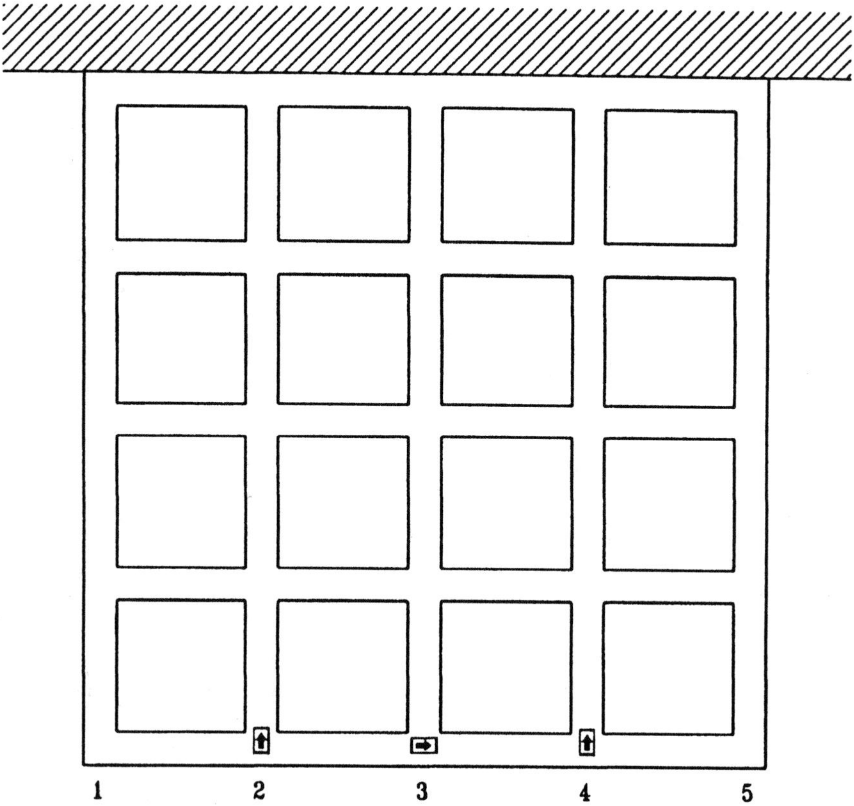

Figure 6.4 Flexible Grid Structure.

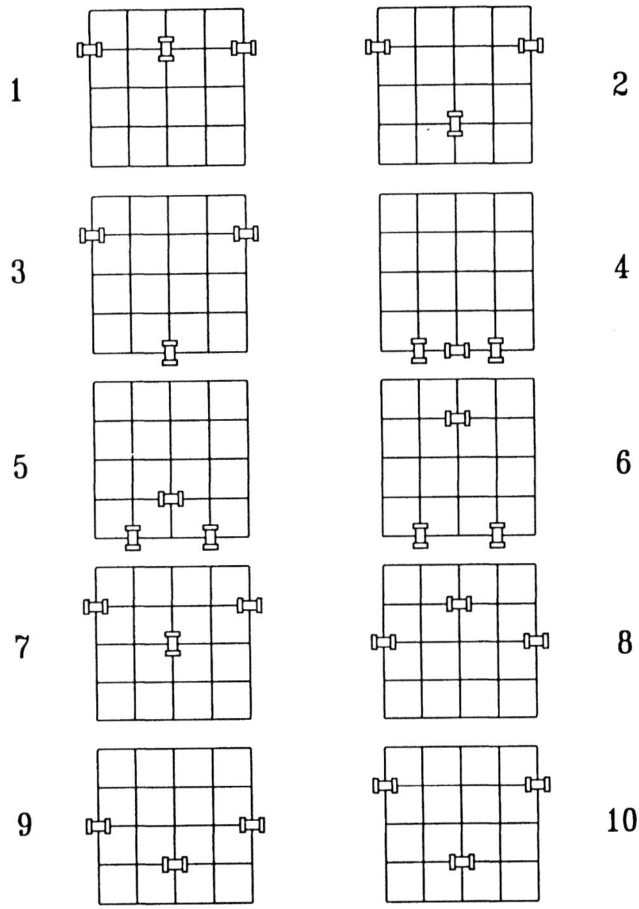

Figure 6.5 Ten Sets of Actuator Locations and Orientations.

preserve symmetry of the grid structure, only symmetrical locations and directions of the actuators are admitted in the present discussion. The admissible actuator locations considered are the 20 grid joints and the torque axis of each actuator is permitted to be either vertical or horizontal.

The locations and the torque directions of the actuators of ten configurations are displayed in Fig. 6.5. In all cases, the actuator torque axis and the active sensor axis are assumed collocated. Since the controllers are designed by using a reduced-order model, there are no additional difficulties introduced by a reduced-order model to evaluate the utility of the α index. For simplicity, we set the disturbance influence matrix equal to the control influence matrix for this example; obviously there are ten different matrices corresponding to the ten actuator configurations of Fig. 6.5. Some physically important variable vector must be selected to construct the cost function. For this illustration, the normal displacement of the lower-left corner is taken to be $\mathbf{y}_d$ in Eq. (6.58). This selection is ad hoc and simply for illustration purposes; more generally, a mission function such as line-of-sight error will dictate the definition of $\mathbf{y}_d$. Even this simple grid structure leads to a fairly high order system, and the well known fact that the finite element method produces a significant number of high-frequency, poorly-approximated modes, provides further practical motivation for order reduction. Motivated by this convergence issue and also to keep the order of the system low for computational purposes, it is desirable to use a reduced-order model to design the controllers. While we do not explicitely address the details of the order reduction here, we assume that a satisfactory subset of the modal coordinates have been selected for control design purposes; we denote the reduced-order dynamical model as

$$\tilde{\mathrm{M}}\ddot{\underline{\eta}}_1 + \tilde{\mathrm{C}}\dot{\underline{\eta}}_1 + \tilde{\mathrm{K}}\underline{\eta}_1 = \tilde{\mathrm{D}}\mathbf{u} \tag{6.94}$$

For the sake of illustration, the particular reduced-order model adopted is the first ten (lowest frequency) modes. Structural damping was also considered while designing the controller, we assumed a modal damping factor of 0.1 %.

6.4.1 Control Design 1 : Symmetric Output Feedback Design

For the above model, we invoke a symmetric output feedback controller design method (Refs. 33, 34) to move the system's first three closed loop eigenvalues to a desired region of the left half-plane and, subject to this condition, minimize a robustness measure (e.g., the sensitivity of the eigenvalues with respect to variation of uncertain parameters). As a measure of eigenvalue sensitivity with respect to system parameter variations, we use the condition number of the closed-loop eigenvector matrix.

The symmetric or "structural" output feedback form of the control law (collocated sensors and actuators) is given by

$$\mathbf{u} = -(\mathrm{G}_1\tilde{\mathrm{D}}^{\mathrm{T}}\underline{\eta}_1 + \mathrm{G}_2\tilde{\mathrm{D}}^{\mathrm{T}}\dot{\underline{\eta}}_1) \tag{6.95}$$

where G_1 and G_2 are $m \times m$ positive definite, symmetric, fully populated gain matrices, and $\underline{\eta}_1$ is the retained subset of modal coordinates.

For any/all choices for the gain matrices from this stable family (i.e., the set of all positive definite G_i matrices), asymptotic stability of the closed-loop system is guaranteed in the Lyapunov sense. Of course, we do not choose the G_i at random. Positive definite gain matrices are parameterized by introducing Cholesky decomposition, and the gain parameter vector is defined as $\mathbf{p}_{gain}$, the distinct elements of the Cholesky factors of G_1 and G_2 (Refs. 33, 34). With the above formulation, our gain design problem (hereafter 'control design 1') becomes a nonlinear parameter optimization problem as follows:

$$\text{Minimize} \qquad \mathcal{K}(\Phi^c(\mathbf{p}_{gain})) \tag{6.96}$$

subject to closed-loop eigenvalue placement constraints. As an example, we adopt the frequency constraints

$$\omega_i^o - \Delta\omega_i \le \omega_i^c \le \omega_i^o + \Delta\omega_i, \qquad \Delta\omega_i = 1\% \text{ of } \omega_i^o, \qquad i = 1, 2, 3$$

and the modal time constant constraints

$$\omega_i^c \zeta_i^c \ge T_i, \qquad i = 1, 2, 3$$

where

$$\omega_i^c(\mathbf{p}_{gain}) = \text{Im}(\lambda_i), \qquad \zeta_i^c(\mathbf{p}_{gain}) = -\frac{\text{Re}(\lambda_i)}{|\lambda_i|},$$

$\mathcal{K}(\Phi^c(\mathbf{p}_{gain}))$ is the condition number of the closed loop eigenvector matrix, and the superscripts c and o denote the closed-loop and the open loop values.

For this example, we take the modal time constants as T_1=0.2, T_2=0.2, and T_3=0.25. The above nonlinear constrained optimization problem can be solved by homotopic nonlinear programming (see Refs. 34, 35 and Chapter 7) in conjunction with a minimum norm gain differential correction strategy, for the ten actuator configurations. The computed output measures [α of Eq. (6.93)] of the system controllability for the ten different configurations and the results of the controllers designed for the corresponding configurations are summarized in Table 6.2. The actuator configuration number is an arbitrary reference number we assigned a priori for each configuration.

The output measure of controllability is conjectured to properly weigh both modal controllability and participation in the output cost function. That means, we expect a configuration possessing a large α index to be more output controllable than one having the small α. We should further expect that more controllable

Table 6.2 The Alpha Measure of Controllability (Output Feedback Controller)

Config. No.	Controllability (α)	$\|\tilde{G}_1\|_f$	$\|\tilde{G}_2\|_f$	Cond. No.
9	0.00333	0.306	6.514	129.864
8	0.00234	0.142	7.434	131.611
10	0.00208	0.444	7.691	161.167
4	0.00158	1.600	7.732	128.707
5	0.00153	6.804	9.827	130.363
6	0.00130	1.595	31.143	147.774
3	0.00123	6.745	19.920	159.009
2	0.00102	13.379	16.483	140.672
7	0.00084	21.576	16.411	140.715
1	0.00078	22.160	25.379	141.402

configurations should require less energy, at least on the average, to control the system and, therefore, we might anticipate the associated optimal controller to have smaller gain matrices. Based upon these heuristic observations, for example, if α [Eq. (6.93)] is an appropriate measure, we should anticipate the large-α designs to correspond to smaller control gains.

As evident in Table 6.2, the magnitudes of the total gain norms of the first four configurations (configuration number 9, 8, 10, and 4) are indeed significantly smaller than those of the remaining six configurations. In fact, designs 9, 8, 10, and 4 appear to be correctly ranked based on the fact that decreasing α corresponds to an increasing control gain norm. All of the above designs also have closed-loop modal matrices with relatively small condition numbers, and it is significant that the small condition numbers appear to be negatively well-correlated with large α values. These results imply that increasing α is highly correlated to the actual controllability and robustness of the system and can apparently serve as a good relative indicator of the desirability of a competing actuator configurations.

6.4.2 Control Design 2: Linear Quadratic Regulator Design

The symmetric output feedback controller design process used in controller design 1 (and implicit in Table 6.2) is perhaps unfamiliar to many readers. Since this is true and in order to confirm the usefulness of the α index, we have also designed control laws for each of the ten actuator configurations using the well-known linear quadratic regulator (LQR), and we compare the results with the α indices as well as the symmetric output feedback designs of Table 6.2. We adopt the following performance index for LQR (hereafter, 'control design 2'):

$$J = \int_0^\infty (\mathbf{z}^T Q \mathbf{z} + \mathbf{u}^T R \mathbf{u}) dt \tag{6.97}$$

where

$$\mathbf{z} = \begin{pmatrix} \underline{\eta}_1 \\ \dot{\underline{\eta}}_1 \end{pmatrix}, \qquad Q = \begin{bmatrix} \Omega & 0 \\ 0 & I \end{bmatrix}, \qquad R = \kappa I$$

and $\Omega = \text{diag}[\omega_1^2, \omega_2^2, \ldots, \omega_{n_r}^2]$, and the scalar κ is chosen in order to satisfy the conditions $\omega_{d_i}^c \zeta_i^c \geq T_i$, $i = 1, \ldots, 3$ with $T_1 = 0.2$, $T_2 = 0.2$, and $T_3 = 0.25$.

The above performance index is an energy type since the first and second terms in the performance index correspond to the state energy and the control energy, respectively. Note that κ is a weight-tuning factor used to place the first three closed-loop eigenvalues to achieve satisfaction of the desired time constant constraints ($\omega_{d_i}^c \zeta_i^c \geq T_i$, $i = 1, 2, 3$). Without judiciously assigning the tuning factor κ, we cannot control the position of the closed-loop eigenvalues of the LQR controller, and the closed-loop behavior for the different configurations becomes dispersed.

The results of the LQR controller designs for the ten configurations and the corresponding configuration's α index are summarized in Table 6.3. Note in Table 6.3 that the magnitude of the gain norm of the first four configurations is again smaller than that of the remaining six configurations. Note also in Table 6.3, that after the first four designs (9, 8, 10, 4), the descending order of the α index is not well-corelated with the gain magnitude of the output feedback controller, and that it also does not correlate as well with the small-α LQR designs. The two gain optimization problems should not be expected to produce identical results; the control designs underlying Table 6.2 are obtained using output feedback whereas those underlying the results in Table 6.3 are full-state feedback. Therefore should not expect that the α index is one-to-one mapped with the results of the two design methods studied (or any other!). In view of this, the α index appears to be a remarkably good measure of controllability for the given configuration, since the results in Tables 6.2 and 6.3 are well-correlated, especially for the most controllable designs. Notice that the first four configurations (9, 8, 10, 4) have the smallest gains and largest αs, in both cases (Tables 6.2 and 6.3), as compared to the other six designs.

To illustrate the consequences of using relatively "good" and "bad" configurations, two of the configurations were adopted for further study. Specifically,

Table 6.3 The Alpha Measure of Controllability (LQR: Full State Feedback)

Config. No.	Controllability (α)	$\|G\|_f$	Cond. No.
9	0.00333	11.270	128.391
8	0.00234	10.510	127.839
10	0.00208	22.114	131.995
4	0.00158	15.157	129.291
5	0.00153	48.807	140.641
6	0.00130	96.378	144.076
3	0.00123	48.977	135.956
2	0.00102	48.830	135.673
7	0.00084	47.401	135.787
1	0.00078	47.097	135.744

configurations 9 and 6 were taken to study the closed-loop performance of both the symmetric output feedback control (design 1) and the LQR full-state feedback control (design 2). The closed-loop response histories of the normal displacement to the grid plane (at three nodes of the bottom of the grid structure) of two configurations that ensues following a typical initial condition disturbance are shown in Figs. 6.6 and 6.7, and the corresponding control input histories are displayed in Figs. 6.8 and 6.9. The initial conditions used were constructed using an imaginary initial static loading applied normal to the lower-left corner of the structure such that the displacement of the lower-left corner was 1 in. (2.54 cm). The static load was removed at the initial time. We imposed control torque saturation bounds [20 oz-in. (0.1412 N-m] make the torques physically realizable and to keep the reaction wheel speeds within a satisfactory range.

First, consider the symmetric output feedback design (control design 1). Note in Figures 6.6a and 6.6b, that the closed-loop performance achieved using actuator configurations No. 6 and No. 9 are almost identical. When you study the Figures 6.8a and 6.8b, however, it is evident by inspection that configuration No. 6 reqiures significantly more control input energy. This behavior is consistent with our expectations based upon the computed output measures of controllability. We reach the same conclusion when we compare the closed-loop responses (Figures 6.7 and 6.9) for the LQR designs with the same two sets of actuator locations. While these are controlled responses given a single set of initial conditions, we are encouraged that the controllability measure predicts the correct trend for the controlled response with two distinct underlying controller design optimizations.

6.4.3 Selection of Optimal Actuator Locations

When the sites available for actuator placement are a continuum, the optimal locations for actuators can be obtained by standard (gradient-based) optimization techniques. In many practical problems, only discrete sites are available. The optimal selection of the locations for these problems becomes an integer programming problem that, if the number of sites is large, usually leads to a difficult and costly-to-solve optimization problem. This problem becomes especially challenging when the number of available sites is very large compared to the number of actuators. This is due to the large number of possible combinations of placement configurations. Skelton et al. (Refs. 16-19) propose a heuristic optimization algorithm that has been found to reliably improve a trial set of actuator locations at a moderate computational cost. The problem is to select m actuator locations from a set of m_{all} available sites. The algorithm is initiated with a configuration in which all m_{all} locations are occupied by imaginary actuators. The least effective location is found by removing one actuator at a time (i.e., remove the actuator having the smallest cost). This least effective actuator is removed, and the process is repeated with $m_{\text{all}} - 1$ actuators. The process is repeated until the number of actuators is reduced to m. The total number of configurations analyzed by this algorithm has a quadratic dependence on the number m_{all} of candidate sites $\left[(m_{\text{all}} - m)(m_{\text{all}} + m + 1)/2\right]$. The major steps of this algorithm are given below.

Algorithm 6.2: Skelton's Heuristic Algorithm

Step 1. Calculate the α index for an initial configuration composed of m_{all} actuators (an actuator at every available location).

Step 2. Calculate α for all possible combinations of actuator locations obtained by removing one actuator from the current configuration.

Step 3. Select the best configuration from Step 2, as the current configuration.

Step 4. Repeat Steps 2 and 3 until the number of remaining actuators is m.

■

For the grid model, we apply Skelton's algorithm to find the optimal actuator locations by using the α index. In our case, m_{all} is 40 since the 20 grid node locations are considered available to place candidate torque actuators, and two directions are considered admissible for each actuator. The optimal configuration obtained (not one of the ten configurations shown in Fig. 6.5) is such that all three actuators' torque directions are the same (transverse direction) and located at the

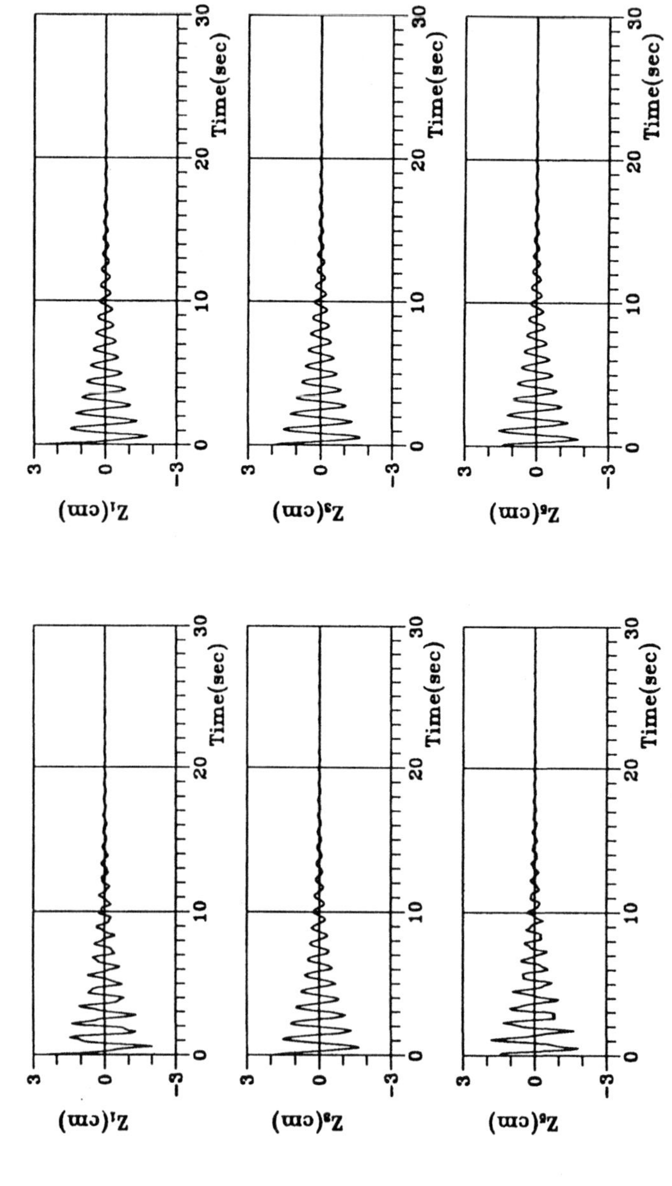

(a) Configuration No. 9 (b) Configuration No. 6

Figure 6.6 Closed Loop Response of the Flexible Grid (Control Design 1)

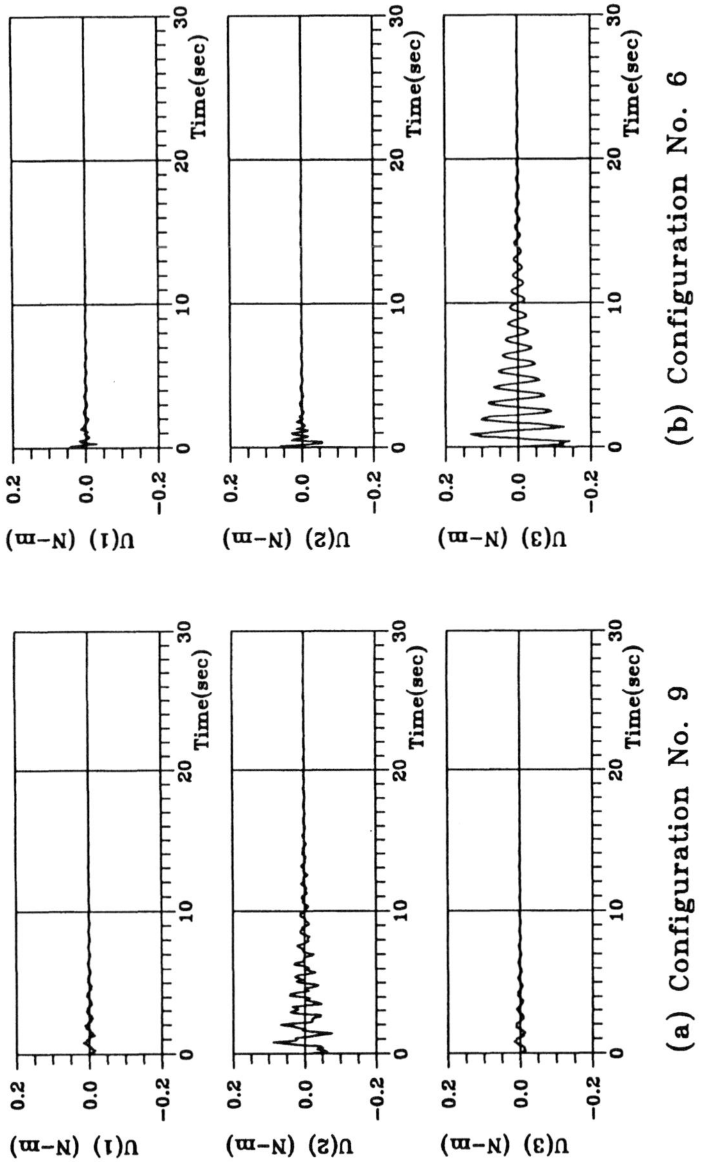

Figure 6.7 Control Torque Input History (Control Design 1)

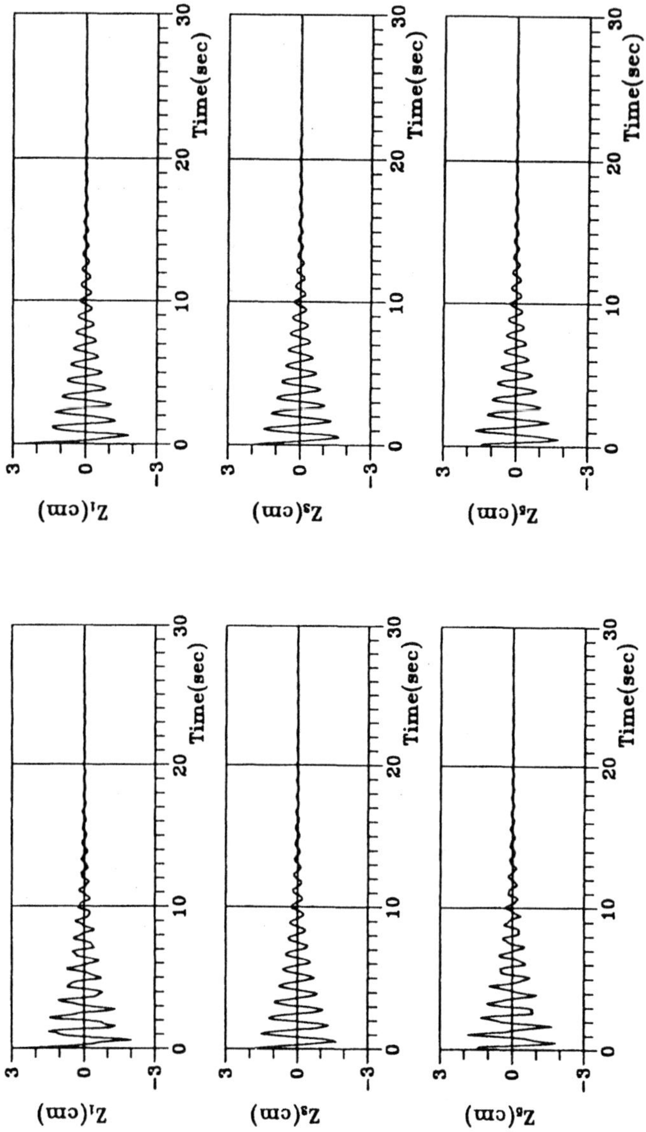

(a) Configuration No. 9

(b) Configuration No. 6

Figure 6.8 Closed Loop Response of the Flexible Grid (Control Design 2)

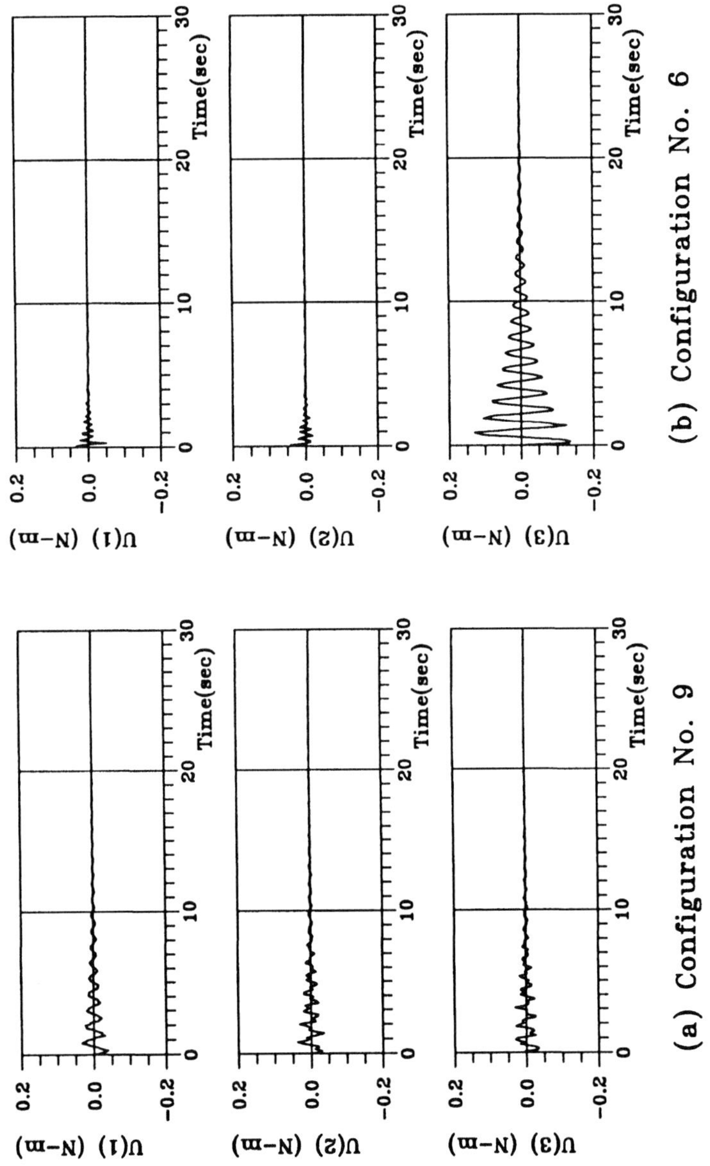

Figure 6.9 Control Torque Input History (Control Design 2)

at the nodes 2, 3, and 4 (nodes in the middle in the lower end). The computed α index is 0.00460. Skelton's algorithm seems to work well for the α index. When we design the controllers (output feedback controller and LQR), however, we find that the gain norms of the resulting controller design are large compared to the best cases (configurations 9, 8, 10, and 4 in Tables 6.5 and 6.6) of the ten configurations discussed above. After examining this case carefully, we found that the second mode's degree of controllability (Hamdan/Nayfeh's measure) of the selected configuration is very low and that the first mode's contribution to the α index is 97.8 %. Had we used a disturbance input matrix which more significantly disturbed this mode, of course, this problem would have been eliminated.

One reason we obtained good correlation between the α index and the designed controllers in section 6.4.2 lies in the preliminary criteria we used to select an admissible subset of ten configurations from all possible combinations. The heuristic process we used to select these admissible ten configurations is following: (1) compute Hamdan and Nayfeh's degree of controllability for all possible combinations of actuator locations, (2) compute the degree of controllability for the modes we wish to control (in our case, the first five modes), (3) select as candidates for further evaluation the best ten configurations whose first five modes are more controllable than those of all other configurations. This means, before attempting optimization over the α index for selecting best actuator configuration, we find it useful to identify an admissible subset of actuator configurations for which *all of the controlled modes are "well controlled" based on Hamdan and Nayfeh's degree of controllability.* In actual applications, we believe this heuristic approach will usually lead to a set of good controller locations, with the final selection made from this set on heuristic grounds.

Therefore, in lieu of directly applying Skelton's optimization process, we suggest an alternative algorithm to accelerate determination of suboptimal location of actuators as follows:

Algorithm 6.3: Alternative Algorithm for Actuator Placement

Step 1. Choose good candidates (e.g., restrict attention to configurations that result in relatively high controllability for important modes), as follows:

- **a.** Compute Hamdan and Nayfeh's degree of controllability for all possible combinations of actuator locations.
- **b.** Based on evaluating the controllability of the "most important modes," select a final subset of candidate actuator configurations.

Step 2. Use the α index to rank the configurations in the final subset (Step b).

As the available actuator locations increase, the combinations of actuator locations also increase and, therefore, more computational time is required to choose the subset of candidates (even before computing the α index). In order to obtain an economic and systematic algorithm for optimal actuator location based on the α index, further research is needed, and we anticipate that some variation of Skelton's heuristic algorithm will prove useful. It is important that the final selections (e.g., Step 2) take into account perturbations of system mass and stiffness variations due to finite actuator inertia. Strictly speaking, this should be done at the onset; however, this greatly increases the expense of the optimization process. We have found (Ref. 36) that the final selection (Step 2) is a reasonable stage at which to include actuator mass and stiffness effects.

In this section, the α measure of controllability for linear time invariant dynamical systems is introduced for guiding the placement of actuators to control vibrating structures. The index α is the sum of the squares of Hamdan and Nayfeh's modal controllability measures weighted by the respective modes' fractional contributions to a quadratic output cost function. The measure of controllability/disturbability is not only a function of the physical system, but also clearly is highly sensitive to actuator placement. We have found that the reachability of the feasible region (for pole assignment) is related to controllability, that is, the degree of controllability affects dramatically the extent to which gain variations can reshape the eigenstructure. The usefulness of the α index for actuator placement has been verified by comparing the results of two control design methods for a grid structure with ten different actuator configurations. The optimal selection of actuator locations can be obtained by integer programming or a heuristic algorithm.

In recent research (Ref. 36), we report an experimental study to further verify some of the actuator placement results on the flexible grid. We concentrated on two of the ten configurations shown in Figure 6.5, in the experimental implementation of our control laws. We implemented linear controllers based on LQR, eigenstructure assignment, and a symmetric positive feedback law as well as Kalman filter and other state estimation algorithms; we imposed time constant constraints on all control designs to make them more readily comparable. We found that the differences between (LQR and eigenstructure assignment) control law design methodlolgies were overwhelmingly less significant than the controllability variations induced by actuator location variations. The key physical difference between the hardware as compared to the finite element model used in the above numerical study, was that our actuator masses totaled about 20% of the total structure mass and therefore could obviously not be neglected as we did in the above optimization study. In other words, the finite element model on which the controllability indices and the control gain design optimization were based would have to be modified for each candidate actuator configuration, because the eigenvalues, eigenvectors, and modal control influence matrices are significantly affected by the mass perturbations. Since we did not account for actuator mass in the above numerical studies, we should expect that these studies would have to be generalized to include these effects, in order to obtain good correlation with the experimental studies (as reported in Ref. 36). Nonideal performance of sensors and actuators played only a modest role in the experimental results, because we restricted the bandwidth (< 25 HZ) to the low frequency range where the sensors and actuators were linear and well-modeled, this was accomplished by filtering both the measurements and the control inputs, as described in Ref. 36.

The actuator mass was indeed found to be significant, as confirmed computationally and experimentally; even the lowest natural frequencies and vibration mode shapes were measurably affected by reconfiguration of the actuators. However, upon accounting for the actuator mass variations, our numerical studies and laboratory experiments indicated that actuator configuration 9 was still highly controllable, and in fact significantly more controllable than configuration 5; the remaining configurations were not studied experimentally. This perhaps fortunate circumstance is not a basis for making broad generalizations, however. As a practical matter, since redoing the finite element model for every actuator configuration is probably too expensive for practical applications (especially if there are hundreds of candidate actuator configurations), the results we obtained suggest that one might gain useful information from a preliminary optimization study that neglects actuator mass and stiffness contributions. A manageable subset of highly controllable configurations (determined on the basis of an approximate model that neglects mass and stiffness variations due to actuator reconfiguration), can then be the subject of detailed studies that fully account for actuator mass and stiffness contributions. This heuristic approach must be viewed as a reasonable path to efficiently select a good feasible design, but it will generally lead to a suboptimal actuator configuration. While mathematicians may be justifiably uncomfortable with such heuristics, most engineers recognize the necessity of compromises; without them, we would never have completed project Apollo!

Problems

Problem 6-1

Consider the first-order system

$$\dot{\mathbf{x}}(t) = \mathbf{A}\mathbf{x}(t) + \mathbf{B}\mathbf{u}(t)$$

The solution of the above system can be represented as

$$\mathbf{x}(t) = e^{\mathbf{A}t}\mathbf{x}(0) + \int_0^t e^{\mathbf{A}(t-\tau)}\mathbf{B}\mathbf{u}(\tau)d\tau \qquad (*)$$

By considering the definition of controllability, we let $\mathbf{x}(t_f) = 0$; then Eq. $(*)$ can be reduced to

$$-\mathbf{x}(0) = \int_0^{t_f} e^{-\mathbf{A}\tau}\mathbf{B}\mathbf{u}(\tau)d\tau \qquad (\dagger)$$

Show that Eq. $(\dagger)$ can be rewritten in the form (using the Cayley-Hamilton theorem: see Problem 2-2)

$$-\mathbf{x}(0) = \mathcal{C}\begin{Bmatrix} \mathbf{c}_0 \\ \mathbf{c}_1 \\ \vdots \\ \mathbf{c}_{n-1} \end{Bmatrix}$$

where $\mathbf{c}_i$ are vectors that include the integration of input signals, and $\mathcal{C}$ is a controllability matrix. What conclusion can be induced?

Problem 6-2

A reaction wheel control system for a spacecraft can be modeled as

$$\dot{\theta} = q$$
$$I_y\dot{q} = N\,i$$
$$J\dot{q}_w = -N\,i$$
$$R\,i = e - N(q - q_w)$$

where θ is the pitch angle, q is the pitch rate, q_w is the wheel pitch rate, I_y is the moment of inertia of the spacecraft, J is the moment of inertia of the wheel plus motor, N is the torque per unit current, R is the armature resistance, i is the current, and e is the voltage (control variable).

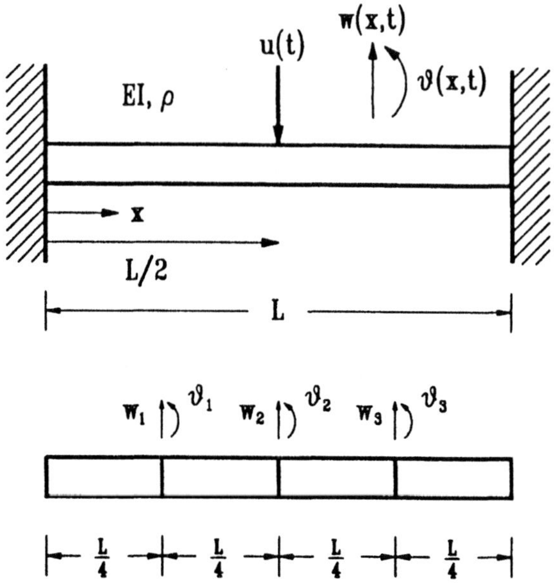

Figure P6.1 Uniform Beam Fixed at Both Ends.

Your tasks are as follows:

(a) Discuss the controllability of the system by examining the controllability matrix.

(b) Obtain the transformed equations of motion by using the following similarity transformation matrix T:

$$T = [\,T_1 \quad T_2\,]$$

where

$$T_1 = [\,b \quad Ab \quad \dots \quad A^{r-1}b\,]$$

where r is the rank of the controllability matrix, and choose T_2 such that its columns are independent of each other and also those of T_1.

If the system is not completely controllable, describe which variables or combinations of variables are uncontrollable.

Problem 6-3

Consider the transverse vibration of uniform beam fixed at both ends and with one actuator located at $x = L/2$ location; the structural data are EI=1, L=4, ρ=420, A=1, as shown in Figure P6.1. By using the finite element method (4 beam elements) the following linearized second-order equation of motion can be obtained:

$$\mathrm{M\ddot{x}(t) + C\dot{x}(t) + Kx(t) = Fu(t)} \qquad (\ddagger)$$

where the state vector is

$$\mathrm{x^T} = \begin{bmatrix} \mathrm{w_1} & \theta_1 & \mathrm{w_2} & \theta_2 & \mathrm{w_3} & \theta_3 \end{bmatrix}$$

and the system matrices are

$$\mathrm{M} = \begin{bmatrix} 312 & 0 & 54 & -13 & 0 & 0 \\ 0 & 8 & 13 & -3 & 0 & 0 \\ 54 & 13 & 312 & 0 & 54 & -13 \\ -13 & -3 & 0 & 8 & 13 & -3 \\ 0 & 0 & 54 & 13 & 312 & 0 \\ 0 & 0 & -13 & -3 & 0 & 8 \end{bmatrix}$$

$$\mathrm{K} = \begin{bmatrix} 24 & 0 & -12 & 6 & 0 & 0 \\ 0 & 8 & -6 & 2 & 0 & 0 \\ -12 & -6 & 24 & 0 & -12 & 6 \\ 6 & 2 & 0 & 8 & -6 & 2 \\ 0 & 0 & -12 & -6 & 24 & 0 \\ 0 & 0 & 6 & 2 & 0 & 8 \end{bmatrix}, \qquad \mathrm{F} = \begin{Bmatrix} 0 \\ 0 \\ 1 \\ 0 \\ 0 \\ 0 \end{Bmatrix}$$

$$\mathrm{C} = \alpha \mathrm{M} + \beta \mathrm{K}, \quad \text{with } \alpha = \beta = 0.001$$

Your tasks are:

(a) Transform the second-order equation, Eq. ($\ddagger$), into state space form. Check the rank of the controllability grammian, and use the PBH rank test (or the PBH eigenvector test) to determine which modes are uncontrollable.

(b) Rewrite Eq. ($\ddagger$) using modal coordinate by using the modal coordinate transformation. Discuss the form of the modal equations in relation to the results from part a.

(c) Explain the results from parts a and b from a physical point of view. Discuss the relevance of the location of the actuator to controlling this flexible structure.

Problem 6-4 (Ref. 13)

The initial conditions constituting the recovery region is the set of all $\mathbf{x}(0) = \delta$ that can be returned to the origin in time T, where [see Eq. (6.19)]:

$$\delta = \sum_{i=1}^{n} \underline{\phi}_i \left[\int_0^T e^{-\lambda_i t} \underline{\psi}_i^T \mathbf{B}\mathbf{u}(t) dt \right] = \sum_{i=1}^{n} \frac{\underline{\phi}_i}{\|\underline{\phi}_i\|} \rho_i^*$$

and where

$$\rho_i^* \equiv \|\underline{\phi}_i\| \left[\int_0^T e^{-\lambda_i t} \underline{\psi}_i^T \mathbf{B}\mathbf{u}(t) dt \right]$$

Assume that [instead of Eq. (6.21)] we choose the following control input:

$$u_j = \text{sign } \left(\underline{\psi}_i^T \mathbf{b}_j\right)$$

to minimize the component of δ in the direction of $\underline{\phi}_i$.

Show that the length ρ_i^* of the semiaxis of the parallelepiped in the direction of the eigenvector $\underline{\phi}_i$ is related to the Hamdan and Nayfeh's gross measure $\rho_i = \|\mathbf{f}_i\|$ of modal controllability, Eq. (6.50), by

$$\rho_i^* = (\cos\ \delta_i)^{-1} \|\mathbf{f}_i\|_1 \int_0^T e^{-\lambda_i t} dt$$

where δ_i is the angle between $\underline{\phi}_i$ and $\underline{\psi}_i$ such that

$$(\cos\ \delta_i)^{-1} = \|\underline{\phi}_i\| \ \|\underline{\psi}_i\|$$

and $\|\mathbf{f}_i\|_1$ is the 1-norm of $\mathbf{f}_i$.

Problem 6-5 (Ref. 16)

Consider the time-invariant linear dynamic system

$$\dot{\mathbf{x}}(t) = \mathbf{A}\mathbf{x}(t) + \mathbf{B}\mathbf{u}(t)$$

The cost function is taken that represents the system performance as follows:

$$V = \sum_{i=1}^{m} \int_0^{\infty} \mathbf{y}_d^{i\,T}(t) Q_v \mathbf{y}_d^i(t) dt$$

$$\mathbf{y}_d^i(t) = C_d \mathbf{x}^i(t)$$

where $\mathbf{x}^i(t)$ is the response due to an impulse input $u_i(t) = u_i\delta(t)$ (with $u_j(t) = 0$, $i \neq j$) applied at $t = 0$ with zero initial conditions, Q_v is a weighting matrix, and C_d is defined so that $\mathbf{y}_d(t)$ models an important output for the design objectives.

The decomposition of V into contributions of control variable u_i is defined by

$$V_{u_i} \equiv \frac{1}{2}\left(\frac{\partial V}{\partial \mathbf{b}_i}\right)\mathbf{b}_i$$

where $\mathbf{b}_i$ denote the ith column vector of input matrix B.

Show that the cost function V and the 'component cost' contribution V_{u_i} of the control variable u_i in the cost function V can be computed as follows:

$$V = \text{trace}\{KBUB^T\}$$
$$V_{u_i} = [B^TKBU]_{ii}$$

where K is the observability grammian that satisfies the following Lyapunov equation:

$$KA + A^TK + C_d^TQ_vC_d = 0$$

and $U \equiv \text{diag}\{u_1^2, u_2^2, \ldots, u_m^2\}$.

References

1. Viswanathan, C.N., Longman, R.W., and Likins, P.W., "A Degree of Controllability Definition: Fundamental Concepts and Application to Modal Systems," *Journal of Guidance, Control, and Dynamics,* Vol. 7, No. 2, 1984, pp. 222-230.

2. Viswanathan, C.N., Longman, R.W., and Likins, P.W., "A Definition of the Degree of Controllability – A Criterion for Actuator Placement," *Proceedings of the Second VPI&SU AIAA Symposium on Dynamics and Control of Large Flexible Spacecraft,* Blacksburg, VA, June 1979.

3. Viswanathan, C.N. and Longman, R.W., "The Determination of the Degree of Controllability for Dynamic Systems with Repeated Eigenvalues," *Advances in the Astronautical Sciences,* Vol. 50, 1983.

4. Longman, R.W. and Alfriend, K.T., "Energy Optimal Degree of Controllability and Observability for Regulator and Maneuver Problems," *Proceedings of the Sixteenth Annual Conference on Information Sciences and Systems,* Princeton, NJ, Mar. 1982.

5. Longman, R.W., Sirlin, S.W., Li, T., and Sevaston, G., "The Fundamental Structure of Degree of Controllability and Degree of Observability," AIAA/AAS Paper 82-1434, San Diego, CA, Aug. 1982.

6. Laskin, R.A., Longman, R.W., and Likins, P.W., "A Definition of the Degree of Controllability for Fuel-Optimal Systems," *Proceedings of the Third VPI&SU AIAA Symposium on Dynamics and Control of Large Flexible Spacecraft,* Blacksburg, VA, June 1981.

7. Laskin, R.A., Longman, R.W., and Likins, P.W., "Actuator Placement in Modal Systems Using Fuel Optimal Degree of Controllability Concepts," *Proceedings of the 20th Annual Allerton Conference on Communication, Control and Computing,* Monticello, IL, Oct. 1982.

8. Klein, G., Lindberg, R.E., and Longman, R.W., "Computation of a Degree of Controllability via System Discretization," *Journal of Guidance, Control, and Dynamics,* Vol. 5, Nov.-Dec. 1982, pp. 583-588.

9. Lindberg, R.E. and Longman, R.W., "The Influence of Time and Normalization on Actuator Placement by Degree of Controllability," *Proceedings of the AAS/AIAA Astrodynamics Specialist Conference,* Lake Placid, NY, Aug. 1983.

10. Lindberg, R.E. and Longman, R.W., "Optimization of Actuator Placement via Degree of Controllability Criteria Including Spillover Considerations," AIAA/AAS Paper 82-1435, San Diego, CA, Aug. 1982.

11. Friedland, B., "Controllability Index Based on Conditioning Number," *Journal of Dynamic Systems, Measurement, and Control, Transactions of the ASME,* Dec. 1975, pp. 444-445.

12. Moore, B.C., "Principal Component Analysis in Linear Systems: Controllability, Observability, and Model Reduction," *IEEE Transactions on Automatic Control,* Vol. AC-26, No. 1, 1981, pp. 17-32.

13. Hamdan, A.M.A. and Nayfeh, A.H., "Measure of Modal Controllability and Observability for First- and Second-Order Linear Systems," *Journal of Guidance, Control, and Dynamics,* Vol. 12, No. 3, 1989, pp. 421-428.
see also... Same authors and title: Erratum, Journal of Guidance, Control, and Dynamics, Vol. 12, No. 5, 1989.

14. Lindner, D.L., Babendreier, J., and Hamdan, A.M.A., "Measures of Controllability and Observability and Residues," *IEEE Transactions on Automatic Control,* Vol. AC-34, No. 6, 1989, pp. 648-650.

15. Vander Velde, W.E. and Carignan, C.R., "A Dynamic Measure of Controllability and Observability for the Placement of Actuators and Sensors on Large Space Structures," NASA-CR-168520, Jan. 1982.

16. Skelton, R.E. and DeLorenzo, M.L., "Selection of Noisy Actuators and Sensors in Linear Stochastic Systems," *Journal of Large Scale Systems, Theory and Applications,* Vol.4, Apr. 1983, pp. 109-136.

17. Skelton, R.E. and Chiu, D., "Optimal Selection of Inputs and Outputs in Linear Stochastic Systems," *Journal of the Astronautical Sciences,* Vol. XXXI, No. 3, Jul.-Sep. 1983, pp. 399-414.

18. Skelton, R.E. and DeLorenzo, M.D., "Space Structure Control Design by Variance Assignment," *Journal of Guidance, Control, and Dynamics,* Vol. 8, No. 4, 1985, pp. 454-462.

19. Norris, G.A. and Skelton, R.E., "Selection of Dynamic Sensors and Actuators in the Control of Linear Systems," *Journal of Dynamic Systems, Measurement, and Control,* Vol. 111, Sept. 1989, pp. 389-397.

20. Skelton, R.E. and Yousuff, A., "Component Cost Analysis of Large Scale Systems," *International Journal of Control,* Vol. 37, No. 2, 1983, pp. 285-304.

21. Skelton, R.E. and Hughes, P.C., "Modal Cost Analysis for Linear Matrix-Second-Order Systems," *Journal of Dynamic Systems, Measurement, and Control,* Vol. 102, Sept. 1980, pp. 151-158.

22. Skelton, R.E., *Dynamic System Control – Linear System Analysis and Synthesis,* John Wiley &, New York, NY, 1988.

23. Skelton, R.E. and Gregory, C.Z., "Measurement Feedback and Model Reduction by Modal Cost Analysis," *Proceedings of Joint Automatic Control Conference,* Denver, CO, 1979.

24. Skelton, R.E., "Cost Decomposition of Linear Systems with Application to Model Reduction," *International Journal of Control,* Vol. 32, No. 6, 1980, pp. 1031-1055.

25. Yousuff, A. and Skelton, R.E., "Controller Reduction by Component Cost Analysis," *IEEE Transactions on Automatic Control,* Vol. AC-29, No. 6, June 1984, pp. 520-530.

26. Skelton, R.E., Singh, R., and Ramakrishnan, J., "Component Model Reduction by Component Cost Analysis," AIAA Guidance, Navigation, and Control Conference, Minneapolis, MN, Aug. , 1988.

27. Kailath, T., *Linear Systems*, Prentice-Hall, Englewood Cliffs, NJ, 1980.

28. Chen, C.T., *Linear System Theory and Design*, Holt, Rinehart and Winston, New York, NY,1984.

29. Carrol, K.A. and Hughes, P.C., "Controller Order Reduction for Flexible Spacecraft Using Closed-Loop Balancing Methods," Symposium on Dynamics of Flexible Structures in Space, Convened at the College of Aeronautics, Cranfield Institute of Technology, Bedford, England, May 15-18, 1990.

30. Laub, A., "A Schur Method for Solving Algebraic Riccati Equations," *IEEE Transactions on Automatic Control,* Vol. AC-24, June 1979, pp. 913-925.

31. Kim, Y. and Junkins, J.L., "A Measure of Controllability for Actuator Placement," *Journal of Guidance, Control, and Dynamics,* Vol. 14, No. 5, Sept.-Oct. 1991, pp. 895-902.

32. James, G.H., "An Experimental and Analytical Study of Structural Identification," Ph.D. Dissertation, Texas A&M University, College Station, TX, Dec. 1989.

33. Junkins, J.L. and Kim, Y., "First and Second Order Sensitivity of the Singular Value Decomposition," *Journal of the Astronautical Sciences,* Vol. 38, No. 1, Jan.-Mar., 1990, pp. 69-86.

34. Junkins, J.L. and Kim, Y., "A Minimum Sensitivity Design Method for Output Feedback Controllers," *Mechanics and Control of Large Space Structures,*, Progress in Aeronautics and Astronautics Series, Vol. 129, Edited by J.L. Junkins, AIAA, New York, NY, 1990, Chapter 15.

35. Dunyak, J.P., Junkins, J.L., and Watson, L.T., "Robust Nonlinear Least Square Estimation Using the Chow–Yorke Homotopy Method," *AIAA Journal of Guidance, Control, and Dynamics,* Vol. 7, No. 6, 1984, pp. 752-755.

36. Hurtado, J., Rahman, Z., and Junkins, J.L., "Vibration Suppression for a Planar Flexible Grid: Analysis and Experiment," *33rd AIAA Stuctures, Structural Dynamics and Materials Conference,* Dallas, TX, Apr. 1992.

CHAPTER 7

Design of Linear Output Feedback Control Systems

Considerable attention has been recently focused upon developing active and passive control methods for suppression of vibration as well as maneuvering large flexible structures. These problems arise in both civil and aerospace applications. The key distinguishing feature of the most challenging problems of current interest is that there is a significant overlap in the desired controller bandwidth and the natural frequencies of the open-loop structure. The difficulty of the control design problem is especially great for the case of the next generation of large space structures, because the per unit mass cost of placing these systems on orbit often leads to extremely flexible structural designs with many modes in the controller bandwidth. Aside from the inherent difficulty of simultaneously carrying out vibration suppression and, for example precision maneuvers and/or pointing of the overall structure, three major problems arise when designing controllers for even the linearized dynamical models of these systems. These difficulties are overviewed as follows:

First, many of the available control design methods assume that the full-state vector (or an estimate of the full-state vector) is available. In most practical situations, the estimation algorithms and associated feedback control laws encounter difficulties when only a low-dimensioned output vector is available. Real time implementation of high-dimensioned state estimators, while perhaps feasible in the coming generation of processors, remains computationally unattractive.

Second, it is computationally expensive to design controllers for extremely high order systems such as a flexible structure. It is usually feasible to design a controller based on a reduced-order model, which includes only the most important subset of the vibration modes, although we typically retain a high-order model for validation purposes. A reduced-order controller may destabilize higher-order modes, due to truncation and spill over effects (Ref. 1). Therefore we should include consideration of truncation and spill over effects when designing controllers based on a reduced-order model for a large flexible structure.

Third, there almost always exists significant modeling errors for mathematical models of large flexible structures. These unknown, unmodeled effects can potentially destabilize the closed-loop system. In order to tolerate model errors, it is desirable to optimize the robustness (tolerance of model errors and disturbances) by judicious design of the control law.

Motivated by the need to overcome these difficulties, a large research literature has evolved over the past two decades. We offer some overview comments on the literature, however, we confine ourselves to an important subset of the methodology for detailed discussion.

In this chapter, an inequality-constrained pole placement technique is introduced for mechanical second-order systems that optimally tunes the gains in a symmetric output form of the feedback control law. To provide a context for these developments, we first review the most pertinent recent literature.

In section 7.1, we review the literature related to the output feedback control. In section 7.2, we provide a theoretical foundation for this chapter and introduce the main control design method to be discussed (that leads to robust, globally stable linear output feedback controllers). This method makes use of design freedom that remains (after regional or point placement of some of the closed-loop eigenvalues) to increase the robustness of the system via an optimization process. As a consequence of the specific parameterization adopted for the feedback law, the structure of the output feedback controller guarantees stability. Therefore, truncation and spillover effects cannot destabilize the system, even though (1) we may not have chosen a good reduced-order model, and (2) finite model errors are present. While the most fundamental property (stability) of the controller can be guaranteed, loss of predicted performance due to model errors is also a significant issue. Stability robustness degradation due to imprecise collocation of sensors and actuators is also discussed.

In section 7.3, a pole placement technique is presented for mechanical second-order systems by using symmetric output feedback (Refs. 2-5). The approach formulated in section 7.3 leads to a nonlinear constrained optimization problem. There are two large classes of iterative algorithms to solve such problems. One method is sequential linear programming (Refs. 6-8) based on an iterative local solution of linear programming problems. A second optimization method attempts to deal with the nonlinearity of the problem via gradient-based nonlinear programming techniques. The former method can be easily coded by using familiar linear programming algorithms such as the simplex method (Ref. 7). This approach has been found to suffer from some difficulties in choosing control parameters and local nonconvexity when the problem is highly nonlinear (Ref. 7). In our numerical studies, we also have experienced occasional convergence oscillations near the optimum, preventing accurate convergence to the solution. Pole placement problems are often highly nonlinear, since the performance index and the constraint equations are evaluated through solutions of eigenvalue problems and using singular value decompositions. Therefore, we enhance the convergence of the latter approach by introducing a two-stage homotopic nonlinear programming algorithm in section 7.4.

So that we do not sacrifice system performance on the altar of global stability, we introduce a method for optimizing over a stable family of gains to minimize a closed-loop sensitivity measure with respect to model parameter variations. Finally (section 7.5), numerical demonstrations illustrate the salient features and practicality of this approach to design of robust output feedback controllers.

7.1 Literature Survey

The most fundamental and widely useful contributions of modern control theory are the linear quadratic (LQ) regulator and pole placement design methods; design of large flexible structures' control systems has been attempted through implementation of both design techniques. Since it is almost always impractical to measure all components of the state vector, a Kalman filter or some other observer must be operated in real time to estimate a high-dimensioned state vector as required to directly implement conventional LQ control laws. This often leads to unacceptably large controller dimension with the associated requirement for excessive real time matrix arithmetic operations. Motivated by the desire to simplify the designs and to reduce the requirements for real-time processing, alternative control structures and design methods (Refs. 9-38) have been investigated. Significant new output feedback design procedures have been established which augment the extensive literature on LQ and pole placement approaches.

In the LQ design approach using output feedback, the problem of finding the optimal, constant output feedback gain matrix for linear time-invariant multivariable systems with quadratic cost has been the subject of research over the past two decades (Refs. 9-15). The necessary conditions (Refs. 9, 10) for these problems take the form of coupled nonlinear matrix equations. The algorithm given in Ref. 9 for computing the optimal feedback gain is of the so-called indirect type: The necessary conditions for optimality are found to be a set of coupled nonlinear equations which are solved iteratively by generalized Newton or homotopic methods. These nonlinear algebraic matrix equations have to be solved on every design iteration, and this is computationally expensive for high-dimensioned applications. The necessary conditions developed in Ref. 10 are more general than those in Ref. 9, so that this approach can be applied, in principle, to time-varying systems and the problem involving output feedback with parameter uncertainty. Various algorithms and numerical studies for solving these necessary conditions have been reported in Refs. 11-15. In order to avoid solving the nonlinear algebraic equations at each iteration, direct algorithms were developed in Refs. 11 and 12, that require the solution of only linear equations and attractive numerical results were obtained.

The optimal feedback gains can be computed by direct minimization of the cost function using nonlinear programming techniques such as Davidon-Fletcher-Powell algorithm (Ref. 39). Several iterative techniques for finding the optimal gain matrix are studied in Refs. 9-12, however, they all require a good initial guess for a stabilizing gain matrix within the domain of attraction of the optimal solution. Establishing a satisfactory initial guess for the control gains is often difficult when the open-loop system is unstable. Therefore, an alternative algorithm is proposed in Ref. 13, that requires only a stabilizing state feedback law to initialize the search. Such a law always exists for a controllable system and, moreover, can be easily found using, e.g., the method of Kleinman (Ref. 40). An output feedback matrix F is said to dominate another $\bar{F}$, if the cost for F is not higher for any initial condition than the cost for $\bar{F}$ and if the cost is lower for some (perhaps all) initial conditions. Two algorithms can be found in Ref. 14, for finding feedback gains that dominate any given stabilizing feedback gain. For dealing with more general performance indices, a sequential programming algorithm is suggested in Ref. 15, which obtains gains minimizing a broad class of performance indices, including a standard LQ cases.

The fundamental ideas underlying pole assignment control design techniques were first presented by Wonham (Ref. 41). Since then, the problem of pole assignment has received wide attention. As mentioned in Ref. 41, if feedback is applied from every state variable, the characteristic poles of a controllable system can be positioned without restriction. Since complete state observation is unlikely in most practical situations, many researchers have studied the conditions (Refs. 16-29) under which the system is pole assignable with incomplete state observation (i.e., output feedback).

Generalization of Wonham's full-state feedback results to address the case of output feedback is presented along with an algorithm in Ref. 16. For a controllable-observable system, m poles of the closed-loop system are assignable almost arbitrarily by gain output feedback, where m is the number of independent outputs (Ref. 16). These results were extended by Ref. 17, in which it is shown that, under the same conditions as in Ref. 16, max (r, m) eigenvalues are assignable, almost arbitrarily, by gain output feedback, where r is the number of independent inputs. When the order n of the system becomes large, however, the algorithm in Ref. 16 usually encounters numerical difficulties as a consequence of ill-conditioning of a matrix that must be inverted. To partially circumvent this problem, an alternate algorithm is presented in Ref. 18, that allows pole assignment to be carried out on large, linear, time-invariant multivariable systems with output feedback.

The necessary and sufficient conditions for pole assignment based on frequency-domain techniques are established in Refs. 19, 20. In Refs. 19 and 20, unity-rank feedback is used to reduce the multivariable system to a much simpler equivalent single-input system. However, exact assignment of all the poles by unit rank feedback is not always possible, whereas it can be achieved using a full-rank feedback matrix, and the feedback gains resulting from the unity feedback may be far from optimal.

A new proof of Davison's theorem (Refs. 16, 17) on pole assignment and an associated design procedure which has certain advantages over Davison's method are provided in Ref. 21. Reference 21 shows that, in some cases, more poles than the maximum of the number of inputs or outputs can be assigned arbitrarily. In Ref. 22, a more specific condition is derived for the pole assignment in linear multivariable systems via output feedback, and it is established that, given a controllable-observable system satisfying [(A, B, C) in $\dot{\mathbf{x}} = \mathbf{A}\mathbf{x} + \mathbf{B}\mathbf{u}$, $\mathbf{y} = \mathbf{C}\mathbf{x}$, in which $A \in R^n$, rank B=m, rank C=r], it can be shown that for almost all (B, C) pairs, min $(n, m+r-1)$ poles can be assigned arbitrarily close to min $(n, m+r-1)$ specified symmetric values by using output feedback.

In the development of numerically reliable eigenvalue placement algorithms, several issues must be considered (Refs. 42, 43), such as conditioning of the open-loop system. In Ref. 23, a reliable numerical algorithm is provided, that uses only orthogonal transformations with output feedback to assign the desired eigenvalues and, therefore, the algorithm has good numerical properties. However, when used in conjunction with a nonlinear performance measure, no known nonlinear programming method can absolutely guarantee that it will always converge to the global optimal solution.

After series of studies (Refs. 16-22) of pole assignment via output feedback, the problem of simultaneous assignment of closed-loop eigenvalues (poles) and eigenvectors (eigenstructure assignment) has received considerable attention (Refs. 24-29). Kimura (Refs. 24, 25) used properties of the eigenspace of the closed-loop system to derive a sufficient condition for pole assignability based on the dimensions of the matrices A, B, C. In Ref. 26, reinterpretation of Srinathkumar and Rhoten's design procedure (Ref. 44) for eigenvalue/eigenvector assignment for state-feedback is given, and applied to the eigenvalue/eigenvector assignment for output feedback control. Srinathkumar (Ref. 27) also derived a different sufficient condition for assigning an almost arbitrary set of *min* $(n, m+r-1)$ distinct eigenvalues. In Ref. 27, the freedom (that exists over and above eigenvalue placement) is used in selecting closed-loop eigenvectors to achieve closed-loop response shaping.

A new method for reducing controller complexity, proposed in Ref. 28, is based on eigenvalue sensitivity concepts. Here, residue analysis (based upon right and left eigenvectors) is used to estimate the effect on the eigenvalues of the constraints imposed on the feedback gains. Most of the previous results for the eigenstructure assignment, however, have some limitations in the sense that eigenvalues of the closed-loop system are distinct or different from eigenvalues of the open-loop system. Generalized results without these assumptions can be found in Ref. 29; the necessary and sufficient conditions reveal that the closed-loop eigenstructure assignment via output feedback is constrained by the requirement that the generalized right and left eigenvectors lie in a certain subspace.

Both the above methods – output feedback using LQ and pole placement via output feedback – can lead to designs that suffer from robustness problems when parameter uncertainties are present. In the LQ approach, the degree of arbitrariness associated with order reduction and selection of weight matrices results in a large space of possible solutions. Likewise, in all pole placement and eigenstructure assignment methods, the user must select target eigenvalue locations and either directly or indirectly, determine the eigenvector locations. All of these choices affect nominal performance and robustness. Motivated by these issues, other researchers have studied robust designs for output feedback controllers. A direct approach is suggested in Ref. 30, that accesses the additional design freedom remaining once the eigenvalue assignment has been accomplished, and this freedom can be used to optimize a robustness measure. In some approaches (Refs. 31, 32), a standard performance index is modified for output feedback by including a quadratic type sensitivity measure. Also, in Ref. 33, the traditional linear quadratic performance index is used with additional state sensitivity terms. Reference 25 shows that the derivative feedback term is effective in improveing the stability robustness. Both approaches – LQ and pole placement – to obtain robust output feedback controller are addressed in Refs. 34 and 35. In order to use a steepest descent method to minimize the performance index, they discussed performance sensitivity with respect to parameter variation for the LQ case and eigenvalue sensitivities for the pole placement case.

In recent developments (Refs. 2-6, 36-38), a symmetric or "structural" output feedback structure is introduced for the control law. As a consequence of the structure of the control law and the choice of symmetric, positive definite gain matrices, global stability of the closed-loop distributed parameter system is guaranteed in the Lyapunov sense. Stability analysis for a closed-loop system is developed in Refs. 36, 37. In Ref. 38, a perturbation approach is used to predict the change in the eigen properties of a closed-loop system. Positive semidefinite gain matrices are parameterized by linear summation of prescribed symmetric, positive semidefinite submatrices in Ref. 6, and used for pole placement techniques with standard linear programming algorithms. The design method of Ref. 6 allows not only eigenstructure constraints, but also actuator saturation inequality conditions to be imposed. In Refs. 2-5, Cholesky decomposition is utilized to parameterize positive definite gain matrices. To improve the robustness of the closed-loop system, a robustness measure is taken as the performance index in Refs. 3-5, and nonlinear programming is used for inequality-constrained pole placement technique. The methods for control gain design may also be generalized to design gains for analogous "structural" dynamic observers to provide real time state estimates (Ref. 6).

This chapter draws upon the above literature to present and extend methods focused on robust output feedback design using eigenstructure assignment techniques. These ideas build directly upon the work reported in Refs. 2-6, 36-38. The algorithms presented allow the output feedback gain matrix to be directly iterated over a stable family to satisfy design constraints, and after satisfactory eigenvalue placement is accomplished, the remaining freedom utilized to optimize a measure of robustness.

7.2 Theoretical Background of the Symmetric Output Feedback

7.2.1 Stability and Robustness Analysis

Stability Analysis

Consider the class of dynamical systems modeled by the equation of motion

$$\mathrm{M}\ddot{\mathbf{x}} + \mathrm{C}\dot{\mathbf{x}} + \mathrm{K}\mathbf{x} = \mathrm{D}\mathbf{u} \tag{7.1}$$

where $\mathbf{x} \in \mathrm{R}^n$ and $\mathbf{u} \in \mathrm{R}^m$ are configuration and control vectors, respectively, M is the $n \times n$ positive definite symmetric mass matrix, C is the $n \times n$ positive semidefinite symmetric structural damping matrix, K is the $n \times n$ positive semidefinite stiffness matrix, D is the $n \times m$ control influence matrix, and overdots denotes differentiation with respect to time.

Motivated by recent developments (Refs. 2, 6, 38, 45, and 46), we introduce the following symmetric or "structural" output feedback form of the control law (collocated sensors and actuators):

$$\begin{aligned} \mathbf{u} &= -(\mathrm{G}_1\mathbf{y} + \mathrm{G}_2\dot{\mathbf{y}}) \\ \mathbf{y} &= \mathrm{D}^{\mathrm{T}}\mathbf{x} \end{aligned} \tag{7.2}$$

where G_1 and G_2 are fully populated, $m \times m$ positive definite symmetric gain matrices.

With the assumption that the sensors and actuators are perfect, i.e., linear and instantaneous, the closed-loop equations can now be written as

$$\mathrm{M}\ddot{\mathbf{x}} + (\mathrm{C} + \mathrm{D}\mathrm{G}_2\mathrm{D}^{\mathrm{T}})\dot{\mathbf{x}} + (\mathrm{K} + \mathrm{D}\mathrm{G}_1\mathrm{D}^{\mathrm{T}})\mathbf{x} = 0 \tag{7.3}$$

Note that, since G_i are positive definite, then the control-induced stiffnes and damping perturbations $\mathrm{DG}_1\mathrm{D}^{\mathrm{T}}$ and $\mathrm{DG}_2\mathrm{D}^{\mathrm{T}}$ are symmetric positive semidefinite. We have already shown (Theorem 3.7) that, for all members of this special class of output feedback controllers, asymptotic stability of the closed-loop system is guaranteed. An important consequence of the stability analysis is that the stability of a controllable symmetric system (having symmetric positive definite output feedback) is maintained: (1) regardless of the number of modes retained in the reduced-order model, and (2) regardless of inaccuracies in the parameter values used in the design; only the predicted performance (and therefore the optimality) may be degraded as a consequence of model errors (Ref. 46). This consequence is attributable to the fact that, since positive definite G_1 and G_2 are chosen, M, $\mathrm{K} + \mathrm{DG}_1\mathrm{D}^{\mathrm{T}}$, and $\mathrm{C} + \mathrm{DG}_2\mathrm{D}^{\mathrm{T}}$ are always positive definite regardless of whether full-order or reduced-order models are used. Thus, by using the symmetric output feedback controller, the potential instability problems due to spillover can be completely avoided, and the parameters do not have to be known accurately to guarantee stability. This closed-loop stability analysis leads us to the conclusion that if the system is controllable, then the closed-loop system is at least asymptotically stable so long as the gain matrices are chosen as positive definite matrices. In practice, these optimistic stability arguments must be tempered, of course, by consideration of nonideal sensors and actuators.

Stability Robustness Analysis

We have established in the previous section that, with ideal conditions (i.e., exact actuator/sensor collocation, perfect actuators and sensors), the stability of structural output feedback control is guaranteed. However, it may not be possible in practice to place sensors and actuators exactly at the same locations. In this section, we investigate the robustness of structural output feedback controllers to imprecise collocation of sensors and actuators.

The closed-loop system (for perfect collocation) is at least asymptotically stable if the gain matrices are chosen as positive definite matrices, as proven in the previous section. Considering a first-order state space form of the system differential equations, that is equivalent to the second-order closed-loop system of Eq. (7.3), we can write

$$\mathrm{B}\dot{\mathbf{z}} = \mathrm{A}\mathbf{z} \tag{7.4}$$

where

$$\mathbf{z} = \begin{Bmatrix} \mathbf{x} \\ \dot{\mathbf{x}} \end{Bmatrix}, \quad \mathrm{A} = \begin{bmatrix} 0 & \mathrm{M} \\ -(\mathrm{K} + \mathrm{D}\mathrm{G}_1\mathrm{D}^{\mathrm{T}}) & -(\mathrm{C} + \mathrm{D}\mathrm{G}_2\mathrm{D}^{\mathrm{T}}) \end{bmatrix}, \quad \mathrm{B} = \begin{bmatrix} \mathrm{M} & 0 \\ 0 & \mathrm{M} \end{bmatrix} \tag{7.5}$$

Consider the Lyapunov function

$$\mathrm{U}(\mathbf{z}) = \mathbf{z}^{\mathrm{T}}\mathrm{P}\mathbf{z} \tag{7.6}$$

where P is some positive definite matrix.

It can be shown [using Eq. (7.4)] that

$$\dot{\mathrm{U}} = \mathbf{z}^{\mathrm{T}}(\mathrm{A}^{\mathrm{T}}\mathrm{B}^{-\mathrm{T}}\mathrm{P} + \mathrm{P}\mathrm{B}^{-1}\mathrm{A})\mathbf{z} \equiv -\mathbf{z}^{\mathrm{T}}\mathrm{Q}\mathbf{z} \tag{7.7}$$

From the Lyapunov theorem, since the closed-loop system, Eq. (7.4), is asymptotically stable, for any given positive definite Q there exists a positive definite P that is the solution of the Lyapunov equation

$$(\mathrm{B}^{-1}\mathrm{A})^{\mathrm{T}}\mathrm{P} + \mathrm{P}(\mathrm{B}^{-1}\mathrm{A}) = -\mathrm{Q} \tag{7.8}$$

Suppose the output vectors due to imprecise collocation of sensors and actuators are given by

$$\mathbf{y} = (\mathrm{D} + \Delta_1)^{\mathrm{T}}\mathbf{x}, \qquad \dot{\mathbf{y}} = (\mathrm{D} + \Delta_2)^{\mathrm{T}}\dot{\mathbf{x}} \tag{7.9}$$

The following theorem, originally proven in Ref. 46, gives an upper bound [on the size of noncollocation perturbation matrices Δ_1 and Δ_2 of the sensor influence matrix] which guarantee stability.

Theorem 7.1

The closed-loop system with imprecisely collocated sensors and actuators is asymptotically stable if

$$\|\mathrm{G}_1\|_s \ \|\Delta_1\|_s + \|\mathrm{G}_2\|_s \ \|\Delta_2\|_s < \frac{\lambda_{\mathrm{m}}(\mathrm{Q}) \ \lambda_{\mathrm{m}}(\mathrm{M})}{2\|\mathrm{D}\|_s \ \lambda_{\mathrm{M}}(\mathrm{P})} \leq \frac{\lambda_{\mathrm{m}}(\mathrm{M})}{2\|\mathrm{D}\|_s \ \lambda_{\mathrm{M}}(\tilde{\mathrm{P}})} \tag{7.10}$$

where $\tilde{\mathrm{P}}$ is the solution of the Lyapunov equation (7.8) when Q is an identity matrix, $\|\cdot\|_s$ denotes the spectral norm (maximum singular value) of a matrix $[\cdot]$, and $\lambda_{\mathrm{M}}(\cdot)$ and $\lambda_{\mathrm{m}}(\cdot)$ denote the largest and the smallest eigenvalues of the matrix $[\cdot]$.

■

Proof of Theorem 7.1

The closed-loop system with imprecise collocation is given by

$$\mathrm{B}\dot{\mathbf{z}} = \mathrm{A}\mathbf{z} - \begin{bmatrix} 0 & 0 \\ \mathrm{D}\mathrm{G}_1\Delta_1^{\mathrm{T}} & \mathrm{D}\mathrm{G}_2\Delta_2^{\mathrm{T}} \end{bmatrix} \mathbf{z} \equiv (\mathrm{A} + \mathrm{E})\mathbf{z} \tag{7.11}$$

Consider the Lyapunov function

$$\mathrm{U}(\mathbf{z}) = \mathbf{z}^{\mathrm{T}}\mathrm{P}\mathbf{z} \tag{7.12}$$

Using Eqs. (7.8) and (7.11), it can be shown that

$$\begin{aligned} \dot{\mathrm{U}} &= \dot{\mathbf{z}}^{\mathrm{T}}\mathrm{P}\mathbf{z} + \mathbf{z}^{\mathrm{T}}\mathrm{P}\dot{\mathbf{z}} \\ &= \mathbf{z}^{\mathrm{T}}\left[\mathrm{A}^{\mathrm{T}}\mathrm{B}^{-\mathrm{T}}\mathrm{P} + \mathrm{P}\mathrm{B}^{-1}\mathrm{A}\right]\mathbf{z} + 2\mathbf{z}^{\mathrm{T}}\mathrm{P}\mathrm{B}^{-1}\mathrm{E}\mathbf{z} \\ &= -\mathbf{z}^{\mathrm{T}}\mathrm{Q}\mathbf{z} + 2\mathbf{z}^{\mathrm{T}}\mathrm{P}\mathrm{B}^{-1}\mathrm{E}\mathbf{z} \end{aligned} \tag{7.13}$$

For $\dot{\mathrm{U}}$ to be negative definite, a sufficient condition is

$$|2\mathbf{z}^{\mathrm{T}}\mathrm{P}\mathrm{B}^{-1}\mathrm{E}\mathbf{z}| < \mathbf{z}^{\mathrm{T}}\mathrm{Q}\mathbf{z} \tag{7.14}$$

Since

$$|z^T P B^{-1} E z| \leq \|P\|_s \; \|B^{-1}\|_s \; \|E\|_s \; \|z\|^2 \tag{7.15}$$

and

$$z^T Q z \geq \lambda_m(Q) \|z\|^2 \tag{7.16}$$

and noting that (since B^{-1} and P are symmetric real matrices), we can set

$$\|P\|_s = \lambda_M(P), \qquad \|B^{-1}\|_s = \left[\lambda_m(B)\right]^{-1} \tag{7.17}$$

Using Eqs. (7.15)-(7.17) in Eq. (7.14), we find $\dot{U}$ is negative definite if

$$\|E\|_s \leq \frac{\lambda_m(Q)\; \lambda_m(B)}{2\lambda_M(P)} \tag{7.18}$$

From the definition of E in Eq. (7.11),

$$\|E\|_s \leq \|DG_1\Delta_1^T\|_s + \|DG_2\Delta_2^T\|_s \leq \left[\|G_1\|_s\|\Delta_1\|_s + \|G_2\|_s\|\Delta_2\|_s\right]\|D\|_s \tag{7.19}$$

Using the upper bound on $\|E\|_s$ [Eq. (7.19)] in Eq. (7.18) gives

$$\|G_1\|_s\|\Delta_1\|_s + \|G_2\|_s\|\Delta_2\|_s < \frac{\lambda_m(Q)\; \lambda_m(B)}{2\|D\|_s\; \lambda_M(P)} \tag{7.20}$$

The inequality in Eq. (7.10) can be obtained by noting that $\lambda_m(B) = \lambda_m(M)$, and this is the original condition for stability in Ref. 46.

It is proven in Appendix B (Lemmas B.1 and B.2) that the least conservative (i.e., maximum) bound for the size of the perturbations $\|E\|_s$ which still guarantee stability results by choosing Q=I [in Eq. (7.8)], so that

$$\frac{\lambda_m(Q)}{\lambda_M(P)} \leq \frac{1}{\lambda_M(\tilde{P})}$$

where $\tilde{P}$ is the Lyapunov solution when Q is an identity matrix.

Using the above upper bound, the rightmost inequality in Eq. (7.10) can be obtained, and this is a less conservative condition than that originally presented in Ref. 46.

Q.E.D.

Note that Theorem 7.1 generally provides conservative estimates, and it is apparent that the system will be guaranteed asymptotically stable for sufficiently small collocation errors. In practice, one often finds that particular systems remain stable for much larger Δ_i collocation errors than those that satisfy the conservative bound of Eq. (7.10). The robustness of stability properties for actuator/sensor placement errors, in the presence of unmodeled actuator/sensor dynamics and nonlinearities, are discussed in Ref. 46.

7.2.2 Relation to LQR Design

Every stabilizing controller is an optimal controller in some sense, and it is This viewpoint is useful, because it reveals the most important arbitrariness which is always present in selecting a performance measure, and it sometimes provides us with new physical insights into the properties of the controller. The following theorem shows that every structural output feedback controller is an optimal LQR that minimizes a generalized quadratic performance index (that includes a state-control cross penalty in addition to the standard state and control quadratic penalties). In Ref. 46, the following theorem is proven by substituting weightmatrices into the necessary conditions of optimality. Our alternative proof utilizes the optimality condition and a given structural feedback law. This alternative proof of the central theorem relating symmetric feedback to the LQR seems qualitatively appealing. This theorem provides anther way to see that structural output feedback guarantees the global stability of the closed-loop system.

Theorem 7.2

For the system given by Eq. (7.1), the symmetric output feedback controller given by Eq. (7.2) minimizes the performance index:

$$J = \int_0^\infty \left(\mathbf{z}^T Q \mathbf{z} + 2\mathbf{z}^T S \mathbf{u} + \mathbf{u}^T R \mathbf{u}\right) dt \tag{7.21}$$

where the weight matrices are parameterized as a function of the symmetric gains as

$$Q = \begin{bmatrix} DG_1G_2^{-1}G_1D^T & 0 \\ 0 & DG_2D^T + 2C \end{bmatrix}, \quad R = G_2^{-1}, \quad S = \begin{bmatrix} DG_1G_2^{-1} \\ 0 \end{bmatrix} \tag{7.22}$$

and $\mathbf{z}$ is defined as in Eq. (7.5).

■

Proof of Theorem 7.2

Defining $\mathbf{z}$ as in Eq. (7.5), Eq. (7.1) can be expressed as

$$\dot{\mathbf{z}} = A_o \mathbf{z} + B_o \mathbf{u} \tag{7.23}$$

where

$$A_o = \begin{bmatrix} 0 & I \\ -M^{-1}K & -M^{-1}C \end{bmatrix}, \qquad B_o = \begin{bmatrix} 0 \\ M^{-1}D \end{bmatrix} \tag{7.24}$$

The problem is to determine a performance index for which the given control law is optimum, where the given control law is a symmetric output feedback law with the structure:

$$\mathbf{u} = -(G_1D^T\mathbf{x} + G_2D^T\dot{\mathbf{x}}) = -\,[\,G_1D^T \quad G_2D^T\,]\,\mathbf{z} \tag{7.25}$$

where G_1 and G_2 are positive definite symmetric gain matrices.

The solution for the optimal control problem [Eqs. (7.21) and (7.23)] can be obtained by using a variational calculus method (Ref. 47). Instead of this standard solution process, we use a device analogous to square-completion, which helps to understand the optimality condition and the role of the weight matrices.

Letting P be any constant symmetric matrix, and with $\mathbf{u}(t)$ and $\mathbf{z}(t)$ as above, we have

$$\begin{aligned}\frac{d}{dt}(\mathbf{z}^T P\mathbf{z}) &= \dot{\mathbf{z}}^T P\mathbf{z} + \mathbf{z}^T P\dot{\mathbf{z}} \\ &= (A_o\mathbf{z} + B_o\mathbf{u})^T P\mathbf{z} + \mathbf{z}^T P(A_o\mathbf{z} + B_o\mathbf{u})\end{aligned}$$

Integrating both sides of this equation from t=0 to t=∞, we obtain

$$\mathbf{z}^T P\mathbf{z}\Big|_{t=\infty} - \mathbf{z}^T P\mathbf{z}\Big|_{t=0} = \int_0^\infty \left\{\mathbf{z}^T(PA_o + A_o^T P)\mathbf{z} + 2\mathbf{u}^T B_o^T P\mathbf{z}\right\}dt$$

Since we consider only stable admissible controllers for which

$$\mathbf{z}(t) \longrightarrow 0 \qquad \text{as} \qquad t \longrightarrow \infty \tag{7.26}$$

it follows that

$$\int_0^\infty \left\{\mathbf{z}^T(PA_o + A_o^T P)\mathbf{z} + 2\mathbf{u}^T B_o^T P\mathbf{z}\right\}dt + \mathbf{z}_o{}^T P\mathbf{z}_o = 0 \tag{7.27}$$

where $\mathbf{z}_o = \mathbf{z}(0)$.

Note that since the system of Eq. (7.23) is assumed a priori to be stabilizable, then the controller implicitly assumed in Eq. (7.27) exists.

Adding Eq. (7.27) to the performance index of Eq. (7.21), we obtain, for any admissible controllers,

$$J = \int_0^\infty \left\{\mathbf{z}^T(PA_o + A_o^T P + Q)\mathbf{z} + 2\mathbf{u}^T(B_o^T P + S^T)\mathbf{z} + \mathbf{u}^T R\mathbf{u}\right\}dt + \mathbf{z}_o{}^T P\mathbf{z}_o$$

By considering the second and third terms in the integrand, we can add and subtract the following term to make square-completion

$$\mathbf{z}^T(B_o^T P + S^T)^T R^{-1}(B_o^T P + S^T)\mathbf{z}$$

Then the performance index becomes

$$\begin{aligned} J = &\int_0^\infty \mathbf{z}^T\left[PA_o + A_o^T P + Q - (B_o^T P + S^T)^T R^{-1}(B_o^T P + S^T)\right]\mathbf{z}dt \\ &+ \int_0^\infty \left\{\mathbf{u} + R^{-1}(B_o^T P + S^T)\mathbf{z}\right\}^T R\left\{\mathbf{u} + R^{-1}(B_o^T P + S^T)\mathbf{z}\right\}dt \\ &+ \mathbf{z_o}^T P \mathbf{z_o} \end{aligned} \tag{7.28}$$

Since R is positive definite, and the symmetric matrix is entirely arbitrary, the integrand here is non-negative and can be made zero (i.e., if we select P to make the first integrand zero and then choose $\mathbf{u}(t)$ to minimize J).

We find $\mathbf{u}(t)$, the optimal feedback law, as follows:

$$\mathbf{u} = -R^{-1}(B_o^T P + S^T)\mathbf{z} \tag{7.29}$$

and

$$A_o^T P + PA_o - SR^{-1}B_o^T P - PB_o R^{-1}S^T - PB_o R^{-1}B_o^T P = -(Q - SR^{-1}S^T) \tag{7.30}$$

Since our system is a mechanical second-order system, it is reasonable to impose a partitioned form for the matrices P, Q, and S as follows (since P and Q are symmetric matrices, matrices P_1, P_3, Q_1, and Q_3 are symmetric):

$$P = \begin{bmatrix} P_1 & P_2 \\ P_2^T & P_3 \end{bmatrix}, \qquad Q = \begin{bmatrix} Q_1 & Q_2 \\ Q_2^T & Q_3 \end{bmatrix}, \qquad S = \begin{bmatrix} S_1 \\ S_2 \end{bmatrix} \tag{7.31}$$

Substituting Eqs. (7.24) and (7.31) into Eq. (7.30) yields the three equations

$$\begin{aligned} -KM^{-1}P_2^T - P_2M^{-1}K - S_1R^{-1}D^TM^{-1}P_2^T - P_2M^{-1}DR^{-1}S_1^T \\ -P_2M^{-1}DR^{-1}D^TM^{-1}P_2^T = -Q_1 + S_1R^{-1}S_1^T \end{aligned} \tag{7.32}$$

$$\begin{aligned} -KM^{-1}P_3 + P_1 - P_2M^{-1}C - S_1R^{-1}D^TM^{-1}P_3 - P_2M^{-1}DR^{-1}S_2^T \\ -P_2M^{-1}DR^{-1}D^TM^{-1}P_3 = -Q_2 + S_1R^{-1}S_2^T \end{aligned} \tag{7.33}$$

$$\begin{aligned} P_2 - CM^{-1}P_3 + P_2^T - P_3M^{-1}C - S_2R^{-1}D^TM^{-1}P_3 - P_3M^{-1}DR^{-1}S_2^T \\ -P_3M^{-1}DR^{-1}D^TM^{-1}P_3 = -Q_3 + S_2R^{-1}S_2^T \end{aligned} \tag{7.34}$$

Since we have a given controller, Eq. (7.25) should be equal to Eq. (7.29). Equating Eq. (7.25) and Eq. (7.29), after substituting Eqs. (7.24) and (7.31) into Eq. (7.29), gives

$$R^{-1}(D^TM^{-1}P_2^T + S_1^T) = G_1D^T \tag{7.35}$$

$$R^{-1}(D^TM^{-1}P_3 + S_2^T) = G_2D^T \tag{7.36}$$

By considering the fact that P_3 is a positive definite symmetric matrix and by using heuristic insights, we can take P_2, P_3, S_1, S_2, and R from Eqs. (7.35) and (7.36) as follows:

$$P_3 = M, \qquad R = G_2^{-1}, \qquad S_2 = 0, \qquad D^TM^{-1}P_2^T + S_1^T = G_2^{-1}G_1D^T \tag{7.37}$$

These are solutions, as can be verified by direct substitutions, but we have not proven their uniqueness.

Substituting Eqs. (7.37) into Eqs. (7.32)-(7.34) yields

$$Q_1 = (KM^{-1}P_2^T + P_2M^{-1}K) + DG_1G_2^{-1}G_1D^T \tag{7.38}$$

$$P_1 = K + DG_1D^T + P_2M^{-1}C - Q_2 \tag{7.39}$$

$$Q_3 = 2C + DG_2D^T - (P_2 + P_2^T) \tag{7.40}$$

Observing that P_1 is a positive definite symmetric matrix and by heuristic insights, we can take P_1, P_2, and Q_2 from Eq. (7.39) as follows:

$$P_1 = K + DG_1D^T, \qquad P_2 = 0, \qquad Q_2 = 0 \tag{7.41}$$

Substituting Eqs. (7.41) into Eqs. (7.37), (7.38) and (7.40) gives

$$Q_1 = DG_1G_2^{-1}G_1D^T, \qquad Q_3 = 2C + DG_2D^T, \qquad S_1 = DG_1G_2^{-1} \tag{7.42}$$

Equations (7.37), (7.41), and (7.42) establish the weight matrices [Eq. (7.22)] in the the performance index [Eq. (7.21)] minimized by using the symmetric output feedback control [Eq. (7.25)].

Q.E.D.

In this section, two central aspects of the theoretical background of symmetric output feedback control have been treated: (1) stability robustness analysis in the presence of sensor/actuator noncolocation, (2) the relationship of symmetric output feedback to the LQR problem. These developments supplement the orginal work of Joshi (Ref. 46), by offering alternative formulation paths and modest generalizations. The latter point is interesting, owing to the degree of arbitrariness in selection of weight matrices. Given the above developments, it is evident that symmetric output feedback and the LQR have an equivalence in the sense that they can both be looked upon as related means for parameterizing an infinite family of stable control laws. To implement either, it is necessary to exercise the implicit design freedom to select, from the infinity of possiblilties, a particular design, using heuristic judgement or numerical studies to directly specify admissible control gains in the symmetric output feedback case, and to indirectly specify them (through weight matrix choices) for the LQR. Since many practical constraints and considerations such as actuator saturation are not captured in either formulation, it is to be anticipated that design iteration will be required to implement either of these approaches. For control gain design using symmetric output feedback, an eigenstructure assignment optimization process is developed in the following section.

7.3 Optimization Problem Formulation Using Pole Placement

From section 7.2, it is evident that judicious selection of the gain matrices from a stable family (i.e. the positive definiteness of the G_i matrices defines an infinity of stable feedback controls) guarantees the global stability of the closed-loop system in the Lyapunov sense. The most familiar member of the family is the so-called *direct feedback* case of diagonal G_1 and G_2. More generally, however, fully populated G_i matrices allow commands to every actuator to be generated by linearly operating on all sensor measurements, and thereby provide significantly increased controller design freedom while maintaining the stability and robustness guarantees. This class of controllers is attractive because a stable family of controllers is guaranteed even in the presence of model errors, and it is not necessary to implement a state observer. However, it is necessary to have all sensors located at actuator stations, and this requirement presents a stumbling block in some applications. On the other hand, as developed by Joshi and extended above, it is relatively staightforward to gain the necessary insight to account for departure from the ideal co-location of sensors and actuators.

It is significant that including sensor and actuator dynamics destroys some of the above-mentioned properties and weakens the stability claims. For these reasons, generalizations and modifications of this approach have recently been proposed (Refs. 6, 46) to circumvent some of these difficulties.

We can admit the most general family of positive semidefinite gain matrices by introducing the Cholesky decomposition gain parameterizations

$$G_1 = L_1 L_1^T, \qquad G_2 = L_2 L_2^T \tag{7.43}$$

Note that G_1 and G_2 are guaranteed to be positive semidefinite for all real elements (q_{ij}, r_{ij}) assigned to the following lower triangular matrices

$$L_1 = \begin{bmatrix} q_{11}^2 & 0 & 0 & \cdots & 0 \\ q_{21} & q_{22}^2 & 0 & \cdots & 0 \\ q_{31} & q_{32} & q_{33}^2 & \cdots & 0 \\ \vdots & \vdots & \vdots & \ddots & \vdots \\ q_{m1} & q_{m2} & q_{m3} & \cdots & q_{mm}^2 \end{bmatrix}, \quad L_2 = \begin{bmatrix} r_{11}^2 & 0 & 0 & \cdots & 0 \\ r_{21} & r_{22}^2 & 0 & \cdots & 0 \\ r_{31} & r_{32} & r_{33}^2 & \cdots & 0 \\ \vdots & \vdots & \vdots & \ddots & \vdots \\ r_{m1} & r_{m2} & r_{m3} & \cdots & r_{mm}^2 \end{bmatrix} \tag{7.44}$$

Let the corresponding gain parameter vector be

$$\mathbf{p} = [q_{11} \quad q_{21} \quad \cdots \quad q_{m1} \quad q_{22} \quad \cdots \quad q_{mm} \quad r_{11} \quad r_{21} \quad \cdots \quad r_{mm}]^T \tag{7.45}$$

Through Eqs. (7.43)-(7.45), it is clear that the gain matrices

$$G_1 = G_1(\mathbf{p}), \qquad G_2 = G_2(\mathbf{p}) \tag{7.46}$$

are continuous, differentiable functions of $\mathbf{p}$. The partial derivatives of the gain matrices $(\partial G_i / \partial p_k)$ with respect to the elements of the parameter vector $\mathbf{p}$ can be constructed easily by direct differentiation, these are given in Appendix F.

The closed-loop system can be rewritten as

$$M\ddot{\mathbf{x}} + \hat{C}\dot{\mathbf{x}} + \hat{K}\mathbf{x} = 0 \tag{7.47}$$

where the closed-loop system damping and stiffness matrices are

$$\hat{C} = C + DG_2D^T, \qquad \hat{K} = K + DG_1D^T \tag{7.48}$$

It is apparent from Eqs. (7.47) and (7.48) that the closed-loop damping and stiffness matrices are simple algebraic functions of the parameter vector $\mathbf{p}$, i.e.,

$$\hat{C} = \hat{C}(\mathbf{p}), \qquad \hat{K} = \hat{K}(\mathbf{p}) \tag{7.49}$$

Considering a first-order statespace form of the system differential equations which is equivalent to the second-order closed-loop system of Eq. (7.47), we write

$$B\dot{\mathbf{z}} = A\mathbf{z} \tag{7.50}$$

where

$$\mathbf{z} = \begin{Bmatrix} \mathbf{x} \\ \dot{\mathbf{x}} \end{Bmatrix}, \qquad A = \begin{bmatrix} 0 & M \\ -\hat{K} & -\hat{C} \end{bmatrix}, \qquad B = \begin{bmatrix} M & 0 \\ 0 & M \end{bmatrix} \tag{7.51}$$

It is evident that

$$A = A(\mathbf{p}) \tag{7.52}$$

The closed-loop right and left eigenvalue problems [associated with $\mathbf{z} = \hat{\underline{\phi}} e^{\lambda t}$ solutions of Eq. (7.50)] are

$$\begin{aligned} \text{right:} \quad & \lambda_i B \hat{\underline{\phi}}_i = A \hat{\underline{\phi}}_i, \qquad i = 1, 2, \ldots, 2n \\ \text{left:} \quad & \lambda_i B^T \hat{\underline{\psi}}_i = A^T \hat{\underline{\psi}}_i, \qquad i = 1, 2, \ldots, 2n \end{aligned} \tag{7.53}$$

where the conventional normalization (Ref. 48) of the biorthogonality conditions for the eigenvectors are adopted as

$$\begin{aligned} & \hat{\underline{\phi}}_i^T B \hat{\underline{\phi}}_i = 1, \qquad i = 1, 2, \ldots, 2n \\ & \hat{\underline{\psi}}_j^T B \hat{\underline{\phi}}_i = \delta_{ij}, \qquad i, j = 1, 2, \ldots, 2n \end{aligned} \tag{7.54}$$

so that

$$\hat{\underline{\psi}}_j^T A \hat{\underline{\phi}}_i = \lambda_i \delta_{ij}, \qquad i, j = 1, 2, \ldots, 2n \tag{7.55}$$

Since $A = A(\mathbf{p})$ is a continuous function of $\mathbf{p}$, it is natural that we consider the eigenvalues and eigenvectors to also be functions of parameter vector $\mathbf{p}$.

Our goal is to find the parameter vector $\mathbf{p}$ that satisfies prescribed design objectives. To focus the discussion, suppose that our design objectives are: (1) to move some of the eigenvalues into an acceptable region of the complex plane and (2) to maximize a robustness measure for the closed-loop system. More generally, we can impose inequality constraints to enforce actuator saturation limits and other admissible control conditions (Ref. 6). The desired region for the eigenvalues can be defined by using either equality or inequality constraints on the eigenvalues and the optimization crateria can, for example be a closed-loop robustness measure. From the developments of Chapter 2, it is clear that conditioning of the eigenvectors is a direct measure of the sensitivity of the corresponding eigenvalues to perturbations in the elements of the closed-loop system matrices. The condition number of the modal matrix of closed-loop eigenvectors can be used to establish finite bounds on the variation of system matrices within which stability can be guaranteed. To define a representative robust control design problem, we adopt as a optimization criterion the condition number $\mathcal{K}(\Phi^c(\mathbf{p}))$ of the closed-loop modal matrix Φ^c (whose columns are the closed-loop right eigenvectors $\hat{\underline{\phi}}_i$, $i = 1, \ldots, 2n$).

The condition number of a general complex matrix Φ^c is related to the largest and least singular values of Φ^c as

$$\mathcal{K}(\Phi^c) = \frac{\overline{\sigma}(\Phi^c)}{\underline{\sigma}(\Phi^c)} \tag{7.56}$$

where $\overline{\sigma}$ and $\underline{\sigma}$ denote the maximum and minimum singular values, respectively.

A general treatment of singular value decomposition and a derivation of the associated sensitivities with respect to parameter variations are derived in section 2.3 and Appendix C.

Our problem can be stated as a nonlinear parameter optimization problem in which we seek to impose specified eigenspace equality and inequality constraints and minimize the condition number of the closed-loop modal matrix. By considering all constraints and the performance index as functions of the parameter vector $\mathbf{p}$, we can formulate our nonlinear programming problem as follows:

$$\underset{\mathbf{p}\in\Omega}{\text{Minimize}} \qquad J(\mathbf{p}) = \mathcal{K}(\Phi^c(\mathbf{p})) = \frac{\overline{\sigma}(\Phi^c(\mathbf{p}))}{\underline{\sigma}(\Phi^c(\mathbf{p}))} \tag{7.57}$$

subject to

$$f_i(\mathbf{p}) = \omega_{di}^o - \omega_{di}(\mathbf{p}) = 0, \qquad i = 1, \dots, m \tag{7.58}$$

$$g_i(\mathbf{p}) = \omega_{di}(\mathbf{p})\zeta_i(\mathbf{p}) \geq T_i, \qquad i = 1, \dots, m \tag{7.59}$$

where

$$\omega_{di} = \Im(\lambda_i) \ \text{(damped natural frequency)}$$

$$\zeta_i(\mathbf{p}) = -\frac{\Re(\lambda_i)}{|\lambda_i|} \ \text{(damping factor)}$$

Ω is the set of admissible gain $\mathbf{p}$ vectors, ω_{di}^o denotes the objective values for the damped natural frequencies, and T_i denotes the objective values for bounding the modal time constants.

When applying gradient-based optimization algorithms, it is useful to compute analytical partial derivatives of the eigenvalues, eigenvectors, and the singular values.

For the case of distinct eigenalues, the partial derivatives of eigenvalues and eigenvectors are as follows (see section 2.3):

$$\begin{aligned} \frac{\partial \lambda_i}{\partial p_k} &= \hat{\underline{\psi}}_i^T \left(\frac{\partial A}{\partial p_k} - \lambda_i \frac{\partial B}{\partial p_k} \right) \hat{\underline{\phi}}_i, \qquad i = 1, 2, \dots, 2n \\ \frac{\partial \hat{\underline{\phi}}_i}{\partial p_k} &= \sum_{j=1}^{2n} a_{ji}^k \hat{\underline{\phi}}_j \\ \frac{\partial \hat{\underline{\psi}}_i}{\partial p_k} &= \sum_{j=1}^{2n} b_{ji}^k \hat{\underline{\psi}}_j \end{aligned} \tag{7.60}$$

where

$$
\begin{aligned}
a_{ji}^{k} &= \frac{1}{\lambda_i - \lambda_j}\hat{\underline{\psi}}_j^T\left(\frac{\partial A}{\partial p_k} - \lambda_i\frac{\partial B}{\partial p_k}\right)\hat{\underline{\phi}}_i\,, & j \neq i \\
&= -\frac{1}{2}\left[\hat{\underline{\phi}}_i^T\frac{\partial B}{\partial p_k}\hat{\underline{\phi}}_i + \sum_{\substack{l=1\\ l\neq i}}^{2n} a_{li}^{k}\hat{\underline{\phi}}_l^T(B + B^T)\hat{\underline{\phi}}_i\right], & j = i \\
b_{ji}^{k} &= \frac{1}{\lambda_i - \lambda_j}\hat{\underline{\psi}}_i^T\left(\frac{\partial A}{\partial p_k} - \lambda_i\frac{\partial B}{\partial p_k}\right)\hat{\underline{\phi}}_j\,, & j \neq i \\
&= -\hat{\underline{\psi}}_i^T\frac{\partial B}{\partial p_k}\hat{\underline{\phi}}_i - a_{ii}^{k}\,, & j = i
\end{aligned}
\tag{7.61}
$$

Except for isolated events of multiple eigenvalues and root bifurcations, we can consider the eigenvalues and eigenvectors to be smooth differentiable functions of the gain parameter vector **p**.

The gradient of the singular values are as follows (see section 2.3):

$$\frac{\partial \sigma_i}{\partial p_k} = \frac{1}{2}\left(U_i^H\frac{\partial \Phi^c}{\partial p_k}V_i + V_i^H\frac{\partial \Phi^{cH}}{\partial p_k}U_i\right) = \Re\left(U_i^H\frac{\partial \Phi^c}{\partial p_k}V_i\right) \tag{7.62}$$

where $\Re(\cdot)$ denotes the real part of $(\cdot)$.

The partial derivatives of the damping factors and damped frequencies with respect to the elements of the parameter vector **p** can be constructed easily via the chain rule (Appendix G).

For high-order systems such as a flexible structure, it is usually desirable and often necessary to develop a reduced-order model to save on computational time when designing control laws. A high-order model is generally retained to verify the resulting design. The most common order reduction method is the modal truncation method which introduces a partition into the modal matrix to select a subset of low-frequency modes (determined to be of most importance in the particular application). The criteria for mode selection can be based on modal cost analysis, for example. To perform the modal coordinate transformation, the open-loop eigenvalue problem of the system (7.1) should be solved first [see section 2.2.1 and Eqs. (4.136)-(4.140)]. The general modal coordinate transformation is

$$\mathbf{x}(t) = \Phi\underline{\eta}(t) = [\,\Phi_1 \quad \Phi_2\,]\begin{Bmatrix}\underline{\eta}_1(t)\\ \underline{\eta}_2(t)\end{Bmatrix} \tag{7.63}$$

where Φ is the modal matrix of open-loop eigenvectors, $\underline{\eta}_1$ is the retained subset of n_r modal coordinates that we wish to include in the reduced-order model, and $\underline{\eta}_2$ is the vector of residual modal coordinates.

Then the reduced-order equation of motion becomes

$$\tilde{M}\ddot{\underline{\eta}}_1 + \tilde{C}\dot{\underline{\eta}}_1 + \tilde{K}\underline{\eta}_1 = \tilde{D}\mathbf{u} \tag{7.64}$$

where

$$\begin{aligned}
\tilde{M} &= \Phi_1^T M \Phi_1 = I \\
\tilde{C} &= \Phi_1^T C \Phi_1 = \mathrm{diag}(2\zeta_1\omega_1, \ldots, 2\zeta_{n_r}\omega_{n_r}) \\
\tilde{K} &= \Phi_1^T K \Phi_1 = \mathrm{diag}(\omega_1^2, \ldots, \omega_{n_r}^2) \\
\tilde{D} &= \Phi_1^T D
\end{aligned}$$

The diagonal structure of $\tilde{C}$ requires that C be a linear combination of M and K (see Ref. 49 for more general conditions); we make this restriction for the sake of convenience. As a practical matter, the ζ_i are usually estimated from experimental identification processes, using the methods of Ref. 45, for example.

For "structural feedback" control, the control force vector takes the form:

$$\mathbf{u} = -(\tilde{G}_1\tilde{D}^T\underline{\eta}_1 + \tilde{G}_2\tilde{D}^T\dot{\underline{\eta}}_1) \tag{7.65}$$

where $\tilde{G}_1$ and $\tilde{G}_2$ are positive gain matrices for the reduced-order model. After manipulating the above equations, we can easily verify that gain matrices G_1 and G_2 for the original high-order system are identical, to within order reduction truncation errors, to the gain matrices $\tilde{G}_1$ and $\tilde{G}_2$ for the reduced-order model. That is, upon introducing the order reduction approximation $\mathbf{x} \approx \Phi_1\underline{\eta}_1$ into Eqs. (7.47) and (7.48), they become identical to Eqs. (7.64) and (7.65) with $G_i = \tilde{G}_i$. Even though we design the gain matrices by using the reduced-order model, we can use them for the original (in the limit, infinite dimensional) system with guaranteed stability.

While some approximations are obviously introduced in any order-reduction process, it is important that the stability of the full-order closed-loop system is theoretically guaranteed by using symmetric output feedback. This is because the order reduction approximations, in this case, merely select particular gain matrices from an inherently globally stable family (G_1 and G_2 must be positive definite). An analogous claim cannot be made, for example, if a conventional linear quadratic regulator control design is derived from a reduced order model. While stability is guaranteed for symmetric output feedback, of course, other performance measures may not be satisfactorily predicted by the particular reduced-order model, and in actual applications, one must address stability characteristics including the effects of external disturbances and sensor/actuator bandwidth limitations. In the following section, we discuss numerical methods for the nonlinear constrained optimization problem formulated in this section.

7.4 Two-Stage Homotopic Nonlinear Programming

We now consider a nonlinear programming algorithm, the central feature of which is the recursive use of a minimum norm differential correction to locally modify the design variables. This is a generalized Newton process for solving a system of underdetermined nonlinear equations. On each iteration, we minimize the norm (sum square) of the correction to the design variables required to satisfy specified equality constraints. The constraint functions are locally linearized to provide a system of linear algebraic equations to solve for the correction vector. The process is iterated with the derivatives (the Jacobian matrix) locally evaluated. We discuss below the procedure used to deal with inequality as well as equality constraints. We also consider a method (a homotopic continuation method) to dramatically enhance convergence and make convergence failures informative so that the optimization problem can be intelligently revised if the constraint set is unreachable.

Of course, the two key requirements underlying reliable convergence of any Newton-like nonlinear programming algorithm are: (1) the existence of a feasible solution and (2) the availability of a sufficiently close starting iterative. As we show below, the homotopic continuation method deforms the constraints by controllable, arbitrarily small amounts and thereby remains controllably close to a neighboring converged solution; this effectively guarantees that "arbitrarily close" starting iteratives are available for each local iteration. An embedding method is used to define a continuous family of neighboring problems that contains two important members: (1) a trivial problem (one for which the solution is available) and (2) the problem of interest.

The continuous one-parameter family of problems is constructed in such a way that an embedding parameter ($0 \leq \gamma \leq 1$) may be set to define any member of the family with $\gamma = 0$ generating the trivial problem and $\gamma = 1$ generating the problem of interest. By sweeping γ and judiciously assigning the γ increments, we control how closely spaced the neighboring solutions lie (in the space of the constraint functions), and the γ increments can be assigned adaptively based on convergence progress. Thus, in the iterative design process, we can remain as near a neighboring converged solution as is necessary to maintain sufficiently good starting iteratives. Therefore, if a local convergence cannot be achieved (for some γ value), we have eliminated the most common problem (having too a poor starting iterative to justify the linearizations), and we can focus on other, more interesting convergence issues, such as the nonexistence of feasible solutions, local linear dependence of the linearized constraint equations, and turning points. For example, comparing the active constraint sets and the corresponding Jacobian rows between the apparently unreachable set (the one for which convergence cannot be achieved) and the neighboring (previous) converged solution will reveal which constraints are in competition and will usually suggest avenues for revision of the problem and/or provide insights as to why the convergence failure occurred.

We now discuss the algorithm, beginning with the statement of the parameter optimization problem we are addressing: We seek to find the $\mathbf{p}^*$ that minimizes a function

$$J = J(\mathbf{p}) \tag{7.66}$$

subject to the constraints

$$\begin{aligned} f_i(\mathbf{p}) &= f_i^o \qquad & i &= 1, \ldots, m \\ g_j(\mathbf{p}) &\geq T_j \qquad & j &= 1, \ldots, m \end{aligned} \tag{7.67}$$

where f_i^o is the "objective" value of the ith equality constraint, and T_j represents boundaries on the feasible region.

First Stage: Seeking a Feasible Solution

Converging from an approximate starting solution $\mathbf{p}_{\mathbf{start}}$ to a feasible solution satisfying the constraints of Eqs. (7.67) may be approached by locally considering all of the equality constraints, but only the *active* (locally violated) subset of inequality constraints. The locally active inequality constraints are simply considered as additional equality constraints of the form

$$f_j(\mathbf{p}) = T_j, \qquad j = m+1, \ldots, r \tag{7.68}$$

The subset of inequality constraints $g_j(\mathbf{p}) \geq T_j$ treated as local equality constraints $f_j(\mathbf{p}) \equiv g_j(\mathbf{p}) = T_j$ in Eqs. (7.68) obviously depends on which constraints are locally violated and thus will typically change during the sequence of differential correction processes. So long as r (the number of locally violated constraints plus the m equality constraints) does not exceed N (the number of elements in $\mathbf{p}$), then we can proceed with the discussion below; otherwise, only the $N - m$ "most important" inequality constraints will be locally imposed, assuming that the original ordering represents the "importance hierarchy" of the inequality constraints. Adopting this approach permits the specification of greater than $N - m$ inequality constraints, although not more than $N - m$ can be active on any iteration, especially the last one. We have observed that it is often the case that final convergence to a feasible solution satisfying Eqs. (7.67) can be achieved even when local iterations encounter more than $N - m$ locally active constraints (i.e., it is typical that only a small subset of the specified inequality constraints are finally active when convergence to a feasible solution is achieved and, only under special circumstances will this finally active subset of constraints be invariant during all local iterations). These ideas must also be viewed in the context of the generalized homotopic continuation procedure that we discuss below.

Our restated local optimization problem is to minimize J of Eq. (7.66) subject to the equality constraint

$$\mathbf{f}(\mathbf{p}) = \mathbf{f}_o \tag{7.69}$$

where $\mathbf{f}(\mathbf{p})$ is the $r \times 1$ vector of m equality constraints and the $r - m$ active inequality constraints.

We initially address the problem of achieving a feasible solution near our starting solution, and delay considering minimization of J until the first feasible design has been achieved. To enhance convergence, we introduce a "portable objective" vector $\mathbf{f_p}(\gamma)$ as the one-parameter family of objective constraint values

$$\mathbf{f_p}(\gamma) = \gamma \mathbf{f}_o + (1 - \gamma)\mathbf{f}(\mathbf{p_{start}}), \qquad 0 \leq \gamma \leq 1 \tag{7.70}$$

where γ is the imbedding parameter. The *linear homotopy map* $\mathbf{H}(\mathbf{p}(\gamma), \gamma)$ is generated by replacing $\mathbf{f}_o$ in Eq. (7.69) by $\mathbf{f_p}$ from Eq. (7.70) to obtain

$$\mathbf{H}(\mathbf{p}(\gamma), \gamma) \equiv \gamma \mathbf{f}_o + (1 - \gamma)\mathbf{f}(\mathbf{p_{start}}) - \mathbf{f}(\mathbf{p}(\gamma)) = 0 \tag{7.71}$$

Notice the γ boundary conditions satisfied by $\mathbf{H}(\mathbf{p}(\gamma), \gamma)$ in Eq. (7.71):

$$\begin{aligned} &\text{at} \quad \gamma = 0: \ \text{trivial problem} \qquad [\mathbf{p}(0) \equiv \mathbf{p_{start}}] \\ &\text{at} \quad \gamma = 1: \ \text{problem of interest} \quad [\mathbf{p}(1) \equiv \mathbf{p_{feasible}}] \\ &\qquad\qquad [\equiv \text{a solution satisfying} \quad \mathbf{f}_o - \mathbf{f}(\mathbf{p}(1)) = 0] \end{aligned} \tag{7.72}$$

It is clear that an arbitrary guess $\mathbf{p_{start}}$ satisfies the homotopy map of Eq. (7.71) for $\gamma = 0$ and that, if we can obtain a solution for $\gamma = 1$, then we have determined a feasible solution $\mathbf{p_{feasible}}$. It is also apparent that sweeping γ at a suitably small increment generates a sequence of neighboring problems. We solve this sequence of problems using the neighboring converged solutions to generate starting iteratives for each subsequent problem.

The local design corrections (for each γ specification) are done by linearizing Eq. (7.71) about the local $\mathbf{p}$ iterative and computing the minimum norm differential correction (Appendix H) that satisfies the linearized constraint equation. The minimum norm differential correction is as follows:

$$\Delta\mathbf{p} = -\mathrm{W}^{-1}\mathrm{A}^{\mathrm{T}}(\mathrm{A}\mathrm{W}^{-1}\mathrm{A}^{\mathrm{T}})^{-1}\mathbf{H}(\mathbf{p}(\gamma), \gamma) \tag{7.73}$$

and we use the differential corrections

$$\mathbf{p_{new}} = \mathbf{p_{old}} + \Delta\mathbf{p} \tag{7.74}$$

to iterate until Eq. (7.71) is satisfied for each local γ value. When a local convergence is acheived, γ is incremented by a prescribed amount. If local convergence is not achieved, the attempted γ increment is reduced and γ is assigned nearer the largest γ value for which the neighboring convergence was previously achieved. The process is halted if (1) convergence to a feasible solution is achieved (at $\gamma = 1$), or (2) local convergence cannot be achieved when we increment γ by some small tolerance $\Delta\gamma \leq \epsilon$. In the latter case the user should reconsider the problem statement in view of the active constraint set and other considerations from evaluation of the final (largest γ) solution achieved.

Second Stage: Descent to the Inequality Constrained Minimum Solution

The first stage of the process is complete when/if a feasible solution is obtained ($\gamma = 1$ achieved in the above algorithm). Suppose that we have achieved convergence to a feasible solution. Having a feasible solution, we now consider minimization of the performance index J subject to the local equality constraint of Eq. (7.71). Assuming there are $r < \mathrm{N}$ active constraints, the above algorithm can be modified to make it seek out the minimum value of J while satisfying the constraints. This modification can readily be accomplished (Ref. 50) by simply introducing an objective value ($\mathrm{J_o}$) for the performance index and treating the performance index as the (r+1)th equality constraint. The corresponding minimum norm differential correction, in lieu of Eq. (7.73), is

$$\Delta\mathbf{p} = -\mathrm{W}^{-1}\tilde{\mathrm{A}}^{\mathrm{T}}(\tilde{\mathrm{A}}\mathrm{W}^{-1}\tilde{\mathrm{A}}^{\mathrm{T}})^{-1}\tilde{\mathbf{H}}(\mathbf{p}(\gamma),\gamma) \tag{7.75}$$

where

$$\tilde{\mathrm{A}} \equiv \begin{bmatrix} & \mathrm{A} & & \\ \cdots & \cdots & \cdots & \cdots \\ \frac{\partial \mathrm{J}}{\partial \mathrm{p}_1} & \frac{\partial \mathrm{J}}{\partial \mathrm{p}_2} & \cdots\cdots & \frac{\partial \mathrm{J}}{\partial \mathrm{p}_\mathrm{N}} \end{bmatrix}, \qquad \tilde{\mathbf{H}}(\mathbf{p}(\gamma),\gamma) \equiv \begin{pmatrix} \mathbf{H}(\mathbf{p}(\gamma),1) \\ \cdots\cdots\cdots \\ \mathrm{J}(\mathbf{p}(\gamma)) - \mathrm{J_o} \end{pmatrix} \tag{7.76}$$

Note in Eq. (7.76) that the upper part of the vector $\tilde{\mathbf{H}}(\mathbf{p}(\gamma),\gamma)$ is $\mathbf{H}(\mathbf{p}(\gamma),1)$ in Eq. (7.71); this means that the feasible solution constraints [satisfaction of Eq. (7.67)] should be retained while minimizing the performance index (seeking the objective value $\mathrm{J_o}$). Observe that the performance objective value $\mathrm{J_o}$ should be assigned with some insight into the physical significance and typical values for J. In the event that convergence difficulties arise, $\mathrm{J_o}$ can be replaced by a portable value $\mathrm{J_o}(\gamma)$, analagous to the portable constraint boundaries [Eq. (7.70)], as discussed below [see Eq. (7.81)].

Instead of computing the explicit inverse indicated in Eqs. (7.73) or Eqs. (7.75), we typically make use of the more numerically robust SVD pseudoinverse method (see section 2.2.2) to solve for the correction $\Delta\mathbf{p}$ from

$$\tilde{\mathbf{H}} = -\tilde{\mathrm{A}}\Delta\mathbf{p} \tag{7.77}$$

Any $(r+1) \times N$ matrix $\tilde{A}$ can be decomposed as follows:

$$\tilde{A} = [U_1 \quad U_2] \begin{bmatrix} \Sigma_1 & 0 \\ 0 & 0 \end{bmatrix} \begin{bmatrix} V_1^H \\ V_2^H \end{bmatrix} \tag{7.78}$$

where, if $\tilde{A}$ has maximum rank, Σ_1 is an $(r+1) \times (r+1)$ diagonal matrix (otherwise, the dimension of Σ_1 is the rank of $\tilde{A}$), (U_1, V_1) are unitary matrices spanning the (row, column) space of $\tilde{A}$, and (U_2, V_2) span the corresponding null spaces of $\tilde{A}$.

The Moore-Penrose pseudoinverse of $\tilde{A}$ is given by (see Application 2.2)

$$\tilde{A}^\dagger = V_1 \Sigma_1^{-1} U_1^H \tag{7.79}$$

Therefore, the minimum norm solution Eq. (7.77) can be computed as

$$\Delta \mathbf{p} = -\tilde{A}^\dagger \tilde{\mathbf{H}} \tag{7.80}$$

If needed, of course, we can easily modify the pseudoinverse solution to accommodate a nonidentity weighting matrix. For moderate-dimensioned and well-conditioned problems, Eq. (7.75) is usually numerically well behaved.

Upon achieving convergence to the feasible region, we reset $\gamma = 0$ and subsequently use Eq. (7.75) or Eq. (7.80) to compute the differential corrections. For the iterations subsequent to the first feasible solution, a new "constrained optimization homotopy process" is established to drive the objective function to its minimum value or, alternatively, drive it to a "goal" value $J_{\mathbf{goal}}$. The objective value $J_o(\gamma)$ is assigned using the one-parameter family of portable values

$$J_o(\gamma) = \gamma J_{\mathbf{goal}} + (1-\gamma) J(\mathbf{p}_{\mathbf{feasible}}) \tag{7.81}$$

where $\mathbf{p}_{\mathbf{feasible}}$ is the *first* feasible solution [satisfying Eq. (7.71) with $\gamma = 1$].

The value for $J_{\mathbf{goal}}$ may be an actual goal, but is more typically interpreted as "the best one could possibly hope for;" it is often taken as zero for minimization of positive performance indices. We increase γ from zero toward unity, with the size of the increments dictated by the convergence progress of the differential corrections for each γ value. When convergence cannot be achieved for a small tolerance increase in γ, we adopt the solution corresponding to the largest γ for which convergence was achieved as the constrained minimum. It is rigorously proven in Ref. 51 that this process is mathematically equivalent to a gradient projection algorithm with appropriate strategies for correction step size control. However, this algorithm is superior to the conventional implementations of the gradient projection method because (1) the correction formulas are more easily programmed, and (2) the one-embedded-parameter homotopic continuation process has been found much more attractive than step size control because it is inherently adaptive and self-starting. Of course, if the problem under consideration has multiple constrained minima, one will find that the above problem formulation does not save us from the usual local versus global convergence difficulties. Turning points are occasionally encountered (characterized by rank deficiency of the local constraint jacobian); these can be accommodated by using the Chow-Yorke homotopy algorithm (Ref. 51). Figure 7.1 provides a logical flowchart showing (the macroscopic details of) a homotopic nonlinear programming algorithm.

In the following section, we design a robust output feedback controller for a simple system by using homotopic nonlinear programming which was discussed above. We also examine the performance of the resulting controller design in the presence of model uncertainty.

7.5 Case Studies: Numerical Applications

7.5.1 Simple Example (Three Degree of Freedom Problem)

To illustrate the simplest application of the robust output feedback controller approach developed in the previous sections, we consider a coupled oscillator system of sixth order having two actuators. The system matrices are

$$\mathrm{M} = \begin{bmatrix} 1 & 0 & 0 \\ 0 & 1 & 0 \\ 0 & 0 & 1 \end{bmatrix}, \qquad \mathrm{K} = \begin{bmatrix} 20 & -10 & 0 \\ -10 & 30 & -20 \\ 0 & -20 & 20 \end{bmatrix}, \qquad \mathrm{D} = \begin{bmatrix} 1 & 0 \\ 0 & 0 \\ 0 & 1 \end{bmatrix}$$

We adopt the condition number of the closed-loop modal matrix as our performance measure and consider two equality and two inequality constraints on eigenvalues placement [Eqs. (7.58) and (7.59)]. The objective values for the closed-loop damped natural frequencies ($\omega^o_{\mathrm{d}_i}$) are taken as their corresponding zero gain values. Inequality constraints are introduced to bound the slowest system time constant, and we adopted $\mathrm{T}_i = 0.3, i = 1, 2$.

Utilizing the two-stage homotopic nonlinear programming developed in the previous section, we successfully solved the above problem. The optimization process was initiated by simply adopting identity matrices for the $\mathrm{G_i}$ starting iteratives. Ten homotopy steps were required with three or four minimum norm differential gain corrections on each step. The final converged position and velocity gain matrices are

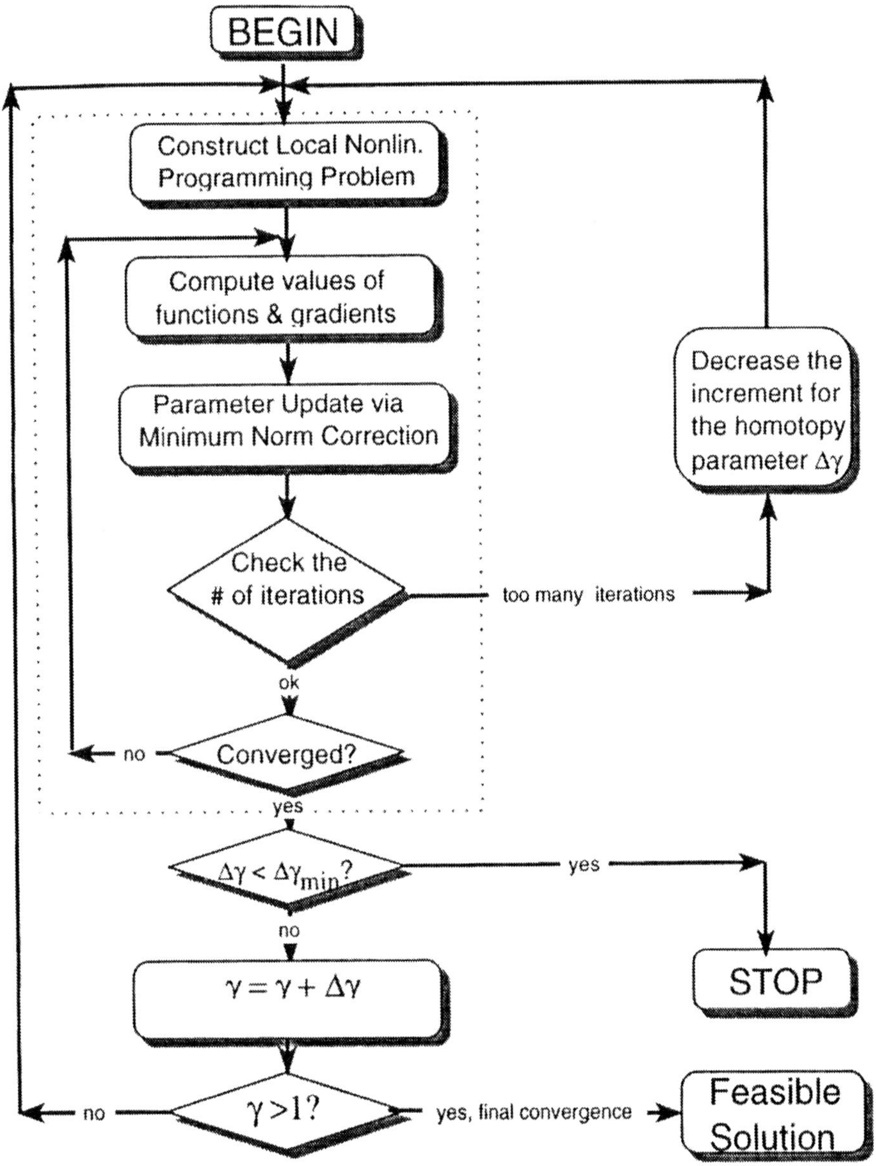

Figure 7.1 Logic Flow of the Minimum Correction Homotopy Algorithm for Nonlinear Programming.

Table 7.1 Characteristics of the Open-Loop and Closed-Loop System: Three Coupled Oscillator System

Open Loop		Feasible Solution	
Eigenvalues	Condition No.	Eigenvalues	Condition No.
$\pm 1.4447i$		$-0.8833 \pm 1.4447i$	
$\pm 4.4721i$	9.4903	$-1.0574 \pm 4.4721i$	8.1291
$\pm 6.9219i$		$-0.5322 \pm 6.7872i$	

$$\begin{aligned} G_1 &= \begin{bmatrix} 1.1591 & 0.0253 \\ 0.0253 & 1.0557 \end{bmatrix}, \qquad \|G_1\|_f = 1.5682 \\ G_2 &= \begin{bmatrix} 2.2539 & 0.2834 \\ 0.2834 & 2.6918 \end{bmatrix}, \qquad \|G_2\|_f = 3.5336 \end{aligned} \tag{7.82}$$

where $\|G\|_f$ denotes the Frobenius (sum-square) norm of the gain matrix G.

The characteristics of the open-loop and closed-loop systems are summarized in Table 7.1. As is evident, we successfully imposed the system time constants and also modestly improved the conditioning of the closed-loop eigenvectors as compared to the open-loop system. In order to evaluate the performance of this controller in comparison to competing robust gain design methods, we design a full-state feedback controller for the same system. It is well known that there are extra degrees of freedom when we design a full-state feedback controller. We use pole placement gain design, which exploits the extra design freedom to reduce the sensitivity of the poles to the perturbations in the closed-loop system (Ref. 52). We take the closed-loop eigenvalues obtained from the above output feedback controller (Table 7.1) as the target eigenvalues of the full-state feedback controller and use the open-loop eigenvector matrix as target closed-loop eigenvectors (see section 5.2). The resulting full-state feedback gain matrix is as follows:

$$\begin{aligned} G &= \begin{bmatrix} 6.2450 & 3.2694 & 7.7150 & 2.2386 & 0.8388 & 0.2675 \\ 6.0412 & 1.7898 & -3.9661 & 0.5140 & -0.2674 & 2.7071 \end{bmatrix} \\ \|G\|_f &= 13.3450 \end{aligned} \tag{7.83}$$

Table 7.2 The Closed-Loop Eigenvalues (Perturbed System)

Output Feedback	Full-State Feedback
$-0.8351 \pm 1.5358i$	$-0.9216 \pm 1.4695i$
$-1.0432 \pm 4.7783i$	$-0.9734 \pm 4.3725i$
$-0.5946 \pm 6.5030i$	$-0.4116 \pm 6.8147i$

The condition number of the closed-loop eigenvector matrix of this case is 8.4026, which is slightly larger than the above optimum output feedback case.

Now, we perturb all elements of the mass and stiffness matrices by about 10%, and use the gains designed for the unperturbed system in order to obtain some insight into the robustness of both controllers. The perturbed mass and stiffness matrices are

$$\mathrm{M} = \begin{bmatrix} 1.1 & 0 & 0 \\ 0 & 0.9 & 0 \\ 0 & 0 & 1.05 \end{bmatrix}, \qquad \mathrm{K} = \begin{bmatrix} 22 & -11 & 0 \\ -11 & 29 & -18 \\ 0 & -18 & 18 \end{bmatrix}$$

It is obvious that the perturbed matrices remain positive definite, and therefore might be interpreted as feasible errors associated with the modeling process. For all physically possible perturbations (so long as the mass and stiffness matrices remain positive definite), we know that the symmetric output feedback controller guarantees stability. Therefore, we can expect that the closed-loop system with the output feedback controller will at least remain stable, even though it was designed for the unperturbed system. We have no such stability guarantees for the full-state feedback controller, however. On the other hand, the performance of the system will definitely be modified by model perturbations.

The closed-loop eigenvalues of the perturbed system with the controllers that were designed for the unperturbed system are summarized in Table 7.2. As we can see in Table 7.2, both controllers remain stable for the perturbed system, and, comparing Tables 7.1 and 7.2, both are fairly insensitive to these particular perturbations. By comparing the norm of gain matrices (7.82) and (7.83), however, we can conjecture that the output feedback controller uses less energy than the full-state feedback controller for most initial conditions. Also, of course, evaluating the performance of the full state feedback controller must take into account the additional performance degradation due to state estimation errors not considered in this discussion. The output feedback controller appears to be a better selection for this particular case, although a detailed study over some distribution of uncertainty and disturbances would be needed to complete this evaluation. Computational studies for large flexible structures of moderate dimensionality are presented in the following sections.

7.5.2 Draper/RPL Model

As a second example, consider the planar rotational/vibrational dynamics of the flexible structure consisting of a rigid hub with four cantilevered flexible appendages (see Chapter 4). The equation of motion can be written as

$$\mathbf{M}\ddot{\mathbf{x}} + \mathbf{K}\mathbf{x} = \mathbf{D}\mathbf{u} \tag{7.84}$$

Using modal truncation [Eqs. (7.63) and (7.64)], the reduced-order equation of motion becomes

$$\tilde{\mathbf{M}}\ddot{\underline{\eta}}_1 + \tilde{\mathbf{K}}\underline{\eta}_1 = \tilde{\mathbf{D}}\mathbf{u} \tag{7.85}$$

The order-reduction process is not central to this discussion, although it is important that a "good" reduced-order model be used to obtain practical results efficiently.

For "structural feedback" control, the control force vector takes the following form [Eq. (7.2)].

$$\mathbf{u} = -\left(\tilde{\mathbf{G}}_1\tilde{\mathbf{D}}^{\mathrm{T}}\underline{\eta}_1 + \tilde{\mathbf{G}}_2\tilde{\mathbf{D}}^{\mathrm{T}}\dot{\underline{\eta}}_1\right) \tag{7.86}$$

For any/all choices of gain matrices from this stable family (i.e., the set of all positive definite $\mathrm{G_i}$ matrices), asymptotic stability of the closed-loop system is guaranteed in the Lyapunov sense as we discussed in section 7.2.1. In the present example, we take N=20 [where N denotes the number of terms retained in the Ritz ("assumed modes") method] and $n_r = 9$ [where n_r denotes the number of the modes considered in the reduced-order model]. Therefore, we design the control gain matrices $\tilde{\mathrm{G}}_1$ and $\tilde{\mathrm{G}}_2$ by using the reduced (18th)-order model and using the full (42nd)-order system to verify the controller design.

We adopt the condition number of the closed-loop modal matrix as our performance measure [Eq. (7.57)] and impose constraints on the first three closed-loop frequencies and time constants [Eqs. (7.58) and (7.59)]. The objective value of the closed-loop damped natural frequency of the rigid body mode is taken as 0.3, and the objective values for the first two flexible modes' closed-loop damped natural frequencies are taken as their corresponding zero gain values. We adopt the following time constants for lower bounds on the first three closed-loop eigenvalues: $\mathrm{T_1} = 0.2$, $\mathrm{T_2} = 0.2$, and $\mathrm{T_3} = 0.25$. The optimization process was initiated by adopting identity matrices for the $\mathrm{G_i}$ starting iteratives. The characteristics of the

Table 7.3 Characteristics of the Open-Loop and Closed-loop System: Draper RPL Model

Open Loop		Optimal Solution	
ω_i	$\zeta_i\omega_i$	ω_{di}	$\zeta_i\omega_{di}$
0	0	0.3000	0.2000
4.3724	0	4.3726	0.2061
7.9070	0	7.9071	0.2359
51.4510	0	52.0041	5.3354
52.8058	0	53.0653	8.0875
157.5566	0	158.0285	14.0467
158.3659	0	158.5273	21.4808
313.7099	0	308.8039	34.8815
314.3402	0	311.7689	22.6602
Cond. No.	∞	Cond. No.	37210.8
		$\|\tilde{G}_1\|_f =$	1.9080
		$\|\tilde{G}_2\|_f =$	7.5612

open-loop and closed-loop systems are summarized in Table 7.3. The converged gain matrices are

$$\tilde{G}_1 = \begin{bmatrix} 1.8342 & 0.2592 & 0.2601 \\ 0.2592 & 0.0409 & 0.0405 \\ 0.2601 & 0.0405 & 0.0403 \end{bmatrix}, \qquad \|\tilde{G}_1\|_f = 1.9080$$
$$\tilde{G}_2 = \begin{bmatrix} 7.4284 & 0.4579 & 0.8335 \\ 0.4579 & 0.2411 & 0.0211 \\ 0.8335 & 0.0211 & 0.3509 \end{bmatrix}, \qquad \|\tilde{G}_2\|_f = 7.5612 \tag{7.87}$$

The small differences between the objective achieved values of the first few flexible mode eigenvalues (Table 7.3) stem from the fact that we designed the controller

Table 7.4 Mathematical Model of A Flexible Grid Structure ($n_r = 10$)

$$\tilde{M} = I_{10\times 10}, \qquad \tilde{C} = \mathrm{diag}[2\zeta_1\omega_1, 2\zeta_1\omega_1, \ldots, 2\zeta_{10}\omega_{10}] \qquad \tilde{K} = \mathrm{diag}[\omega_1^2, \omega_2^2, \ldots, \omega_{10}^2]$$

$$\tilde{D} = \begin{bmatrix} -0.0031 & -0.2298 & 0.0031 \\ 0.4376 & 0.0000 & 0.4376 \\ 0.0066 & 0.8139 & -0.0066 \\ 0.8004 & 0.0107 & -0.8004 \\ 0.3858 & 0.0000 & 0.3858 \\ -0.8465 & 0.0597 & 0.8465 \\ 0.2701 & 1.3708 & -0.2701 \\ -0.2938 & 0.0000 & -0.2938 \\ 0.3110 & 0.0000 & 0.3110 \\ -0.4646 & 0.0000 & -0.4646 \end{bmatrix}, \qquad \underline{\omega} = \begin{Bmatrix} 5.6491 \\ 14.6826 \\ 30.4589 \\ 37.7848 \\ 48.8524 \\ 80.4699 \\ 82.9575 \\ 83.6582 \\ 102.4535 \\ 125.7395 \end{Bmatrix}$$

and

$$\zeta_i = 0.001, \qquad i = 1, \ldots, 10$$

based on a reasonably accurate reduced-order model. It is obvious that our reduced-order model was sufficiently accurate for control design purposes. Because of the structure of symmetric output feedback, all modes of the full-order model are stabilized; this form of the control *guarantees stability*, even if we had chosen a poor reduced-order model. In this case it is obvious that we imposed the desired constraints on the eigenvalues and also improved the conditioning of the closed-loop eigenvectors.

7.5.3 Vibration Suppression for a Flexible Grid Structure

As a third example, consider a 60 degree of freedom model of a flexible grid structure (see section 6.4). The particular position and orientation of the sensors and actuators shown in Fig 6.4 were chosen for this example. Even this simple structure leads to a fairly high-order dynamical model, and as usual, it is desirable to use a reduced-order model to design the controller. The reduced-order equation of motion becomes

$$\tilde{M}\ddot{\underline{\eta}}_1 + \tilde{C}\dot{\underline{\eta}}_1 + \tilde{K}\underline{\eta}_1 = \tilde{D}\mathbf{u} \tag{7.88}$$

In this case, we adopted the first ten lowest-frequency modes. Structural damping was also considered, we assumed a 0.1% structural damping factor. Table 7.4 lists the data for the reduced order model of the flexible grid structure.

In this example, the formulation of the problem is different from the previous example. We adopt the condition number of the closed-loop modal matrix as our performance measure, and seek to impose constraints on the first three closed-loop frequencies and time constants. Thus all the constraint equations are of the inequality type as follows:

$$\underset{\mathbf{p} \in \Omega}{\text{minimize}} \;\; \mathrm{J}(\mathbf{p}) = \mathcal{K}(\Phi^{c}(\mathbf{p})) = \frac{\sigma_1(\Phi^{c}(\mathbf{p}))}{\sigma_{2n}(\Phi^{c}(\mathbf{p}))} \tag{7.89}$$

subject to

$$\begin{aligned} \omega_{oi} - \Delta\omega_i \le \omega_{di}(\mathbf{p}) \le \omega_{oi} + \Delta\omega_i, \qquad & i = 1, \ldots, m \\ \omega_{di}(\mathbf{p})\zeta_i(\mathbf{p}) \ge T_i, \qquad & i = 1, \ldots, m \end{aligned} \tag{7.90}$$

where $\Delta\omega_i$ is adopted arbitrarily as 1% of the open-loop natural frequency ω_{oi}.

We can solve the above problem by using a technique similar to that discussed in section 7.4. The above formulation has some practical advantages over the previous example, since we do not require the pole placements to be exactly at prescribed point locations (this is usually an unnecessarily hard constraint). So, we seek to place selected eigenvalues anywhere within an acceptable region of the complex plane; this is obviously more attractive from an engineering design point of view, because it better reflects reality and provides increased design freedom which can be exploited to increase the robustness of the closed-loop system.

We adopt the following time constants for lower bounds on the first three closed-loop eigenvalues: $T_1 = 0.2$, $T_2 = 0.2$, and $T_3 = 0.25$. The optimization process was initiated by adopting the following vector for the parameter starting iteratives:

$$\mathbf{p_{start}} = [1 \quad 0 \quad 0 \quad 1 \quad 0 \quad 1 \quad 1 \quad 0 \quad 0 \quad 2 \quad 0 \quad 1]^{T}$$

The feasible gains, obtained upon completing the first stage of the homotopic differential correction algorithm, are

$$\begin{aligned} \tilde{G}_1 &= \begin{bmatrix} 0.4793 & -0.0033 & -0.0632 \\ -0.0033 & 0.0795 & 0.0015 \\ -0.0632 & 0.0015 & 0.4691 \end{bmatrix}, \qquad \|\tilde{G}_1\|_f = 0.6812 \\ \tilde{G}_2 &= \begin{bmatrix} 1.1215 & -0.0317 & 0.0163 \\ -0.0317 & 11.0473 & 0.3169 \\ 0.0163 & 0.3169 & 1.1269 \end{bmatrix}, \qquad \|\tilde{G}_2\|_f = 10.6028 \end{aligned} \tag{7.91}$$

Table 7.5 Characteristics of the Open-Loop and Closed-Loop System: Cantilevered Flexible Grid Model

Open Loop		Feasible Solution		Optimal Solution	
ω_i	$\zeta_i\omega_i$	ω_{di}	$\zeta_i\omega_{di}$	ω_{di}	$\zeta_i\omega_{di}$
5.649	0.006	5.659	0.282	5.661	0.200
14.683	0.015	14.687	0.233	14.688	0.232
30.459	0.030	30.545	3.545	30.563	2.493
37.785	0.038	37.795	0.745	37.769	0.778
48.852	0.049	48.852	0.219	48.852	0.218
80.470	0.080	80.455	0.897	80.489	0.755
82.958	0.083	81.334	9.746	81.987	7.639
83.658	0.084	83.658	0.182	83.657	0.176
102.454	0.102	102.453	0.213	102.450	0.211
125.740	0.126	125.736	0.372	125.732	0.370
Cond. No.	125.7396	Cond. No.	129.5280	Cond. No.	128.7071
		$\|\tilde{G}_1\|_f$ =	0.6812	$\|\tilde{G}_1\|_f$ =	1.5999
		$\|\tilde{G}_2\|_f$ =	10.6028	$\|\tilde{G}_2\|_f$ =	7.7324

and the final converged position and velocity gain matrices are

$$\begin{aligned} \tilde{G}_1 &= \begin{bmatrix} 0.4468 & 0.0009 & -0.0665 \\ 0.0009 & 1.4699 & -0.0027 \\ -0.0665 & -0.0027 & 0.4364 \end{bmatrix}, \qquad \|\tilde{G}_1\|_f = 1.5999 \\ \tilde{G}_2 &= \begin{bmatrix} 1.1366 & 0.2102 & -0.0030 \\ 0.2102 & 7.3446 & -1.2589 \\ -0.0030 & -1.2589 & 1.1389 \end{bmatrix}, \qquad \|\tilde{G}_2\|_f = 7.7324 \end{aligned} \tag{7.92}$$

The characteristics of the open-loop and feasible solution are summarized in Table 7.5. In this case, it is obvious that we have successfully imposed the

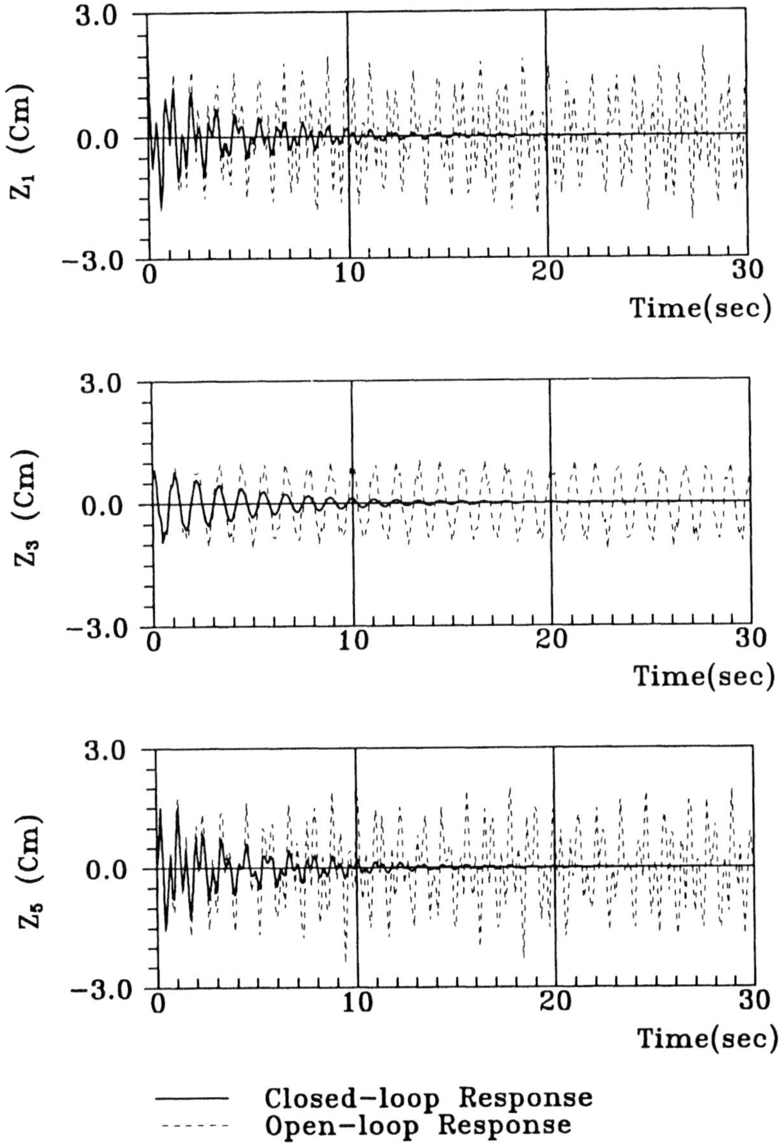

Figure 7.2 Open- and Closed-Loop Response of the Flexible Grid Due to an Initial Disturbance.

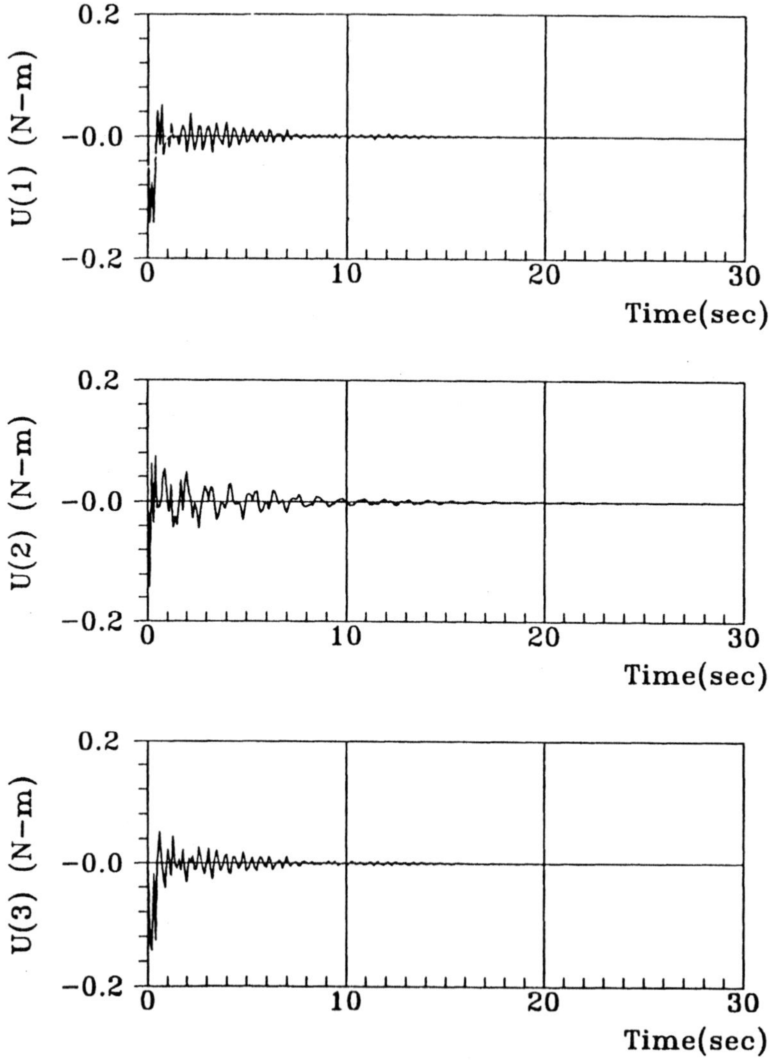

Figure 7.3 Control Torque History.

specified constraints on the closed-loop eigenvalues, with only a modest change in the eigenvector condition number. The open-loop and closed-loop response histories of the normal displacement to the grid plane (at three nodes on the horizontal member of the bottom of the grid structure), which ensue from a typical set of initial conditions, are shown in Fig. 7.2, and the corresponding control torque input histories are displayed in Fig. 7.3. The Newmark direct (implicit) numerical integration method (Ref. 53) was used to compute the time response of the system. Note that the time step for integration must in general be selected carefully (which is a fraction of the highest significantly participating mode's natural period) in order to obtain accurate results. On the other hand, implicit integration methods usually result in a much more graceful accuracy degradation as a function of step size than for explicit methods such as Runge-Kutta algorithms. In the above examples, the Newmark integrations were done using both the reduced-order and full-order system, with negligible differences. In the simulations, have imposed control saturation bounds (20 oz-in ; 0.1412 N-m) to keep the reaction wheel speeds to modest levels. All the states of the full order system are successfully controlled, and the amplitudes of the normal displacement of the closed-loop (Figure 7.2) are almost zero after 15 s.

In this section, a robust output feedback controller design method for second-order dynamical systems was applied to three numerical examples. These examples provide a demonstration of numerical methods tuning symmetric, fully populated, positive definite gain matirces for a stable family of controllers using collocated sensors and actuators. The gains were designed by a nonlinear programming algorithm and made use of a reduced-order model, yet the stability of the corresponding full-order system was guaranteed.

Problems

Problem 7-1 (Ref. 6)

Instead of using the nonlinear Cholesky decomposition to ensure a positive definite gain parameterization for the symmetric output feedback controller, Eq. (7.43), use the following linear parameterization of gain matrices:

$$G_1 = \sum_{k=1}^{l} g_k^1 G_k^1, \qquad G_2 = \sum_{k=1}^{l} g_k^2 G_k^2$$

where G_k^1 and G_k^2 are *prescribed* $(m \times m)$ positive semidefinite gain submatrices, g_k^1 and g_k^2 are non-negative scalar gains (thereby guaranteeing positivity of theG_i), and l is the number of parameters $[l \leq m(m+1)/2]$.

Discuss the advantages and disadvantages of this linear parameterization relative to the Cholesky parameterization. Can you think of a physical/mechanical spring/dashpot analog that would allow one to efficiently construct intuitively meaningful choices for the prescribed matrices G_k^1 and G_k^2?

Problem 7-2

The output vectors due to imprecise collocation of sensors and actuators can be represented as

$$\mathbf{y} = (D + \Delta_1)^T \mathbf{x}$$
$$\dot{\mathbf{y}} = (D + \Delta_2)^T \dot{\mathbf{x}}$$

Assume that the colocation errors have the simple form

$$\Delta_1 = \varepsilon_1 D, \qquad \Delta_2 = \varepsilon_2 D$$

From Theorem 7.1, the robust stability bound is

$$\varepsilon_1 \|G_1\|_s + \varepsilon_2 \|G_2\|_s < \frac{\lambda_m(M)}{2\|D\|_s^2 \lambda_M(\tilde{P})} \qquad (\dagger)$$

Construct the mass, stiffness, damping, and control influence matrices for the flexible grid structure by using the data in Table 7.4. Using the gain matrices G_1 and G_2 given in Eq. (7.92), examine the above inequality condition (†) for the following three cases:

(1) $\varepsilon_1 = 1$, $\varepsilon_2 = 1$

(2) $\varepsilon_1 = -1$, $\varepsilon_2 = -0.5$

(3) $\varepsilon_1 = -1$, $\varepsilon_2 = -1.2$.

Compute the closed-loop eigenvalues for the above cases. What conclusions can be drawn from these results?

Problem 7-3 (Ref. 8)

Consider the general nonlinear constrained minimization problem:

Find the parameter vector $\mathbf{p}^*$ that minimizes a differentiable function

$$J = J(\mathbf{p})$$

subject to the differentiable constraints

$$\mathbf{f}(\mathbf{p})\ \{\leq, =, \geq\}\ \mathbf{f}^o$$

Using a homotopic continuation process (at step i, $\gamma = \gamma^i$; where γ is swept from 0 to 1 at increment $\Delta\gamma$), and linearization about the previous best solution $\mathbf{p}^{i-1}$, show that the above nonlinear programming problem can be transformed, on each correction step, to the following local linear programming problem to determine the correction $\Delta\mathbf{p} = \mathbf{y}$:

$$\text{minimize} \quad \left[\left. \frac{\partial J}{\partial \mathbf{p}} \right|_{\mathbf{p}^{i-1}} \right] \mathbf{y}$$

subject to

$$\left[\left. \frac{\partial \mathbf{f}}{\partial \mathbf{p}} \right|_{\mathbf{p}^{i-1}} \right] \mathbf{y}\ \{\leq, =, \geq\}\ (1 - \gamma^i)\mathbf{f}(\mathbf{p}^s) + \gamma^i \mathbf{f}^o - \mathbf{f}(\mathbf{p}^{i-1}) + \left[\left. \frac{\partial \mathbf{f}}{\partial \mathbf{p}} \right|_{\mathbf{p}^{i-1}} \right] \underline{\epsilon}$$

$$0 \leq \mathbf{y} \leq 2\underline{\epsilon}$$

$$\mathbf{p}^i = \mathbf{p}^{i-1} + \mathbf{y} - \underline{\epsilon}; \qquad \mathbf{p}^0 = \mathbf{p}^s$$

Note that the above linear programming problem can be solved using standard Simplex codes.

What is the role of the constraint $0 \leq \mathbf{y} \leq 2\underline{\epsilon}$?

Comment upon pitfalls implicit in the above formulation which might inhibit convergence in practical applications. Can you suggest refinements of this algorithm to overcome these anticipated pitfalls?

References

1. Balas, M.J., "Trends in Large Space Structures Control Theory : Fondest Hopes, Wildest Dreams," *IEEE Transactions on Automatic Control,* Vol. AC-27, No. 3, June 1982, pp. 522-535.

2. Rew, D.W. and Junkins, J.L., "Multi-Criterion Approaches to Optimization of Linear Regulators," AIAA Paper 86-2198-CP, *Proceedings of the AIAA Guidance, Navigation, and Control Conference,* Williamsburg, VA, Aug. 1986.

3. Junkins, J.L. and Kim, Y., "First and Second Order Sensitivity of the Singular Value Decomposition," *Journal of the Astronautical Sciences,* Vol. 38, No. 1, Jan.-Mar., 1990, pp. 69-86.

4. Junkins, J.L. and Kim, Y., "A Minimum Sensitivity Design Method for Output Feedback Controllers," *Mechanics and Control of Large Space Structures,* edited by J.L. Junkins, Vol. 129, Progress in Astronautics and Aeronautics, AIAA, Washington, DC, 1990.

5. Kim, Y. and Junkins, J.L., "A Measure of Controllability for Actuator Placement," *Journal of Guidance, Control, and Dynamics,* Vol. 14, No. 5, Sept.-Oct. 1991, pp. 895-902.

6. Creamer, N.G. and Junkins, J.L., "Low-Authority Eigenvalue Placement for Second-Order Structural Systems," *Journal of Guidance, Control, and Dynamics,* Vol. 14, No. 3, May-June 1991, pp. 698-701.

7. Vanderplatts, G.N., *Numerical Optimization Techniques for Engineering Design with Applications*, McGraw-Hill, New York, NY 1984.

8. Lim, K.B. and Junkins, J.L., "Optimal Design of Dynamic Structures via Sequential Linear Programming," 4th International Modal Analysis Conference, Los Angeles, CA, Feb. 3-6, 1986.

9. Levine, W.S. and Athans, M., "On the Determination of the Optimal Constant Output Feedback Gains for Linear Multivariable Systems," *IEEE Transactions on Automatic Control,* Vol. AC-15, No. 1, 1970, pp. 44-48.

10. Knapp, C.H. and Basuthakur, S., "On Optimal Output Feedback," *IEEE Transactions on Automatic Control,* Vol. AC-17, Dec. 1972, pp. 823-825.

11. Choi, S.S. and Sirisena, H.R., "Computations of Optimal Output Feedback Gains for Linear Multivariable Systems," *IEEE Transactions on Automatic Control,* Vol. AC-19, June 1974, pp. 257-258.

12. Horisberger, H.P. and Belanger, P., "Solution of the Optimal Constant Output Feedback Problem by Conjugate Gradients," *IEEE Transactions on Automatic Control,* Vol. AC-19, Aug. 1974, pp. 434-435.

13. Choi, S.S. and Sirisena, H.R., "Computation of Optimal Output Feedback Controls for Unstable Linear Multivariable Systems," *IEEE Transactions on Automatic Control,* Vol. AC-22, Feb. 1977, pp. 134-136.

14. Allwright, J.C., "LQP: Dominant Output Feedbacks," *IEEE Transactions on Automatic Control,* Vol. AC-27, Feb. 1982, pp. 915-921.

15. Moerder, D.D. and Calise, A.J., "Convergence of a Numerical Algorithm for Calculating Optimal Output Feedback Gains," *IEEE Transactions on Automatic Control,* Vol. AC-30, Sept. 1985, pp. 900-903.

16. Davison, E.J., "On Pole Assignment in Linear Systems with Incomplete State Feedback," *IEEE Transactions on Automatic Control,* Vol. AC-15, June 1970, pp. 348-351.

17. Davison, E.J. and Chatterjee, R., "A Note on Pole Assignment in Linear Systems with Incomplete State Feedback," *IEEE Transactions on Automatic Control,* Vol. AC-16, Feb. 1971, pp. 98-99.

18. Davison, E.J. and Chow, S.G., "An Algorithm for the Assignment of Closed-Loop Poles Using Output Feedback in Large Linear Multivariable Systems," *IEEE Transactions on Automatic Control,* Vol. AC-18, Feb. 1973, pp.74-75.

19. Fallside, F. and Seraji, H., "Pole-Shifting Procedure for Multivariable Systems Using Output Feedback," *Proceedings of IEEE,* Vol. 118, No. 11, Nov. 1971, pp. 1648-1654.

20. Seraji, H., "Restrictions on Attainable Poles and Methods for Pole Assignment with Output Feedback," *Proceedings of IEEE,* Vol. 121, No. 3, Nov. 1974, pp. 205-212.

21. Sridhar, B. and Lindorff, D.P., "Pole-Placement with Constant Gain Output Feedback," *International Journal of Control,* Vol. 18, No. 5, 1973, pp. 993-1003.

22. Davison, E.J. and Wang, S.H., "On Pole Assignment in Linear Multivariable Systems Using Output Feedback," *IEEE Transactions on Automatic Control,* Vol. AC-20, Aug. 1975, pp. 516-518.

23. Misra, P. and Patel, R.V., "Numerical Algorithm for Eigenvalue Assignment by Constant and Dynamic Output Feedback," *IEEE Transactions on Automatic Control,* Vol. AC-34, Aug. 1989, pp. 579-588.

24. Kimura, H., "Pole Assignment by Gain Output Feedback," *IEEE Transactions on Automatic Control,* Vol. AC-20, Aug. 1975, pp. 509-516.

25. Kimura, H., "A Further Result on the Problem of Pole Assignment by Output Feedback," *IEEE Transactions on Automatic Control,* Vol. AC-22, June 1977, pp. 458-463.

26. Shah, S.L., Fisher, D.G., and Seborg, D.E., "Eigenvalue/Eigenvector Assignment for Multivariable Systems and Further Results for Output Feedback Control," *Electronics Letters,* Vol. 11, No. 16, Aug. 1975, pp. 388-389.

27. Srinathkumar, S., "Eigenvalue/Eigenvector Assignment Using Output Feedback," *IEEE Transactions on Automatic Control,* Vol. AC-23, No. 1, Feb. 1978, pp. 79-81.

28. Calvo-Ramon, J.R., "Eigenstructure Assignment by Output Feedback and Residue Analysis," *IEEE Transactions on Automatic Control,* Vol. AC-31, No. 3, Mar. 1986, pp. 247-249.

29. Kwon, B. and Youn, M., "Eigenvalue-Generalized Eigenvector Assignment by Output Feedback," *IEEE Transactions on Automatic Control,* Vol. AC-32, No. 5, May 1987, pp. 417-421.

30. Kabamba, P.T. and Longman, R.W., "Exact Pole Assignment Using Direct or Dynamic Output Feedback," *IEEE Transactions on Automatic Control,* Vol. AC-27, No. 6, Dec. 1982, pp. 1244-1246.

31. Subbayyan, R. and Vaithilingam, M.C., "Sensitivity – Reduced Design of Linear Regulators," *International Journal of Control,* Vol. 29, No. 3, 1979, pp. 435-440.

32. O'Reilly, J., "Low-Sensitivity Feedback Controllers for Linear Systems with Incomplete State Information," *International Journal of Control,* Vol. 29, No. 6, 1979, pp. 1047-1058.

33. Haraldsdottir, A., Kabamba, P.T., and Ulsoy, A.G., "Control of Linear Systems by Output Proportional Plus Derivative Feedback," *Proceedings of ACC.,* WP4, Minneapolis, MN, June 10-12, 1987.

34. Longman, R.W. and Gu. A.Q., "Pole Placement with Optimization," *Proceedings of the Conference on Optimal Control and Variational Calculus,* Oberwolfach, West-Germany, June 15-21, 1986.

35. Gu, Z.Q. and Longman, R.W., "Design of Dynamic Output Feedback Controllers with Parameter Variation Insensitivity," AIAA Guidance, Navigation and Control Conference, Williamsburg, VA, Aug. 18-20, 1986.

36. Canavin, J.R., "The Control of Spacecraft Vibrations Using Multivariable Output Feedback," AIAA/AAS Astrodynamics Conference, Palo Alto, CA, Aug. 7-9, 1978.

37. Goh, C.J. and Caughey, T.K., "On the Stability Problem Caused by Finite Actuator Dynamics in the Collocated Control of Large Space Structures," *International Journal of Control,* Vol. 41, No. 3, 1985, pp. 787-802.

38. Aubrun, J.N., "Theory of Control of Structures by Low-Authority Controllers," *Journal of Guidance, Control, and Dynamics,* Vol. 3, No. 5, 1980, pp. 444-451.

39. Fletcher, R. and Powell, M.J.D., "A Rapidly Convergent Descent Method for Minimization," *Computer Journal,* Vol. 6, 1963, pp. 163-168.

40. Kleinman, D.L., "An Easy Way to Stabilize a Linear Constant System," *IEEE Transactions on Automatic Control,* Vol. AC-15, Dec. 1970, p. 692.

41. Wonham, W.M., "On Pole Assignment in Multi-Input Controllable Linear Systems," *IEEE Transactions on Automatic Control,* Vol. AC-12, 1967, pp. 660-665.

42. Stewart, G.W., *Introduction to Matrix Computations*, Academic Press, New York, NY, 1973.

43. Wilkinson, J.H., *The Algebraic Eigenvalue Problem*, Oxford University Press, New York, NY, 1965.

44. Srinathkumar, S. and Rhoten, R.P., "Eigenvalue/Eigenvector Assignment for Multivariable Systems," *Electronics Letters,* Dec. 1975, pp. 124-125.

45. Juang, J.N. and Pappa, R.S., "An Eigensystem Realization Algorithm (ERA) for Model Parameter Identification and Model Reduction," *Journal of Guidance, Control, and Dynamics,* Vol. 8, No. 5, Sept.-Oct. 1985.

46. Joshi, S.M., *Control of Large Flexible Space Structures*, Lecture Notes in Control and Information Sciences, Vol. 131, Springer-Verlag, New York, NY 1989.

47. Athans, M. and Falb, P.L., *Optimal Control, an Introduction to the Theory and Its Applications*, McGraw-Hill, New York, NY 1966.

48. Lim, K.B., Junkins, J.L., and Wang, B.P., "Re-examination of Eigenvector Derivatives," *Journal of Guidance, Control, and Dynamics,* Vol. 10, No. 6, 1987, pp. 581-587.

49. Caughey, T.K., "Classical Normal Modes in Damped Linear Dynamic Systems," *Journal of Applied Mechanics,* Transactions of the ASME, Vol. 27, June 1960, pp. 269-271.

50. Junkins, J.L., "Equivalence of the Minimum Norm and Gradient Projection Constrained Optimization Techniques," *AIAA Journal,* Vol. 10, No. 7, July 1972, pp. 927-929.

51. Dunyak, J.P., Junkins, J.L., and Watson, L.T., "Robust Nonlinear Least Square Estimation Using the Chow–Yorke Homotopy Method," *Journal of Guidance, Control, and Dynamics,* Vol. 7, No. 6, 1984, pp. 752-755.

52. Laub, A.J. and Little, J.N., *MATLAB User's Guide - Control System Toolbox*, The Mathworks, Inc., Sherborn, MA, 1986.

53. Bathe, K.J., *Finite Element Procedures in Engineering Analysis*, Prentice-Hall, Englewood Cliffs, NJ, 1982.

APPENDIX A

Description of MATLAB® Operators

The following table shows a list of operators grouped by subject area. The remainder of this appendix provides "users' reference manual" styled descriptions. These operators were developed by the authors for the applications discussed in this text; these teaching aides do not constitute a professional toolbox. MATLAB is distributed by MathWorks, Inc., 24 Prime Parkway, Natick, MA 01760. A PC-AT 386 version of MATLAB was used to develop these operators (diskette appended, see the 'readme' file for installation information). With minor modifications, these operators should be compatible with most modern implementations of MATLAB.

	Eigenvalue and Singular Value Decomposition Problems
eign	Eigenvalue solver with specified normalizations
eigsen	Partial derivatives of eigenvalues and eigenvectors
svsen	Partial derivatives of singular values
co2mod	Coordinate transformation to modal coordinate
	Modeling of Flexible Structures
beamasm	Beam model using assumed modes method
beamfem	Beam model using finite element method
rplasm	RPL model using assumed modes method
rplfem	RPL model using finite element method
	Design of Linear Full-State Feedback Control
lqrmsf	LQR design utilizing matrix sign function
assign	Robust eigenstructure assignment
	Controllability of Linear Systems
doc	Degree of controllability for lightly damped system
moc	Measure of modal controllability
modcst2	Modal cost analysis for modal space system
indcon	Measure of controllability for actuator placement

assign

Purpose:

Determine full-state gain matrix to exactly place eigenvalues and approximately place eigenvectors.

Synopsis:

G=assign(A,B,Lambda,T)

Description:

assign is used for pole placement gain selection for the multi-input systems

$$\dot{\mathbf{x}} = \mathbf{A}\mathbf{x} + \mathbf{B}\mathbf{u}$$

with full-state feedback law of

$$\mathbf{u} = \mathbf{G}\mathbf{x}$$

Input argument Lambda is a vector including the desired closed-loop poles (closed loop eigenvalues), and matrix T contains as columns the corresponding target eigenvectors for eigenstructure assignment.

Using specified eigenvalues (Lambda) and a well-conditioned target eigenvector matrix (T), the feedback gains can be computed. The selection of nearly achievable, well-conditioned target eigenvectors is important to reliably obtain an attractive control design, as is choosing good target locations for the closed-loop eighevalues. Note that this algorithm assigns the eigenvalues (Lambda) exactly and the eigenvectors (T) approximately.

Algorithm:

assign solves a Sylvester equation (which is obtained from the closed-loop eigenvalue problem), to obtain the feedback gain.

References:

1. Bhattacharyya, S.P. and deSouza, E., "Pole Assignment via Sylvester's Equation," *Systems and Control Letters,* Vol. 1, No. 4, Jan. 1982, pp.261-263.
2. Cavin, R.K. III and Bhattacharyya, S.P., "Robust Well-Conditioned Eigenstructure Assignment via Sylvester's Equation," *Journal of Optimal Control Applications and Methods,* Vol. 4, 1983, pp. 205-212.
3. Rew, D.W., Junkins, J.L., and Juang, J.N., "Robust Eigenstructure Assignment by a Projection Method: Application Using Multiple Optimization Criteria," *Journal of Guidance, Control, and Dynamics,* Vol. 12, No. 3, May-June 1989, pp.396-403.

beamasm, beamfem

Purpose:

Mathematical model of a fixed-free uniform beam undergoing transverse vibration. Specifically, compute the mass and stiffness matrices using the assumed modes method (**beamasm**) and the finite element method (**beamfem**).

Synopsis:

[M,K]=beamasm(n1,rho,L,E,I)
[M,K]=beamfem(n2,rho,L,E,I)

Description:

beamasm and **beamfem** generate mass matrix M and stiffness matrix K for fixed-free beam (length, L; mass density, rho; elastic modulus, E; cross-sectional area moment of inertia I) undergoing transverse vibration.

Algorithm:

beamasm uses assumed modes method, wherein n1 is the number of admissible functions retained in the approximation.
beamfem uses finite element method, wherein n2 is the number of finite elements.

References:

1. Meirovitch, L., *Computational Methods in Structural Dynamics*, Sijthoff & Noordhoff, Leyden, The Netherlands, Rockville, MD, 1980.

2. Craig, R.R. Jr., *Structural Dynamics - an Introduction to Computer Methods,* John Wiley & Sons, New York, NY, 1981.

3. Reddy, J.N., *An Introduction to the Finite Element Method,* McGraw-Hill Book Company, New York, NY, 1984.

co2mod

Purpose:

Coordinate transformation to modal coordinates for second-order undamped mechanical system. Specifically, given fully populated, symmetric mass and stiffness matrices (M,K), and corresponding control influence matrix (D), compute the diagonal modal space matrices (Mtil, Ktil) and corresponding control influence matrix (Dtil) are computed by similarity transformations.

Synopsis:

[Mtil,Ktil,Dtil]=co2mod(M,K,D)
[Mtil,Ktil,Dtil]=co2mod(M,K,D,nr)

Description:

Consider the second-order system: $M\ddot{\mathbf{x}} + K\mathbf{x} = D\mathbf{u}$. For $M = M^T > 0$, $K = K^T \geq 0$ $n \times n$ matrices, the modal coordinate transformation is the $n \times n$ modal matrix Φ of eigenvectors, computed by solving the open-loop eigenvalue problem: $\omega_i^2 M\underline{\phi}_i = K\underline{\phi}_i$, with $\Phi = [\underline{\phi}_1\ \underline{\phi}_2\ \cdots\ \underline{\phi}_n]$ and the specific biorthogonality normalizations: $\Phi^T M \Phi = I = \tilde{M}$, and $\Phi^T K \Phi = \text{diag}(\omega_i^2) = \tilde{K}$. The transformed equation of motion becomes uncoupled in the form $\tilde{M}\ddot{\underline{\eta}} + \tilde{K}\underline{\eta} = \tilde{D}\mathbf{u}$. For the specified nomalizations, observe that $\tilde{M}$ is an identity matrix, $\tilde{K}$ is diagonal matrix containing the squares of the natural frequencies, and $\underline{\eta}$ is the modal coordinate vector (length n). More generally, $\tilde{M}$ is the *modal mass matrix.* By specifying $\text{nr} \leq n$, an nrth reduced-order model (modal truncation method) is obtained, wherein $\tilde{M}$ and $\tilde{K}$ are nr $\times$ nr diagonal matrices.

Algorithm:

co2mod uses eigenvalue solver **eign**, which imposes the biorthogonality normalization $[\tilde{M} = I,\ \tilde{K} = \text{diag}(\omega_i^2)]$ for computing the modal matrix Φ.

References:

1. Meirovitch, L., *Computational Methods in Structural Dynamics*, Sijthoff & Noordhoff, Leyden, The Netherlands, 1980.

2. Meirovitch, L., *Dynamics and Control of Structures,* John Wiley & Sons, New York, NY, 1990.

doc

Purpose:

Computes a degree of controllability index for lightly damped systems. Specifically, an approximate measure of the size of the *recovery region* (see Ref. 1) which can be returned to the origin in specified time T is determined.

Synopsis:

ind=doc(M,C,K,D,T,Ni,Ki)

Description:

doc computes degree of controllability (Refs. 1, 2) for lightly damped second-order systems using modal coordinates.

$$M\ddot{\underline{\eta}} + C\dot{\underline{\eta}} + K\underline{\eta} = \mathbf{Du}$$

Note that M, C, K are diagonal matrices, using the transformation matrix Φ (the eigenvector matrix obtained by solving the eigenvalue problem of original mass and stiffness matrices).

Input arguments Ni denotes normalization factors for modal coordinates, whereas Ki denotes unnormalized actuator strength, and T is the duration of time for controlling the system.

Algorithm:

doc utilizes the concepts of recovery region and its approximation (Refs. 1,2). The recovery region is defined as the set of initial states that can be returned to the origin in a given time T using bounded controls. The degree of controllability is a scalar measure of the size of the region, where the scalar is chosen as the shortest distance from the origin to an initial state that cannot be returned to the origin in time T.

References:

1. Viswanathan, C.N., Longman, R.W., and Likins, P.W., "A Degree of Controllability Definition: Fundamental Concepts and Application to Modal Systems, *Journal of Guidance, Control, and Dynamics,* Vol. 7, No. 2, 1984, pp. 222-230.

2. Viswanathan, C.N. and Longman, R.W., "The Determination of the Degree of Controllability for Dynamic Systems with Repeated Eigenvalues," *Advances in the Astronautical Sciences,* Vol. 50, 1983.

eign

Purpose:

Eigenvalue solver with specified normalization

Synopsis:

[Lambda,Phi,Psi]=eign(A,B)

Description:

The generalized eigenvalue problem is to determine the nontrivial eigenvalues λ_i and eigenvectors $(\underline{\phi}_i, \underline{\psi}_i)$ satisfying the pair of equations:

$$\lambda_i B\underline{\phi}_i = A\underline{\phi}_i, \quad i = 1, 2, \ldots, n$$
$$\lambda_i B^T\underline{\psi}_i = A^T\underline{\psi}_i, \quad i = 1, 2, \ldots, n$$

with the specified normalization of the *biorthogonality conditions*

$$\underline{\phi}_i^T B\underline{\phi}_i = 1, \qquad i = 1, 2, \ldots, n$$
$$\underline{\psi}_j^T B\underline{\phi}_i = \delta_{ij}, \qquad \underline{\psi}_j^T A\underline{\phi}_i = \lambda_i\delta_{ij}, \qquad i, j = 1, 2, \ldots, n$$

Algorithm:

eign uses the built-in MATLAB function **eig** for eigenvalue solution.

Limitations:

(1) From **eig** if the limit of $30n$ iterations is exhausted while iterating for an eigenvalue $\Longrightarrow$ **Solution will not converge**.
(2) For the complex repeated eigenvalue case, sometimes we encounter $\underline{\phi}_i^T B\underline{\phi}_i \cong 0$ or $\underline{\psi}_i^T B\underline{\phi}_i \cong 0$. This may cause problems in normalizing the right or left eigenvectors, respectively.

References:

1. Wilkinson, J.H., *The Algebraic Eigenvalue Problem,* Oxford University Press, New York, NY., 1965.

2. Meirovitch, L., *Dynamics and Control of Structures,* John Wiley & Sons, New York, NY, 1990.

eigsen

Purpose:

Partial derivatives of eigenvalues and eigenvectors.

Synopsis:

[EVAD,REVTD,LEVTD]=eigsen(Lambda,Phi,Psi,B,DADP,DBDP)

Description:

eigsen calculates the partial derivatives of eigenvalues (EVAD), partial derivatives of right eigenvector matrix (REVTD) and partial derivatives of the left eigenvector matrix (LEVTD) with respect to the parameter p for the following generalized eigenvalue problem:

$$A(p)\underline{\phi}_i = \lambda_i B(p)\underline{\phi}_i$$

Before this program is used, the eigenvalue solver **(eign)** with biorthogonality normalization should be pre-executed to determined the input variables Lambda, Phi and Psi. Note that arbitrarily normalized input eigenvectors will usually lead to incorrect results; the derivatives of the eigenvectors are based on the specific normalization employed in **eign** (see above description). Partial derivatives of A and B matrices (DADP and DBDP) are computed from analytical expressions or finite differences if analytical derivatives are untractable.

Algorithm:

eigsen uses the formulas derived in Ref. 1 for computing partial derivatives.

Limitations:

This program appilies only to the distinct eigenvalue problem. If there exists repeated eigenvalues, use the generalized methods discussed in Refs. 2, 3.

References:

1. Lim, K.B., Junkins, J.L., and Wang, B.P., "Re-examination of Eigenvector Derivatives," *Journal of Guidance, Control, and Dynamics,* Vol. 10, No. 6, 1987, pp. 581-587.
2. Lim, K.B., Juang, J.N., and Ghaemmaghami, P., "Eigenvector Derivatives of Repeated Eigenvalue Using Singular Value Decompositions," *J. of Guidance, Control, and Dynamics,* Vol 12, No 2, Mar-Apr 1989, pp.282-283.
3. Juang, J.N., Ghaemmaghami, P., and Lim, K.B., "Eigenvalue and Eigenvector Derivatives of a Non-Defective Matrix," *Journal of Guidance, Control, and Dynamics,* Vol. 12, No. 4, July-August 1989, pp.480-486.

indcon

Purpose:

Compute the α measure of controllability for actuator placement (for modal second order systems).

Synopsis:

alpha=indcon(M,C,K,D,Phi,Cx,Qv)

Description:

indcon returns alpha, which is a scalar measure of controllability and disturbability for the system. The second-order "structural form" equations of motion should be transformed (see Chapter 6) using the open-loop modal matrix before this program is used.

Algorithm:

indcon uses formulas derived in Ref. 1 for computing the alpha controllability index useful for actuator placement.

References:

1. Kim, Y. and Junkins, J.L., "A Measure of Controllability for Actuator Placement," AIAA Guidance, Navigation and Control Conference, AIAA Paper 90-3464-CP, Portland, OR, Aug. 1990.

lqrmsf

Purpose:

Linear quadratic regulator design.

Synopsis:

[G,P]=lqrmsf(A,B,Q,R)

Description:

lqrmsf solves for the continuous-time linear quadratic regulator (LQR) optimal control gain and the solution of the associated Riccati equation. **lqrmsf** returns the optimal feedback gain matrix G such that the feedback law

$$\mathbf{u} = -\mathrm{G}\mathbf{x}$$

minimizes the quadratic performance index

$$J = \frac{1}{2}\int_{t_0}^{\infty}\left[\mathbf{x}^T Q\mathbf{x} + \mathbf{u}^T R\mathbf{u}\right]dt$$

subject to the system differential equation constraints

$$\dot{\mathbf{x}} = A\mathbf{x} + B\mathbf{u}$$

P is the unique positive definite solution (for controllable systems) of the associated matrix Riccati equation

$$A^T P + PA - PBR^{-1}B^T P + Q = 0$$

Algorithm:

lqrmsf utilizes the matrix sign function for solving the matrix Riccati equation.

Limitations:

Certain conditions must be satisfied in order for a unique positive definite solution of the Ricatti equation (and therefore the solution for the optimal LQR gain) to exist:

(1) Matrix Q must be symmetric and positive semidefinite
(2) Matrix R must be symmetric and positive definite
(3) The (A,B) pair must be controllable.

References:

1. Bierman, G.J., "Computational Aspects of the Matrix Sign Function to the ARE," *Proceedings of 23rd Conference on Decision and Control,* Las Vegas, NV, Dec. 1984, pp. 514-519.

2. Gardiner, J.D. and Laub, A.J., "A Generalization of the Matrix Sign Function Solution for the Algebraic *Proceedings of 24th Conference on Decision and Control,* Ft. Lauderdale, FL, Dec. 1985, pp. 1233-1235.

3. Shieh, L.S., Lian, S.R., and Mcinnis, B.C., "Fast and Stable Algorithms for Computing the Principal Square Root of a Complex Matrix," *IEEE Transactions on Automatic Control,* Vol. AC-32, No. 9, 1987, pp. 820-822.

moc

Purpose:

Compute the Hamdan-Nayfeh measure of modal controllability.

Synopsis:

[MC,GMC]=moc(A,B)

Description:

moc calculates matrix MC containing the measure of modal controllability of the ith mode from the jth input of the given system

$$\dot{\mathbf{x}} = \mathbf{A}\mathbf{x} + \mathbf{B}\mathbf{u}$$

and a vector GMC that includes each mode's gross measure of controllability.

Algorithm:

moc uses the formulas derived in Ref. 1. MC is a measure of orthogonality of the open loop left eigenvectors to the control influence matrix. This measure of controllability is essentially a norm of the modal space control influence matrix and provides a dynamic/geometric interpretation and generalization of the PBH eigenvector test.

References:

1. Hamdan, A.M.A. and Nayfeh, A.H., "Measure of Modal Controllability and Observability for First- and Second-Order Linear Systems," *Journal of Guidance, Control, and Dynamics,* Vol. 12, No. 3, 1989, pp. 421-428. See also the related Erratum, *Journal of Guidance, Control, and Dynamics,* Vol. 12, No. 5, 1989.

modcst2

Purpose:

Modal cost analysis for the second modal coordinate system.

Synopsis:

[V,Vi]=modcst2(M,C,K,D,Phi,Cx,Cxd,Qx,Qxd)

Description:

modcst2 returns the total cost V and the vector Vi containing the modal cost of each mode for a second-order system (represented in a modal coordinate system).

$$M\ddot{\underline{\eta}} + C\dot{\underline{\eta}} + K\underline{\eta} = Du$$

Note that M, C, K are diagonal mass damping and stiffness matrices (i.e., $\underline{\eta}$ is a vector of modal coordinates). The transformation matrix Φ (obtained by solving the eigenvalue problem of the original fully populated mass and stiffness matrices), maps $\underline{\eta}$ into the configuration vector as $\mathbf{x} = \Phi\underline{\eta}$.

For measuring (ideally zero) system performance that represents a physically important quantity to the control designer, the following cost function is used:

$$V = \sum_{i=1}^{m} \int_0^\infty \mathbf{y}_d^{i\,T}(t) Q_v \mathbf{y}_d^i(t) dt$$

where $\mathbf{y}_d(t)$ is the vector of most important outputs, defined as

$$\mathbf{y}_d(t) = Cd \begin{Bmatrix} \mathbf{x}(t) \\ \dot{\mathbf{x}}(t) \end{Bmatrix}, \qquad Cd \equiv \begin{bmatrix} C_{dx} & 0 \\ 0 & C_{d\dot{x}} \end{bmatrix}$$

and Q_v is a suitable weighting matrix defined as

$$Q_v = \begin{bmatrix} Q_x & 0 \\ 0 & Q_{\dot{x}} \end{bmatrix}$$

Algorithm:

modcst2 uses the analytical solution for the Lyapunov equation (Refs. 1-4, as discussed in Ref. 4). The equation of motion should be transformed into modal coordinates to diagonalize the system matrices before this program is used.

References:

1. Skelton, R.E. and DeLorenzo, M.D., "Space Structure Control Design by Variance Assignment," *Journal of Guidance, Control, and Dynamics,* Vol. 8, No. 4, 1985, pp. 454-462.

2. Skelton, R.E. and Hughes, P.C., "Modal Cost Analysis for Linear Matrix-Second-Order Systems," *Journal of Dynamic Systems, Measurement, and Control,* Vol. 102, Sept. 1980, pp. 151-158.

3. Skelton, R.E., *Dynamic System Control – Linear System Analysis and Synthesis,* John Wiley & Sons, New York, NY, 1988.

4. Kim, Y. and Junkins, J.L., "A Measure of Controllability for Actuator Placement," AIAA Guidance, Navigation and Control Conference, AIAA Paper 90-3464-CP, Portland, OR, Aug. 1990.

rplasm, rplfem

Purpose:

Mathematical model of the flexible hub-appendage-tip mass structure.

Synopsis:

[M,K,D]=rplasm(n1,rho,L,E,I,r,Jh,mt,Jt)
[M,K,D]=rplfem(n2,rho,L,E,I,r,Jh,mt,Jt)

Description:

The **rplasm** and **rplfem** operators generate the mass matrix M, stiffness matrix K, and input influence matrix D for the "Draper/RPL model" (also known as the "TAMU/AFOSR maneuver experimental structure", see Chapter 4).

'RPL' acknowledges the Air Force Rocket Propulsion Laboratory's historical sponsorship of an experiment invloving a similar configuration at Charles Stark Draper Laboratory during the early 1980's. TAMU/AFOSR denotes a recent configuration developed at Texas A&M University under the sponsorship of the Air Force Office of Scientific Research.

The arguement list of the **rplasm** and **rplfem** operators inclued the geometric and mass properties:

rho : mass density of appendages

E : elastic modulus of the appendages

I : cross – sectional area moment of inertia

L : appendage length

r : hub radius

Jh : rotary inertia of hub

mt : tip mass

Jt : rotary inertia of tip masses

Note that the control input influence matrix D is generated for the assumed control inputs: A torque u_1 applied on the hub, a torque u_2 applied on the end of both appendages 1 and 3, and a torque u_3 applied on the end of both appendages 2 and 4; all torques are about the vertical axis and all vibration is assumed to occur in the horizontal plane, and the vibrational deformations are restriced to occur in an antisymmetric fashion which preserves the mass center at its inertially fixed, undeformed location, the hub center.(see Figure 4.4).

Algorithm:

rplasm uses the assumed modes method, wherein n1 is the number of admissible functions retained in the approximation.
rplfem uses the finite element method, wherein n2 is the number of finite elements.

References:

1. Meirovitch, L., *Computational Methods in Structural Dynamics*, Sijthoff & Moordhoff, Leyden, The Netherlands,1980.

2. Craig, R.R. Jr., *Structural Dynamics - an Introduction to Computer Methods,* John Wiley & Sons, New York, NY, 1981.

3. Reddy, J.N., *An Introduction to the Finite Element Method,* McGraw-Hill Book Company, New York, NY, 1984.

4. Bathe, K.J., *Finite Element Procedure in Engineering Analysis,* Prentice-Hall, Inc., Englewood Cliffs, NJ, 1982.

5. Bodden, D.S. and Junkins, J.L., "Eigenvalue Optimization Algorithms for Structural/Controller Design Iterations," *Journal of Guidance, Control, and Dynamics,* Vol. 8, No. 6., Nov.-Dec. 1985.

6. Junkins, J.L. and Turner, J.D., *Optimal Spacecraft Rotational Maneuvers,* Studies in Astronautics, Vol. 3, Elsevier Scientific Publishing Company, New York, NY, 1985.

7. Kim, Y., Junkins, J.L., and Kurdila, A.J., "On the Consequences of Certain Modeling Approximations in Dynamics and Control of Flexible Space Structures," Preprint, *33th SDM Conference,* Dallas, TX, April 1992.

svsen

Purpose:

Partial derivatives of singular values.

Synopsis:

svdev=svsen(A,DADP)

Description:

svsen calculates the first partial derivatives of singluar values (and stores them in vector svdev) with respect to the parameter p for the following singular value problem:

$$A(p) = U\Sigma V^H$$

Note that the first partial derivative of matrix A (matrix DADP) is needed, this can be determined by explicit differentiation or by finite difference approximations prior to executing **svsen**.

Algorithm:

svsen uses the formula derived in Ref. 1 for computing singular value sensitivities.

References:

1. Junkins, J.L. and Kim, Y., "First and Second Order Sensitivity of the Singular Value Decomposition," *Journal of the Astronautical Sciences,* Vol. 38, No. 1, Jan.-Mar., 1990, pp. 69-86.

APPENDIX B

Proof of Theorem 2.7 (Patel and Toda)

The proof is based on the Lyapunov approach from Ref. 1.

Consider the system

$$\dot{\mathbf{x}}(t) = A\mathbf{x}(t) + \mathbf{f}(\mathbf{x}(t), t) \tag{B.1}$$

where $\mathbf{x}(t) \in R^n$, A is a time-invariant, asymptotically stable matrix and has distinct eigenvalues so that there exists a set of n linearly independent eigenvectors.

Consider the eigenvalue problem of the open-loop system of Eq. (B.1)

$$A\Phi = \Phi\Lambda \tag{B.2}$$

$$A^T\Psi = \Psi\Lambda \tag{B.3}$$

with the biorthonormality conditions

$$\Psi^T\Phi = I \tag{B.4}$$

$$\Psi^T A\Phi = \Lambda \tag{B.5}$$

where $\Lambda = \text{diag}\{\lambda_1, \ldots, \lambda_n\}$, Φ and Ψ denote modal matrices corresponding to right and left eigenvalue problem, and λ_i are eigenvalues of A.

Let us introduce the modal transformation

$$\mathbf{x}(t) = \Phi\underline{\eta}(t) \tag{B.6}$$

By using Eq. (B.6), we can rewrite Eq. (B.1) in modal coordinates as

$$\dot{\underline{\eta}}(t) = \Lambda\underline{\eta}(t) + \underline{\mathbf{g}}(t) \tag{B.7}$$

where $\underline{\mathbf{g}}(t) = \Psi^T\underline{\mathbf{f}}(t)$.

Choose as the Lyapunov function

$$U(\underline{\eta}) = \underline{\eta}^H(t)P\underline{\eta}(t) \tag{B.8}$$

where P is a constant positive definite matrix.

Since A is an asymptotically stable matrix, Λ is also an asymptotically stable matrix, and P always exists by the Lyapunov theorem such that

$$\Lambda^H P + P\Lambda = -2Q \tag{B.9}$$

where Q is *any* positive definite matrix.

Note that $U(\underline{\eta}) > 0$ for all $\underline{\eta} \neq 0$ and $U(\underline{\eta}) \to \infty$ as $\|\underline{\eta}\| \to \infty$. We require that $\dot{U}(\underline{\eta}) \leq 0$ for the stability of the system (B.7). Differentiating Eq. (B.8) with respect to t, along with the solution of Eq. (B.7), yields

$$\begin{aligned}\dot{U}(\underline{\eta}) &= \dot{\underline{\eta}}^H P \underline{\eta} + \underline{\eta}^H P \dot{\underline{\eta}} \\ &= \underline{\eta}^H (\Lambda^H P + P\Lambda)\underline{\eta} + \underline{\mathbf{g}}^H P \underline{\eta} + \underline{\eta}^H P \underline{\mathbf{g}}\end{aligned}$$

Using Eq. (B.9) in the above equation and noting that P is a positive definite matrix

$$\dot{U}(\underline{\eta}) = -2\underline{\eta}^H Q \underline{\eta} + 2\Re\{\underline{\eta}^H P \underline{\mathbf{g}}\} \tag{B.10}$$

where $\Re(\cdot)$ denotes the real part of $(\cdot)$. Observe that the second term of Eq. (B.10) is generally sign variable, hence we can guarantee that $\dot{U}(\underline{\eta}) \leq 0$, if the first term of Eq. (B.10) is dominate, that is, if

$$\Re\{\underline{\eta}^H P \underline{\mathbf{g}}\} \leq \underline{\eta}^H Q \underline{\eta} \tag{B.11}$$

Notice that the terms in Eq. (B.11) have the upper and lower bounds

$$\Re\{\underline{\eta}^H P \underline{\mathbf{g}}\} \leq \|\underline{\eta}\|_2 \|P\|_s \|\underline{\mathbf{g}}\|_2$$

and

$$\underline{\eta}^H Q \underline{\eta} \geq \min\ \lambda(Q) \|\underline{\eta}\|_2^2$$

where $\|\cdot\|_2$ denotes the Euclidean norm, and the spectral norm $\|[\cdot]\|_s$ = maximum singular value of matrix $[\cdot]$.

Note that the maximum singular value of the symmetric real matrix is identically the maximum eigenvalue of that matrix. Utilizing the above equations, Eq. (B.11) is satisfied, if the perturbation in modal coordinates, $\underline{\mathbf{g}}$, satisfies the following inequality equation

$$\|\underline{\eta}\|_2 \|P\|_s \|\underline{\mathbf{g}}\|_2 \leq \min\ \lambda(Q) \|\underline{\eta}\|_2^2$$

or

$$\frac{\|\underline{\mathbf{g}}\|_2}{\|\underline{\eta}\|_2} \leq \frac{\min\ \lambda(Q)}{\|P\|_s} = \frac{\min\ \lambda(Q)}{\max\ \lambda(P)} \equiv \mu \tag{B.12}$$

where P and Q are related by Eq. (B.9).

Note that the above results hold for an infinity of Q matrices, and that each Q generates a different bound on $\underline{g}$. Clearly, we seek the largest value of μ as a bound on $\underline{g}$. In order to choose the best Q matrix, note the following lemmas (Ref. 2).

Lemma B.1

Let $\bar{P}$ and P be the unique positive definite solution of the Lyapunov equation

$$A^T\bar{P} + \bar{P}A = -2\bar{Q} \tag{B.13}$$

$$A^T P + PA = -2Q \tag{B.14}$$

where

$$\bar{Q} = qQ \tag{B.15}$$

and q is a positive scalar. Then,

$$\frac{\min\ \lambda(\bar{Q})}{\max\ \lambda(\bar{P})} = \frac{\min\ \lambda(Q)}{\max\ \lambda(P)} = \mu \tag{B.16}$$

Proof of Lemma B.1

Using Eq. (B.15), Eq. (B.13) can be written as

$$A^T\left(\frac{1}{q}\bar{P}\right) + \left(\frac{1}{q}\bar{P}\right)A = -2Q \tag{B.17}$$

Because of the uniqueness of solutions of Eqs. (B.14) and (B.17), observe that $\frac{1}{q}\bar{P} = P$. Therefore we see that: $\max\ \lambda(\bar{P}) = \max\ \lambda(qP) = q\ \max\ \lambda(P)$, so we conclude that

$$\frac{\min\ \lambda(\bar{Q})}{\max\ \lambda(\bar{P})} = \frac{\min\ \lambda(qQ)}{q\ \max\ \lambda(P)} \equiv \frac{\min\ \lambda(Q)}{\max\ \lambda(P)}$$

Q.E.D.

The above lemma, which proves that μ is independent of scaling, will be used in the following to show that μ is maximum when we choose Q as an identity matrix [i.e., the choice on Q leading to the largest (or least conservative) of μ is Q = I].

Lemma B.2

The bound μ in Eq. (B.16) is maximum when Q=I in Eq. (B.14), where I is the $n \times n$ identity matrix.

Proof of Lemma B.2

Let Q be any arbitrary positive definite matrix, and

$$\bar{Q} = qQ, \qquad q = \frac{1}{\min\ \lambda(Q)}$$

Then,

$$\min\ \lambda(\bar{Q}) = 1 \tag{B.18}$$

The solution of the Lyapunov equation

$$A^T\bar{P} + \bar{P}A = -2\bar{Q} \tag{B.19}$$

can be written as (Ref. 3)

$$\bar{P} = 2\int_0^\infty e^{A^Tt}\bar{Q}e^{At}dt \tag{B.20}$$

and by using Eq. (B.18),

$$\bar{\mu} \equiv \frac{\min\ \lambda(\bar{Q})}{\max\ \lambda(\bar{P})} = \frac{1}{\max\ \lambda(\bar{P})}$$

Consider a second solution $\tilde{P}$ of the following Lyapunov equation:

$$A^T\tilde{P} + \tilde{P}A = -2I \tag{B.21}$$

From Eq. (B.20), the solution of Eq. (B.21) is

$$\tilde{P} = 2\int_0^\infty e^{A^Tt}Ie^{At}dt \tag{B.22}$$

and we also observe that

$$\tilde{\mu} \equiv \frac{\min\ \lambda(I)}{\max\ \lambda(\tilde{P})} = \frac{1}{\max\ \lambda(\tilde{P})}$$

From Eqs. (B.20) and (B.22), we see $\bar{P} - \tilde{P} = 2\int_0^\infty e^{A^Tt}(\bar{Q} - I)e^{At}dt$. Since $\lambda(\bar{Q}) \geq 1$ from Eq. (B.18), it follows that $\bar{Q} - I \geq 0$. Noting that e^{A^Tt} and e^{At} are nonsingular, the positivity of $\int_0^\infty e^{A^Tt}(\bar{Q} - I)e^{At}dt$ yields the conclusion that

$$\bar{P} - \tilde{P} \geq 0$$

and

$$\max \lambda(\bar{P}) \geq \max \lambda(\tilde{P})$$

Now, by Lemma B.1, $\bar{\mu}$ is invariant under any choice of the positive scalar q; hence, the result $\tilde{\mu} \geq \bar{\mu}$ holds for all positive definite $\bar{Q}$ [and not just for those satisfying $[\min \lambda(\bar{Q})] = 1$].

Q.E.D.

From Lemma B.2, the bound in Eq. (B.12) is maximum when Q is the identity matrix. Also, since the modal forces and coordinates are related to the physical forces and coordinates by

$$\mathbf{x}(t) = \Phi\underline{\eta}(t)$$
$$\underline{\mathbf{g}}(t) = \Psi^T\underline{\mathbf{f}}(t) = \Phi^{-1}\underline{\mathbf{f}}(t)$$

and because

$$\|\mathbf{x}\|_2 \leq \|\Phi\|_s\|\underline{\eta}\|_2$$
$$\|\underline{\mathbf{g}}\|_2 \leq \|\Phi^{-1}\|_s\|\underline{\mathbf{f}}\|_2$$

we can establish the following inequality:

$$\frac{\|\underline{\mathbf{g}}\|_2}{\|\underline{\eta}\|_2} \leq \frac{\|\Phi^{-1}\|_s\|\underline{\mathbf{f}}\|_2}{\|\mathbf{x}\|_2/\|\Phi\|_s}$$

The stability condition (B.12) is now satisfied if

$$\frac{\|\Phi^{-1}\|_s\|\underline{\mathbf{f}}\|_2}{\|\mathbf{x}\|_2/\|\Phi\|_s} \leq \frac{1}{\max \lambda(P)}$$

or if

$$\frac{\|\underline{\mathbf{f}}\|_2}{\|\mathbf{x}\|_2} \leq \frac{1}{\|\Phi^{-1}\|_s\|\Phi\|_s \max \lambda(P)} \tag{B.23}$$

where P satisfies the Lyapunov equation

$$\Lambda^H P + P\Lambda = -2I \tag{B.24}$$

The solution of P in Eq. (B.24) can be easily obtained if written in index notation,

$$\sum_{k=1}^{n} \Lambda_{ik}^H P_{kj} + \sum_{k=1}^{n} P_{ik}\Lambda_{kj} = -2\delta_{ik}, \qquad i,j = 1,\ldots,n$$

Since Λ is a diagonal matrix that consists of eigenvalues (of A), the above equation becomes

$$P_{ij}(\lambda_i^H + \lambda_j) = -2\delta_{ij}$$

or

$$\begin{aligned} P_{ij} &= 0, && i \neq j \\ &= -\frac{1}{\Re\{\lambda_i\}}, && i = j \end{aligned}$$

Note that the Lyapunov solution P is also a diagonal matrix and, therefore, the eigenvalues of P are $\left\{-\frac{1}{\Re\{\lambda_i\}}, \quad i = 1, \ldots, n\right\}$.

Clearly,

$$\frac{1}{\max\ \lambda(P)} = \min\ \left[-\Re\{\lambda_i\}\right] = \min\ \left[-\Re\{\lambda_i(A)\}\right]$$

Therefore, the stability condition of Eq. (B.23) becomes

$$\frac{\|\underline{\mathbf{f}}\|_2}{\|\mathbf{x}\|_2} \leq \frac{\min\ \left[-\Re\{\lambda_i(A)\}\right]}{\mathcal{K}(\Phi)} \tag{B.25}$$

where $\mathcal{K}(\Phi) = \|\Phi^{-1}\|_s \|\Phi\|_s =$ condition number of Φ.

For linear perturbations (proportional to $\mathbf{x}$) of the form $\underline{\mathbf{f}}(\mathbf{x}) = E\mathbf{x}$,

$$\|\underline{\mathbf{f}}\|_2 = \|E\mathbf{x}\|_2 \leq \|E\|_s \|\mathbf{x}\|_2$$

and the stability condition of Eq. (B.25) becomes

$$\|E\|_s \leq \frac{\min\ \left[-\Re\{\lambda_i(A)\}\right]}{\mathcal{K}(\Phi)} \tag{B.26}$$

Q.E.D.

References

1. Lim, K.B., and Junkins, J.L., "Robustness Optimization of Structural and Controller Parameters," AIAA Paper 87-0791-CP, AIAA 28th SDM Conference, Monterey, CA, 1987.

2. Patel, R.V. and Toda, M., "Quantitative Measures of Robustness for Multivariable Systems," TP-8A, *Proceedings of JACC,* San Francisco, CA, 1980.

APPENDIX C

Derivation of Singular Value Sensitivities

Consider any $n \times n$ complex-valued matrix Φ, factored using the singular value decomposition as:

$$\Phi = U\Sigma V^H \tag{C.1}$$

where U and V are $n \times n$ unitary matrices normalized so that

$$U^H U = I \tag{C.2}$$

$$V^H V = I \tag{C.3}$$

The matrix $\Sigma = \text{diag}(\sigma_1, \sigma_2, \ldots, \sigma_n)$, contains as its elements the real *singular values* of Φ ordered as $\sigma_1 \geq \sigma_2 \geq \ldots \geq \sigma_n \geq 0$. The superscript $()^H$ denotes the conjugate transpose.

Since U and V are unitary, we can use Eqs. (C.2) and (C.3) to rearrange Eq. (C.1) as

$$\Sigma = U^H \Phi V \tag{C.4}$$

The above four equations can be expressed as conditions on the $n \times 1$ complex column vectors of $U = [U_1 U_2 \ldots U_n]$ and $V = [V_1 V_2 \ldots V_n]$ as follows

$$\Phi V_i = U_i \sigma_i, \qquad i = 1, 2, \ldots, n \tag{C.5}$$

$$\Phi^H U_i = V_i \sigma_i, \qquad i = 1, 2, \ldots, n \tag{C.6}$$

$$U_i^H U_j = \delta_{ij}, \qquad i, j = 1, 2, \ldots, n \tag{C.7}$$

$$V_i^H V_j = \delta_{ij}, \qquad i, j = 1, 2, \ldots, n \tag{C.8}$$

$$\sigma_i \delta_{ij} = U_i^H \Phi V_j, \qquad i, j = 1, 2, \ldots, n \tag{C.9}$$

Let us define the $N \times 1$ parameter vector $\mathbf{p}$ as

$$\mathbf{p} = [\,p_1 \quad p_2 \quad \ldots \quad p_N\,] \tag{C.10}$$

and we assume that the matrix Φ is a twice differentiable function of $\mathbf{p}$, i.e.,

$$\Phi = \Phi(\mathbf{p}) \tag{C.11}$$

Differentiating Eqs. (C.5) and (C.6) with respect to a typical element p_k of $\mathbf{p}$, on premultiplying the resulting two equations by U_i^H and V_i^H, gives

$$U_i^H \frac{\partial \Phi}{\partial p_k} V_i + U_i^H \Phi \frac{\partial V_i}{\partial p_k} = U_i^H \frac{\partial U_i}{\partial p_k} \sigma_i + U_i^H U_i \frac{\partial \sigma_i}{\partial p_k} \tag{C.12}$$

$$V_i^H \frac{\partial \Phi^H}{\partial p_k} U_i + V_i^H \Phi^H \frac{\partial U_i}{\partial p_k} = V_i^H \frac{\partial V_i}{\partial p_k} \sigma_i + V_i^H V_i \frac{\partial \sigma_i}{\partial p_k} \tag{C.13}$$

Adding together Eqs. (C.12) and (C.13) gives

$$\begin{aligned} U_i^H \frac{\partial \Phi}{\partial p_k} V_i + V_i^H \frac{\partial \Phi^H}{\partial p_k} U_i + (U_i^H \Phi - V_i^H \sigma_i) \frac{\partial V_i}{\partial p_k} \\ + (V_i^H \Phi^H - U_i^H \sigma_i) \frac{\partial U_i}{\partial p_k} = \frac{\partial \sigma_i}{\partial p_k} (U_i^H U_i + V_i^H V_i) \end{aligned} \tag{C.14}$$

Using Eqs. (C.5)-C.9) in Eq. (C.14), we obtain the following result (identical to Ref. 2).

$$\frac{\partial \sigma_i}{\partial p_k} = \frac{1}{2} (U_i^H \frac{\partial \Phi}{\partial p_k} V_i + V_i^H \frac{\partial \Phi^H}{\partial p_k} U_i) = \Re (U_i^H \frac{\partial \Phi}{\partial p_k} V_i) \tag{C.15}$$

where $\Re(\cdot)$ denotes the real part of $(\cdot)$.

Upon differentiating Eq. (C.15) with respect to p_l, we obtain the following expression for the second partial derivatives of the singular values

$$\frac{\partial^2 \sigma_i}{\partial p_k \partial p_l} = \Re (\frac{\partial U_i^H}{\partial p_l} \frac{\partial \Phi}{\partial p_k} V_i + U_i^H \frac{\partial^2 \Phi}{\partial p_k \partial p_l} V_i + U_i^H \frac{\partial \Phi}{\partial p_k} \frac{\partial V_i}{\partial p_l}) \tag{C.16}$$

Since Eq. (C.16) involves the singular vector gradients $\frac{\partial U_i}{\partial p_l}$ and $\frac{\partial V_i}{\partial p_l}$, we must either evaluate them or eliminate them. We choose the latter approach.

Notice that either set of unitary vectors (the columns of U or V) form a complete set of basis vectors spanning the $n \times 1$ complex vector space, so we know that any other vector can be formed as a linear combination of either one of these two sets of basis vectors. We choose to project $\frac{\partial U_i}{\partial p_l}$ and $\frac{\partial V_i}{\partial p_l}$ onto the corresponding sets of unitary vectors as

$$\begin{aligned} \frac{\partial U_i}{\partial p_l} &= \sum_{j=1}^{n} \alpha_{ji}^{l} U_j \\ \frac{\partial V_i}{\partial p_l} &= \sum_{j=1}^{n} \beta_{ji}^{l} V_j \end{aligned} \tag{C.17}$$

where α_{ji}^{l} and β_{ji}^{l} are scalar complex constant "components" of the singular vector gradients.

Differentiating Eq. (C.5) with respect to the typical parameter p_l, upon premultiplying the result by U_r^H and making use of Eqs. (C.17), gives

$$U_r^H \frac{\partial \Phi}{\partial p_l} V_i + \sum_{j=1}^{n} \beta_{ji}^{l} U_r^H \Phi V_j = \sigma_i \sum_{j=1}^{n} \alpha_{ji}^{l} U_r^H U_j + U_r^H U_i \frac{\partial \sigma_i}{\partial p_l} \tag{C.18}$$

Using Eqs. (C.7) and (C.9) in Eq. (C.18), we obtain the following equation:

$$\sigma_i \alpha_{ri}^{l} - \sigma_r \beta_{ri}^{l} = U_r^H \frac{\partial \Phi}{\partial p_l} V_i - \frac{\partial \sigma_i}{\partial p_l} \delta_{ri} \tag{C.19}$$

Similarly, differentiating Eq. (C.6) with respect to p_l, upon premultiplying the resulting equation by V_r^H and making use of Eqs. (C.5), (C.7), (C.8), and (C.17), gives

$$\sigma_r \alpha_{ri}^{l} - \sigma_i \beta_{ri}^{l} = -V_r^H \frac{\partial \Phi^H}{\partial p_l} U_i + \frac{\partial \sigma_i}{\partial p_l} \delta_{ri} \tag{C.20}$$

In the case of the "diagonal" terms for which $r = i$, Eqs. (C.19) and (C.20) become

$$\begin{aligned} \alpha_{ii}^{l} - \beta_{ii}^{l} &= \frac{1}{\sigma_i} \left(U_i^H \frac{\partial \Phi}{\partial p_l} V_i - \frac{\partial \sigma_i}{\partial p_l} \right) \\ \alpha_{ii}^{l} - \beta_{ii}^{l} &= \frac{1}{\sigma_i} \left(-V_i^H \frac{\partial \Phi^H}{\partial p_l} U_i + \frac{\partial \sigma_i}{\partial p_l} \right) \end{aligned} \tag{C.21}$$

It is evident that subtraction of the first of Eqs. (C.21) from the second provides another verification of Eq. (C.15), but we have another use for Eqs. (C.21) in evaluating the second partials of Eq. (C.16). In the case of $r \neq i$, we have the following result from Eqs. (C.19) and (C.20):

$$\begin{aligned} \sigma_i \alpha_{ri}^{l} - \sigma_r \beta_{ri}^{l} &= U_r^H \frac{\partial \Phi}{\partial p_l} V_i \\ \sigma_r \alpha_{ii}^{l} - \sigma_i \beta_{ii}^{l} &= -V_r^H \frac{\partial \Phi^H}{\partial p_l} U_i = -\left(U_i^H \frac{\partial \Phi}{\partial p_l} V_r \right)^H \end{aligned} \tag{C.22}$$

Except for the $r = i$ case, and the case of repeated singular values, we can solve Eqs. (C.22) for the singular vector gradient projection coefficients as

$$\begin{aligned} \alpha_{ri}^{l} &= \frac{1}{\sigma_i^2 - \sigma_r^2} \left[\sigma_i \left(U_r^H \frac{\partial \Phi}{\partial p_l} V_i \right) + \sigma_r \left(U_i^H \frac{\partial \Phi}{\partial p_l} V_r \right)^H \right] \\ \beta_{ri}^{l} &= \frac{1}{\sigma_i^2 - \sigma_r^2} \left[\sigma_r \left(U_r^H \frac{\partial \Phi}{\partial p_l} V_i \right) + \sigma_i \left(U_i^H \frac{\partial \Phi}{\partial p_l} V_r \right)^H \right] \end{aligned} \tag{C.23}$$

The singular vector sensitivities must be consistent with the normalization of Eqs. (C.7) and (C.8). Upon setting $j = i$ and differentiating with respect to p_l, then substituting Eqs. (C.17) and making use of Eqs. (C.7) and (C.8), we find that the diagonal projection coefficients are purely imaginary.

$$
\begin{aligned}
(\alpha_{ii}^{l})^{H} + (\alpha_{ii}^{l}) &= \Re(\alpha_{ii}^{l}) = 0 \\
(\beta_{ii}^{l})^{H} + (\beta_{ii}^{l}) &= \Re(\beta_{ii}^{l}) = 0
\end{aligned} \tag{C.24}
$$

Using Eqs. (C.21) in Eqs. (C.24), and substituting Eqs. (C.15) into the result, gives

$$
(\alpha_{ii}^{l})^{H} + (\beta_{ii}^{l}) = \frac{-1}{\sigma_i}\Big(U_i^{H}\frac{\partial \Phi}{\partial p_l}V_i - \frac{\partial \sigma_i}{\partial p_l}\Big) = \frac{-i}{\sigma_i}\Im\Big(U_i^{H}\frac{\partial \Phi}{\partial p_l}V_i\Big), \qquad i \equiv \sqrt{-1} \tag{C.25}
$$

where $\Im(\cdot)$ denotes the imaginary part of $(\cdot)$.

Substitution of Eqs. (C.17) into Eq. (C.16) gives

$$
\frac{\partial^2 \sigma_i}{\partial p_k \partial p_l} = \Re\Big[\sum_{j=1}^{n}\Big((\alpha_{ji}^{l})^{H}U_j^{H}\frac{\partial \Phi}{\partial p_k}V_i + (\beta_{ji}^{l})U_i^{H}\frac{\partial \Phi}{\partial p_k}V_j\Big) + U_i^{H}\frac{\partial^2 \Phi}{\partial p_k \partial p_l}V_i\Big]
$$

or

$$
\frac{\partial^2 \sigma_i}{\partial p_k \partial p_l} =
$$

$$
\Re\Big[\sum_{\substack{j=1\\ j\neq i}}^{n}\Big((\alpha_{ji}^{l})^{H}U_j^{H}\frac{\partial \Phi}{\partial p_k}V_i + (\beta_{ji}^{l})U_i^{H}\frac{\partial \Phi}{\partial p_k}V_j\Big) + U_i^{H}\frac{\partial^2 \Phi}{\partial p_k \partial p_l}V_i + [(\alpha_{ii}^{l})^{H} + (\beta_{ii}^{l})]U_i^{H}\frac{\partial \Phi}{\partial p_k}V_i\Big] \tag{C.26}
$$

Finally, eliminating $(\alpha_{ii}^{l})^{H} + (\beta_{ii}^{l})$ using Eq. (C.25) and $\alpha_{ji}^{l}, \beta_{ji}^{l}$ for $j \neq i$ using Eq. (C.23) and after straightforward algebraic operations, we obtain the final expression for the second partial derivatives of the singular values:

$$
\frac{\partial^2 \sigma_i}{\partial p_k \partial p_l} =
$$

$$
\begin{aligned}
\Re\Big[\sum_{\substack{j=1\\ j\neq i}}^{n}\frac{1}{\sigma_i^2 - \sigma_j^2}\Big\{&\sigma_i[(U_j^{H}\frac{\partial \Phi}{\partial p_l}V_i)^{H}(U_j^{H}\frac{\partial \Phi}{\partial p_k}V_i) + (U_i^{H}\frac{\partial \Phi}{\partial p_l}V_j)^{H}(U_i^{H}\frac{\partial \Phi}{\partial p_k}V_j)] \\
&+ \sigma_j[(U_i^{H}\frac{\partial \Phi}{\partial p_l}V_j)(U_j^{H}\frac{\partial \Phi}{\partial p_k}V_i) + (U_j^{H}\frac{\partial \Phi}{\partial p_l}V_i)(U_i^{H}\frac{\partial \Phi}{\partial p_k}V_j)]\Big\} \\
&+ U_i^{H}\frac{\partial^2 \Phi}{\partial p_k \partial p_l}V_i\Big] + \frac{1}{\sigma_i}\Im(U_i^{H}\frac{\partial \Phi}{\partial p_l}V_i)\Im(U_i^{H}\frac{\partial \Phi}{\partial p_k}V_i)
\end{aligned} \tag{C.27}
$$

This expression has been carefully validated by second finite difference studies, analogous to those reported in Ref. 3 for validating analytical partial derivatives for eigenvalue and eigenvector derivatives.

Note that the first- and second-order sensitivities of the singular values of the eigenvector need the first and second partial derivatives of eigenvalues and eigenvectors that are presented in section 2.3.1.

Q.E.D.

References

1. Chen, C.T., *Linear System Theory and Design*, Holt, Rinehart and Winston, New York, NY, 1984.

2. Mukhopadhyay, V. and Newsom, J.R., "Application of Matrix Singular Value Properties for Evaluating Gain and Phase Margins of Multiloop Systems," NASA TM-84524, July 1982.

3. Lim, K.B., Junkins, J.L., and Wang, B.P., "Re-examination of Eigenvector Derivatives," *Journal of Guidance, Control, and Dynamics*, Vol. 10, No. 6, 1987, pp. 581-587.

APPENDIX D

Work Energy Rate Principle

Consider a mechanical (natural) system that consists of N multiply connected substructures. For simplicity, consider the simplest possible case in which each substructure is idealized as a particle. It is straightforward to generalize this simplest case to consider systems of rigid bodies and flexible bodies. The motion of the mass m_k of the kth substructure is governed by Newton's second law as

$$m_k \ddot{\mathbf{x}}_k = \mathbf{F}_k \tag{D.1}$$

where $\mathbf{x}_k$ is the physical position vector, and $\mathbf{F}_k$ denotes the applied and constrained forces acting on the kth substructure.

The virtual work principle is stated as (Refs. 1-3)

$$\left(\mathbf{F}_k - m_k \ddot{\mathbf{x}}_k\right)^T \delta \mathbf{x}_k = 0 \tag{D.2}$$

where $\delta \mathbf{x}_k$ indicates the virtual displacement vector.

Suppose n independent generalized coordinates $(q_1, q_2, \ldots, q_n)$ are introduced; the coordinate transformations between the physical and the generalized coordinates are given by

$$\mathbf{x}_k = \mathbf{x}_k(q_1, q_2, \ldots, q_n) \tag{D.3}$$

and the virtual displacement vector can be represented using the generalized coordinates as follows:

$$\delta \mathbf{x}_k = \frac{\partial \mathbf{x}_k}{\partial \mathbf{q}} \delta \mathbf{q} \tag{D.4}$$

where $\mathbf{q}$ denotes the generalized coordinate vector.

Using Eq. (D.4) in Eq. (D.1), the virtual work principle can be rewritten as

$$\left[\mathbf{F}_k^T \frac{\partial \mathbf{x}_k}{\partial \mathbf{q}} - m_k \ddot{\mathbf{x}}_k^T \frac{\partial \mathbf{x}_k}{\partial \mathbf{q}}\right] \delta \mathbf{q} = 0 \tag{D.5}$$

The force-related term in Eq. (D.5) can be written by using the definition of generalized forces as follows (Refs. 1, 2):

$$\mathbf{F}_k^T \frac{\partial \mathbf{x}_k}{\partial \mathbf{q}} = \mathbf{Q}_k^T \tag{D.6}$$

Specifically, by considering the conservative forces, Eq. (D.6) can be rewritten as

$$\mathbf{F}_k^T \frac{\partial \mathbf{x}_k}{\partial \mathbf{q}} = \mathbf{Q}_k^T - \frac{\partial V}{\partial \mathbf{q}} \tag{D.7}$$

Note that the first term, $\mathbf{Q}_k^T$, of the above equation includes the generalized forces arising from external forces, constraint forces, and boundary forces acting on the kth substructure; the second term includes the conservative generalized forces arising from the gradient of a potential energy function; and V denotes the potential energy function into which all of these conservative forces are absorbed.

Using Eq. (D.7), Eq. (D.5) can be rewritten as

$$\left[\mathbf{Q}_k^T - \frac{\partial V}{\partial \mathbf{q}} - m_k \ddot{\mathbf{x}}_k^T \frac{\partial \mathbf{x}_k}{\partial \mathbf{q}}\right] \delta \mathbf{q} = 0 \tag{D.8}$$

or, as a consequence of the independence of the generalized coordinates, we have

$$m_k \ddot{\mathbf{x}}_k^T \frac{\partial \mathbf{x}_k}{\partial \mathbf{q}} + \frac{\partial V}{\partial \mathbf{q}} = \mathbf{Q}_k^T \tag{D.9}$$

Postmultiplying the vector $\dot{\mathbf{q}}$ on both sides of Eq. (D.9) leads to

$$m_k \ddot{\mathbf{x}}_k^T \frac{\partial \mathbf{x}_k}{\partial \mathbf{q}} \dot{\mathbf{q}} + \frac{\partial V}{\partial \mathbf{q}} \dot{\mathbf{q}} = \mathbf{Q}_k^T \dot{\mathbf{q}} \tag{D.10}$$

Noting the following relations which come from the chain rule for diferentiation:

$$\dot{\mathbf{x}}_k = \frac{\partial \mathbf{x}_k}{\partial \mathbf{q}} \dot{\mathbf{q}}, \qquad \frac{dV}{dt} = \frac{\partial V}{\partial \mathbf{q}} \dot{\mathbf{q}} \tag{D.11}$$

So we can write Eq. (D.10) as follows

$$m_k \ddot{\mathbf{x}}_k^T \dot{\mathbf{x}}_k + \frac{dV}{dt} = \mathbf{Q}_k^T \dot{\mathbf{q}} \tag{D.12}$$

We note also that the first term of the left-hand side in Eq. (D.12) can be manipulated as

$$m_k \ddot{\mathbf{x}}_k^T \dot{\mathbf{x}}_k = \frac{d}{dt}\left(\frac{1}{2} m_k \dot{\mathbf{x}}_k^T \dot{\mathbf{x}}_k\right) \equiv \frac{dT_k}{dt} \tag{D.13}$$

using the definition of the kinetic energy of a kth substructure (of the natural system),

$$T_k \equiv \frac{1}{2} m_k \dot{\mathbf{x}}_k \dot{\mathbf{x}}_k \tag{D.14}$$

Finally, substituting Eq. (D.13) into Eq. (D.12) yields

$$\frac{dE_k}{dt} = \frac{d}{dt}\left(T_k + V_k\right) = \mathbf{Q}_k^T \dot{\mathbf{q}} = \mathbf{F}_k^T \dot{\mathbf{x}}_k \tag{D.15}$$

where E_k $(= T_k + V_k)$ is the total energy of the kth substructure.

Equation (D.15) represents the *work/energy rate principle* (Ref. 3). It is observed that *the rate of total energy is equal to the rate of change of work done by the generally nonconservative applied forces and constraint forces* $\mathbf{Q}_k$. Observe further that the work rate can easily be expressed in terms of the inner product of generalized forces acting on generalized coordinate time derivatives (e.g. $\mathbf{Q}_k^T \dot{\mathbf{q}}$), or as an inner product of actual forces acting on the inertial velocities of the point of application of the forces (e.g., the sum of all $\mathbf{F}_k^T \dot{\mathbf{x}}_k$ acting on the kth substructure). Having made these observations, it is easy to use kinematics to determine the work rate for rather general systems. This principle is of most significant importance in efficiently applying Lyapunov stability theory for the case in which the candidate Lyapunov function is expressible as a function of the total energy of substructures. Through the use of this principle, we can often avoid most of the tedious algebra and calculus required to determine the final expressions for the Lyapunov functions time derivative; this expression is most often simply a weighted power expression that can immediately be written from the *work/energy rate principle.* In Ref. 3, this principle is generalized for the case of multirigid and distributed parameter systems, which include working moments as well as forces, and its utility in deriving Lyapunov function rates is shown.

References

1. Meirovitch, L., *Analytical Methods in Vibrations,* Macmillan Company, Toronto, 1967.

2. Junkins, J.L. and Turner, J.D., *Optimal Spacecraft Rotational Maneuvers,* Elsevier, Amsterdam, 1986.

3. Oh, H.S., Vadali, S.R., and Junkins, J.L., "On the Use of the Work-Energy Rate Principle for Designing Feedback Control Laws," *Journal of Guidance, Control, and Dynamics,*Vol. 15, No. 1, Jan.-Feb. 1992, pp. 275-277.

APPENDIX E

Analytical Solution of the Lyapunov Equation for Modal Second-Order Systems

An analytical solution of the Lyapunov equation for the modal second-order system is presented in Ref. 1. We present in this appendix a novel derivation of this solution using index notation.

Consider the modal second-order system

$$\ddot{\underline{\eta}} + \tilde{\mathrm{C}}\dot{\underline{\eta}} + \tilde{\mathrm{K}}\underline{\eta} = \tilde{\mathrm{D}}\mathbf{u} \tag{E.1}$$

where

modal coordinate vector $\underline{\eta} \in \mathrm{R}^n$
input vector $\mathbf{u} \in \mathrm{R}^m$
$\tilde{\mathrm{C}} = \mathrm{diag}(2\zeta_1\omega_1, 2\zeta_2\omega_2, \ldots, 2\zeta_n\omega_n)$
$\tilde{\mathrm{K}} = \mathrm{diag}(\omega_1^2, \omega_2^2, \ldots, \omega_n^2)$
$\tilde{\mathrm{D}}$ is an $n \times m$ input influence matrix

For control applications, the system dynamics are usually modeled in first-order differential equations. By introducing the $2n \times 1$ modal state vector

$$\mathbf{z} = \begin{pmatrix} \underline{\eta} \\ \dot{\underline{\eta}} \end{pmatrix} \tag{E.2}$$

Eq. (E.1) can be written as the first-order system

$$\dot{\mathbf{z}} = \mathrm{A}\mathbf{z} + \mathrm{B}\mathbf{u} \tag{E.3}$$

where

$$\mathrm{A} = \begin{bmatrix} 0 & \mathrm{I} \\ -\tilde{\mathrm{K}} & -\tilde{\mathrm{C}} \end{bmatrix}, \qquad \mathrm{B} = \begin{bmatrix} 0 \\ \tilde{\mathrm{D}} \end{bmatrix} \tag{E.4}$$

Let us introduce the controllability grammian X,

$$\mathrm{X} \equiv \int_0^\infty e^{\mathrm{A}t}\mathrm{B}\mathrm{B}^{\mathrm{T}}e^{\mathrm{A}^{\mathrm{T}}t}dt \tag{E.5}$$

which satisfies the following algebraic Lyapunov equation:

$$\mathrm{X}\mathrm{A}^{\mathrm{T}} + \mathrm{A}\mathrm{X} + \mathrm{B}\mathrm{B}^{\mathrm{T}} = 0 \tag{E.6}$$

In order to obtain an analytical solution for the above second-order modal system, let the Lyapunov solution be partitioned as follows

$$X = \begin{bmatrix} X_{\eta\eta} & X_{\eta\dot{\eta}} \\ X_{\eta\dot{\eta}}^T & X_{\dot{\eta}\dot{\eta}} \end{bmatrix} \tag{E.7}$$

where $X_{\eta\eta}$ and $X_{\dot{\eta}\dot{\eta}}$ are symmetric matrices since X is a symmetric matrix.

Substitution of Eqs. (E.4) and (E.7) into Eq. (E.6) gives the following three matrix equations:

$$X_{\eta\dot{\eta}} + X_{\eta\dot{\eta}}^T = 0 \tag{E.8}$$

$$X_{\dot{\eta}\dot{\eta}} - \tilde{K}X_{\eta\eta} - \tilde{C}X_{\eta\dot{\eta}}^T = 0 \tag{E.9}$$

$$-X_{\eta\dot{\eta}}^T\tilde{K} - X_{\dot{\eta}\dot{\eta}}\tilde{C} - \tilde{K}X_{\eta\dot{\eta}} - \tilde{C}X_{\dot{\eta}\dot{\eta}} + \tilde{D}\tilde{D}^T = 0 \tag{E.10}$$

Note from Eq. (E.8) that $X_{\eta\dot{\eta}}$ is a skew-symmetric matrix; therefore, the diagonal elements are zero.

Using Eqs. (E.8) and (E.9) in Eq. (E.10), we obtain the following equation:

$$(X_{\eta\dot{\eta}}\tilde{K} - \tilde{K}X_{\eta\dot{\eta}}) - (\tilde{K}X_{\eta\eta}\tilde{C} + \tilde{C}\tilde{K}X_{\eta\eta}) + \tilde{C}(X_{\eta\dot{\eta}}\tilde{C} + \tilde{C}X_{\eta\dot{\eta}}) + \tilde{D}\tilde{D}^T = 0 \tag{E.11}$$

Equation (E.9) can be rewritten by using Eq. (E.8) as

$$X_{\dot{\eta}\dot{\eta}} = \tilde{K}X_{\eta\eta} - \tilde{C}X_{\eta\dot{\eta}} \tag{E.12}$$

Since $X_{\dot{\eta}\dot{\eta}}$ is a symmetric matrix, subtraction of Eq. (E.12) from the transpose of Eq. (E.12) gives

$$X_{\eta\dot{\eta}}\tilde{C} + \tilde{C}X_{\eta\dot{\eta}} = \tilde{K}X_{\eta\eta} - X_{\eta\eta}\tilde{K} \tag{E.13}$$

Substituting Eq. (E.13) into Eq. (E.11) yields

$$(X_{\eta\dot{\eta}}\tilde{K} - \tilde{K}X_{\eta\dot{\eta}}) - (\tilde{K}X_{\eta\eta}\tilde{C} + \tilde{C}X_{\eta\eta}\tilde{K}) + \tilde{D}\tilde{D}^T = 0 \tag{E.14}$$

We can obtain the solution easily by writing Eqs. (E.12)–(E.14) in index notation. Note that $i, j = 1, \ldots, n$ respectively in the following equations. Equation (E.13) can be written in index notation as follows:

$$\sum_{k=1}^{n} [X_{\eta\dot{\eta}}]_{ik}[\tilde{C}]_{kj} + \sum_{k=1}^{n} [\tilde{C}]_{ik}[X_{\eta\dot{\eta}}]_{kj} = \sum_{k=1}^{n} [\tilde{K}]_{ik}[X_{\eta\eta}]_{kj} - \sum_{k=1}^{n} [X_{\eta\eta}]_{ik}[\tilde{K}]_{kj}$$

Noting that $\tilde{K}$ and $\tilde{C}$ are diagonal matrices, the above equation can be written as

$$[X_{\eta\dot{\eta}}]_{ij}[\tilde{C}]_{jj} + [\tilde{C}]_{ii}[X_{\eta\dot{\eta}}]_{ij} = [\tilde{K}]_{ii}[X_{\eta\eta}]_{ij} - [X_{\eta\eta}]_{ij}[\tilde{K}]_{jj}$$

Therefore,

$$[X_{\eta\dot{\eta}}]_{ij} = \frac{[\tilde{K}]_{ii} - [\tilde{K}]_{jj}}{[\tilde{C}]_{ii} + [\tilde{C}]_{jj}}[X_{\eta\eta}]_{ij} \tag{E.15}$$

The (i,j)th element of Eq. (E.12) can be written as follows:

$$\begin{aligned}[X_{\dot{\eta}\dot{\eta}}]_{ij} &= \sum_{k=1}^{n}[\tilde{K}]_{ik}[X_{\eta\eta}]_{kj} - \sum_{k=1}^{n}[\tilde{C}]_{ik}[X_{\eta\dot{\eta}}]_{kj} \\ &= [\tilde{K}]_{ii}[X_{\eta\eta}]_{ij} - [\tilde{C}]_{ii}[X_{\eta\dot{\eta}}]_{ij}\end{aligned}$$

Substituting Eq. (E.15) into the above equation gives

$$[X_{\dot{\eta}\dot{\eta}}]_{ij} = \frac{[\tilde{C}]_{ii}[\tilde{K}]_{jj} + [\tilde{C}]_{jj}[\tilde{K}]_{ii}}{[\tilde{C}]_{ii} + [\tilde{C}]_{jj}}[X_{\eta\eta}]_{ij} \tag{E.16}$$

Equation (E.14) can be written in index notation as follows:

$$\begin{aligned}&\sum_{k=1}^{n}[X_{\eta\dot{\eta}}]_{ik}[\tilde{K}]_{kj} - \sum_{k=1}^{n}[\tilde{K}]_{ik}[X_{\eta\dot{\eta}}]_{kj} \\ &\quad - \sum_{k=1}^{n}\sum_{l=1}^{n}[\tilde{K}]_{ik}[X_{\eta\eta}]_{kl}[\tilde{C}]_{lj} - \sum_{k=1}^{n}\sum_{l=1}^{n}[\tilde{C}]_{ik}[X_{\eta\eta}]_{kl}[\tilde{K}]_{lj} + [\tilde{D}\tilde{D}^T]_{ij} = 0\end{aligned}$$

The above equation can be re-written as

$$[X_{\eta\dot{\eta}}]_{ij}[\tilde{K}]_{jj} - [\tilde{K}]_{ii}[X_{\eta\dot{\eta}}]_{ij} - [\tilde{K}]_{ii}[X_{\eta\eta}]_{ij}[\tilde{C}]_{jj} - [\tilde{C}]_{ii}[X_{\eta\eta}]_{ij}[\tilde{K}]_{jj} + [\tilde{D}\tilde{D}^T]_{ij} = 0$$

Substituting Eq. (E.15) into the above equation gives

$$[X_{\eta\eta}]_{ij} = \frac{[\tilde{C}]_{ii} + [\tilde{C}]_{jj}}{([\tilde{K}]_{ii} - [\tilde{K}]_{jj})^2 + ([\tilde{C}]_{ii} + [\tilde{C}]_{jj})([\tilde{C}]_{ii}[\tilde{K}]_{jj} + [\tilde{C}]_{jj}[\tilde{K}]_{ii})}[\tilde{D}\tilde{D}^T]_{ij} \tag{E.17}$$

Now, substitution of Eq. (E.17) into Eqs. (E.15) and (E.16) yields

$$[X_{\eta\dot{\eta}}]_{ij} = \frac{[\tilde{K}]_{ii} - [\tilde{K}]_{jj}}{([\tilde{K}]_{ii} - [\tilde{K}]_{jj})^2 + ([\tilde{C}]_{ii} + [\tilde{C}]_{jj})([\tilde{C}]_{ii}[\tilde{K}]_{jj} + [\tilde{C}]_{jj}[\tilde{K}]_{ii})}[\tilde{D}\tilde{D}^T]_{ij} \tag{E.18}$$

$$[X_{\dot{\eta}\dot{\eta}}]_{ij} = \frac{[\tilde{C}]_{ii}[\tilde{K}]_{jj} + [\tilde{C}]_{jj}[\tilde{K}]_{ii}}{([\tilde{K}]_{ii} - [\tilde{K}]_{jj})^2 + ([\tilde{C}]_{ii} + [\tilde{C}]_{jj})([\tilde{C}]_{ii}[\tilde{K}]_{jj} + [\tilde{C}]_{jj}[\tilde{K}]_{ii})}[\tilde{D}\tilde{D}^T]_{ij} \tag{E.19}$$

Equations (E.17)-(E.19) give us a complete analytical solution of the Lyapunov Equation (E.6).

For the modal second-order system, note in Eq. (E.1) that

$$\begin{aligned}[\tilde{C}]_{ii} &= 2\zeta_i\omega_i \\ [\tilde{K}]_{ii} &= \omega_i^2\end{aligned} \tag{E.20}$$

By considering the above physical properties, the solution of The Lyapunov equation can be summarized as follows:

$$X = \begin{bmatrix} X_{\eta\eta} & X_{\eta\dot{\eta}} \\ X_{\eta\dot{\eta}}^T & X_{\dot{\eta}\dot{\eta}} \end{bmatrix} \tag{E.21}$$

where the elements of $n \times n$ block matrices are

$$\begin{aligned}[X_{\eta\eta}]_{ij} &= \frac{2(\zeta_i\omega_i + \zeta_j\omega_j)}{\Delta}[\tilde{D}\tilde{D}^T]_{ij}, & i,j &= 1,2,\ldots,n \\ [X_{\eta\dot{\eta}}]_{ij} &= \frac{\omega_i^2 - \omega_j^2}{\Delta}[\tilde{D}\tilde{D}^T]_{ij}, & i,j &= 1,2,\ldots,n \\ [X_{\dot{\eta}\dot{\eta}}]_{ij} &= \frac{2\omega_i\omega_j(\zeta_i\omega_j + \zeta_j\omega_i)}{\Delta}[\tilde{D}\tilde{D}^T]_{ij}, & i,j &= 1,2,\ldots,n\end{aligned} \tag{E.22}$$

with

$$\Delta = (\omega_i^2 - \omega_j^2)^2 + 4\omega_i\omega_j(\zeta_i\omega_i + \zeta_j\omega_j)(\zeta_i\omega_j + \zeta_j\omega_i)$$

Reference

1. Skelton, R.E., Singh, R., and Ramakrishnan, J., "Component Model Reduction by Component Cost Analysis," AIAA Guidance, Navigation, and Control Conference, Minneapolis, MN, Aug., 1988.

APPENDIX F

Partial Derivatives of Symmetric Gain Matries

In Chapter 7, we admit a general positive definite gain matrix by introducing the Cholesky decomposition parameterization

$$G = LL^T \tag{F.1}$$

Gain matrix G is guaranteed to be positive semidefinite for all real elements q_{ij} assigned to the lower triangular matrix.

$$L = \begin{bmatrix} q_{11}^2 & 0 & 0 & \cdots & 0 \\ q_{21} & q_{22}^2 & 0 & \cdots & 0 \\ q_{31} & q_{32} & q_{33}^2 & \cdots & 0 \\ \vdots & \vdots & \vdots & \ddots & \vdots \\ q_{m1} & q_{m2} & q_{m3} & \cdots & q_{mm}^2 \end{bmatrix} \tag{F.2}$$

Note that the diagonal elements of L are quadratic, and index i is always greater than or equal to index j in q_{ij}.

The elements q_{ij} are often taken as design parameters in an optimization problem; therefore, it is useful to derive the analytical partial derivative of the gain matrix with respect to these parameters.

Differentiating Eq. (F.1) with respect to a typical element q_{ij} of L gives

$$\frac{\partial G}{\partial q_{ij}} = \frac{\partial L}{\partial q_{ij}} L^T + L \frac{\partial L^T}{\partial q_{ij}} \tag{F.3}$$

Taking partial derivatives of Eq. (F.2) gives

$$\begin{aligned} \left[\frac{\partial L}{\partial q_{ii}}\right]_{kl} &= 2q_{ii}, \quad && \text{for } k = i,\ l = i \\ &= 0, && \text{otherwise} \\ \left[\frac{\partial L}{\partial q_{ij}}\right]_{kl} &= 1, && \text{for } k = i,\ l = j \\ &= 0, && \text{otherwise} \end{aligned} \tag{F.4}$$

where $[\cdot]_{kl}$ is the (k,l) element of matrix $[\cdot]$.

By using Eqs. (F.4) in Eq. (F.3), we obtain the following results:

$$
\begin{aligned}
\left[\frac{\partial G}{\partial q_{ii}}\right]_{kl} &= 4q_{ii}^3, && \text{for } k = i,\ l = i \\
&= 2q_{ii}q_{ki}, && \text{for } k = i,\ l > i \\
&= 2q_{ii}q_{ki}, && \text{for } k > i,\ l = i \\
&= 0, && \text{otherwise} \\
\left[\frac{\partial G}{\partial q_{ij}}\right]_{kl} &= 2q_{ij}, && \text{for } k = i,\ l = i \\
&= q_{jj}^2, && \text{for } k = i,\ l = j \\
&= q_{jj}^2, && \text{for } k = j,\ l = i \\
&= q_{kj}, && \text{for } k = i,\ \ l > j(l \neq i) \\
&= q_{kj}, && \text{for } k > j(k \neq i),\ l = i \\
&= 0, && \text{otherwise}
\end{aligned}
\qquad (F.5)
$$

APPENDIX G

Derivatives of Damping Factors and Damped Frequencies

The eigenvalues of the system A(**p**) can be written as

$$\begin{aligned} \lambda_i &= \Re\{\lambda_i\} + i\Im\{\lambda_i\} \\ &= -\zeta_i\omega_i + i\omega_{di} \end{aligned} \tag{G.1}$$

where $\omega_{di} \equiv \omega_i\sqrt{1-\zeta_i^2}$: damped natural frequency, $i \equiv \sqrt{-1}$, and $\Re\{\cdot\}$ and $\Im\{\cdot\}$ denote the real part and imaginary part of $\{\cdot\}$, respectively.

The purpose of this appendix is to establish the analytical sensitivity of the eigenvalues' damped natural frequencies and damping factors with respect to the parameter vector **p**. We assume that A(**p**) is continuous and differentiable, and we note that excluding the obvious isolated cases of root bifurcations and multiple eigenvalues, the eigenvalues of A(**p**) can be expected to be differentiable with respect to the elements of **p**

Note that ω_i, ζ_i represent the undamped natural frequency and damping factor of the ith mode, respectively.

The derivative of the damping factors can be derived from the relation

$$\zeta_i(\mathbf{p}) = -\frac{\Re\{\lambda_i\}}{|\lambda_i|} \tag{G.2}$$

$$|\lambda_i| = \sqrt{\left(\Re\{\lambda_i\}\right)^2 + \left(\Im\{\lambda_i\}\right)^2} \tag{G.3}$$

By taking the partial derivative of Eqs. (G.2) and (G.3) with respect to a typical element p of parameter vector **p**, we get

$$\frac{\partial\zeta_i}{\partial p} = -\frac{\frac{\partial}{\partial p}\left(\Re\{\lambda_i\}\right)}{|\lambda_i|} + \frac{\Re\{\lambda_i\}}{|\lambda_i|^2}\frac{\partial|\lambda_i|}{\partial p} \tag{G.4}$$

$$\frac{\partial|\lambda_i|}{\partial p} = \frac{1}{|\lambda_i|}\left[\Re\{\lambda_i\}\frac{\partial}{\partial p}\left(\Re\{\lambda_i\}\right) + \Im\{\lambda_i\}\frac{\partial}{\partial p}\left(\Im\{\lambda_i\}\right)\right] \tag{G.5}$$

Note that

$$\frac{\partial}{\partial p}\left(\Re\{\lambda_i\}\right) = \Re\left(\frac{\partial\lambda_i}{\partial p}\right)$$

and

$$\frac{\partial}{\partial p}\left(\Im\{\lambda_i\}\right) = \Im\left(\frac{\partial\lambda_i}{\partial p}\right)$$

Substituting the above Eqs. (G.3) and (G.5) into Eq. (G.4) gives the damping factor sensitivity as

$$\frac{\partial \zeta_i}{\partial p} = \frac{\Im\{\lambda_i\}}{|\lambda_i|^3}\left[\Re\{\lambda_i\}\Im\left(\frac{\partial \lambda_i}{\partial p}\right) - \Im\{\lambda_i\}\Re\left(\frac{\partial \lambda_i}{\partial p}\right)\right] \tag{G.6}$$

From Eq. (G.1), the damped natural frequency is simply

$$\omega_{di} = \Im\{\lambda_i\}$$

Therefore, the derivative of damped frequency can be obtained directly by simply taking the imaginary part of the eigenvalue derivative:

$$\frac{\partial \omega_{di}}{\partial p} = \Im\left(\frac{\partial \lambda_i}{\partial p}\right) \tag{G.7}$$

APPENDIX H

Minimum Norm Differential Correction Algorithm

The derivation is based on the results originally developed in Ref. 1.

Consider the underdetermined root-solving problem

$$\mathbf{f}(\mathbf{p}) = 0 \tag{H.1}$$

where $\mathbf{p} \in \mathrm{R}^{\mathrm{N}}$ is a parameter vector and $\mathbf{f}(\mathbf{p})$ is the $r \times 1$ vector of equality equation (note that $\mathrm{N} > r$ for the underdetermined problem).

Given a starting estimate for $\mathbf{p}$ that does not satisfy Eq. (H.1), the correction $\Delta\mathbf{p}$ is determined by linearizing Eq. (H.1) about the current $\mathbf{p}$ iterative and computing the differential correction that will satisfy the linearized constraint equation with the minimum correction norm.

In order to seek the solution $\mathbf{p} = \mathbf{p}_{\mathrm{old}} + \Delta\mathbf{p}$ nearest the current state ($\mathbf{p}_{\mathrm{old}}$), we can state the above problem as the following local minimization problem:

$$\text{Minimize} \quad \Delta\mathbf{p}^{\mathrm{T}}\mathrm{W}\Delta\mathbf{p} \tag{H.2}$$

$$\text{Subject to} \quad \mathbf{f}(\mathbf{p}_{\mathrm{old}}) + \mathrm{A}\Delta\mathbf{p} = 0 \tag{H.3}$$

where $\mathrm{A} \equiv \left[\dfrac{\partial\mathbf{f}}{\partial\mathrm{p}}\right]_{\mathbf{p}=\mathbf{p}_{\mathrm{old}}}$ and W is an $\mathrm{N} \times \mathrm{N}$ positive definite weighting matrix.

Let us introduce the Lagrange multiplier $\underline{\lambda} \in \mathrm{R}^{r}$ to define the Lagrangian function

$$\mathcal{L}(\Delta\mathbf{p}, \underline{\lambda}) = \Delta\mathbf{p}^{\mathrm{T}}\mathrm{W}\Delta\mathbf{p} + \underline{\lambda}^{\mathrm{T}}\{\mathbf{f}(\mathbf{p}_{\mathrm{old}}) + \mathrm{A}\Delta\mathbf{p}\} \tag{H.4}$$

The necessary conditions for constrained minimization require the vanishing of the gradients of $\mathcal{L}$ with respect to $\mathbf{p}$ and $\underline{\lambda}$ as

$$\nabla_{\Delta\mathbf{p}}\mathcal{L}(\Delta\mathbf{p}, \underline{\lambda}) = 0 \tag{H.5}$$

$$\nabla_{\underline{\lambda}}\mathcal{L}(\Delta\mathbf{p}, \underline{\lambda}) = 0 \tag{H.6}$$

Using Eq. (H.4) to carry out the gradient operations in Eqs. (H.5) and (H.6) gives

$$\mathrm{W}\Delta\mathbf{p} + \mathrm{A}^{\mathrm{T}}\underline{\lambda} = 0 \tag{H.7}$$

$$\mathbf{f}(\mathbf{p}_{\mathrm{old}}) + \mathrm{A}\Delta\mathbf{p} = 0 \tag{H.8}$$

From Eqs. (H.7) and (H.8), we obtain the minimum norm differential correction vector

$$\Delta\mathbf{p} = -\mathrm{W}^{-1}\mathrm{A}^{\mathrm{T}}(\mathrm{A}\mathrm{W}^{-1}\mathrm{A}^{\mathrm{T}})^{-1}\mathbf{f}(\mathbf{p}_{\mathrm{old}}) \tag{H.9}$$

and we use the recursion formula

$$\mathbf{p}_{new} = \mathbf{p}_{old} + \Delta\mathbf{p} \tag{H.10}$$

until Eq. (H.1) is satisfied.

In some problems, the inverse of Eq. (H.9) is poorly conditioned. For these cases, we can re-arrange Eqs. (H.7) and (H.8) as

$$\begin{bmatrix} W & A^T \\ A & 0 \end{bmatrix} \begin{Bmatrix} \Delta\mathbf{p} \\ \underline{\lambda} \end{Bmatrix} = \begin{Bmatrix} 0 \\ -\mathbf{f}(\mathbf{p}_{old}) \end{Bmatrix} \tag{H.10}$$

and determine the solution for $\Delta\mathbf{p}$ using either the QR or SVD algorithms (see Chapter 2 for discussion of these algorithms).

References

1. Junkins, J.L., "Equivalence of the Minimum Norm and Gradient Projection Constrained Optimization Techniques," *AIAA Journal,* Vol. 10, No. 7, July 1972, pp. 927-929.

APPENDIX I

TAMU/AFOSR Maneuver Experiment Hardware Description

Figures. 3.1 and I.1 show sketches of an experimental structure at Texas A&M University; the configuration parameters are given in Table I.1. The structure has a central hub and four identical cantilevered flexible appendages, each having a finite tip mass. This structure (the "TAMU/AFOSR maneuver experiment") was developed recently at Texas A&M University under the sponsorship of the Air Force Office of Scientific Research. It is a modification of a design developed originally at the C. S. Draper Laboratories sponsored by the Air Force Rocket Propulsion Laboratory and is therefore often referred to as the "Draper/RPL" model. The original Draper/RPL structure used on/off thrusters at the beam tips and was supported on an air bearing, whereas the TAMU/AFOSR structure is supported by a combination of a ceramic needle bearing and a thin "piano wire," and control actuation is accomplished using a reaction wheel. These experimental facilities are discussed, in Ref. 1, in the context of other recent flexible structure experiments.

The reaction wheel is driven by a DC motor. The stator of the motor is rigidly connected to the top of the hub, keeping the axis of rotation aligned with the vertical. The whole assembly is pivoted on a ceramic bearing so that it can rotate freely in the horizontal plane; the vertical wire supports the structure along its undeformed geometric centerline (colinear with the fixed steel needle on which the ceramic bearing rotates) and is used to reduce the nominal bearing load to less than 1% of the weight of the structure. Thus the wire supports the weight of the structure, whereas the needle in the inverted conical ceramic bearing serves to eliminate hub translation and pin the structure's geometric center. Tipping motion is restrained by an adjustable slip collar bearing. An angle encoder, two full-bridge strain gauges near the root of an appendage, and a motor tachometer are employed to measure, respectively, the hub angular position, the appendage bending moment and shear forces, and the reaction wheel speed as functions of time. Based on these measurements and appropriate signal processing and feedback control law, a control signal is provided to a current amplifier that drives the DC motor to produce the necessary control torque. The motor bandwidth (0 to about 100 Hz) and accurately commandable torque (from about 1 up to about 600 oz-in. (depending on rotor speed), linear with respect to current over a wide range of rotor speeds) are sufficient, in principle, to control both overall rigid body motion and to suppress vibration. We note in passing that the bearing friction is known to be between 5 and 20 oz-in., depending on the adjustable tension in the wire support, based on our experimental studies.

Figure I.2 shows a block diagram of the overall system. The user interface and overall data management for the system is accomplished in a Packard Bell 286 (PB286) PC-AT compatible computer and a Data Translation DT2841 input/output (I/O) board. The DT2841 performs digital I/O, analog-to-digital (A/D)

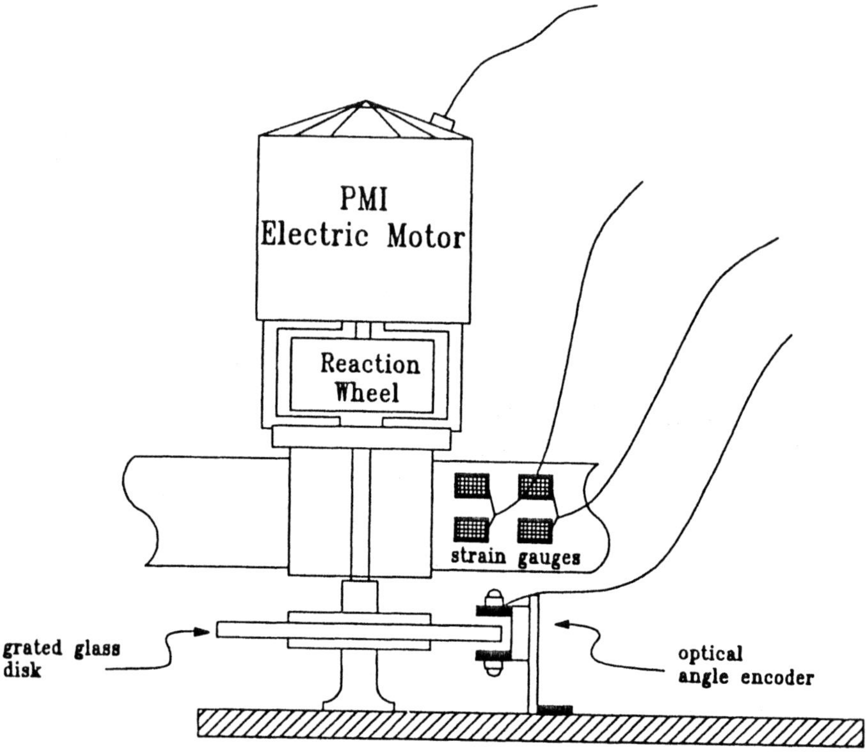

Figure I.1 Hub Assembly: Motor, Reaction Wheel, and Sensors.

conversion, digital-to-analog (D/A) conversion, and timing operations. Under the control of appropriate software, the Packard Bell 286 computer reads in data from the I/O board, does the necessary computations in its own CPU, and then sends out the resulting control signal to the amplifier through the I/O board. Depending on the complexity of the signal processing and the control law, the feedback loop can easily be closed with control loop update rates exceeding 200 Hz. Since the bandwidth of the existing hub motion sensor (the angle encoder's resolution of 0.01 deg. results in a defacto bandwidth restriction, because the vibratory modes above 20 HZ seldom have associated hub motions with amplitudes of 0.01 deg.) restricts us to controlling the dynamics having frequencies less than 20 Hz, we usually find that update rates of less than 100 Hz are sufficient to research the issues under study.

The angle encoder (Teledyne Gurley model 8708) output is decoded and

Table I.1 Hub/Appendage Experiment Configuration Parameters

Hub and Appendages	
Total undeformed system inertia, I	2128 oz-s^2-in.
Hub center to gauge center, L_0	5.5470 in.
Hub center to tip mass, L	51.07 in.
Tip mass, m_t	0.15627 oz-s^2/in.
Appendage modules of elasticity, E	161.6 million oz/in.^{2}
Inertia of bending section, I	0.000813 in.^{4}
Mass density of appendage/length, ρ	0.003007 oz-s^2/in.^{2}
Distance between two-gauge set	1.365 in.
Motor (PMI, servodisk DC motor: JR16M 4 CH)	
Torque constant, K_t	52.77 oz-in./amp
Back electromotive force constant, K_e	39.77 V/1000 rpm
Tachometer constant	3 V/1000 rpm
Power Supply (KEPCO BOP 36-12M)	
Current gain	1.2 amp/commanded V

converted into a 16-bit number in hundredths of a degree by a logic converter (Teledyne Gurley model SCA CC) and counter (C-Tek model Lin-101-50-21-41-BE) combination. As mentioned, the angle encoder output is accurate to about 0.01 deg, before smoothing. The raw angle information is read into the PB286 through the 16-bit digital port of the DT2841 I/O board. We utilize a sequential least squares filter to extract smoothed angle and angular velocity estimates from the angle encoder measurements. Least squares filters are also used to smooth the shear and moment measurements. When operating these digital filters, the sensors were sampled about two orders of magnitude faster than the frequency at which the control loop was closed to provide suitable redundancy for noise rejection without introducing significant phase lags.

The motor used to apply the control torque is a PMI servodisk DC motor equipped with a tachometer. The motor produces a torque proportional to the current in the armature as sent by the current amplifier. The commanded motor

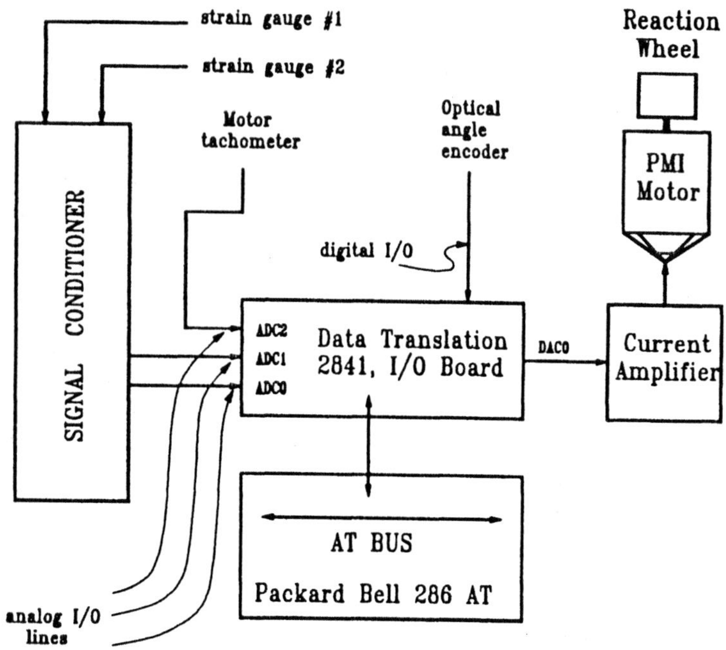

Figure I.2 System Functional Block Diagram.

current is sent out to the amplifier by PB286 through the D/A line (DAC0) of the DT2841. The tachometer output is read by the computer through one A/D line (ADC2) and used in our control laws to compensate for the motor friction torque. A separate study has been carried out to characterize the motor bearing friction. Except near zero wheel speed, the motor friction is well modeled as a quadratic function of the wheel speed, which could be estimated given the tachometer output. Likewise, the friction of the support bearing was found to have a similar nonlinear model with stiction torque near zero hub rotational speed. The friction torque discontinuity characterizing the nonlinear stiction region at zero angular velocity of the hub has a magnitude that is approximately linearly dependent on the tension in the wire support, and varies between 5 and 20 oz-in over the tension range of practical interest. The wire tension had to be maintained less than some critical value (about 95% of the total weight of the structure) to ensure that the ceramic needle bearing functioned properly.

The two full-bridge strain gauges are employed near the root of one of the appendages at a finite distance apart from each other to determine the instantaneous bending moment and shear force. The strain-gauge signals are read with the PB286 through two additional A/D channels (ADC0, ADC1).

Four KEPCO bipolar-operational power supplies (BOPs) in series are used to supply current to the motor, these BOPs have the feature that one can be named the master and the others slaved to reproduce its output, thus one-fourth of the requied current is commanded to the master BOP. These power supplies can supply up to 12 A of current for up to a compliance voltage of 150 V. The BOPs are remotely programmed by PB286 through one of the D/A channels (DAC0) of DT2841.

The DT2841 board from DATA Translation performs all of the I/O (digital, A/D, and D/A) operations for PB286. The DT2841 has two programmable clocks (DACLK and ADCLK). DACLK and ADCLK are programmed to run at 75 (or 100) and 4000 Hz, respectively; DACLK runs continuously during a whole control operation, and ADCLK runs in a burst mode (i.e., intermittently).

Every DACLK timing signal initiates the following:

(1) Converts the digital data into an analog signal to be used by the BOP to push the corresponding current through the motor.

(2) Generates a hardware interrupt to initiate a new control cycle.

The new control cycle proceeds as follows:

(1) Read angle encoder (through a digital port).

(2) Start ADCLK, complete reading the two strain gauges and the tachometer values (through A/D channels 0, 1, and 2) during the next three ADCLK cycles, and then disable ADCLK.

(3) Filter the data digitally.

(4) Predict the current required at the beginning of the next cycle.

(5) Store the data in random access memory (RAM) for time history and make the predicted data available to the D/A converter to be converted at the beginning of the next cycle.

(6) Wait until the next DACLK signal, which converts the digital data into an analog signal and generates a hardware interrupt to initiate the next control cycle.

The control software is written in FORTRAN, except for the PB286's interrupt controller setting, the I/O board functions, and the I/O routines, which are all written in assembly code.

The control cycle was typically updated at the rate of 75 Hz or 100 Hz (DACLK rate), depending on the experiment. The practical bandwidth of the closed-loop system was found to be less than 20 Hz for all experimental studies and typically about 10 Hz; we found that the limitation to low frequencies is a direct consequence of the errors introduced by deriving the angular rate approximation from the angle encoder measurements. Note that flexibility-induced oscillations of the hub having an amplitude of less than 0.01 deg could not be detected, and because the rotational/vibrational coupling effects of higher frequency modes seldom exhibit hub oscillations of this amplitude, this gives rise to a defacto, motion dependent, bandwith limitiation to control those significantly excited modes having measureable hub motion. As described in Ref. 2, it is necessary to take care in the design of the signal processing and control law that modes outside of the controller bandwith are not de-stabilized. Even with this relatively low bandwith, we have been able to do significant experiments on rotational/vibrational maneuvers of flexible structures, because it is easy to vary the configuration parameters to arrange for several highly disturbable and controllable flexible modes to be in the 0-10 Hz bandwith of this controller. The primary focus of this experiment is to allow evaluation of various control design methods for simultaneous large angle maneuvers and vibrational control, and controlling a limited number of significantly participating flexible modes is sufficient for this purpose. The strain guages provide visiblitiy of the higher frequency modes and permit post experiment estimation of the high frequency dynamics of the system to validate the control system designs. This experimental facility is an example of an configuration that is (1) fairly easy to implement, (2) relatively inespensive, and therefore, (3) appropriate for research within the constraints of university-based analytical/experimental studies.

References

1. Das, A., "Large-Angle Maneuver Experiments in Ground-Based Laboratories," *Mechanics and Control of Large Flexible Structures*, edited by J.L.Junkins, Vol. 129, *Progress in Astronautics and Aeronautics,* AIAA, 1990, pp. 491-506.

2. Junkins, J.L., Rahman, Z., and Bang, H., "Near-Minimum-Time Maneuvers of Flexible Vehicles: A Liapunov Control Law Design Method," *Mechanics and Control of Large Flexible Structures*, edited by J.L.Junkins, Vol. 129, *Progress in Astronautics and Aeronautics,* AIAA, 1990, pp. 565-593.

INDEX

U

V

W